Michael Dellas

Laser Remote Sensing

Useful Quantities

	Page
Electron's mass, $m_e = 9.108 \times 10^{-31}$ kg	31
Electron's charge, $-e = -1.602 \times 10^{-19}$ C	29
Classical electron radius, $r_e = \dfrac{e^2}{4\pi\varepsilon_0 m_e c^2} = 2.818 \times 10^{-15}$ m	31
Vacuum permittivity, $\varepsilon = 8.854 \times 10^{-12}\ \mathrm{C^2\,N^{-1}\,m^{-2}}$	12
Vacuum permeability, $\mu_0 = 4\pi \times 10^{-7}\quad \mathrm{N\,s^2\,C^{-2}}$	12
Vacuum light speed, $c = (\mu_0\varepsilon_0)^{-1/2} = 2.9979 \times 10^8\ \mathrm{ms^{-1}}$	13
Refractive index of air at 0 K, $\eta_{\mathrm{air}} = 1.000293$	14
Refractive index of water at 20 K, $\eta_w = 1.333$	14
Refractive index of fused silica, $\eta = 1.55$	14
Planck's constant, $h = 6.626 \times 10^{-34}$ Js (and $\hbar \equiv h/2\pi$)	20
Boltzmann's constant, $k = 1.381 \times 10^{-23}\ \mathrm{J\,K^{-1}}$	60
Loschmidt's number, $N_L = 2.69 \times 10^{25}\ \mathrm{m^{-3}}$ (STP)	42
Air density at sea level and 0 K $\cong 2.55 \times 10^{25}\ \mathrm{m^{-3}}$	42
1 eV $= 1.602 \times 10^{-19}$ J ($\cong$ 1239 nm)	62
One Kayser or wavenumber $= 1\ \mathrm{cm^{-1}}$	74
Rayleigh backscattering cross section for air at 694.3 nm, $\sigma_\pi^R \cong 2.15 \times 10^{-28}\ \mathrm{cm^2\,sr^{-1}}$	42

Laser Remote Sensing

FUNDAMENTALS AND APPLICATIONS

RAYMOND M. MEASURES

Professor of Applied Science and Engineering
University of Toronto
Institute for Aerospace Studies

KRIEGER PUBLISHING COMPANY
MALABAR, FLORIDA

Original Edition 1984
Reprint Edition 1992

Printed and Published by
KRIEGER PUBLISHING COMPANY
KRIEGER DRIVE
MALABAR, FLORIDA 32950

Library of Congress Cataloging-In-Publication Data
Measures, Raymond M.
Laser remote sensing : fundamentals and applications / Raymond M. Measures.
p. cm.
Reprint. Originally published: New York : Wiley, 1984.
Includes biliographical references and index.
ISBN 0-89464-619-2 (alk. paper)
1. Remote sensing--Equipment and supplies. 2. Optical radar.
I. Title.
G70.6.M4 1991
621.36'78--dc20 91-20352
CIP

10 9 8 7 6 5 4

To my parents,
sisters,
wife,
and children.

Preface

And the Lord shed light upon the earth, and saw that it was good. . . .
Genesis 1 : 4

The *Laser* is one of the most versatile tools ever invented. Lasers are likely to revolutionize computers and communications, replacing electronics with optoelectronics; they have performed delicate microsurgery on single cells, and might one day be used from space battle stations to destroy missiles. It therefore should not come as a great surprise that they have an important role in the field of remote sensing, for their special properties make them ideal for probing the environment. Indeed, the laser's capability of performing an analysis at a distance effectively adds a new dimension to remote sensing. The range and kind of applications made possible are quite extraordinary. For example, a ground-based laser remote sensor has been able to distinguish sodium atoms from the much larger concentration of other species at an altitude of 90 km. What is even more remarkable is that this system was able to effectively count the atoms at this altitude when their density was less than 1000 per cubic centimeter.

In writing this book I was faced with the usual problems encountered in compiling a work on a rapidly advancing field of research. In response I have endeavored to view the subject matter as one might a building. I have attempted to lay a strong foundation of basic material that is intended to have an extended life. Upon this foundation I have built a structure with many windows that are designed to provide the reader with an almost unobstructed view of the research field. This should enable the reader to see the evolution of the subject and the major lines of its advance.

This book is intended for the broad class of scientists, researchers, and students who are interested in the environment and wish to become acquainted with the impact that lasers are having on remote sensing. I have attempted to make the book as self-contained as is reasonable and have therefore, in the early chapters, reviewed much of the basic physics needed to understand the subject.

The outline of *electromagnetic theory* provided in Chapter 2 serves amongst other purposes to prepare the reader for the concepts of *elastic* (*Rayleigh and Mie*) *scattering*. The *quantum description* of atoms and molecules given in Chapter 3 lays the basis for introducing *spectroscopy*, the basic *radiation processes*, and *inelastic* (*Raman*) *scattering*. The material in these two chapters prepares the reader for the *interaction and propagation of radiation*, treated in Chapter 4. This includes a discussion of the *radiative transfer equation* with and without scattering and also deals with laser-induced fluorescence. In Chapter 5, I have reviewed the *fundamentals of lasers* and have included a brief description of the types of lasers relevant to remote sensing. Chapter 6 examines the *basic methods* by which lasers are used in remote sensing, provides a description of the *systems* involved, and considers the problem of *signal-to-noise ratio*. The basic *laser remote-sensing equations* are derived in Chapter 7, where particular care is taken to show the difference between the equations used for *scattering* and for *laser-induced fluorescence*. The *analysis* and *interpretation* of the signals obtained through laser remote sensing are treated in Chapter 8. Broad reviews of *atmospheric* and *hydrospheric* laser remote-sensing applications form the basis of Chapters 9 and 10, respectively. These chapters have been written with an eye to providing the reader with some insight into the enormous breadth of measurements that can be undertaken remotely with lasers as well as indicating the state of the art. Although I have endeavored to be as comprehensive as possible, the literature is already so large that there are bound to be some omissions. For these I apologize.

The material of this book could serve as the basis of a two-semester course on laser remote sensing, or to enhance any more general course on remote sensing.

SI units have been used throughout much of this book and in the formation of the equations. Nevertheless, there are many instances where I have chosen to use more convenient units. For example, irradiance is generally given in $W\ cm^{-2}$ rather than $W\ m^{-2}$.

I am indebted to many researchers in the field who have contributed material and offered encouragement in preparing this book. I would like to offer special thanks to Mrs. Winifred Dillon for her diligent and patient typing of the manuscript, Mrs. Laura Quintero for her clear and beautiful art work, and my wife for her patience, help, and understanding in very many ways. I also appreciate the sacrifice of my children and the help in proofreading given by my brother-in-law.

RAYMOND M. MEASURES

Thornhill, Ontario

Contents

Laser Remote Sensing

1

Introduction

The concurrence of high technology with man's expanding need to study his environment is timely. Within the past two decades we have witnessed the invention of the laser and its remarkable development, while at the same time we have grown acutely aware of the finite nature of the earth and the fragile balance of its ecosystems. Environmental issues such as the influence of fluorocarbons and nitric oxide on the earth's protective shield of ozone, the effect of carbon dioxide and volcanic dust on the climate, the formation of photochemical smog, oil pollution, and acid rain have drawn to our attention the ease with which the biosphere can be perturbed.

Solar radiation is the principal source of energy incident upon the earth and consequently plays a key role in determining both the structure and the composition of the atmosphere. The kinds of interaction permissible with the constituents of the atmosphere are essentially controlled by the wavelength of the radiation. Short-wavelength radiation ($\lambda < 200$ nm or 0.2 μm, corresponding to a photon energy in excess of 6 eV) is capable of dissociating, and for short enough wavelengths ($\lambda < 100$ nm) ionizing, the major components nitrogen (N_2) and Oxygen (O_2). Ozone can be dissociated for wavelengths less than 320 nm. Less-violent interactions occur for longer-wavelength radiation, and in the infrared ($\lambda > 760$ nm) molecular vibration is the main result of excitation. This state of affairs is reflected in the change in the solar spectral irradiance brought about by its passage through the atmosphere: see Fig. 1.1. The upper curve gives the coarse spectral irradiance beyond the earth's atmosphere. The maximum is seen to occur at a wavelength of about 470 nm, (Lintz and Simonett, 1976). About 20% of the sun's energy falls in the spectral band $\lambda < 470$ nm, while 44% lies in the visible band from 400 to 760 nm.

The dashed curve of Fig. 1.1 represents the spectral irradiance of a black body at $T = 5900$ K, and the lowest curve the spectral irradiance of direct sunlight at sea level. The difference between the two continuous curves represents the attenuation by scattering and absorption, while the stippled

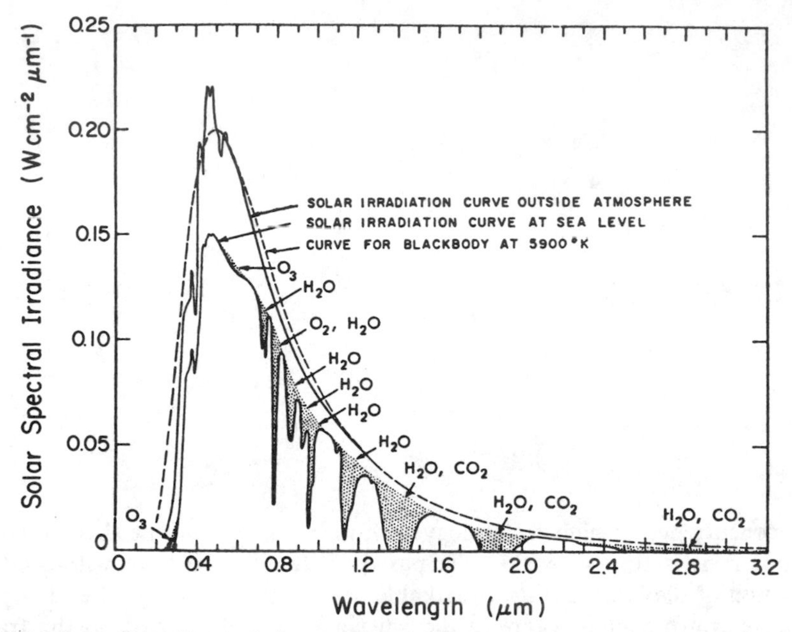

Fig. 1.1. Spectral irradiance of direct sunlight before and after it passes through the earth's atmosphere. The stippled portion gives the atmospheric absorption. The sun is at the zenith (Valley, 1965).

portion indicates the loss by absorption. Aside from the narrow absorption band of O_2 at 760 nm (0.76 μm), H_2O and CO_2 are responsible for the major losses.

The spectral radiance of the earth and its atmosphere as seen from a meteorological satellite (NIMBUS 4) under clear-sky conditions is presented as Fig. 1.2. The smooth curves give the computed spectral radiance that would be emitted by black bodies at temperatures of 260, 280, and 300 K (Lintz and Simonett, 1976). It is evident that the bulk of the energy radiated away from the earth lies between 8 and 14 μm.

The thermal balance between the incident solar radiant flux and the emission from the earth leads to the thermal structure of the atmosphere portrayed in Fig. 1.3. The lowest, and therefore most dense, part of the atmosphere is termed the *troposphere*. Its boundary is defined to coincide with the lowest temperature minimum and ranges from about 10 km at the poles to about 15 km in the tropics. Above the troposphere lies the *stratosphere*—a region owing its characteristics to its ozone content. As seen from Fig. 1.3, the classification of the atmosphere, above the troposphere, depends on whether it is the density or temperature structure that is of interest. The temperature profile shown in this figure is only intended as a guide, and in fact is more representative of a polar region than a tropical zone (Thrush, 1977).

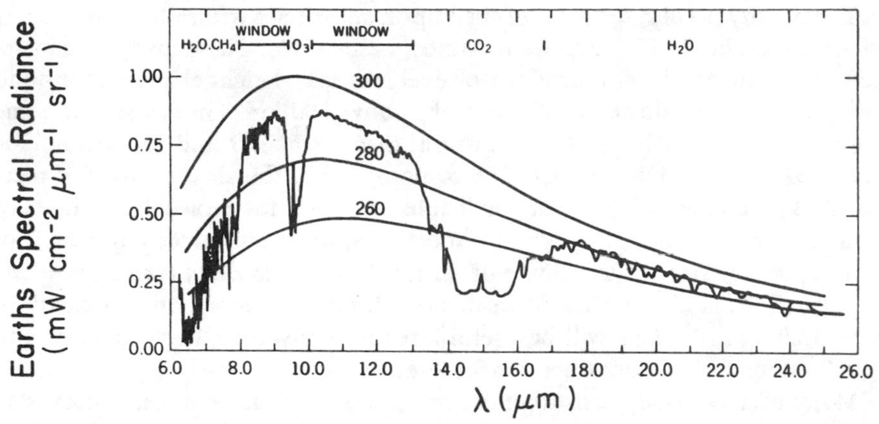

Fig. 1.2. Nadir radiance measured from a satellite (curve with structure). The smooth curves give values of blackbody radiances at the indicated temperature in kelvins (Lintz and Simonett, 1976).

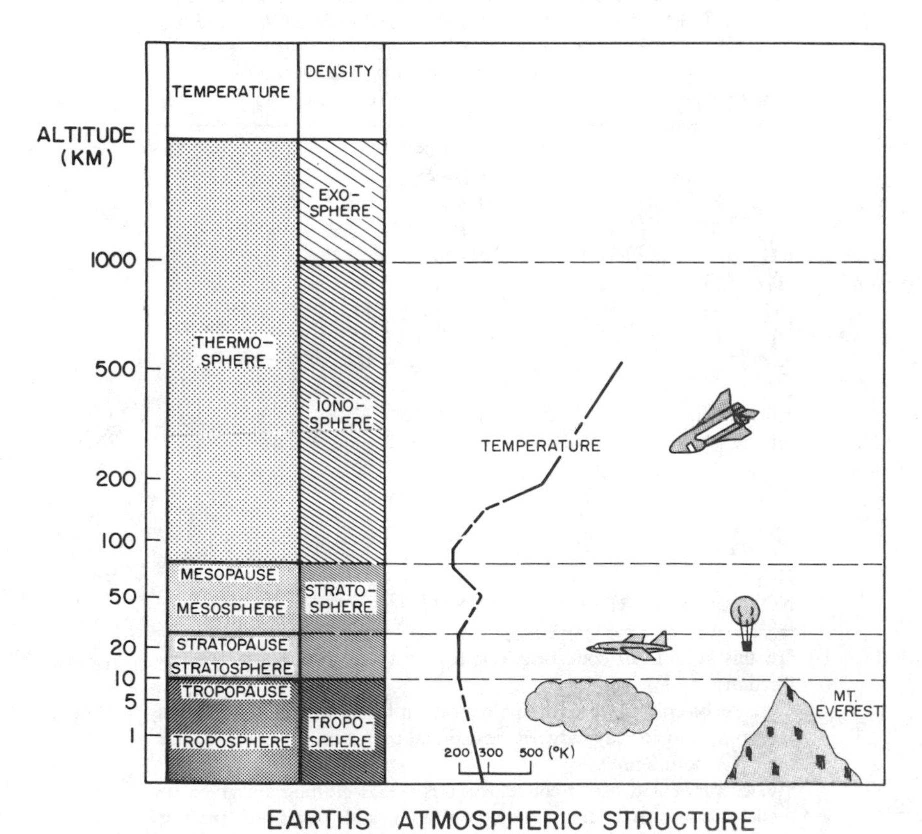

Fig. 1.3. Schematic of earth's atmospheric structure. Adapted from Schriever (1960).

The density profiles for most of the important species within the atmosphere are described in the U.S. Standard Atmosphere (1976). The density in general falls off exponentially with height. However, the scale height changes at around 100 km so that the decrease is less severe above 100 km. The density of some constituents (primarily those that are photoactive like O and O_3) cannot be described in this simple manner. The composition of the air (exclusive of inert gases) at ground level is provided in Table 1.1. Since the molecular weights of N_2 and O_2 are 28 and 32 kg/kmol respectively, the mean molecular weight of air is 28.96 kg/kmol. The volume of an ideal gas at *standard temperature and pressure* (STP) is 22.4 m^3/kmol. In this condition the average mass density of air is 1.29 kg m^{-3}. This will be useful later for converting between ppm and $\mu g\ m^{-3}$ when we consider trace constituents.

Many of these trace constituents can be classified as *pollutants*; that is to say, they can be regarded as harmful to mankind. Some affect man

TABLE 1.1. COMPOSITION OF THE AIR EXCLUSIVE OF THE INERT GASES AT GROUND LEVEL (KELLY ET AL., 1976)

Molecule	Average Concentration (Volume Fraction)[a]
N_2	0.78084[b]
O_2	0.20946[b]
H_2O	1.3×10^{-7} to 4.5×10^{-2}[c]
CO_2	3.18×10^{-4} [3.30×10^{-4}][d]
CH_4	$(1–1.4) \times 10^{-6}$ [1.6×10^{-6}][d]
H_2	5×10^{-7}[b]
CO	$(0.5–2.5) \times 10^{-7}$ [0.75×10^{-7}][d]
O_3	$(2–7) \times 10^{-8}$[c]
N_2O	$(2.7–3.5) \times 10^{-7}$ [2.8×10^{-7}][d]
NO	$10^{-8}–10^{-6}$
NO_2	$10^{-9}–10^{-6}$
HNO_3	2.8×10^{-9}
NH_3	$\leq 10^{-4}$
SO_2	$(0.5–7.2) \times 10^{-9}$
H_2S	$(1.6–16) \times 10^{-9}$
HCHO	$\leq 10^{-7}$
HCl	$(1–2.6) \times 10^{-9}$
NO_3, OH, HO_2, CH_3O	5×10^{-11}?

[a] In this section all concentrations are reported in terms of volume fractions.

[b] These molecules have zero dipole moment for vibrational transitions. However, collisionally induced vibrational transition are observed (i.e. the "N_2 continuum").

[c] Water vapor and ozone concentrations versus altitude are given for model atmospheres for several latitudes and times of year in McClatchey et al. (1972).

[d] Values in square brackets are those used in atmospheric transmittance calculations based on McClatchey et al. (1973).

directly—carbon monoxide for example—while the influence of others is much more subtle. Chlorofluorocarbons (CFC, also known by their trade name Freon) were once thought to be completely innoxious due to their chemical inertness; however, a calculation by Molina and Rowland (1974) suggested that because of their stability these molecules would be carried into the stratosphere, where intense solar UV would break them up and release chlorine atoms, which would then participate in the catalytic annihilation of ozone molecules. This is schematically illustrated in Fig. 1.4 (Thrush, 1977).

A 1979 report of the U.S. National Academy of Sciences projected that continued release of CFC at the 1977 rate would lead to a 16% reduction in the ozone concentration within a few decades. This decline of stratospheric ozone would allow about 44% more of the so-called *damaging ultraviolet* (290- to 320-nm) radiation from the sun to reach the ground. The report suggests that this would lead to very many thousands of additional cases of skin cancer per year in the U.S. alone and that many other forms of life, including crops, could be adversely affected (Maugh 1980). Although these predictions are somewhat

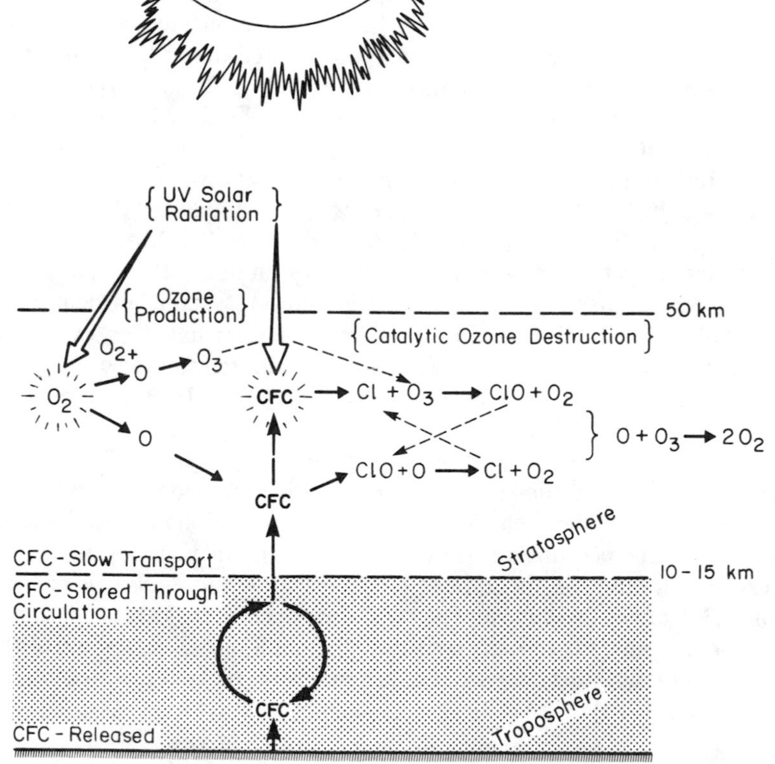

Fig. 1.4. Simplified view of the production of ozone and its catalytic destruction by chlorofluorocarbons, primarily $CFCl_3$ (or CFC-11) and CF_2Cl_2 (or CFC-12).

controversial, there is no argument about the need for a careful and continuing study of the stratosphere and its constituents, particularly in view of the fact that other by-products of man's activities such as NO and CCl_4 can also contribute to the destruction of the O_3 layer (Stedman, 1976; Thrush, 1977).

Another long-term problem facing mankind stems from the increasing amount of carbon dioxide (CO_2) being produced as a result of burning fossil fuels. It has been estimated that the atmospheric CO_2 burden has been increased by about 15% over the past century by man's activities (Revelle, 1982). The importance of CO_2 comes from its optical properties, for it is transparent to visible radiation but relatively opaque to thermal infrared radiation. Consequently, it gives rise to the so-called *greenhouse effect* whereby solar radiation passes relatively unimpeded through the atmosphere while the thermal radiation from the earth's surface is trapped within the CO_2 blanket. Other molecules can also contribute to this phenomena—for example, Varansi and Ko (1977) suggested that the fluorocarbons CFC-11 and CFC-12 mentioned earlier could raise the global surface temperature by as much as 0.9 K when present in a concentration as low as a few parts per 10^9 (ppb) due to their strong absorption bands in the 8 to 12-μm interval—which includes the region of strongest emission from the earth (see Fig. 1.2).

Both nitric oxide and sulfur dioxide have been implicated in the growing environmental problem of *acid rain*, for it is suspected that NO is converted to nitric acid (HNO_3) and SO_2 to sulfuric acid (H_2SO_4) through some chain of chemical reactions within the atmosphere (Likens, 1975). SO_2 has of course been a pollutant over the centuries, back to the start of the industrial revolution in Europe. The early fogs of cities like London were primarily due to oxides of sulfur (SO_x). Most European countries today have opted for an SO_2 standard[†] of 100 μg m^{-3} annual level (Middelton, 1976). Around the end of the forties, the effect of a new kind of air pollution—PAN, causing eye irritation, plant damage, and visibility degradation, became evident in Los Angeles. Subsequently, this form of oxidizing pollution has been experienced in many large cities throughout the world; it is attributed to the action of solar ultraviolet radiation upon mixtures of hydrocarbons (HC) and oxides of nitrogen (NO_x) (Pitts, 1977).

The principal constituents of the troposphere (apart from N_2, O_2, H_2O, and the inert gases) are listed in Table 1.2 together with some idea of their sources and an estimate of their emission rates, background concentrations, and characteristic residence times (Wright et al., 1975). An analysis of the main hydrocarbons observed in the untreated exhaust of an automobile is presented in Table 1.3. Besides the atmospheric constituents listed in Tables 1.2 and 1.3, there are many others that are present in very small quantities. Some of these have an importance that is way out of line with their concentration. The

[†] The molecular weight of SO_2 is 64 kg/kmol. Consequently its density is 2.86 kg m^{-3}, as 1 kmol occupies 22.4 m^3. Thus the volume taken up by 100 μg (or 10^{-7} kg) of SO_2 is 3.5×10^{-8} m^3, and the concentration corresponding to 100 μg m^{-3} is then 35 ppb.

hydroxyl free radical OH is an excellent example, for it has been implicated in the global conversion of CO to CO_2 and as a key intermediary in the photochemical formation of smog, yet its concentration has been measured to be less than a few parts per 10^{12} (ppt) (Davis et al., 1979).

In this book we shall attempt to show how lasers are playing an ever increasing role in studying the environmental problems mentioned above as well as opening new vistas with regard to remote sensing of the atmosphere, oceans, lakes, and rivers of our planet. Although Schawlow and Townes (1958) and Maiman (1960) deserve recognition for their roles in creating the first laser, it was the giant-pulse technique, invented by McClung and Hellwarth (1962) that made remote optical probing really attractive. The first laser studies of the atmosphere were undertaken by Fiocco and Smullin (1963), who recorded laser echoes from the upper regions of the atmosphere, and by Ligda (1963), who probed the troposphere.

In the period that followed this pioneering work great strides have been made, both in the development of *lidar* (*li*ght *d*etection *a*nd *r*anging) systems and in the sophistication of their use. It was immediately appreciated that, as with radar, lidar systems could provide spatially resolved measurements in "real time." An important advance was the recognition that the radiation detected at wavelengths different from that of the laser's output contained highly specific information that could be used to determine the composition of the target region. The specificity of this approach was further enhanced by the broad selection of laser wavelengths that became available and, in the case of resonance excitation, by the precise tuning made possible with certain classes of laser. This ability of lidar systems to perform an effective spectral analysis of a distant target has added a new dimension to remote sensing and has made possible an extraordinary variety of applications. These range from ground-based probing of the trace-constituent distribution in the tenuous outer reaches of the atmosphere (Granier and Megie, 1982) to airborne chlorophyll mapping of the oceans to establish rich fishing areas (Hoge and Swift, 1981).

The atmosphere was one of the first arenas in which the laser's special properties of high power, monochromaticity, short duration, and beam collimation were put to the test. The resulting triumph of the laser has been well documented in the many review articles cited throughout the text. Direct laser measurement of the atmospheric parameters important in evaluating the composition, structure, properties, and dynamic behavior of the atmosphere have been undertaken. The results have led to speculation that, in the future, lidar systems could play a key role in providing the basic information pertinent to man's understanding of the atmosphere and his ability to both predict and modify the weather. Laser probing of the atmosphere has also revealed that lidar systems can detect and quantify trace constituents that arise both naturally and as by-products of our technological society.

Towards the end of the laser's first decade, some consideration was given to the development of earth-oriented laser sensors that could be used from mobile platforms, such as aircraft and helicopters. In the initial applications, these

TABLE 1.2. CHARACTERISTICS OF ATMOSPHERIC TRACE GASES (WRIGHT ET AL., 1975)

Pollutant	Major Sources		Estimated Emission Rates		Polluted-Atmosphere Concentrations	Atmospheric Background Concentrations	Calculated Atmospheric Residence Time
	Anthropogenic	Natural	Anthropogenic (10^9 kg/yr)	Natural (10^9 kg/yr)			
CO_2	Combustion	Biological decay; release from oceans	13,000	10^6	350 ppm	320 ppm	2–4 years
CO	Auto exhaust; combustion	CH_4 oxidation	250	3000	5 ppm	0.1 ppm	0.1 year
SO_2	Combustion of coal and oil	Volcanoes	133	Small	1 ppm	0.2–2 ppb	4 days
H_2S	Chemical processes; sewage treatment	Volcanoes, biological action in swamps	2.7	90	4 ppb	0.2 ppb	2 days
O_3	Photochemical smog	Photolysis of O_2 (25–50 km)	Uncertain	Uncertain	0.3 ppm	0.01 ppm	1 day
NO	Combustion	Bacterial action in soil	48	460	0.2 ppm	0.2–2 ppb	1 day
NO_2	None	Conversion of NO	Negligible	Negligible	0.1 ppm	0.5–4 ppb	5 days
N_2O	None	Biological action in soil	Small	540	0.25 ppm	0.25 ppm	4 years
NH_3	Waste treatment	Biological decay	3.6	1000	0.01 ppm	6–20 ppb	7 days
Hydro-carbons	Combustion exhaust; chemical processes	Biological processes	CH_4: 80 Others: uncertain	CH_4: 2000 Others: uncertain	CH_4: 3 ppm Others: 2 ppm	CH_4: 1.5 ppm Others: < 1 ppb	CH_4: 1 year Others: unknown
HCHO	Combustion exhaust; atmospheric reactions	Biological processes	100	1000	0.05 ppm	1 ppb	1–5 days
HCl	Chemical processes; rocket engine exhaust	Unknown	Uncertain	Unknown	1–5 ppm	Unknown	Unknown

TABLE 1.3. PRINCIPAL HYDROCARBONS IN EXHAUST—ORIGIN AND CONTRIBUTION TO PHOTOCHEMICAL REACTIVITY (HURN, 1968)

Most Prominent on Basis of Concentration	Percent of Total Hydrocarbons
Ethylene	19.0
Methane	13.8
Propylene	9.1
Toluene[a]	7.9
Acetylenc	7.8
1-Butene, *i*-butene, and 1,3-butadiene	6.0
p-, *m*-, and *o*-xylene[a]	2.5
i-Pentane[a]	2.4
n-Butane[a]	2.3
Ethane	2.3
Total	73.1

[a]Fuel components—others are products of fuel cracking or rearrangement in the engine.

downward-pointing laser systems were operated in a mode somewhat analogous to radar, where surface scattering and reflection represented the dominant form of interaction. Surface-wave studies and bathymetric measurements in coastal waters were the first topics to be given serious consideration. The possibility of undertaking studies of water turbidity grew naturally from the latter series of experiments.

Another notable advance was made with the realization that use of a short-wavelength laser could broaden the spectrum of applications, as a result of laser-induced fluorescence, and led to the development of a new form of remote sensor, termed a "laser fluorosensor." The fluorescence induced in thin oil films was of particular interest, as it provided the motivation to develop this new instrument in the hope that it would be capable of detecting and classifying oil slicks. Recent flight tests with these instruments have demonstrated that in addition to such functions they can also undertake measurements of the oil film thickness in the 1 to 10-μm range.

Early in these studies laser-induced fluorescence from natural bodies of water was considered to constitute a source of background emission that could interfere with the oil fluorescence signal. Further studies have not only diminished this worry, but have discovered that the water fluorescence signal might serve to indicate the presence of high organic contamination and enable the dispersion of various kinds of effluent plumes to be remotely mapped. The fluorescence of chlorophyll has long been known, and recently airborne laser fluorosensors have been shown to be capable of remotely mapping the chlorophyll concentration of natural bodies of water.

It is clear that the scope of lasers in environmental remote sensing is extensive. We may generalize by stating that they can be used to undertake:

1. Concentration measurements including both major and minor constituents, and are therefore well suited to pollution surveillance and monitoring.
2. Evaluation of thermal, structural, and dynamic properties.
3. Threshold detection of specific constituents, and are therefore well suited for alarm purposes.
4. Mapping of effluent plume dispersal as a function of time.
5. Spectral fingerprinting of a specific target, such as an oil slick.

Furthermore, these observations can be made remotely from the ground or from ships, helicopters, aircraft, or satellites with both spatial and temporal resolution in most instances.

It is worth remembering that the subject is still young and that the current rapid development taking place in lasers, detectors, data-processing electronics, and computers is sure to be reflected in more sensitive, reliable, and accurate laser remote sensors. This in turn will probably stimulate their use in an even wider array of applications—some of which we cannot even conceive at this point in time.

2

Electromagnetic Theory of Radiation

Classical electromagnetic theory is completely described by the elegant set of equations due to J. C. Maxwell. We shall see that these equations lead naturally to *wavelike* propagation of electromagnetic energy, and we shall study the properties and nature of these electromagnetic waves, including their classical interaction with matter.

In the subsequent chapter we shall allow for the quantized nature of atoms (and molecules) but retain the classical description of the electromagnetic fields —the so called *semiclassical* treatment of the interaction of radiation with matter. Although this treatment is adequate for our purposes, it should be recognized that from certain standpoints the radiation field must also be considered to be quantized into packets of energy called *photons*. Reconciliation of these apparently inconsistent views is possible if we think of Maxwell's equations as describing the statistical properties of large numbers of photons.

2.1. MAXWELL'S EQUATIONS

In the presence of a medium, Maxwell's equations can be expressed in the form

$$\nabla \cdot \mathbf{E} = \frac{q}{\epsilon} \tag{2.1}$$

$$\nabla \cdot \mathbf{H} = 0 \tag{2.2}$$

$$\nabla \times \mathbf{E} = -\mu \frac{\partial \mathbf{H}}{\partial t} \tag{2.3}$$

$$\nabla \times \mathbf{H} = \epsilon \frac{\partial \mathbf{E}}{\partial t} + \mathbf{j} \tag{2.4}$$

Here **E** and **H** represent the *electric* and *magnetic* field vectors, ϵ and μ represent the *electric permittivity* and *magnetic permeability* of the medium, and q and **j** represent the *free charge* and *current densities* respectively. In general we can write the permittivity of the medium in terms of the value associated with a vacuum, $\epsilon_0 = 8.854 \times 10^{-12}\ \mathrm{C^2\ N^{-1}\ m^{-2}}$ and the dielectric constant (relative permittivity) K_e of the medium, namely,

$$\epsilon = K_e \epsilon_0 \tag{2.5}$$

If the medium has a *polarization* **P** (which is defined as the dipole moment per unit volume), then we can write

$$\epsilon = \epsilon_0 + \frac{\mathbf{P}}{\mathbf{E}} \tag{2.6}$$

In a similar manner, the permeability of the material can be expressed in terms of the vacuum value, $\mu_0 = 4\pi \times 10^{-7}\ \mathrm{N\ s^2\ C^{-2}}$, and the relative permeability K_m, namely,

$$\mu = K_m \mu_0. \tag{2.7}$$

Equation (2.1) is derived from Coulomb's law and in essence describes the force between electric charges. For two isolated charges, $Q_1(C)$ and $Q_2(C)$, we can write the force on the first due to the second as

$$\mathbf{F} = Q_1 \mathbf{E}$$

where the electric field at the location of the first charge and arising from the presence of the second is

$$\mathbf{E} = \frac{Q_2}{4\pi\epsilon_0 r^3}\mathbf{r}$$

Here **r** is the vector displacement between the charges, and r is its magnitude.

Equation (2.2) states that no magnetic charges exist, and from this we can deduce that **H** is *solenoidal*. That is to say, **H** can be described in terms of a *magnetic potential* **A**:

$$\mu\mathbf{H} \equiv \nabla \times \mathbf{A}$$

Equation (2.3) is derived from Faraday's law of magnetic induction. If $\partial\mathbf{H}/\partial t = 0$ this leads to a *conservative* electric field, and under these conditions the electric field is *irrotational* and can be described in terms of a *scalar potential* ϕ:

$$\mathbf{E} = -\nabla\phi.$$

The more general description of the electric field involves both ϕ and **A** in the form

$$\mathbf{E} = -\nabla\phi - \frac{\partial \mathbf{A}}{\partial t}$$

Although equation (2.4) is based upon Ampère's law, it is the inclusion of the displacement current term, $\epsilon\, \partial\mathbf{E}/\partial t$, that represents one of Maxwell's greatest achievements, for not only did this enable equation (2.4) to be consistent with the continuity of charge, but it leads naturally to the propagation of electromagnetic energy. A more detailed discussion of Maxwell's equations and electromagnetic theory is provided by Born and Wolf (1964). It should be noted that although SI units will be used in the development of formulae, more convenient units will often be used, particularly where numerical quantities are involved; for example, cross sections will be given as cm^2, densities in cm^{-3}, and irradiance in $W\ cm^{-2}$.

2.2. WAVE EQUATION AND PLANE-WAVE SOLUTIONS

If we assume $q = \mathbf{j} = 0$ (i.e., no free charges and zero current density) and we can take the *curl* of equation (2.3), we obtain

$$\nabla \times \nabla \times \mathbf{E} = -\mu \frac{\partial(\nabla \times \mathbf{H})}{\partial t} \tag{2.8}$$

If we then use the identity

$$\nabla \times \nabla \times \mathbf{E} \equiv \nabla(\nabla \cdot \mathbf{E}) - \nabla^2 \mathbf{E}$$

and take account of the fact $q = 0$ so $\nabla \cdot \mathbf{E} = 0$, then equation (2.8) with equation (2.4) can be cast in the form of a wave equation

$$\nabla^2 \mathbf{E} = \mu\epsilon \frac{\partial^2 \mathbf{E}}{\partial t^2} \tag{2.9}$$

An identical wave equation can be obtained for the magnetic field $\mathbf{H}$, where in both cases

$$v \equiv (\epsilon\mu)^{-1/2} \tag{2.10}$$

represents the phase velocity of propagation for the electromagnetic fields within the medium. In free space this velocity reduces to

$$c = (\epsilon_0 \mu_0)^{-1/2} = 2.9979 \times 10^8\ \mathrm{m\ s^{-1}} \tag{2.11}$$

the characteristic vacuum velocity of light. In general the phase velocity within the medium can be expressed in terms of c through the introduction of the *refractive index* n of the medium,

$$n \equiv \frac{c}{v} = (K_e K_m)^{1/2} \tag{2.12}$$

For air at 0°C and one atmosphere, $n_{\text{air}} = 1.000293$, while at 20°C, $n_{\text{water}} = 1.333$ and $n_{\text{fused silica}} = 1.55$ (Hecht and Zajac, 1974, p. 38). It should be noted (as we shall see later) that n, as defined by (2.12), can be a complex quantity and under these circumstances it is the *real* component η that corresponds to the refractive index.

It is possible to show by substitution that sinusoidal solutions of the form

$$\mathbf{E}(\mathbf{r}, t) = \mathbf{E}_0 \mathbf{e}^{i(\mathbf{k}\cdot\mathbf{r} - \omega t)} \tag{2.13}$$

satisfy the wave equation for the electric field. Here ω represents the angular frequency of the electric-field disturbance and $\mathbf{k}$ is termed the *propagation vector* (or sometimes the wave vector) and assigned a magnitude $k = 2\pi/\lambda$ (the propagation constant), where λ represents the wavelength of the disturbance. $\mathbf{E}_0$ is taken to be a constant vector (i.e., one that is independent of space or time) that represents the electric-field disturbance at $\mathbf{r} = 0$ and $t = 0$. The quantity $\mathbf{k} \cdot \mathbf{r} - \omega t$ is termed the *phase* of the disturbance.

It should be noted that although complex representation of field quantities simplifies much of the mathematical manipulation, *the physically measured fields are obtained from the real part of the complex quantities at the end of the calculation*. However, as shown by Yariv (1976), errors can arise in using this procedure where products (or powers) of sinusoidal functions are involved.

It is evident from (2.13) that $\mathbf{E}(\mathbf{r}, t)$ is the same at any instant of time t, over a surface described by the relation

$$\mathbf{k} \cdot (\mathbf{r} - \mathbf{r}_0) = 0 \tag{2.14}$$

for $\mathbf{r} - \mathbf{r}_0$ represents a vector that lies in a plane that is perpendicular to $\mathbf{k}$. Furthermore, since $\mathbf{E} = \mathbf{E}_0$, whenever,

$$\mathbf{k} \cdot \mathbf{r} - \omega t = 0 \tag{2.15}$$

it is apparent that the electric-field disturbance described by (2.13) can be viewed as propagating in the $\mathbf{k}$-direction with a *phase* velocity

$$v = \frac{\omega}{k} \tag{2.16}$$

which when combined with (2.12) yields

$$k = \frac{n\omega}{c} \tag{2.17}$$

Thus (2.13) represents a plane traveling-wave solution of (2.9) wherein the wavefront (a surface of constant phase) is a plane perpendicular to the direction of propagation and traveling with a velocity c in a vacuum; see Fig. 2.1. By contrast a solution with a phase factor of the form $\mathbf{e}^{i(\mathbf{k}\cdot\mathbf{r}+\omega t)}$ describes a plane wave propagating in a direction opposite to **k**.

If the electric field disturbance is given by (2.13), then it is reasonable to assume

$$\mathbf{H}(\mathbf{r}, t) = \mathbf{H}_0\mathbf{e}^{i(\mathbf{k}\cdot\mathbf{r}-\omega t)} \tag{2.18}$$

where $\mathbf{H}_0$ is a constant vector that represents the magnetic field disturbance at $\mathbf{r} = 0$ and $t = 0$.

If we use $\nabla \cdot \mathbf{E} = 0$ (in the absence of free charges) and $\nabla \cdot \mathbf{H} = 0$, then we can write

$$\mathbf{E}_0 \cdot \mathbf{k} = 0 \quad \text{and} \quad \mathbf{H}_0 \cdot \mathbf{k} = 0 \tag{2.19}$$

which leads us to conclude that both $\mathbf{E}_0$ and $\mathbf{H}_0$ are perpendicular to **k**, the direction of propagation.

If we also substitute (2.13) and (2.18) into Maxwell's form of Faraday's equation, we can write

$$i\mathbf{k} \times \mathbf{E}_0 = i\omega\mu\mathbf{H}_0$$

which means that

$$\mathbf{H} = \epsilon v(\hat{\mathbf{k}} \times \mathbf{E}) \tag{2.20}$$

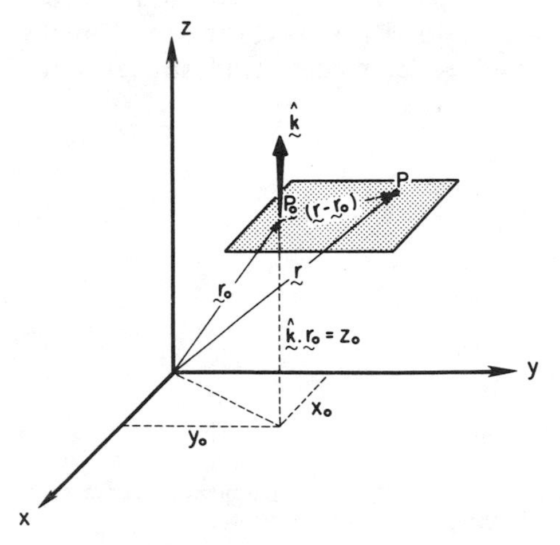

Fig. 2.1. The point P clearly lies in a plane that is perpendicular to the vector **k** if $\mathbf{k} \cdot (\mathbf{r} - \mathbf{r}_0) = 0$.

or in terms of the *magnetic induction* $\mathbf{B} \equiv \mu\mathbf{H}$ we can write

$$\mathbf{B} = \frac{1}{v}(\hat{\mathbf{k}} \times \mathbf{E}) \tag{2.21}$$

where $\hat{\mathbf{k}}$ is the unit vector in the $\mathbf{k}$ direction (i.e. the direction of propagation). It is clear from (2.19) and (2.20) that in a vacuum the electric and magnetic fields $\mathbf{E}$ and $\mathbf{H}$ of a plane traveling electromagnetic wave are perpendicular to each other and each is perpendicular to the direction of propagation—which constitutes a transverse wave; see Fig. 2.2.

The usual treatment of the propagation of light including superposition, reflection, refraction, and dispersion (to be discussed later) presumes a linear relationship between the electromagnetic field and the responding atomic (or molecular) system. Although this is justified for most sources of light, the power density level achieved by certain lasers can lead to *nonlinear* effects such as the generation of *harmonics*. Nonlinear optics in fact embraces a wide variety of phenomena, and it is hoped that the following brief discussion will stimulate the curiosity of the reader.

We saw earlier, in equation (2.6), that the electric polarization induced by an electric field of modest strength can be written in the form

$$\mathbf{P} = \epsilon_0(K_e - 1)\mathbf{E} = \mathscr{P}\mathbf{E}$$

where $\mathscr{P}$ represents the volume *polarizability* of the medium. Clearly, as $\mathbf{E}$ increases, some point is reached at which $\mathbf{P}$ saturates for any given medium. Under these circumstances we are forced to introduce higher-order terms into the relationship, and we can write

$$\mathbf{P} = \mathscr{P}_1 E + \mathscr{P}_2 E^2 + \mathscr{P}_3 E^3 + \cdots \tag{2.22}$$

for an isotropic medium in which $\mathbf{P}$ is parallel to $\mathbf{E}$. The usual linear polarizability (written here $\mathscr{P}_1$ as opposed to $\mathscr{P}$) is much greater than the *nonlinear polarizabilities* $\mathscr{P}_2, \mathscr{P}_3, \ldots,$ and so at modest field strengths the nonlinear effects are negligible.

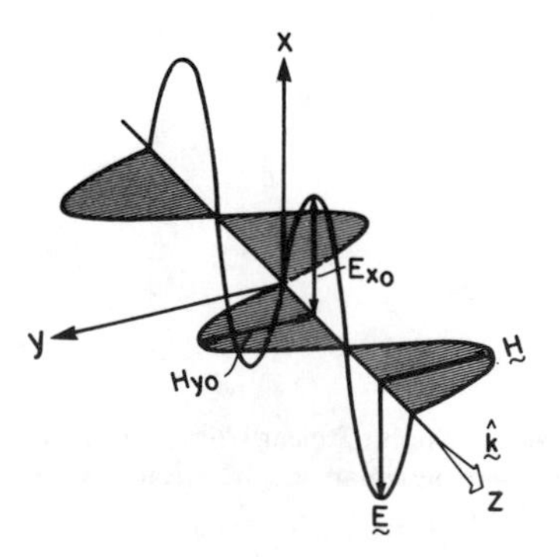

Fig. 2.2. E and H fields of a harmonic electromagnetic wave propagating in the *z*-direction.

In the case of a *harmonic* electromagnetic disturbance we can express the magnitude of the electric-field component incident upon the medium in the form

$$\mathbf{E}(\mathbf{r}, t) = E(\mathbf{r})\cos \omega t$$

the resulting electric polarization

$$\mathsf{P}(\mathbf{r}, t) = \mathscr{P}_1 E(\mathbf{r})\cos \omega t + \mathscr{P}_2 E^2(\mathbf{r})\cos^2 \omega t + \mathscr{P}_3 E^3(\mathbf{r})\cos^3 \omega t + \cdots$$

can be rewritten as

$$\mathsf{P}(\mathbf{r}, t) = \mathscr{P}_1 E(\mathbf{r})\cos \omega t + \tfrac{1}{2}\mathscr{P}_2 E^2(\mathbf{r})\{\cos 2\omega t + 1\}$$
$$+ \tfrac{1}{4}\mathscr{P}_3 E^3(\mathbf{r})\{\cos 3\omega t + 3\cos \omega t\} + \cdots \qquad (2.23)$$

The interesting terms in this expansion involve the second and third harmonics of the incident frequency ω. These terms represent dipole moments per unit volume induced in the medium at twice and three times the incident frequency. These oscillating dipoles in turn drive electric fields at frequencies of 2ω and 3ω. This *second-* and *third-harmonic generation* of electromagnetic fields is greatly enhanced in certain crystalline materials. These are often, however, *anisotropic*, and consequently the nonlinear polarizability coefficients depend upon the direction of propagation, the polarization of the incident electric field, and the orientation of the optic axis of the crystal. In short, the constant $\mathscr{P}_2$ is replaced by the *tensor* $\mathscr{P}_{ijk}(2\omega)$, and the appropriate relation for the second-harmonic polarization is

$$\mathsf{P}_i(2\omega) = \mathscr{P}_{ijk}(2\omega) E_j(\omega) E_k(\omega)$$

where i, j, k can be either x, y, or z, and the usual tensor convention is adopted where summation is implied over any index which is repeated.

The second-order term in the expansion for the induced polarization (2.22) can also lead to *optical mixing*, that is to say the generation of sum and difference frequencies of the two incident frequencies. Suppose two harmonic electromagnetic fields $E_{01}\cos \omega_1 t$ and $E_{02}\cos \omega_2 t$ are incident upon a suitable medium; then the second-order contribution to the induced polarization is

$$\mathscr{P}_2\{E_{01}^2\cos^2\omega_1 t + E_{02}^2\cos^2\omega_2 t + 2E_{01}E_{02}\cos \omega_1 t \cos \omega_2 t\}$$

As before, the first and second terms can be expressed as functions of $2\omega_1$ and $2\omega_2$, respectively, while the last term can be rewritten in the form

$$E_{01}E_{02}\{\cos(\omega_1 + \omega_2)t + \cos(\omega_1 - \omega_2)t\}$$

which clearly reveals the generation of fields at the *sum* $(\omega_1 + \omega_2)$ and *difference* $(\omega_1 - \omega_2)$ frequencies—a process called *parametric amplification*.

For the interested reader there is a rich literature available on this subject; a few examples are Bloembergen (1965), Yariv (1976), Shen (1976), and Bloembergen (1980).

2.3. POLARIZATION

The electric-field disturbance given by (2.13) represents a *linearly* polarized, or *plane*-polarized, electromagnetic wave, since its orientation is constant even though its magnitude and sign vary with time (Fig. 2.2). The plane of polarization is defined by $\mathbf{E}_0$ and $\mathbf{k}$, from which is apparent that there can only be two independent directions of polarization for each direction of propagation $\mathbf{k}$.

Consider a monochromatic (single frequency) light wave propagating in the z-direction of a Cartesian coordinate system

$$\mathbf{E}(z,t) = \mathrm{Re}\{\mathbf{E}_0 \mathrm{e}^{i(kz-\omega t)}\} \tag{2.24}$$

where Re represents the *real* operator that extracts the real part of the expression within the braces. $\mathbf{E}_0$ is a constant vector in the xy plane having components $(E_x, E_y, 0)$, where we can write

$$E_x = E_{x0}\mathrm{e}^{i\delta_x} \quad \text{and} \quad E_y = E_{y0}\mathrm{e}^{i\delta_y}$$

E_{x0} and E_{y0} are positive amplitudes, while δ_x and δ_y are the appropriate phase angles. Thus we can write

$$E_x(z,t) = E_{x0}\cos(kz - \omega t + \delta_x)$$

$$E_y(z,t) = E_{y0}\cos(kz - \omega t + \delta_y)$$

$$E_z(z,t) = 0$$

Plane-polarized light results when the phase difference is

$$\delta \equiv \delta_y - \delta_x = m\pi \tag{2.25}$$

where m is an integer. That is to say, E_x and E_y have the same phase. Under these circumstances the magnitude of the electric field is

$$E_0 = \{E_{x0}^2 + E_{y0}^2\}^{1/2} \tag{2.26}$$

and the polarization vector makes an angle

$$\phi = \tan^{-1}\left\{(-1)^m \frac{E_{x0}}{E_{y0}}\right\} \tag{2.27}$$

with the y-axis.

Circularly polarized light results when

$$E_{x0} = E_{y0} \quad \text{and} \quad \delta \equiv \delta_y - \delta_x = \pm\frac{\pi}{2} \tag{2.28}$$

Under these conditions

$$E_x = E_{x0}\cos(kz - \omega t + \delta_x) \tag{2.29}$$

$$E_y = E_{x0}\cos\left(kz - \omega t + \delta_x \pm \frac{\pi}{2}\right) = \mp E_{x0}\sin(kz - \omega t + \delta_x) \tag{2.30}$$

The minus sign goes with the $+\pi/2$, and in this instance the electric-field vector appears to rotate with time in a clockwise direction when viewed in the direction of propagation (Fig. 2.3). This is termed *right circularly polarized* (RCP) light.

Conversely, if the phase difference $\delta \equiv \delta_y - \delta_x = -\pi/2$, we have *left circularly polarized* (LCP) light. It is also easy to show that if $E_{x0} \neq E_{y0}$ we have *elliptically polarized light*; see Hecht and Zajac (1974, p. 222). The four possible states of polarization are summarized in Table 2.1.

Natural light which results from the spontaneous emission of a very large number of atoms or molecules is often referred to as *unpolarized*, since its polarization is rapidly and randomly varying with time. If the variation of the electric-field vector is neither totally regular or totally irregular, we refer to the

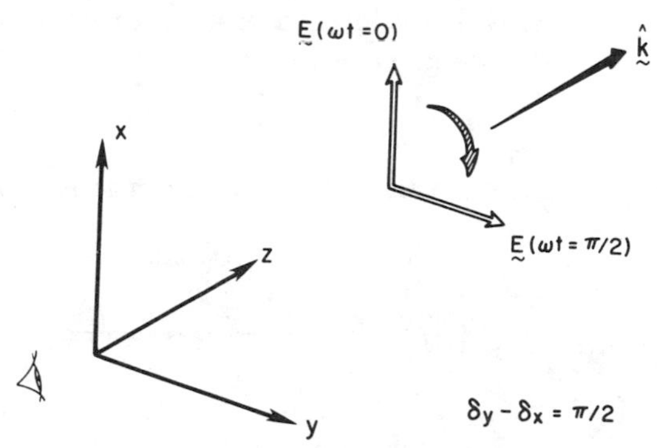

Fig. 2.3. The electric-field vector **E** propagating in the z-direction is seen to rotate in a clockwise direction if $\delta_y - \delta_x = \pi/2$. This constitutes a *right circularly polarized* (RCP) wave.

TABLE 2.1

Description	Criterion
Linearly or plane polarized	$\delta_y = \delta_x$, i.e., $\delta = 0$
Right circularly polarized (RCP)	$\delta = \delta_y - \delta_x = \pi/2$ } $E_{x0} = E_{y0}$
Left circularly polarized (LCP)	$\delta = \delta_y - \delta_x = -\pi/2$ } $E_{x0} = E_{y0}$
Elliptically polarized	$E_{x0} \neq E_{y0}$ and $\delta_y \neq \delta_x$

beam as being *partially polarized*. Such a beam behaves as if some fraction of it were polarized and the remainder unpolarized (Klein, 1970, p. 507).

According to quantum mechanics an electromagnetic wave of frequency ν can be thought of as comprising of a very large number of photons with energy $h\nu$ (where h is Planck's constant and equals 6.626×10^{-34} J s). The *intrinsic* or *spin* angular momentum of each photon is either $+\hbar$ or $-\hbar$. In a purely *left circularly polarized* beam of light the photons have been found to all have their spins aligned in the direction of propagation (see Fig. 2.4). Thus although the electric vector would appear to rotate *counterclockwise* as seen by an observer looking in the direction of propagation, the angular momentum of the photons corresponds to a *clockwise spin*.

A linearly polarized beam of light will interact with matter as if it were composed, at any instant, of an equal number of left- and right-handed photons. This is consistent with the fact that such a beam can be represented classically by a superposition of right- and left-circularly polarized beams.

There are many methods of producing light in any given state of polarization, and the reader is referred to Klein (1970) or Hecht and Zajac (1974) for more details of this topic.

2.4. ENERGY FLOW AND RADIATIVE TERMS

If a point charge Q moves with a velocity $\mathbf{u}$ in the presence of an electromagnetic field, the force on the charge takes the form

$$\mathbf{F} = Q\{\mathbf{E} + \mathbf{u} \times \mathbf{B}\} \tag{2.31}$$

The corresponding rate at which the external electromagnetic field does work

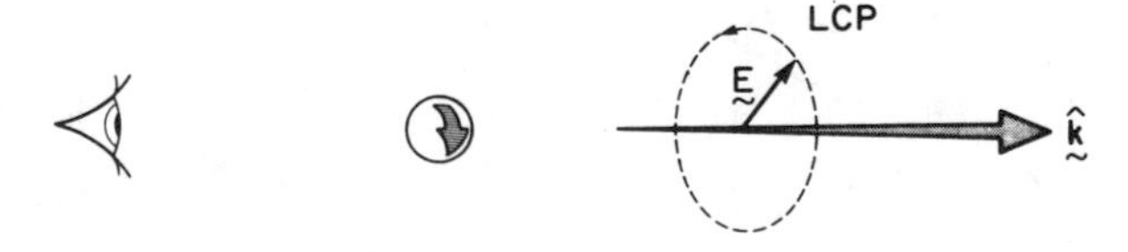

Fig. 2.4. An $\mathscr{L}$-(left circularly) polarized beam is seen to comprise photons with right-handed spin.

on the charge is

$$\mathbf{u} \cdot \mathbf{F} = Q\mathbf{u} \cdot \mathbf{E} \tag{2.32}$$

since $\mathbf{u} \cdot \mathbf{u} \times \mathbf{B} \equiv 0$ and hence no contribution arises from the magnetic field.

The total rate of change of the energy of a continuous distribution of charges within a volume V is given by $\int_V \mathbf{E} \cdot \mathbf{j}\, dV$. If we substitute for $\mathbf{j}$ from equation (2.4) we obtain

$$\int_V \mathbf{E} \cdot \mathbf{j}\, dV = \int_V \left\{ \mathbf{E} \cdot \nabla \times \mathbf{H} - \epsilon \mathbf{E} \cdot \frac{\partial \mathbf{E}}{\partial t} \right\} dV \tag{2.33}$$

If we further use the identity

$$\nabla(\mathbf{E} \times \mathbf{H}) = \mathbf{H} \cdot \nabla \times \mathbf{E} - \mathbf{E} \cdot \nabla \times \mathbf{H} \tag{2.34}$$

and Faraday's law (2.3) for the right-hand side of (2.33), we obtain an equation that expresses the conservation of electromagnetic energy for the volume V bounded by the surface area S, namely,

$$\int_V \frac{\partial}{\partial t} \left\{ \frac{\epsilon E^2}{2} + \frac{\mu H^2}{2} \right\} dV + \int_S (\mathbf{E} \times \mathbf{H}) \cdot \hat{\mathbf{n}}\, dS + \int_V \mathbf{E} \cdot \mathbf{j}\, dV = 0 \tag{2.35}$$

Clearly, $\int_S (\mathbf{E} \times \mathbf{H}) \cdot \hat{\mathbf{n}}\, dS$ represents the flow of electromagnetic energy through the bounding surface S ($\hat{\mathbf{n}}$ is the unit vector normal to the surface), and $(\epsilon E^2 + \mu H^2)/2$ represents the energy density stored respectively in the electric and magnetic fields. The vector

$$\mathfrak{J} = \mathbf{E} \times \mathbf{H} \tag{2.36}$$

is termed the *Poynting vector* and can be interpreted as the instantaneous electromagnetic power flow per unit area. It should be noted that strictly speaking only the surface integral of $\mathfrak{J}$ has physical significance (Born and Wolf, 1964, p. 3). Using (2.20) we can write

$$\mathfrak{J} = \epsilon_0 c(\mathbf{E} \cdot \mathbf{E})\hat{\mathbf{k}} \tag{2.37}$$

In reality detectors of electromagnetic energy at optical frequencies only respond to some time average $\langle\ \rangle$ of this energy flow and so it is more appropriate to introduce the term *irradiance* I, which is defined by the relation[†]

$$I\hat{\mathbf{k}} = \langle \mathfrak{J} \rangle = \epsilon_0 c \langle \mathbf{E} \cdot \mathbf{E} \rangle \hat{\mathbf{k}} \qquad (\mathrm{W\ cm^{-2}}) \tag{2.38}$$

[†]As noted earlier, although in SI units we should use W $\mathrm{m^{-2}}$ it is more convenient in this text to use W $\mathrm{cm^{-2}}$.

It should be noted that if complex notation is used for the **E** field in (2.38), then

$$\langle \mathbf{E} \cdot \mathbf{E} \rangle = \langle \mathrm{Re}\,\mathbf{E} \cdot \mathrm{Re}\,\mathbf{E} \rangle \tag{2.39}$$

With this in mind we shall evaluate the irradiance of a harmonic, linearly polarized plane wave (2.13) propagating through free space in a direction given by $\hat{\mathbf{k}}$:

$$I = \epsilon_0 c \langle \mathbf{E} \cdot \mathbf{E} \rangle = \epsilon_0 c E_0^2 \langle \cos^2(\mathbf{k} \cdot \mathbf{r} - \omega t) \rangle \tag{2.40}$$

Evaluating $\langle \cos^2(\mathbf{k} \cdot \mathbf{r} - \omega t) \rangle$ over a time much greater than the period $2\pi\omega^{-1}$ yields

$$I = \frac{\epsilon_0 c E_0^2}{2} \tag{2.41}$$

The term *irradiance* applies to the flow of electromagnetic energy per unit time per unit area and is certainly applicable to describe the illuminating power density of a collimated (plane-wave approximation) beam of light. However, if we are interested in the radiating power density of a source, it is necessary to take into account its emission properties, and this is done by introducing the term *radiance* J (W cm^{-2} sr^{-1}), defined as the flow of electromagnetic energy per unit time per unit solid angle per unit *projected* area.

Consider a small surface element dS of a source. $\hat{\mathbf{n}}$ represents the unit vector perpendicular to this surface element. The energy emitted from this surface element per second, into solid angle $d\Omega$, centered about an angle θ from the normal (Fig. 2.5) is

$$d^2P = J\,dS \cos\theta\,d\Omega \tag{2.42}$$

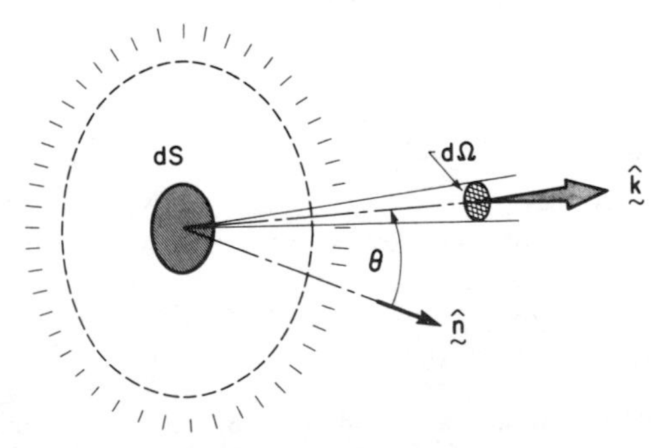

Fig. 2.5. Energy emitted into solid angle $d\Omega$ centered about an angle θ to the normal.

Thus the total power emitted per unit area of the surface, that is to say the irradiance, is

$$I = \int_{\theta=0}^{\pi/2} \int_{\phi=0}^{2\pi} J(\theta, \phi) \cos\theta \, d\Omega(\theta, \phi) \tag{2.43}$$

In general

$$d\Omega(\theta, \phi) = \sin\theta \, d\theta \, d\phi \tag{2.44}$$

and so

$$I = \int_{\theta=0}^{\pi/2} \int_{\phi=0}^{2\pi} J(\theta, \phi) \cos\theta \sin\theta \, d\theta \, d\phi \tag{2.45}$$

If the emission is *isotropic*, we can write $J(\theta, \phi) = J$, independent of θ or ϕ. Under these circumstances

$$I = \pi J$$

More often than not we shall be interested in the electromagnetic energy within a specific bandwidth, and so we introduce the term *spectral irradiance* $I(\nu)$ to designate the optical energy flowing per unit time per unit area per unit frequency interval in a light beam. Consequently

$$I = \int_{-\infty}^{\infty} I(\nu) \, d\nu \tag{2.46}$$

In a like manner we can introduce the *spectral radiance* $J(\nu)$ and write

$$J = \int_{-\infty}^{\infty} J(\nu) \, d\nu \tag{2.47}$$

TABLE 2.2

Term	Symbol	Unit
Energy	$\mathscr{E}$	J
Energy density	ρ	J cm^{-3}
Power	P	W
Irradiance	I	W cm^{-2}
Radiance	J	W cm^{-2} sr^{-1}
Spectral energy density	$\rho(\nu)$	J cm^{-3} Hz^{-1}
Spectral irradiance	$I(\nu)$	W cm^{-2} Hz^{-1}
Spectral radiance	$J(\nu)$	W cm^{-2} sr^{-1} Hz^{-1}

Also frequently we shall use the term *spectral energy density* $\rho(\nu)$—defined as the energy density per unit frequency interval. Consequently, we can write

$$I(\nu) = \rho(\nu)c \quad \text{and} \quad J(\nu) = c\frac{d\rho(\nu)}{d\Omega} \tag{2.48}$$

In Table 2.2 we summarize these radiative terms.

2.5. REFLECTION AND REFRACTION OF LIGHT

We shall now consider what happens to a monochromatic, linearly polarized plane wave of light that is incident upon an interface between two media of differing refractive index.

The *incident*, *reflected*, and *transmitted* waves can be expressed in the form

$$\mathbf{E}_i = \mathbf{E}_{i0}\mathbf{e}^{i(\mathbf{k}_i \cdot \mathbf{r} - \omega_i t)} \tag{2.49}$$

$$\mathbf{E}_r = \mathbf{E}_{r0}\mathbf{e}^{i(\mathbf{k}_r \cdot \mathbf{r} - \omega_r t + \delta_r)} \tag{2.50}$$

and

$$\mathbf{E}_t = \mathbf{E}_{t0}\mathbf{e}^{i(\mathbf{k}_t \cdot \mathbf{r} - \omega_t t + \delta_t)} \tag{2.51}$$

respectively, where δ_r and δ_t are the reflected and transmitted *phase constants*, relative to $\mathbf{E}_i$. Here $\mathbf{k}_i$, $\mathbf{k}_r$, and $\mathbf{k}_t$ are the respective wave vectors, and ω_i, ω_r and ω_t are the respective angular frequencies.

Since the temporal variation of $\mathbf{E}$ cannot change discontinuously at the interface, we can write immediately

$$\omega_i = \omega_r = \omega_t$$

In addition the laws of electromagnetic theory also state that the *tangential* component of both the electric field $\mathbf{E}$ and the magnetic field $\mathbf{H}$ must be continuous across the interface. This enables us to write

$$\hat{\mathbf{n}} \times \mathbf{E}_i + \hat{\mathbf{n}} \times \mathbf{E}_r = \hat{\mathbf{n}} \times \mathbf{E}_t \tag{2.52}$$

where $\hat{\mathbf{n}}$ is the unit vector normal to the interface. Now since $\mathbf{E}_{i0}$, $\mathbf{E}_{r0}$, and $\mathbf{E}_{t0}$ are constant vectors and (2.52) must hold for light incident anywhere on the interface, all the spatial phase factors must be equal, which in turn implies

$$\mathbf{k}_i \cdot \mathbf{r} = \mathbf{k}_r \cdot \mathbf{r} + \delta_r = \mathbf{k}_t \cdot \mathbf{r} + \delta_t \tag{2.53}$$

at any point on the interface. This enables us to write

$$(\mathbf{k}_i - \mathbf{k}_r) \cdot \mathbf{r} = \delta_r, \tag{2.54}$$

and

$$(\mathbf{k}_i - \mathbf{k}_t) \cdot \mathbf{r} = \delta_t \tag{2.55}$$

Now, since δ_r is a constant for any given $\mathbf{k}_i$, irrespective of the location on the boundary, equation (2.54) takes the form

$$(\mathbf{k}_i - \mathbf{k}_r) \cdot \mathbf{r} = \text{constant} \tag{2.56}$$

in which case any two arbitrary points on the interface $(\mathbf{r}_1, \mathbf{r}_2)$ satisfy the relation

$$(\mathbf{k}_i - \mathbf{k}_r) \cdot (\mathbf{r}_1 - \mathbf{r}_2) = 0 \tag{2.57}$$

It immediately follows that since $\mathbf{r}_1 - \mathbf{r}_2$ is a vector that lies in the plane of the interface, $\mathbf{k}_i - \mathbf{k}_r$ is a vector that is parallel to $\hat{\mathbf{n}}$, or

$$k_i \sin\theta_i = k_r \sin\theta_r$$

However, $k_i = k_r$, as both the incident and the reflected waves propagate in the same medium [see equation (2.17)], and consequently

$$\theta_i = \theta_r \tag{2.58}$$

or *the angle of incidence equals the angle of reflection.*

A similar line of reasoning leads us to conclude that equation (2.55) can be interpreted to mean that $\mathbf{k}_i - \mathbf{k}_t$ also lies normal to the surface and therefore

$$k_i \sin\theta_i = k_t \sin\theta_t \tag{2.59}$$

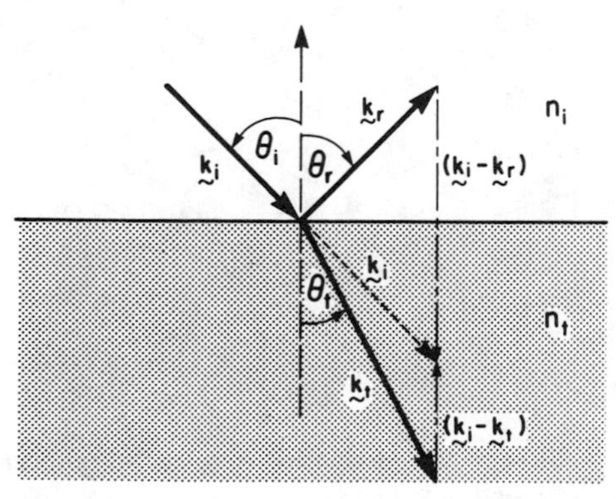

Fig. 2.6. Snell's *law of refraction* states $n_i \sin\theta_i = n_t \sin\theta_t$.

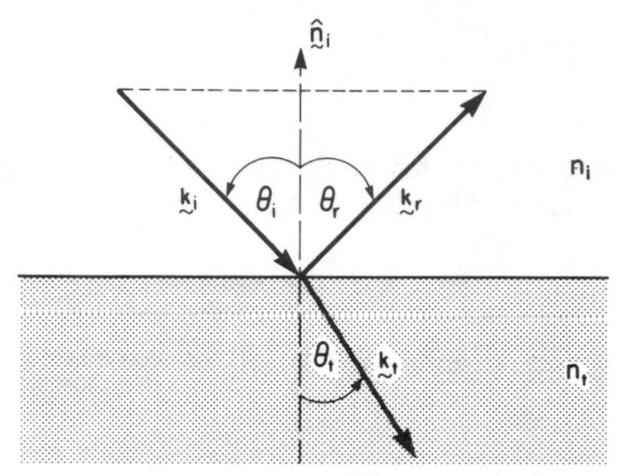

Fig. 2.7. For a plane monochromatic light wave incident on an interface between two media, the *law of reflection* states $\theta_i = \theta_r$.

(see Fig. 2.6). If we draw upon equation (2.17) and the fact that the frequency is unaltered across the interface, we arrive at Snell's law for the *refraction of light* at the boundary of two media, viz.,

$$n_i \sin \theta_i = n_t \sin \theta_t \tag{2.60}$$

where n_i and n_t are the refractive indices of the incident and transmitted media, respectively. Moreover, since $\mathbf{k}_i - \mathbf{k}_r$ and $\mathbf{k}_i - \mathbf{k}_t$ are parallel to $\hat{\mathbf{n}}$, we conclude that $\mathbf{k}_i$, $\mathbf{k}_r$, $\mathbf{k}_t$, and $\hat{\mathbf{n}}$ are *coplanar*: see Fig. 2.7.

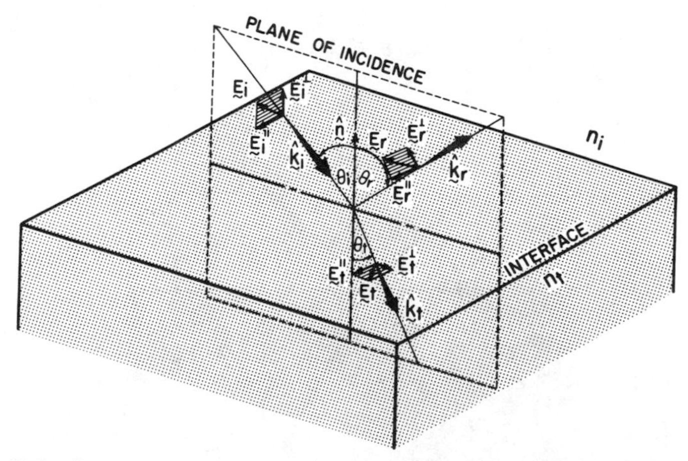

Fig. 2.8. Reflection and refraction of a plane electromagnetic wave incident upon an interface between two media of different refractive index.

The laws of reflection and refraction derived above have been obtained by considering the phase relationship amongst the incident, reflected, and transmitted waves at a boundary. A further study of the boundary conditions leads to the development of reflection and transmission coefficients. We shall simply state the results. Detailed derivation of these Fresnel equations are provided in the excellent texts of Hecht and Zajac (1974) and Born and Wolf (1964).

Whatever the polarization of the incident wave, we shall resolve its electric field **E** into components that are parallel and perpendicular to the plane of incidence—defined by $\mathbf{k}_i$ and $\hat{\mathbf{n}}$ (see Fig. 2.8). In the case of dielectric media Fresnel's equations take the form

$$r_\perp \equiv \left\{\frac{E_{r0}}{E_{i0}}\right\}_\perp = -\frac{\sin(\theta_i - \theta_t)}{\sin(\theta_i + \theta_t)} \tag{2.61}$$

$$t_\perp \equiv \left\{\frac{E_{t0}}{E_{i0}}\right\}_\perp = +\frac{2\sin\theta_t\cos\theta_i}{\sin(\theta_i + \theta_t)} \tag{2.62}$$

$$r_\parallel \equiv \left\{\frac{E_{r0}}{E_{i0}}\right\}_\parallel = +\frac{\tan(\theta_i - \theta_t)}{\tan(\theta_i + \theta_t)} \tag{2.63}$$

$$t_\parallel \equiv \left\{\frac{E_{t0}}{E_{i0}}\right\}_\parallel = +\frac{2\sin\theta_t\cos\theta_i}{\sin(\theta_i + \theta_t)\cos(\theta_i - \theta_t)} \tag{2.64}$$

where $r_\perp$ and $r_\parallel$ are respectively the perpendicular and parallel reflection coefficients and $t_\perp$ and $t_\parallel$ are respectively the perpendicular and parallel transmission coefficients.

The corresponding reflection and transmission coefficients appropriate to irradiance are

$$R_\perp = r_\perp^2, \qquad R_\parallel = r_\parallel^2 \tag{2.65}$$

$$T_\perp = \left\{\frac{\tan\theta_i}{\tan\theta_t}\right\} t_\perp^2 \quad \text{and} \quad T_\parallel = \left\{\frac{\tan\theta_i}{\tan\theta_t}\right\} t_\parallel^2 \tag{2.66}$$

It can also be shown that

$$R_\parallel + T_\parallel = 1 \quad \text{and} \quad R_\perp + T_\perp = 1 \tag{2.67}$$

as might be expected, since absorption within the media has not been taken into account. Examples of the variation in the *reflectance* R and *transmittance* T with incident angle θ (degrees) for both $\parallel$ and $\perp$ components are presented

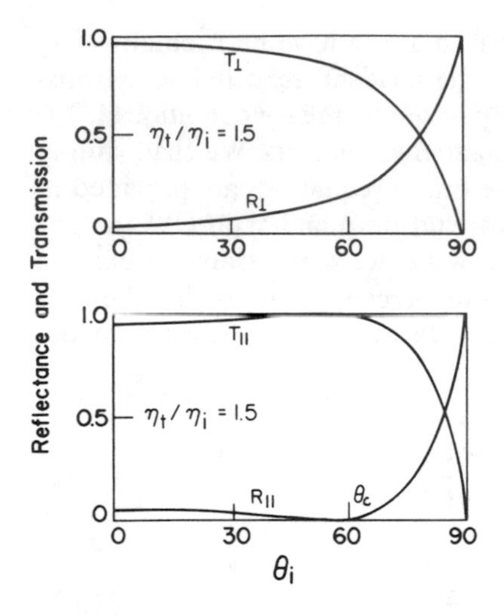

Fig. 2.9. Reflectance and transmittance versus incident angle for two perpendicular planes of polarization (Hecht and Zajac, 1974).

in Fig. 2.9. For the purpose of this illustration $n_t/n_i = 1.5$, corresponding to an air–glass interface.

It is evident from Fig. 2.9 that both $R_{\parallel}$ and $R_{\perp}$ approach unity as $\theta_i \rightarrow 90°$. This implies that any dielectric surface becomes mirrorlike at a glancing angle. At *normal incidence* ($\theta_i = 0$) and using Snell's laws we can write

$$R_{\parallel} = R_{\perp} = \left(\frac{\eta_t - \eta_i}{\eta_t + \eta_i}\right)^2 \tag{2.68}$$

$$T_{\parallel} = T_{\perp} = \frac{4\eta_t\eta_i}{(\eta_t + \eta_i)^2} \tag{2.69}$$

where η_i and η_t represent the *real* refractive indices for the incident and transmitted media respectively. As a result about 4% of the light incident on an air–glass surface is reflected. If $\eta_i > \eta_t$, then *total internal reflection* occurs beyond some critical angle

$$\theta_c \equiv \sin^{-1}\left\{\frac{\eta_t}{\eta_i}\right\} \tag{2.70}$$

This leads to the *guiding* principle of fiber optics and allows light to be efficiently transmitted over a large distance through optical fibers (see Fig. 2.10). In the case of a glass–air interface where $\eta_i/\eta_t = 1.55$, $\theta_c = 40.18°$. The

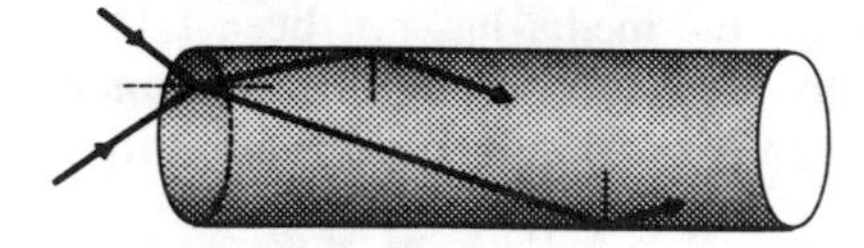

Fig. 2.10. Total internal reflection allows very efficient transmission of light through optical fiber.

light-collection efficiency of a fiber is determined by its *numerical aperture*

$$\mathrm{NA} = \left(\eta_f^2 - \eta_c^2\right)^{1/2} \tag{2.71}$$

where η_f is the refractive index of the fiber core and η_c that of protective cladding. Reasonable values for η_f and η_c are 1.62 and 1.52 respectively, giving NA $\approx$ 0.56. This gives rise to a fiber maximum half angle of acceptance (Hecht and Zajac, 1974) of $\theta_{max} \equiv \sin^{-1}(\mathrm{NA}) = \sin^{-1}(0.56) = 34°$. This means that the fiber totally internally reflects all light incident within a cone of half angle 34°.

Careful inspection of Fig. 2.9 reveals that an angle of incidence exists for which $R_{\parallel} \to 0$ and $T_{\parallel} \to 1.0$. At this *Brewster angle* θ_b, only light with its E-field perpendicular to the plane of incidence is reflected. Consequently, if an unpolarized beam of light strikes a flat surface at this angle, the reflected component is totally polarized parallel to the surface. Under these circumstances $\theta_t = 90° - \theta_b$ and it follows from 2.60 that

$$\theta_b \equiv \tan^{-1}\left(\frac{\eta_t}{\eta_i}\right) \tag{2.72}$$

This is why sunlight reflected at 53° from a smooth lake or other body of water is essentially plane polarized.

2.6. CLASSICAL THEORY OF EMISSION, ABSORPTION, AND DISPERSION

Electromagnetic theory predicts that an accelerating charge such as an electron will radiate electromagnetic energy. Any good text on the subject [e.g. Born and Wolf (1964) or Corney (1977)] will show that the electric field at a position **r** from an accelerating electron of charge $-e$ (-1.602×10^{-19} C) can be written as the sum of the components

$$\mathbf{E}(\mathbf{r}, t) = \frac{-e}{4\pi\epsilon_0 r^2}\hat{\mathbf{k}} + \frac{-e}{4\pi\epsilon_0 c^2 r}\left[\hat{\mathbf{k}} \times (\hat{\mathbf{k}} \times \dot{\mathbf{u}})\right]_{\mathrm{ret}} \tag{2.73}$$

The first term represents the *near-field* Coulomb component. It depends only upon the sign and magnitude of the charge and the distance between the charge and the point of observation. The second term represents the *far-field* component and arises from the acceleration of the charge. At large distances ($r \gg \lambda/2\pi$) only the second term is important, and it constitutes the radiated field. It should be noted that the quantity within the brackets of the far-field component is evaluated at the *retarded time* $(t - r/c)$. As before, the propagation vector $\mathbf{k} \equiv \hat{\mathbf{k}}\, 2\pi/\lambda$, and $\dot{\mathbf{u}}$ represents the instantaneous acceleration of the electron at this retarded time.

The total instantaneous power radiated by the accelerating electron can be evaluated using (2.37):

$$P = \int \epsilon_0 c |E|^2 \, dS \tag{2.74}$$

Reference to the vector diagram of Fig. 2.11 (insert) shows that we can write

$$|\hat{\mathbf{k}} \times (\hat{\mathbf{k}} \times \dot{\mathbf{u}})| = -|\dot{\mathbf{u}}| \sin \psi, \tag{2.75}$$

so that the magnitude of the far-field component is

$$E(\mathbf{r}, t) = \frac{-e}{4\pi\epsilon_0 c^2 r} \left\{ \left| \dot{u}\left(t - \frac{r}{c}\right) \right| \sin \psi \right\} \tag{2.76}$$

where ψ is the angle between direction of acceleration and the direction of propagation. Then the increment of power radiated across an element dS of a spherical surface of radius r and centered on the charge is

$$dP(r, t) = \frac{1}{16\pi^2\epsilon_0} \frac{e^2}{c^3 r^2} \dot{u}_{\text{ret}}^2 \sin^2 \psi \, dS \tag{2.77}$$

where $dS = r^2 \sin \psi \, d\psi \, d\beta$ (see Fig. 2.11), and $\dot{u}_{\text{ret}}$ is the *retarded* value of the acceleration (i.e. the value evaluated at $t - r/c$). Thus the total power radiated from the accelerating electron is

$$P(r, t) = \frac{1}{6\pi\epsilon_0} \cdot \frac{e^2}{c^3} \left| \dot{u}\left(t - \frac{r}{c}\right) \right|^2 \tag{2.78}$$

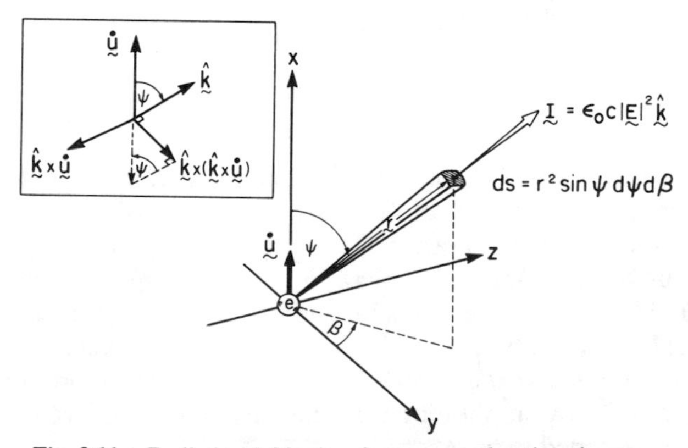

Fig. 2.11. Radiating field arises from an accelerating charge *e*.

This power loss leads to the introduction of a *radiation damping* term in the equation of motion for the bound electrons. Historically this energy loss represented a considerable impediment to the orbital model of the atom proposed by Rutherford in the early days of atomic physics. Bohr circumvented this difficulty by introducing the concept of quantized orbits, and eventually this led to the more complete theory of quantum mechanics.

In order to account for the interaction of electromagnetic fields with atoms and molecules from a classical perspective we shall consider the atom (or molecule) to be modeled by a simple one-dimensional harmonic oscillator. Although this represents a considerable oversimplification, it enables us to derive a set of very useful relations that are in close agreement with those obtained through a more elaborate analysis. In our model of the atom, an electron displaced from its equilibrium position experiences a linear restoring force. Consequently, the equation of motion appropriate to our picture of the atom, in which a single electron is instantly *excited* through some impulse, can be written in the form

$$\ddot{x} + \gamma\dot{x} + \omega_0^2 x = 0 \tag{2.79}$$

where x is the *displacement* of the electron from its equilibrium position (the dots indicate time derivatives; two dots mean the second derivative), $\omega_0 = (\kappa/m_e)^{1/2}$ the *resonant frequency* of the system, κ the appropriate elastic restoring-force coefficient for the electron, and m_e its mass (9.108×10^{-31} kg). The mass of the nucleus is assumed to be effectively infinite in comparison with that of the electron. $\gamma\dot{x}$ represents a damping term which accounts for the *radiative* loss of energy associated with the acceleration experienced by the electron. If we assume that $\gamma\dot{x} \ll \omega_0^2 x$ (valid almost to X-ray frequencies), then to *first order* the solution of (2.79) takes the form

$$x = x_0 \mathrm{e}^{-i\omega_0 t} \tag{2.80}$$

which is tantamount to assuming that the amount of energy lost per cycle is negligible. In this case, if (2.80) is used in conjunction with (2.78), we can write for the time rate of decay of the energy of the system

$$\frac{dW}{dt} = -\frac{1}{6\pi\epsilon_0}\frac{e^2}{c^3}\ddot{x}^2 = -\frac{2}{3}\frac{r_e\omega_0^2}{c}W, \tag{2.81}$$

where $W = m_e\omega_0^2 x^2$ and represents the energy of our simple harmonic-oscillator model of the atom (or molecule). In (2.81) we have introduced the *electron classical radius*

$$r_e \equiv \frac{e^2}{4\pi\epsilon_0 m_e c^2} \approx 2.818 \times 10^{-15}\ \mathrm{m} \tag{2.82}$$

Equation (2.81) gives rise to a simple exponential decay of the excited atom through the emission of radiation, which can be interpreted as the classical analogue of spontaneous emission. The corresponding classical (exponential) lifetime of the excited atom,

$$\tau_0 = \frac{3c}{2r_e\omega_0^2} \tag{2.83}$$

is remarkbly close to the experimentally observed resonance-state lifetimes of a number of atoms.

We shall now consider the classical theory of absorption, which in turn will lead to the subject of dispersion. We begin by again treating the atom as a simple harmonic oscillator, but in this instance the atom is assumed to be irradiated by a plane monochromatic electromagnetic wave. Under these circumstances the relevant equation of motion takes the form

$$\ddot{x} + \gamma\dot{x} + \omega_0^2 x = -\frac{e}{m_e}E_0\mathrm{e}^{-i\omega t} \tag{2.84}$$

where $E = E_0\mathrm{e}^{-i\omega t}$ represents the applied electric field of angular frequency ω. If we assume that the electron displacement achieves the same harmonic dependence, we can write

$$x = x_0\mathrm{e}^{-i\omega t} \tag{2.85}$$

and on substitution back into (2.84) we arrive at the *steady-state* solution

$$x = -\frac{e}{m_e}\cdot\frac{E_0\mathrm{e}^{-i\omega t}}{\omega_0^2 - \omega^2 - i\omega\gamma} \tag{2.86}$$

For a nonmagnetic medium $K_m = 1$, and we saw in (2.12) that the refractive index $n = (\epsilon/\epsilon_0)^{1/2}$, where from (2.6) ϵ/ϵ_0 is determined by the *density of dipole moments*, N_0. If we assume that only a *fraction* f_0 ($\lesssim 1$) of the atoms present are capable of being excited at the frequency ω_0, then we can write

$$N_0 = Nf_0 \tag{2.87}$$

where N represents the atom density of the medium.

The *dipole moment* [or polarizability $\not{p}(\omega)$ times the inducing field times ϵ_0] of each atom having the characteristic frequency ω_0 is $-ex$ and so using (2.87) and (2.6) we can write

$$\frac{\epsilon}{\epsilon_0} = 1 + N_0\not{p}(\omega) = 1 + \left(-\frac{Nexf_0}{\epsilon_0 E}\right) \tag{2.88}$$

which when combined with (2.86) yields the *dispersion equation*

$$\frac{\epsilon(\omega)}{\epsilon_0} = 1 + \frac{Ne^2}{m_e\epsilon_0}\left\{\frac{f_0}{\omega_0^2 - \omega^2 - i\omega\gamma}\right\} \tag{2.89}$$

The form of equation (2.89) suggests that (2.12) would lead us to a *complex refractive index* $n(\omega)$, and so we divide it into real and imaginary parts, namely,

$$n(\omega) \equiv \eta(\omega) + \frac{i\chi(\omega)}{2}, \tag{2.90}$$

where $\eta(\omega)$ represents the experimentally measured refractive index, and as we shall see, $\chi(\omega)$ is related to the atom's ability to absorb electromagnetic energy. Combining (2.89) with (2.90) yields

$$\eta(\omega) + \frac{i\chi(\omega)}{2} = \left[1 + \frac{Ne^2}{m_e\epsilon_0}\left\{\frac{f_0}{\omega_0^2 - \omega^2 - i\omega\gamma}\right\}\right]^{1/2} \tag{2.91}$$

In general the second term within the square root is much smaller than unity, and so we can expand the right-hand side of (2.91), and by equating the real and imaginary parts of both sides of the resultant equation we arrive at expressions for both $\eta(\omega)$ and $\chi(\omega)$.

The relation for the *absolute refractive index* takes the form

$$\eta(\omega) = 1 + \frac{Ne^2 f_0}{2m_e\epsilon_0}\left\{\frac{\omega_0^2 - \omega^2}{\left(\omega_0^2 - \omega^2\right)^2 + \omega^2\gamma^2}\right\} \tag{2.92}$$

We further recognize that at optical frequencies $\omega_0 + \omega \approx 2\omega$, because, in general, the frequency of light is not too far removed from the resonant frequency of the medium within which it is propagating. With this in mind (2.92) can be simplified to give

$$\eta(\omega) = 1 + \frac{Ne^2 f_0}{4m_e\epsilon_0\omega}\left\{\frac{\omega_0 - \omega}{\left(\omega_0 - \omega\right)^2 + \left(\gamma/2\right)^2}\right\} \tag{2.93}$$

Under these same conditions

$$\chi(\omega) = \frac{Ne^2 f_0}{2m_e\epsilon_0\omega}\left\{\frac{\gamma/2}{\left(\omega_0 - \omega\right)^2 + \left(\gamma/2\right)^2}\right\} \tag{2.94}$$

The behavior of both $\eta(\omega)$ and $\chi(\omega)$ around the resonant frequency is portrayed in Fig. 2.12. The increase in the refractive index evident for

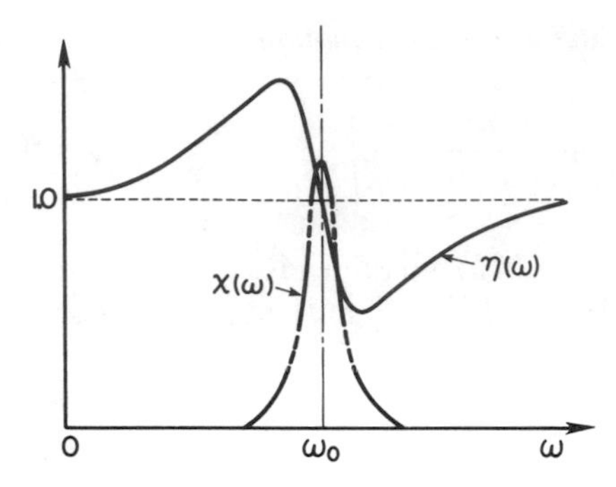

Fig. 2.12. Variation of $\eta(\omega)$ and $\chi(\omega)$ around ω_0.

$\omega_0 - \omega \gg \gamma$ is observed for most transparent media (see Fig. 2.13) and is described by a simplification known as *Sellmeir's dispersion formulae* (Born and Wolf, 1964, p. 96).

It is also apparent from (2.93) and (2.94) that the separation between the maximum and minimum values of $\eta(\omega)$ and the width of $\chi(\omega)$ depend upon γ. The rapid decrease of $\eta(\omega)$ with increasing ω in the immediate vicinity of ω_0 is termed *anomalous dispersion*.

For any real atom (or molecule) many modes of excitation are possible. In terms of the quantum treatment, an atom which happens to exist in a *quantum state* $|n\rangle$ (the integer n is termed a *quantum number* and will be elaborated on later) can undergo one of an almost infinite number of transitions to another quantum state $|m\rangle$ through the absorption or emission of a photon of ap-

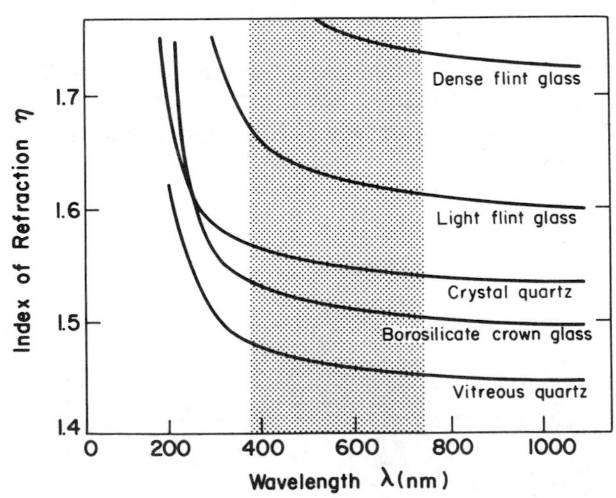

Fig. 2.13. The wavelength dependence of the index of refraction for various materials (Hecht and Zajac, 1974).

propriate energy. The transition from $|n\rangle$ to $|m\rangle$ is characterized by an angular frequency ω_{nm}, a decay rate Γ_{nm}, and an *oscillator strength* f_{nm}, which can be interpreted as the fraction of a classical harmonic oscillator represented by the transition. Since there cannot be more oscillators than atoms in state $|n\rangle$, we are led to the Thomas–Kuhn sum rule

$$\sum_{m}^{(m \neq n)} f_{nm} = 1 \tag{2.95}$$

This quantum treatment leads to a more general relation for the *polarizability* $\not{p}(\omega)$ (defined as the ratio of the induced dipole moment to the strength of the applied field) of the atom or molecule, viz.,

$$\not{p}(\omega) = \frac{e^2}{m_e \epsilon_0} \sum_n \sum_m^{(m \neq n)} \left[\frac{f_{nm}}{\omega_{nm}^2 - \omega^2 - i\omega\Gamma_{nm}} \right] \tag{2.96}$$

This in turn leads to a rather complex frequency structure for the refractive index $\eta(\omega)$. A very much simplified version of which is presented in Fig. 2.14 for the purpose of illustration.

In a cold (and therefore unexcited) medium almost all of the atoms or molecules reside in the ground (i.e. lowest-energy) state. As a consequence the *anomalous dispersion* regime for almost any cold medium lies in, or beyond, the ultraviolet part of the spectrum. The exception to this is found in alkali vapors, where anomalous dispersion can be observed in the visible region of the spectrum.

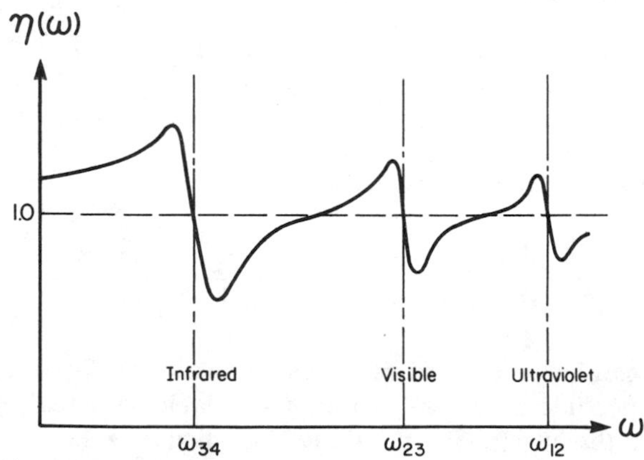

Fig. 2.14. Idealized representation of the refractive-index variation with frequency over a range that includes several transition frequencies.

In the above discussion we have made an implicit assumption regarding the density of the medium through which the light is propagating. In our analysis each atom was treated as if it were in complete isolation from its neighbors and so the equations developed above apply only to low-density gases. When the density is sufficient for each atom (or molecule) to experience the polarization field of its neighbors in addition to the applied electric field, the equation of motion is modified and takes the form

$$\ddot{x} + \gamma\dot{x} + \omega_0^2 x = -\left(E + \frac{\mathsf{P}}{3\epsilon_0}\right)\frac{e}{m_e} \tag{2.97}$$

where $\mathsf{P}/3\epsilon_0[= N\alpha(\omega)E(\omega)/3]$ represents the effective polarization field to which each atom is exposed and $\alpha(\omega)$ represents the atom's *polarizability*.

If we again assume a solution of the form $x = x_0 e^{-i\omega t}$, then we can write

$$\alpha(\omega) = \frac{-ex_0}{\epsilon_0 E_0} \tag{2.98}$$

and if this is used with (2.97) we have

$$\left[-\omega^2 - i\gamma\omega + \omega_0^2\right]\alpha(\omega) = \frac{e^2}{m_e\epsilon_0}\left[1 + \frac{Nf_0\alpha(\omega)}{3}\right]. \tag{2.99}$$

This gives

$$\alpha(\omega) = \frac{e^2}{m_e\epsilon_0}\left[\{\omega_0^2 - \omega^2 - i\omega\gamma\} - \frac{Ne^2 f_0}{3m_e\epsilon_0}\right]^{-1}, \tag{2.100}$$

which when combined with (2.6) yields

$$n^2 - 1 = \frac{Ne^2 f_0}{m_e\epsilon_0}\left[\{\omega_0^2 - \omega^2 - i\gamma\omega\} - \frac{Ne^2 f_0}{3m_e\epsilon_0}\right]^{-1} \tag{2.101}$$

This can be recast in the more usual form

$$\frac{n^2 - 1}{n^2 + 2} = \frac{Ne^2}{3m_e\epsilon_0}\left[\frac{f_0}{\omega_0^2 - \omega^2 - i\gamma\omega}\right] \tag{2.102}$$

which in turn can be generalized in a manner similar to (2.96) by accounting for the many possible modes of excitation available to a real atom. Clearly (2.102) tends to the form (2.89) of a dilute gas when $n \to 1$.

We shall now endeavor to arrive at the physical meaning for the imaginary component of the complex refractive index. The phase velocity of a traveling plane monochromatic wave is

$$v = \frac{\omega}{k} = c\left(\frac{\epsilon_0}{\epsilon}\right)^{1/2}$$

according to (2.10) and (2.16). Thus the electric-field component of such a wave propagating in the z-direction is

$$E(z,t) = E_0 e^{i\omega[(\epsilon/\epsilon_0)^{1/2} z/c - t]} \tag{2.103}$$

Substitution of (2.90) for the complex refractive index yields

$$E(z,t) = E_0 e^{i\omega[\eta(\omega)z/c - t] - \omega\chi(\omega)z/2c} \tag{2.104}$$

The *irradiance* of this wave as a function of position within the medium can be written, using (2.40), in the form

$$I(z) = \tfrac{1}{2}\epsilon_0 c E_0^2 e^{-\kappa(\omega)z} \tag{2.105}$$

where

$$\kappa(\omega) \equiv \frac{\omega\chi(\omega)}{c} \tag{2.106}$$

can be seen to represent the *absorption coefficient* of the medium. If we introduce

$$I_0 \equiv \frac{\epsilon_0 c E_0^2}{2} \tag{2.107}$$

as the irradiance incident on the $z = 0$ interface, then (2.105) can be written in the familiar Beer–Lambert form

$$I(z,\omega) = I_0 e^{-\kappa(\omega)z} \tag{2.108}$$

We see that the irradiance of a plane monochromatic light wave is attenuated exponentially by an absorbing medium. Although this is certainly true at low levels of irradiance and is valid under most conditions of interest in remote sensing, the Beer–Lambert law can break down at sufficiently high levels of irradiance, (Measures and Herchen, 1983).

If we combine (2.94) with (2.106), we arrive at a relation for the absorption coefficient in terms of the characteristic properties of the attenuating medium, viz.,

$$\kappa(\omega) = \frac{Ne^2 f_0}{2m_e\epsilon_0 c}\left\{\frac{\gamma/2}{(\omega_0 - \omega)^2 + (\gamma/2)^2}\right\} \tag{2.109}$$

This absorption coefficient can be written in terms of a so-called *Lorentzian* line profile function

$$\mathscr{L}(\omega) \equiv \frac{1}{\pi}\left\{\frac{\gamma/2}{(\omega_0 - \omega)^2 + (\gamma/2)^2}\right\} \tag{2.110}$$

to give

$$\kappa(\omega) = \kappa_0 \mathscr{L}(\omega) \tag{2.111}$$

where κ_0 is the *line strength* of the absorption transition and is given by

$$\kappa_0 = \frac{Ne^2 f_0 \pi}{2m_e \epsilon_0 c} \tag{2.112}$$

The total absorption coefficient per atom, σ_A, integrated over the entire spectrum [in reality; integration over a frequency interval equal to several times the bandwidth $(\pi\gamma)^{-1}$ is adequate] yields

$$\sigma_A = \int_{-\infty}^{\infty} \frac{\kappa(\omega)\, d\omega}{N} = 2\pi^2 c f_0 r_e \tag{2.113}$$

We see that the *absorption oscillator strength* f_0 can be related directly to this integrated absorption coefficient.

2.7. MOLECULAR RAYLEIGH SCATTERING

We shall again consider a plane, linearly polarized monochromatic wave propagating in the z-direction with its plane of polarization at an angle ϕ to the y-axis. The electric-field component of this wave will induce a dipole moment in the molecules (or atoms) in its path. This dipole moment will then radiate into 4π steradians and this reemission of electromagnetic radiation can be interpreted as a form of *elastic* scattering of the incident wave (the word *elastic* signifying no change of wavelength).

Consider a single molecule (or atom) with only one valance electron (i.e. only one electron that is bound weak enough to be influenced by the applied field) located at the origin of the coordinate system shown in Fig. 2.15. If we can again treat this molecule or atom as a classical harmonic oscillator, then the electron's equation of motion in terms of its displacement from its equilibrium position (at the origin) $\boldsymbol{\xi}$ takes the form

$$\ddot{\boldsymbol{\xi}} + \gamma\dot{\boldsymbol{\xi}} + \omega_0^2\boldsymbol{\xi} = -\frac{e}{m_e}\mathbf{E} \tag{2.114}$$

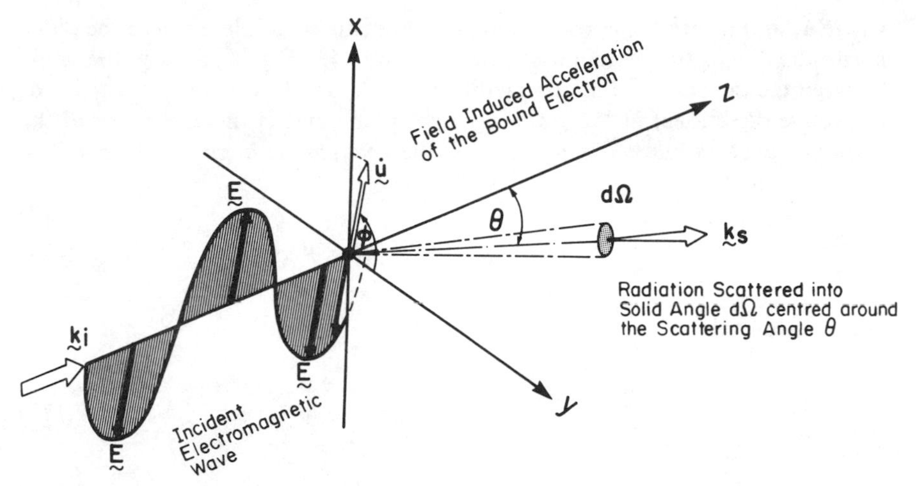

Fig. 2.15. Angular configuration for scattering of electromagnetic radiation from a bound electron.

where again ω_0 represents the resonant angular frequency of the electron and γ the damping constant. The electric field $\mathbf{E}$ for the incident wave can be expressed in terms of its angular frequency ω and a unit polarization vector $\hat{\boldsymbol{\epsilon}}$:

$$\mathbf{E} = \hat{\boldsymbol{\epsilon}} E_0 \mathbf{e}^{-i\omega t} \tag{2.115}$$

If we assume a harmonic steady-state solution for (2.114) of the form

$$\boldsymbol{\xi} = \hat{\boldsymbol{\epsilon}} \xi_0 \mathbf{e}^{-i\omega t} \tag{2.116}$$

it follows that the induced acceleration of the bound electron is

$$\dot{\mathbf{u}} = \ddot{\boldsymbol{\xi}} = -\left[\frac{\omega^2}{\omega_0^2 - \omega^2 - i\omega\gamma}\right]\frac{e}{m_e}\mathbf{E} \tag{2.117}$$

The radiated component of the electric field produced by this accelerating electron—the *scattered field* at a position $Q(r, \theta)$—can be divided into two components

$$\mathbf{E}_1^s = -\frac{e}{4\pi\epsilon_0 c^2 r}\dot{\mathbf{u}}_1 \tag{2.118}$$

and

$$\mathbf{E}_2^s = -\frac{e}{4\pi\epsilon_0 c^2 r}\dot{\mathbf{u}}_2 \tag{2.119}$$

where $\dot{\mathbf{u}}_1$ and $\dot{\mathbf{u}}_2$ are the perpendicular components of acceleration in the plane perpendicular to the direction of scattering $\mathbf{k}_s$, (see Fig. 2.16). θ is the angle between the incident and scattering directions ($\hat{\mathbf{k}}_i$ and $\hat{\mathbf{k}}_s$ are unit vectors in the respective directions) in the *scattering* (yz) plane and is therefore termed the *scattering angle*, while ϕ is referred to as the *polarization angle*. From Fig. 2.16 we can see that

$$\dot{u}_1 = \dot{u} \sin \phi \quad \text{and} \quad \dot{u}_2 = \dot{u} \cos \phi \cos \theta \tag{2.120}$$

and it follows that the *scattered field* at Q is

$$\mathbf{E}_s = -\frac{e\dot{u}}{4\pi\epsilon_0 c^2 r}\left[\hat{\mathbf{u}}_2 \cos \theta \cos \phi + \hat{\mathbf{u}}_1 \sin \phi\right] \tag{2.121}$$

where $\hat{\mathbf{u}}_2$ and $\hat{\mathbf{u}}_1$ represent unit vectors in the $\dot{\mathbf{u}}_2$ and $\dot{\mathbf{u}}_1$ directions.

The corresponding element of power radiated into the element of solid angle $d\Omega$ can be expressed in the form

$$dP(\theta, \phi) = \tfrac{1}{2}\epsilon_0 c |E_s|^2 r^2 \, d\Omega \tag{2.122}$$

If this is combined with (2.121), we arrive at the *scattered power radiated per unit solid angle*,

$$\frac{dP(\theta, \phi)}{d\Omega} = \frac{e^2 |\dot{u}|^2}{2^5 \pi^2 \epsilon_0 c^3}\left[\cos^2\phi \cos^2\theta + \sin^2\phi\right] \tag{2.123}$$

where we have made use of the fact that $\hat{\mathbf{u}}_1 \cdot \hat{\mathbf{u}}_2 = 0$ as the vectors are at right

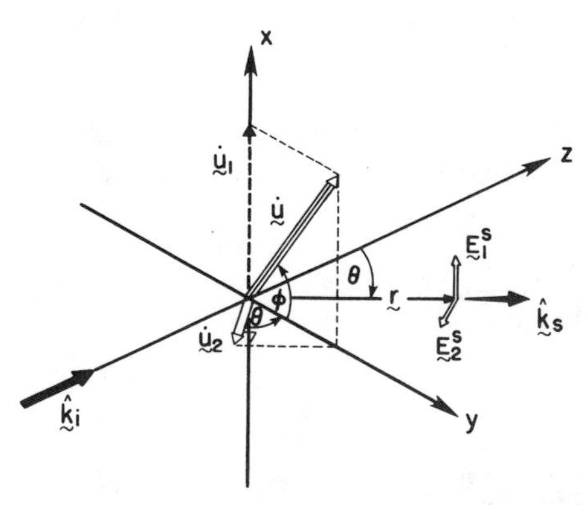

Fig. 2.16. Angular configuration of the scattered electric-field components.

angles to each other. If we draw upon (2.115) and (2.117) we can write

$$\frac{dP(\theta,\phi)}{d\Omega} = \tfrac{1}{2}\epsilon_0 c|E_0|^2 r_e^2 \left[\frac{\omega^2}{\omega_0^2 - \omega^2 - i\omega\gamma}\right]^2 \{\cos^2\phi\cos^2\theta + \sin^2\phi\} \tag{2.124}$$

If we introduce the *differential scattering cross section* of an individual molecule or atom $d\sigma/d\Omega$ and recognize that in general we can write

$$\frac{dP(\theta,\phi)}{d\Omega} = I_0 \frac{d\sigma(\theta,\phi)}{d\Omega} \tag{2.125}$$

then a direct comparison of (2.124) and (2.125) reveals

$$\frac{d\sigma(\theta,\phi)}{d\Omega} = r_e^2\{\cos^2\phi\cos^2\theta + \sin^2\phi\}\left[\frac{\omega^2}{\omega_0^2 - \omega^2 - i\omega\gamma}\right]^2 \tag{2.126}$$

where $I_0 = \frac{1}{2}\epsilon_0 c|E_0|^2$ represents the incident irradiance. At optical frequencies we can usually assume that $\omega_0 + \omega \approx 2\omega$. Consequently, we can write

$$\frac{d\sigma(\theta,\phi)}{d\Omega} \approx \tfrac{1}{4} r_e^2\{\cos^2\phi\cos^2\theta + \sin^2\phi\}\left[\frac{\omega^2}{(\omega_0 - \omega)^2 + (\gamma/2)^2}\right] \tag{2.127}$$

This leads to a total *Rayleigh scattering* cross section per isolated molecule (or atom)

$$\sigma_R(\omega) = \tfrac{1}{4} r_e^2 \left[\frac{\omega^2}{(\omega_0 - \omega)^2 + (\gamma/2)^2}\right] \int_{\theta=0}^{\pi}\int_{\phi=0}^{2\pi} \{\cos^2\phi\cos^2\theta + \sin^2\phi\}\sin\theta \, d\theta \, d\phi \tag{2.128}$$

integrated over all angles. The double integral is easily evaluated to yield

$$\sigma_R(\omega) = \frac{2\pi r_e^2}{3}\left[\frac{\omega^2}{(\omega_0 - \omega)^2 + (\gamma/2)^2}\right] \tag{2.129}$$

In much of the literature on Rayleigh scattering, the frequency dependence is eliminated in terms of the square of the complex refractive index of the medium (2.89), viz.,

$$n^2 = 1 + \frac{Ne^2}{\epsilon_0 m_e}\left[\frac{1}{\omega_0^2 - \omega^2 - i\omega\gamma}\right] \tag{2.130}$$

where N is the number density of the scatterers, assumed to have an oscillator strength of unity. In this case the *Rayleigh differential scattering cross section* is given by

$$\frac{d\sigma_R(\theta, \phi)}{d\Omega} = \frac{\pi^2(n^2 - 1)^2}{N^2\lambda^4}\{\cos^2\phi \cos^2\theta + \sin^2\phi\} \tag{2.131}$$

where λ $(= 2\pi c/\omega)$ represents the wavelength of both the incident and scattered radiation (the photons neither gain nor lose energy during elastic scattering). A typical value for $d\sigma_R/d\Omega$ in the lower atmosphere, at around 700 nm, is 2×10^{-28} cm^2 sr^{-1}.

The *total* Rayleigh scattering cross section $\sigma_R(\lambda)$ can be evaluated from (2.131) by integrating over 4π steradians:

$$\sigma_R(\lambda) = \frac{8\pi}{3}\left[\frac{\pi^2(n^2 - 1)^2}{N^2\lambda^4}\right] \tag{2.132}$$

The λ^{-4} dependence of the cross section (and that for small-particle scattering) is responsible for much of the visual beauty that we too often take for granted. This includes our blue sky and the soft red sun which is sometimes observed at dawn and dusk. In the field of laser remote sensing the *backscattering* cross section

$$\sigma_\pi^R \equiv \frac{d\sigma_R(\theta = \pi)}{d\Omega} = \frac{\pi^2(n^2 - 1)^2}{N^2\lambda^4} \tag{2.133}$$

is of prime importance.

For the mixture of atmospheric gases which occur below 100 km, Collis and Russell (1976, p. 89) have indicated that the molecular Rayleigh backscattering cross section

$$\sigma_\pi^R(\lambda) = 5.45\left[\frac{550}{\lambda(\text{nm})}\right]^4 \times 10^{-28}\ \text{cm}^2\ \text{sr}^{-1} \tag{2.134}$$

which corresponds to an effective cross section of 2.15×10^{-28} cm^2 sr^{-1} at 694.3 nm. This can be compared with the corresponding cross section for a number of gases by reference to Table 2.3. At sea level,† where $N \approx 2.55 \times 10^{19}$ molecules cm^{-3}, the atmospheric volume *backscattering coefficient* is

$$\beta_\pi^R(\lambda) \equiv N\sigma_\pi^R(\lambda) = 1.39\left[\frac{550}{\lambda(\text{nm})}\right]^4 \times 10^{-8}\ \text{cm}^{-1}\ \text{sr}^{-1} \tag{2.135}$$

†Loschmidt's number at standard temperature and pressure (STP) equals 2.69×10^{19} molecules cm^{-3}.

TABLE 2.3. RAYLEIGH BACKSCATTERING CROSS SECTION σ_π^R AT 694.3 nm

Gas	Formula	σ_π^R (10^{-28} cm^2 sr^{-1})	Ref.
Hydrogen	H_2	0.44	*a, b*
Deuterium	D_2	0.43	*a*
Helium	He	0.03	*a, b*
Oxygen	O_2	1.80	*b*
Nitrogen	N_2	2.14	*a, b*
Carbon dioxide	CO_2	6.36	*b*
Methane	CH_4	4.60	*a, b*
Nitrous oxide	N_2O	6.40	*a*
Neon	Ne	0.09	*b*
Argon	Ar	2.00	*a, b*
Xenon	Xe	11.60	*a*
Freons—important to stratospheric studies:			
Freon-12	CCl_2F_2	36.08	*b*
Freon-13B1	$CBrF_3$	24.87	*b*
Freon-14	CB_4	4.91	*b*
Freon-22	$CHClF_2$	21.90	*b*

[a]Rudder and Bach (1968).
[b]Shardanand and Prasad Rao (1977).

In arriving at the λ^{-4} dependence for the elastic scattering cross section the influence of atmospheric dispersion has been neglected. Elterman (1968) has shown that allowance for dispersion leads to a deviation from the -4 power by no more than about 3%, and Shardanand and Prasad Rao (1977) verified the λ^{-4} law for a number of gases over the visible range of wavelengths. Sepucha (1977) experimentally demonstrated that the Rayleigh cross section at 10.6 μm agrees with the value extrapolated from the measurements in the visible using the λ^{-4} dependence.

According to (2.135) the molecular backscattering coefficient at the ruby-laser wavelength is

$$\beta_\pi^R(694.3 \text{ nm}) \approx 5.47 \times 10^{-9} \text{ cm}^{-1} \text{ sr}^{-1}$$

which compares well with the value of 5.38×10^{-9} cm^{-1} sr^{-1}, estimated by McCormick (1971). Atmospheric measurements for β_π are in general much larger [for example, Cooney (1968) suggests that $\beta_\pi \approx 3 \times 10^{-8}$ cm^{-1} sr^{-1} at 694.3 nm], as scattering from particulates and aerosols tend to dominate at the longer wavelengths.

If the incident radiation is *unpolarized*, we average (2.131) over ϕ to obtain

$$\frac{d\sigma_R(\theta)}{d\Omega} = \frac{\pi^2(n^2-1)^2}{N^2\lambda^4}\left\{\frac{\cos^2\theta + 1}{2}\right\}$$

or more succinctly

$$\frac{d\sigma_R(\theta)}{d\Omega} = \tfrac{1}{2}\sigma_\pi^R(\cos^2\theta + 1) \tag{2.136}$$

It is of interest to note that we can obtain the same result by resolving the unpolarized light into two linearly polarized components, one polarized parallel, the other perpendicular to the *yz* scattering plane. This is also equivalent to an incident beam that is plane polarized at 45° to the *x*-axis (i.e., $\phi = \pi/4$). For the component polarized parallel to the *x*-axis (i.e., $\phi = \pi/2$) the appropriate differential scattering cross section is

$$\frac{d\sigma_R(\theta, \pi/2)}{d\Omega} = \sigma_\pi^R \tag{2.137}$$

and the *power scattered* by the molecule into unit solid angle centered around θ is

$$\frac{dP(\theta, \pi/2)}{d\Omega} = \tfrac{1}{2}\sigma_\pi^R I_i \tag{2.138}$$

I_i represents the total incident irradiance propagating in the *z*-direction. The factor of $\frac{1}{2}$ in (2.138) is required for the comparison with the unpolarized beam, that is to say, half of the incident radiation is assumed polarized parallel and half perpendicular to the scattering plane. It is also evident from (2.73) that the scattered electric field cannot have a component perpendicular to $\hat{\mathbf{u}}$, and from (2.117) $\hat{\mathbf{u}}$ is parallel to the dipole inducing field $\mathbf{E}_i$. Consequently, the component of the incident radiation that is polarized perpendicular to the scattering plane cannot give rise to elastically scattered radiation that is polarized parallel to the scattering plane; see Fig. 2.17.

The incident component of light polarized parallel to the *y*-axis ($\phi = 0$) will have a differential scattering cross section

$$\frac{d\sigma_R(\theta,0)}{d\Omega} = \tfrac{1}{2}\sigma_\pi^R\cos^2\theta \tag{2.139}$$

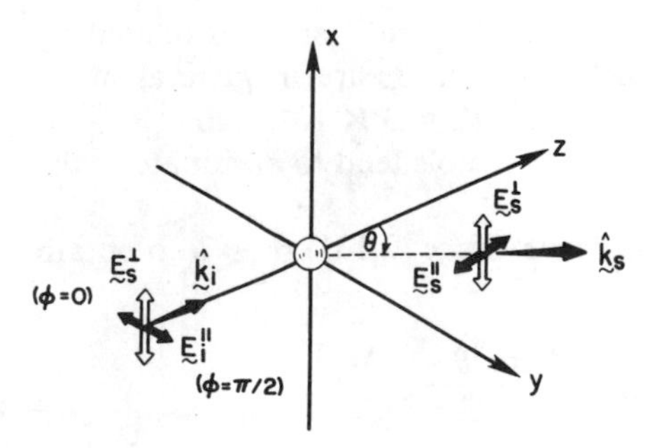

Fig. 2.17. Polarization configuration for scattering of radiation incident in the *z*-direction and either polarized at 45° to the *x*-axis ($\phi = \pi/4$) or unpolarized.

which will lead to power scattered into unit solid angle centered about θ:

$$\frac{dP(\theta,0)}{d\Omega} = \tfrac{1}{2}\sigma_\pi^R I_i \cos^2\theta$$

and consequently the total power scattered into unit solid angle about θ will be

$$\frac{dP(\theta)}{d\Omega} = \tfrac{1}{2}\sigma_\pi^R I_i\{\cos^2\theta + 1\}$$

which can be interpreted in terms of the same cross section as evaluated for unpolarized light; see equation (2.136). The variation in the scattering cross section with θ for each of the planes of polarization is presented in Fig. 2.18.

Each of the two components of the incident beam gives rise to a different angular distribution of scattered light. The incident light polarized perpendicular to the scattering plane ($\phi = \pi/2$) gives rise to a *circular* polar diagram for the distribution of scattered light, because the scattered irradiance at a distance r from the molecule can be related to the scattered power per unit solid angle through the relation

$$\frac{dP_s^\perp}{d\Omega} = I_s^\perp \frac{dS}{d\Omega} = r^2 I_s^\perp$$

Combining this with (2.138) enables us to see that

$$I_s^\perp = \frac{1}{r^2}\sigma_\pi^R \frac{I_i}{2} \tag{2.140}$$

is independent of θ.

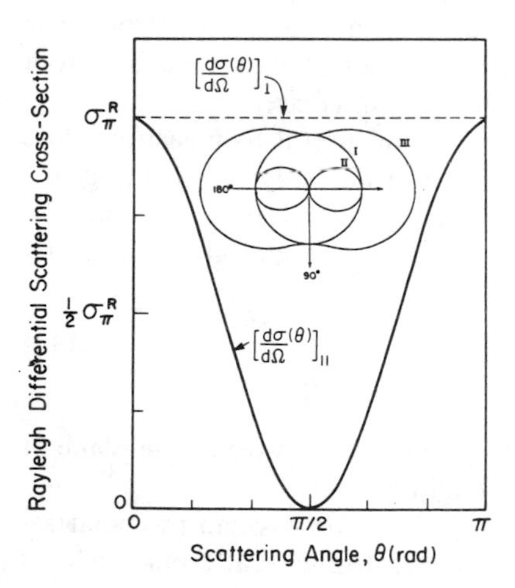

Fig. 2.18. Variation of cross section for molecular Rayleigh scattering with scattering angle θ, for incident radiation polarized $\parallel$ and $\perp$ to the *yz* scattering plane. The insert displays the corresponding polar scattering diagram, with I and II representing the $\perp$ and $\parallel$ components, respectively and III the resultant.

On the other hand the incident light that is polarized parallel to the scattering plane leads to a *figure-8* scattering diagram, as in this case the scattered irradiance at a distance r from the molecule is

$$I_s^{\parallel} = \frac{1}{r^2}\sigma_\pi^R \frac{I_i}{2}\cos^2\theta \tag{2.141}$$

The *degree of polarization* of a beam of light of irradiance I is defined by the relation

$$\mathcal{T} \equiv \frac{I^{\parallel} - I^{\perp}}{I^{\parallel} + I^{\perp}} \tag{2.142}$$

and consequently the degree of polarization for an incident unpolarized beam

$$\mathcal{T}_i = \frac{I_i^{\parallel} - I_i^{\perp}}{I_i^{\parallel} + I_i^{\perp}} = 0$$

as $I_i^{\parallel} = I_i^{\perp} = I_i/2$. Alternately, the degree of polarization of the light scattered at θ can be ascertained from (2.140) and (2.141), namely,

$$\mathcal{T}_s = \frac{I_s^{\parallel} - I_s^{\perp}}{I_s^{\parallel} + I_s^{\perp}} = \frac{1 - \cos^2\theta}{1 + \cos^2\theta} \tag{2.143}$$

and so $\mathcal{T}_s = 0$ for either $\theta = 0$ or π. However, equation (2.143) reveals that molecular scattering at 90° to the direction of propagation leads to polarization of otherwise unpolarized light, since $\mathcal{T}_s = 1$. This explanation of the polarization of 90°-scattered sunlight was first proposed by Lord Rayleigh. In a like manner the scattering of light from a colloidal suspension within a transparent fluid, such as water, is also polarized. This effect is observed when a beam of vertically polarized light is passed through a fine suspension—the path of the beam is visible when viewed horizontally but is invisible from above or below. This is sometimes termed the *Tyndall effect*.

Another scattering parameter that is often found useful to measure is the *depolarization ratio* δ_p. However, since there are several ways of defining this entity (Heller and Nakagaki, 1974), some care is required in reading the literature. The most common definition for the depolarization ratio is given by

$$\delta_p \equiv \frac{I_s^{\perp}}{I_s^{\parallel}} \tag{2.144}$$

where in this instance the perpendicular and parallel signs refer to the plane of polarization of the *incident linearly polarized* light beam.

For single molecular scattering δ_p is related to the *anisotropy* in the polarizability of the molecules, and in the case of isotropic scatterers such as

TABLE 2.4. DEPOLARIZATION OF LIGHT SCATTERING BY GASES[a]

Gas	Formula	δ_p
Argon	Ar	0
Methane	CH_4	0
Nitrogen	N_2	0.036
Oxygen	O_2	0.065
Chlorine	Cl_2	0.041
Air	—	0.042[b]
Nitric oxide	NO	0.027
Ethane	C_2H_6	0.005
Carbon monoxide	CO	0.013
Hydrogen chloride	HCl	0.007
Hydrogen bromide	HBr	0.008
Carbon dioxide	CO_2	0.097
Carbon disulfide	CS_2	0.115
Water	H_2O	0.02
Hydrogen sulfide	H_2S	0.003
Sulfur dioxide	SO_2	0.031
Ammonia	NH_3	0.01

[a] Condon and Odishaw (1967).
[b] Recent measurements for air have indicated that $\delta_{p\ \mathrm{air}} = 0.035$ (Zuev, 1976).

monatomic gases like argon, $\delta_p = 0$. If allowance is made for the anisotropic properties of the scattering molecules, the Rayleigh scattering cross section is modified to take the more general form

$$\sigma_R(\lambda) = \frac{8\pi}{3}\left[\frac{\pi^2(n^2-1)^2}{N^2\lambda^4}\right]\left\{\frac{6+3\delta_p}{6-7\delta_p}\right\} \tag{2.145}$$

The depolarization ratio δ_p for a number of atmospheric constituents are provided by Condon and Odishaw (1967) and are listed in Table 2.4. Depolarization also arises if multiple scattering occurs. Consequently, it is of considerable interest to those studying the lower atmosphere and in particular clouds (Liou and Schotland, 1971; Cohen and Graber, 1975; McNeil and Carswell, 1975; Pal and Carswell, 1976, 1978). We shall return to this topic shortly.

2.8. RAYLEIGH–MIE SCATTERING

As indicated in the introduction of this book, our atmosphere contains, in addition to a variety of molecules, a rich brew of particulates and aerosols that include dust, ice crystals, fog, haze, and clouds. The size range of these *aeroticulates* spans many orders of magnitude, as indicated in Fig. 2.19, and their scattering properties become extremely complex when their dimensions become comparable to the wavelength λ of the light illuminating them. This leads to the introduction of a scattering *size parameter*

$$\alpha \equiv ka = \frac{2\pi a}{\lambda} \tag{2.146}$$

where a is some representative dimension of the aeroticulate (aerosol or particulate).

For *isotropic* dielectric scatterers that are much smaller than the wavelength of the incident light ($\alpha < 0.5$), the scattering is somewhat similar to molecular Rayleigh scattering. This can be seen by expressing the differential scattering cross section, equation (2.126), in terms of the *polarizability* of the scatterer:

$$\frac{d\sigma(\theta,\phi)}{d\Omega} = \frac{\pi^2}{\lambda^4}|\wp(\omega)|^2[\cos^2\phi\cos^2\theta + \sin^2\phi] \tag{2.147}$$

If the scatterer is now taken to be a small dielectric sphere of radius a and refractive index n (relative to the medium outside the sphere) then it can be shown (Kerker, 1969) that the polarizability $\wp$ of this sphere is

$$\wp = 4\pi a^3\left\{\frac{n^2-1}{n^2+2}\right\} \tag{2.148}$$

Substitution of this relation into (2.147) yields

$$\frac{d\sigma(\theta,\phi)}{d\Omega} = a^2\left(\frac{2\pi a}{\lambda}\right)^4\left\{\frac{n^2-1}{n^2+2}\right\}^2[\cos^2\phi\cos^2\theta + \sin^2\phi] \tag{2.149}$$

for the differential elastic scattering cross section. The corresponding *elastic*

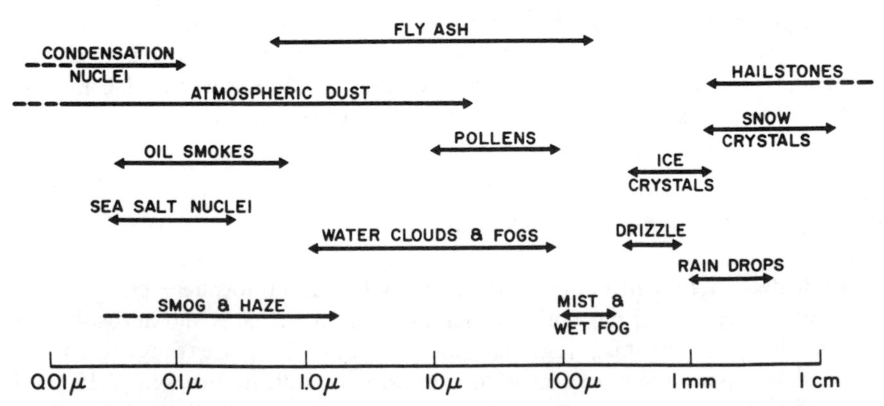

Fig. 2.19. Representative diameters of common atmospheric particles (Johnson, 1969).

backscatterin[illegible]tion for the isotropic dielectric sphere is given by

$$[illegible] \equiv \frac{d\sigma(\theta = \pi)}{d\Omega} = a^2(ka)^4\left\{\frac{n^2 - 1}{n^2 + 2}\right\}^2 \tag{2.150}$$

Both of the[illegible]ctions are seen to have the λ^{-4} dependence associated with molecu[illegible]gh scattering. The angular dependence of the differential scattering cross s[illegible]n in the case of very small dielectric scatterers is also the same as for molecules. The major difference between elastic scattering from molecules and from aeroticulates is the magnitude of the cross section. Clearly, dielectric spheres of radius much greater than the size of molecules will have a scattering cross section that is correspondingly larger.

As the scatterer's size parameter α increases beyond the value of 0.5, the scattering pattern loses its symmetry and the differential scattering cross section becomes a very complicated function of ka, n, and θ. The first comprehensive treatment of scattering by dielectric spheres of size comparable to the wavelength of the incident light was undertaken by Mie in 1908. This theory states that in general we can write

$$\frac{d\sigma_M(\theta, \phi)}{d\Omega} = \frac{\lambda^2}{4\pi^2}\left[i_2(\theta, \alpha, n)\cos^2\phi + i_1(\theta, \alpha, n)\sin^2\phi\right] \tag{2.151}$$

where the perpendicular and parallel Mie intensity functions, $i_1(\theta, \alpha, n)$ and $i_2(\theta, \alpha, n)$ respectively, describe the angular distribution of the scattered radiation in terms of θ, α, and n, (Van de Hulst, 1957; Kerker, 1969). The angular distribution of scattered irradiance at a distance r from a small spherical dielectric scatterer is

$$I_s(\theta, \phi) = \frac{I_i}{r^2}\frac{\lambda^2}{4\pi^2}\left[i_2(\theta)\cos^2\phi + i_1(\theta)\sin^2\phi\right] \tag{2.152}$$

where we have simplified the notation by writing $i_{1,2}(\theta, \alpha, n) = i_{1,2}(\theta)$. It should be noted that throughout this section the various cross sections have an implicit dependence on the wavelength of the light. However, for the sake of notational brevity this will not be indicated. The scattered irradiance can be divided into components that are polarized parallel ($I_s^{\parallel}$) and perpendicular ($I_s^{\perp}$) to the scattering plane (defined by the unit propagation vectors $\hat{\mathbf{k}}_i$ and $\hat{\mathbf{k}}_s$):

$$I_s^{\parallel} = \frac{I_i}{k^2r^2}i_2(\theta)\cos^2\phi \tag{2.153}$$

and

$$I_s^{\perp} = \frac{I_i}{k^2r^2}i_1(\theta)\sin^2\phi \tag{2.154}$$

In the case of *natural* (unpolarized) light, or linearly polarized light with $\phi = \pi/4$, the *normalized* components of scattered irradiance on the surface of the dielectric sphere (i.e., $r = a$) can be directly related to the Mie intensity functions, that is to say

$$i_1(\theta) = (ka)^2 \mathscr{I}_s^{\perp} = \frac{(ka)^2 I_s^{\perp}}{I_i/2} \tag{2.155}$$

and

$$i_2(\theta) = (ka)^2 \mathscr{I}_s^{\parallel} = \frac{(ka)^2 I_s^{\parallel}}{I_i/2} \tag{2.156}$$

The polar diagrams for $i_1(\theta)$ and $i_2(\theta)$, in the limit $ka \ll 1$, have the same form as for Rayleigh scattering (see Fig. 2.18). However, as ka increases

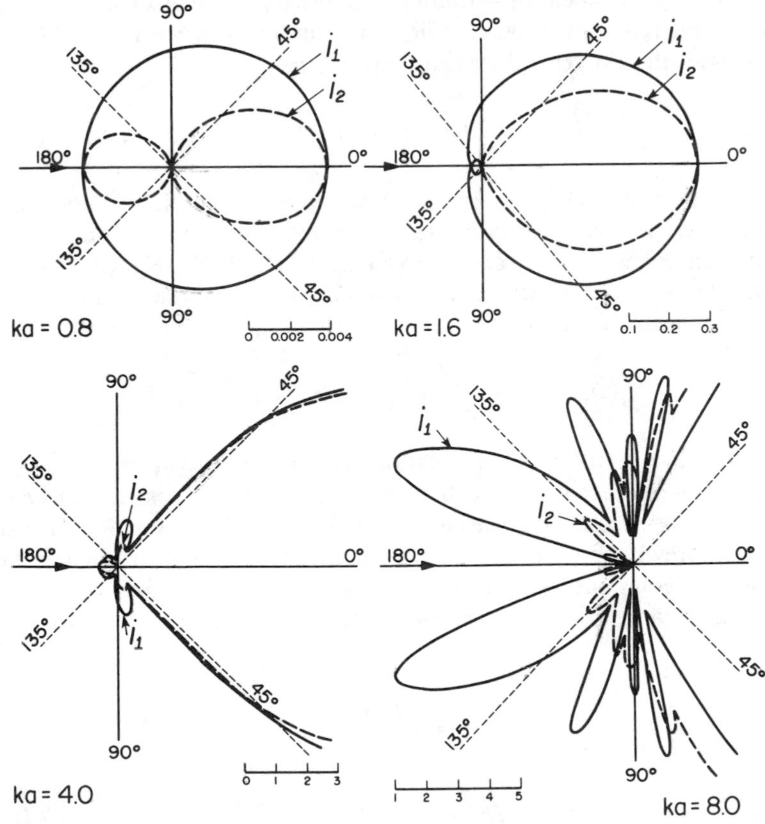

Fig. 2.20. Polar diagrams displaying the angular distribution of the ∥ and ⊥ components of scattered light from a dielectric sphere of refractive index $n = 1.25$. Incident light is assumed unpolarized or with $\phi = \pi/4$. Here $i_1(\theta) = (ka)^2 \mathscr{I}_s^{\perp}$ and $i_2(\theta) = (ka)^2 \mathscr{I}_s^{\parallel}$ (Born and Wolf, 1964).

beyond 0.5 a clear asymmetry develops, leading eventually ($ka > 1.5$) to the pronounced *forward scattering pattern* displayed in Fig. 2.20.

This progressive deviation from the Rayleigh scattering pattern with its enhanced forward scattering and growing angular mode structure is further illustrated in Fig. 2.21, where the logarithm of the Mie intensity functions, $i_1(\theta)$ and $i_2(\theta)$ (each ordinate division corresponds to a factor of 10 increasing upwards), are plotted against scattering angle θ for several values of the size parameter α for refractive-index values of 1.55 (corresponding to glass spheres) and 1.33 (corresponding to water droplets) (Van de Hulst, 1957).

For unpolarized light we see that $(\lambda^2/8\pi^2)[i_1(\theta) + i_2(\theta)]$ represents the radiant energy scattered per second per unit solid angle, centered around θ, for an incident plane wave of unit irradiance and the degree of polarization

$$\mathscr{T}_s = \frac{i_2 - i_1}{i_2 + i_1} \tag{2.157}$$

A more general treatment of Mie scattering theory can be found in several texts (Van der Hulst, 1957; Kerker, 1969; and Deirmendjian, 1969) in addition to the original work of Mie (1908). Extension of this work to nonspherical dielectric scatterers has been undertaken by Barber and Yeh (1975). Recently, Wiscombe (1980) has described an improved Mie-scattering algorithm.

There are two scattering parameters that are of prime significance in the lidar field. These are the *volume backscattering coefficient* β_π and the *total*

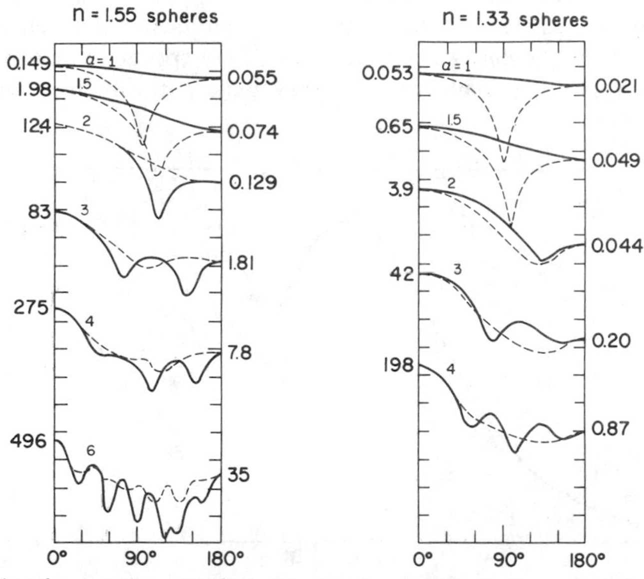

Fig. 2.21. Angular variation of $i_1(\theta)$ (solid curves) and $i_2(\theta)$ (broken curves) for a range of α-values and two refractive indices. The logarithms of the Mie intensity functions are plotted against θ; each ordinate division corresponds to a factor of 10 (Van de Hulst, 1957).

scattering coefficient β_s. In order to evaluate these quantities we need to introduce the Mie *backscattering efficiency*

$$Q_\pi(\alpha, n) \equiv \frac{1}{\pi a^2} \frac{d\sigma_M(\theta = \pi)}{d\Omega} \tag{2.158}$$

and the total *scattering efficiency*

$$Q_s(\alpha, n) \equiv \frac{\sigma_s^M}{\pi a^2} = \frac{1}{\pi (ka)^2} \int_0^\pi \int_0^\pi \left[i_2(\theta)\cos^2\phi + i_1(\theta)\sin^2\phi \right] \sin\theta \, d\theta \, d\phi \tag{2.159}$$

For the backscattering case $i_1 = i_2$ and $Q_\pi(\alpha, n)$ tends to increase with α until a maximum is reached, whereupon $Q_\pi(\alpha, n)$ undergoes a damped oscillation. This behavior is illustrated in Fig. 2.22, for the case of a dielectric sphere of refractive index $n = 1.33$ and a wavelength of 700 nm (Twomey and Howell, 1965). The total scattering efficiency $Q_s(\alpha, n)$ is found to exhibit a similar dependence upon α, although the larger the refractive index of the scatterer (compared to that of the background medium), the greater the amplitude and frequency of the ripple structure on the curves, (Kerker, 1969, p. 105; Penndorf, 1957).

Since the total scattering efficiency represents the ratio of the energy scattered by the particle to the total energy physically intercepted by the particle, values of $Q_s(\alpha, n)$ greater than unity may appear to be something of an enigma. This is especially true for very large particles, inasmuch as the

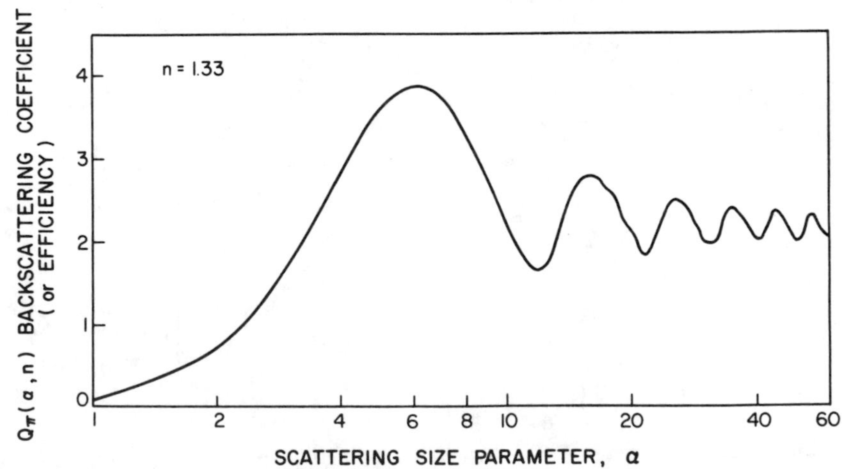

Fig. 2.22. Variation of Mie backscattering efficiency $Q_\pi(\alpha, n)$ with α for $n = 1.33$, $\lambda = 700$ nm based on calculations of (Twomey and Howell, 1965).

limiting value of $Q_s(\alpha, n)$ is two. The explanation lies in the perturbation of the electromagnetic field beyond the physical presence of the scatterer that is required to satisfy the boundary conditions on the surface of the scatterer (Kerker, 1969, p. 106).

Another notable effect stems from the relative independence of the total scattering efficiency upon α for large scatterers. This implies that almost all wavelengths are scattered with equal efficiency if the scatterers are large enough (and free of absorption bands)—consequently, clouds appear white, unless they are so opaque as to appear dark.

Volume Scattering Coefficients and Phase Functions

Scattering of light within the atmosphere invariably involves a distribution of scatterers that vary in *composition*, *size*, and *shape*. The subsequent scattering properties are obviously extremely complex. On many occasions, however, the situation is greatly simplified through the domination of one class of aeroticulates. With this in mind we shall consider the scattering properties of a *polydispersion of homogeneous spheres* illuminated by a linearly polarized plane wave with $\phi = \pi/4$ (the results will also apply to unpolarized light). If we assume that $N(a)\,da$ represents the number of dielectric spheres with refractive index n having a radius that lies in the interval $(a, a + da)$, then the *differential volume scattering coefficients* $\beta_j(\theta, n, \lambda)$ at wavelength λ are given by

$$\beta_j(\theta, n, \lambda) \equiv \frac{1}{2k^3}\int_{\alpha_1}^{\alpha_2} i_j(\theta, \alpha, n, \lambda) N(\alpha)\, d\alpha \qquad (\mathrm{cm^{-1}\,sr^{-1}}) \quad (2.160)$$

where $j = 1, 2$ corresponds to polarization perpendicular or parallel to the scattering plane. The size-parameter limits of integration α_2 and α_1 define the effective range of radius for the distribution of scatterers and correspond to radii a_2 and a_1 respectively.

The *total volume scattering coefficient* is

$$\beta_s(n, \lambda) \equiv \frac{\pi}{k^3}\int_{\alpha_1}^{\alpha_2} \alpha^2 Q_s(\alpha, n, \lambda) N(\alpha)\, d\alpha \qquad (\mathrm{cm^{-1}}) \qquad (2.161)$$

and the *phase functions* for the polydispersion of spherical particles (or aerosols) are

$$P_j(\theta, n, \lambda) \equiv 4\pi \frac{\beta_j(\theta, n, \lambda)}{\beta_s(n, \lambda)} \qquad (\mathrm{sr^{-1}}) \qquad (2.162)$$

where again $j = 1, 2$. In essence the phase function (the name is something of a misnomer, as it has nothing to do with the phase of the wave) indicates the angular distribution of radiation scattered by such a medium under illumination. The *backscattering phase functions* are

$$P_j(\pi, \lambda) \equiv \frac{4\pi\beta_j(\pi, \lambda)}{\beta_s(\lambda)} \qquad (\text{for} \quad j = 1, 2)$$

where $\beta_j(\pi, \lambda)$ represents the *backscattering volume coefficient* corresponding to the appropriate plane of polarization. When depolarization is negligible, the backscattered radiation has essentially the same polarization as that of the incident radiation and we need only consider one *backscattering coefficient* $\beta_\pi(\lambda)$. Alternatively, if the photodetector of a lidar system is insensitive to the polarization of the scattered light, then we can also write

$$\beta_\pi(\lambda) = \frac{1}{2k^3} \int_{\alpha_1}^{\alpha_2} \{i_1(\alpha, n, \pi, \lambda) + i_2(\alpha, n, \pi, \lambda)\} N(\alpha)\, d\alpha \qquad (2.163)$$

It should also be noted that for a single constituent medium and in the absence of absorption, the *volume extinction (attenuation) coefficient* is

$$\kappa_\epsilon = \beta_s \qquad (2.164)$$

(Kerker, 1969, p. 49).

As one might expect, there is no single distribution function that can adequately describe all of the various kinds of particles that exist in the atmosphere. There are, however, two models that seem to be capable of fitting most of the important kinds of atmospheric particle distributions. The first model was proposed by Junge (1963), and takes the form

$$N(a) = \frac{J}{a^{\nu_p + 1}} \qquad (\text{cm}^{-3}\,\mu\text{m}^{-1}) \qquad (2.165)$$

ν_p tends to lie between 2.0 and 5.0 for most particulates found in the atmosphere, and J is a normalization constant that follows from continuity, that is,

$$\int_0^\infty N(a)\, da = N_{\text{total}} \qquad (\text{cm}^{-3}) \qquad (2.166)$$

Junge's power law (2.165) has been found to represent reasonably the distribution of fine water aerosols classified as *continental haze*, or Haze C. In general the radii of these droplets is less than or equal to 0.1 μm, as seen by reference to Fig. 2.23.

The particle size distribution that is found more appropriate to clouds or fogs (maritime haze) is discussed in detail by Deirmendjian (1964, 1969) and Liou and Schotland (1971) and takes the form

$$N\left(\frac{a}{a_m}\right) = C\left(\frac{a}{a_m}\right)^{\nu_p} e^{-\nu_p/a_m} \qquad (\text{cm}^{-3}\,\mu\text{m}^{-1}) \qquad (2.167)$$

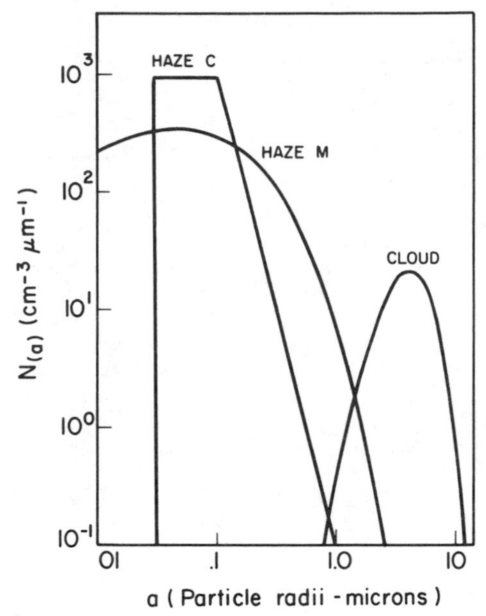

Fig. 2.23. Three kinds of distribution of atmospheric water droplets used in computing Mie functions. Total concentration = 100 cm^{-3} (Deirmendjian, 1964).

where a_m is the mode radius at which the concentration is a maximum and C is a normalization constant which is again given by (2.166); see Fig. 2.23. $\nu_p = 6$ according to Liou and Schotland (1971). They also suggest that a bimodal distribution might better represent the observed distributions found in certain classes of clouds where the population in the tail of the distribution function given by (2.167) appears to be inadequate. It also appears that for most clouds $N_{\text{total}} \approx 10^2$ cm^{-3}.

Liou and Schotland (1971) have evaluated the angular variation in the phase functions for several cloud models and at several wavelengths. Figure 2.24 is representative of their results, showing again the pronounced forward scattering (diffraction peak) that would be expected from a low-density fair-weather cumulus cloud with a mode radius of 4 μm. It should be noted that the vertical scale shown applies to the lowest curve—the scales for the other curves may be obtained by multiplication by a power of 10 such that the horizontal bar on each curve occurs at unity.

McCormick (1971) has predicted the variation of both the aerosol and molecular backscattering coefficients [$\beta_\pi^M(\lambda, z)$ and $\beta_\pi^R(\lambda, z)$ respectively—i.e., M for Mie and R for Rayleigh] with altitude z at wavelengths corresponding to the fundamental and second harmonic of a ruby laser. A Junge power-law distribution was assumed for the aerosols with $n = 1.5$, $\nu_p = 3$, and aerosol-radius integration limits of $a_1 = 0.125$ μm and $a_2 = 10$ μm. The aerosol altitude profile was based on the sampling data of Rosen (1966), with a

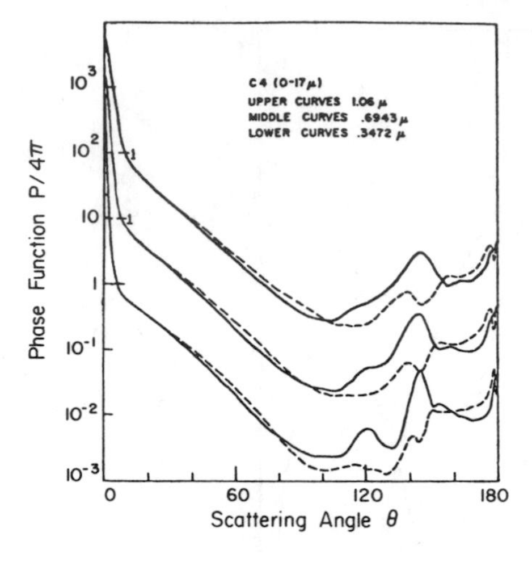

Fig. 2.24. Single-scattering phase functions perpendicular (solid lines) and parallel (dotted lines) to the scattering plane for water drops illuminated by 1.06-μm (upper curves), 0.6943-μm (middle curves), and 0.3472-μm light (lower curves). The size distribution has the mode radius at 4 μm. For scales see text (Liou and Schotland, 1971).

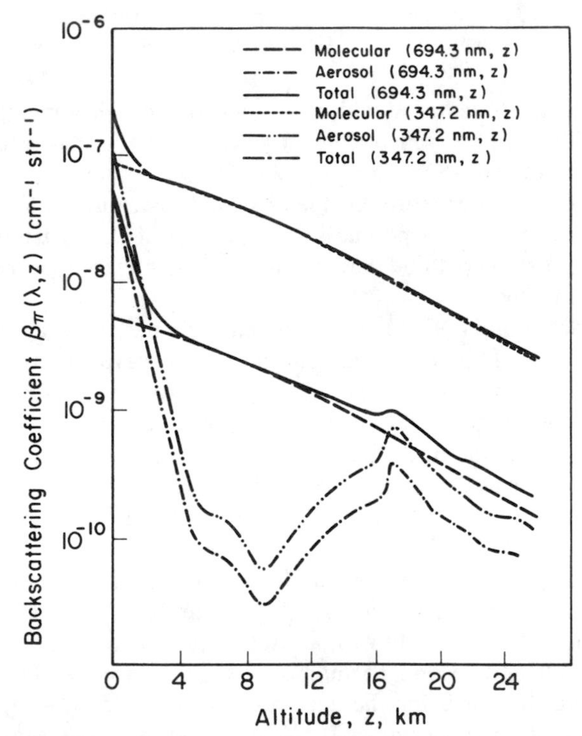

Fig. 2.25. Variation of backscattering coefficient $\beta_\pi(\lambda, z)$ with altitude at fundamental and second harmonic frequencies of ruby laser (McCormick, 1971).

sea-level value of 450 particles cm^{-3} for N_{total}. Recently Gibson (1976) has also made balloon measurements of the aerosol scattering properties of the atmosphere to an altitude of 25 km. McCormick's results are presented as Fig. 2.25 and should be understood to provide only a guide to the relative importance of aerosol and molecular scattering under light haze conditions, since substantial changes in aerosol concentration can occur within a fraction of a kilometer. A table of aerosol scattering coefficients *per particle* based on a similar analysis has previously been published (McCormick et al., 1968).

Deirmendjian (1969) has developed scattering models for several kinds of atmospheric conditions. Figures 2.26 and 2.27 present his results for the volume backscattering and attenuation coefficients as modified by Wright et al. (1975).

Although most Mie-scattering computations assume a polydispersion of homogeneous *spherical* scatterers, Chylek et al. (1976) have attempted to extend these calculations to distributions of *irregular-shaped* and *randomly oriented* particles. The good agreement found between their analysis and experiments appears to vindicate their assertion that the surface waves present in spherical particles (and responsible for both the *glory* and the ripple structure of the scattering coefficients) is absent in scattering from irregular particles.

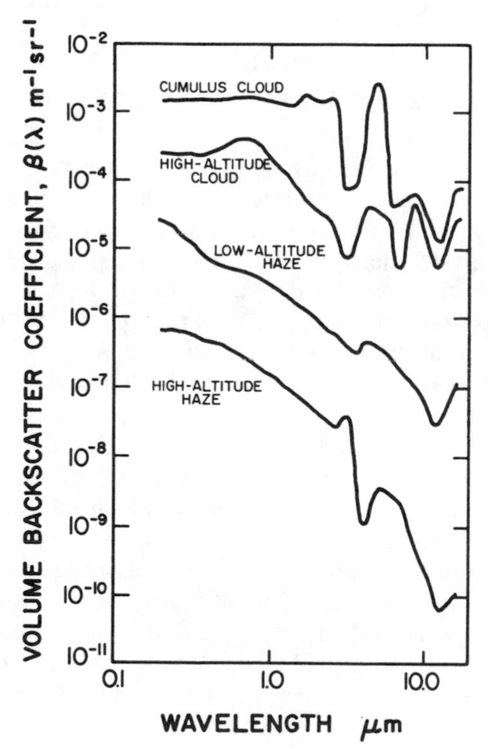

Fig. 2.26. Aerosol volume backscattering coefficient as a function of wavelength (Wright et al, 1975).

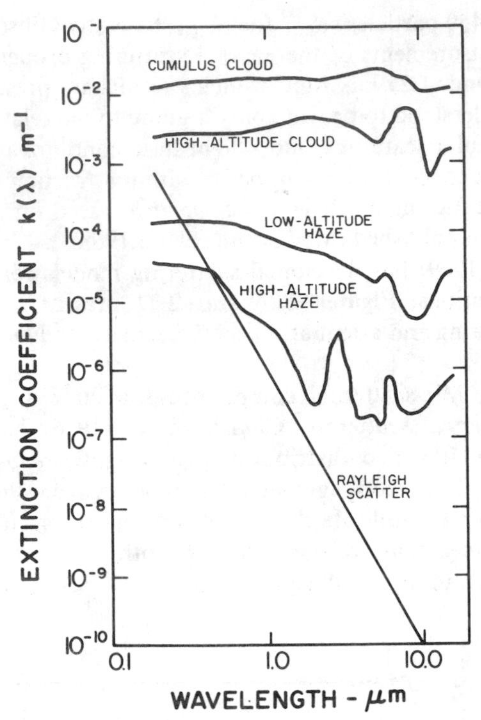

Fig. 2.27. Aerosol extinction coefficient as a function of wavelength (Wright et al, 1975).

The treatment given so far has assumed single scattering of the radiation, and although this is reasonable under clear sky or light haze conditions, at higher values of turbidity (clouds, fogs, and smoke plumes) multiple scattering becomes important. In the case of a linearly polarized lidar pulse, multiple scattering leads to an appreciable depolarization of the scattered radiation, and the rate of attenuation of the laser pulse is reduced under these circumstances due to beam replenishment by forward scattering. We shall return to this point later, in the discussion of radiative transfer (Section 4.4).

The theoretical treatment of multiple scattering in lidar work is extremely complex. Nevertheless, Liou and Schotland (1971) have attempted to formulate an analytical approach to the problem of double scattering, and Kattawar and Plass (1976) have undertaken several studies which include the diffusion (or asymptotic) limit of multiple scattering. To date the agreement between the theoretical predictions of multiple-scattering lidar and the experimental observations leaves a lot to be desired (Collis and Russell, 1976).

The *inversion* of lidar data in order to reconstruct the particle distributions responsible for the scattered signal is beyond the scope of the present text, and the reader is directed to the review on the subject by Fymat (1976) and the recent paper by Heintzenberg et al. (1981).

3

Quantum Physics and Radiation Processes

In the previous chapter we studied the nature of electromagnetic radiation and the form of its interaction with matter that can be accounted for in terms of a classical picture of atoms and molecules. This mechanical view of the microscopic world came under suspicion around the turn of the century through the startling revelations of Planck, Einstein, and Bohr. The monumental work of Heisenberg, DeBroglie, Schrödinger and Dirac (to name only a few) served to formulate the subsequent new branch of physics, termed *quantum mechanics*. This mathematical and somewhat more abstract view of nature proved to be very successful in accounting for the sophisticated experiments in physics that were undertaken during this period, including those that involve the interaction of radiation with matter.

We shall first review those aspects of quantum mechanics that are germane to our interests. This treatment will, for obvious reasons, be somewhat brief and will not include the rigor found in many textbooks on the subject (Eisberg and Resnick, 1974; Liboff, 1980; Loudon, 1973; Heitler, 1954). Nevertheless, we shall attempt to provide a reasonably self-consistent basis for understanding those interactions of radiation with matter that are important to the field of laser remote sensing. These range from spontaneous emission to Raman scattering.

3.1. REVIEW OF QUANTUM MECHANICS

3.1.1. Historical Perspective

The branch point from classical physics was *blackbody radiation*, the emission from an idealized cavity in thermal equilibrium. In reality it is approximated by the radiation emanating from a small hole in an oven. Classical physics (essentially Maxwell's electromagnetic theory and statistical mechanics) could not account for the spectral distribution of this radiation; worse, it predicted the *ultraviolet catastrophe* (the total blackbody emission should be infinite, due to the Rayleigh–Jeans law, which stated that the radiant energy density should increase as the cube of the frequency), with which nature refused to comply. Planck derived the correct expression to describe the observed thermal emission through the introduction of his famous quantization

$$\mathscr{E} = nh\nu \tag{3.1}$$

In effect this states that the energy of an atomic oscillator $\mathscr{E}$ in the walls of the blackbody is restricted to integer values of $h\nu$. Here ν is the frequency of the oscillator, n is an integer $(1, 2, 3, \ldots)$, and h is the important constant named after Planck ($h = 6.626 \times 10^{-34}$ J s). According to Planck the *radiant energy density* $\rho(\nu)$ within a cavity in thermal equilibrium at temperature T is given by the expression

$$\rho(\nu) = \frac{8\pi\nu^2}{c^3} \cdot \frac{h\nu}{e^{h\nu/kT} - 1} \tag{3.2}$$

where k is Boltzmann's constant ($k = 1.381 \times 10^{-23}$ J K^{-1}) (Eisberg and Resnick, 1974). The first factor of (3.2) represents the number density of radiation modes per unit volume and per unit frequency interval that can exist within a cavity. The second factor represents the mean energy of the atomic oscillators; it approaches the classical value of kT when the energy spacing between the quantized states becomes much less than the mean thermal energy of the system, that is, $h\nu \ll kT$.

It is important to understand that Planck's quantization applied specifically to the atomic oscillators within the walls of the blackbody. Full appreciation of the quantized nature of the electromagnetic field only came with Einstein's explanation of the photoelectric effect and Compton's study of X-ray scattering by light elements (Eisberg and Resnick, 1974). As a result of this work the particle-like properties of the electromagnetic field (the energy $\mathscr{E}_\nu$ and momentum p_ν) are related to the wavelike properties (the frequency ν and wavelength λ) through the relations

$$\mathscr{E}_\nu = h\nu \quad \text{and} \quad p_\nu = h/\lambda \tag{3.3}$$

Spectroscopic studies of low-pressure discharges led Bohr to develop a simple model of the atom that involved the *quantization* of the *orbital angular momentum* of the circulating electron. Although this so-called *Bohr atom* could account for some of the basic features of the hydrogen atom spectra, it was too limited to be of general use. Nevertheless, certain aspects of it led de Broglie to make one of man's boldest intellectual leaps and suggest that electrons (and subsequently matter in general) also possessed a dual (wave–particle) nature. This idea met with immediate success in explaining the strange experimental results of Davisson and Germer and of Thomson in scattering electrons from crystals.

The duality relations given by (3.3) consequently apply equally well to both matter and radiation. In order to make the duality of nature more palatable let me present a simple analogy: *A pane of glass when held up to a light acts as a window, yet the same sheet of glass appears as a mirror when viewed at grazing incidence*. The window and the mirror are by nature opposite, *but clearly a sheet of glass can appear as either, depending on the experiment*.

3.1.2. Quantized States of Atoms

As we have seen in the wave picture of light [equation (2.41)], the irradiance of a light beam is proportional to the square of the electric field vector, averaged over one cycle of the wave. In the photon (or particle) picture Einstein proposed that the irradiance of a beam is given by

$$I = \Phi c h \nu \tag{3.4}$$

where Φ represents the density of photons of energy $h\nu$. A reconciliation of the wave and particle aspects of light is possible if we view the wavelike behavior as guiding (in a probabilistic manner) the photons. Thus we can think of the square of the electric field, $|E(\mathbf{r}, t, \nu)|^2$, as describing the *probability* of observing photons of frequency ν at position $\mathbf{r}$ and time t.

Max Born postulated that the duality of matter might be viewed in a similar manner, and Schrödinger proposed that the state of an atomic particle is described by a wave equation of the form

$$\left[-\frac{\hbar^2}{2m}\nabla^2 + V(\mathbf{r}, t)\right]\Psi(\mathbf{r}, t) = i\hbar\frac{\partial}{\partial t}\Psi(\mathbf{r}, t) \tag{3.5}$$

where $V(\mathbf{r}, t)$ represents the potential energy of the particle of mass m and $\Psi(\mathbf{r}, t)$ represents the *wave function of the particle*. The physical meaning of the wave function can be appreciated by recognizing that $\Psi^*(\mathbf{r}, t)\Psi(\mathbf{r}, t)$ provides the probability of observing the particle at position $\mathbf{r}$ and time t. The asterisk signifies the complex conjugate. The two terms within the square brackets constitute the *Hamiltonian operator* $\mathscr{H}$. If the particle described by (3.5) is in a *stationary state* of frequency ω, we shall assume that we can write

$$\Psi(\mathbf{r}, t) = \psi(\mathbf{r})e^{-i\omega t} \tag{3.6}$$

where $\psi(\mathbf{r})$ represents the spatial part of the wave function. Then the Schrödinger wave equation (3.5) is simplified to the form

$$\left[-\frac{\hbar^2}{2m}\nabla^2 + V(\mathbf{r})\right]\psi(\mathbf{r}) = \hbar\omega\psi(\mathbf{r}) \tag{3.7}$$

where $\hbar\omega$ can be viewed as the energy associated with the oscillation of the particle, and $\hbar \equiv h/2\pi$.

In the case of a single electron in the field of a nuclear charge Ze, the potential-energy operator is

$$V(\mathbf{r}) = -\frac{Ze^2}{4\pi\epsilon_0 r} \tag{3.8}$$

and we can express the time-independent Schrödinger equation for this hydrogenlike atom in the form

$$\left[-\frac{\hbar^2}{2m}\nabla^2 - \frac{Ze^2}{4\pi\epsilon_0 r}\right]\psi(r,\theta,\phi) = \mathscr{E}\psi(r,\theta,\phi) \tag{3.9}$$

Although we shall not go through the analysis leading to the solutions of this equation [this is available in any good quantum-mechanics text, e.g. Liboff (1980) or Eisberg and Resnick (1974)] we shall summarize the results. In the first place, finite solutions to (3.9) are found to exist (for $\mathscr{E} < 0$) only for values of energy that are quantized according to the relation

$$\mathscr{E}_n = -\frac{1}{(4\pi\epsilon_0)^2}\frac{Z^2e^4m_e}{2\hbar^2}\left\{\frac{1}{n^2}\right\} = -\frac{13.6 \text{ eV}}{n^2} \tag{3.10}$$

where n is an integer that can range from 1 to ∞. The *ionization energy* of hydrogen (where $Z = 1$) is evidently 13.6 eV, as $n = 1$ corresponds to the ground state. Here one electron volt (eV) represents the kinetic energy of an electron that has fallen through a potential drop of one volt (1 eV $\approx 1.602 \times 10^{-19}$ J). The wave-function solutions are specified by a set of integers n, l, and m, called *quantum numbers*:

$$\psi(r,\theta,\phi) = \psi_{n,l,m}(r,\theta,\phi)$$

or more succinctly

$$\psi = \psi(n,l,m) \tag{3.11}$$

As we have seen, n—the *principal* quantum number ($n = 1, 2, \ldots, \infty$)—determines the energy of the state. Likewise l—the *azimuthal* quantum number ($l = 0, 1, 2, \ldots, n-1$)—specifies the orbital angular momentum of the state. Clearly there are n different l-states for each value of n. Finally, m—the *magnetic* quantum number ($m = 0, \pm 1, \pm 2, \ldots, \pm l$)—is related to the inclination of the orbital angular momentum vector relative to some reference axis (often the direction of a magnetic field). If we allow for the two spin states possible for an electron, then there are $2 \times \sum_{l=0}^{n-1}(2l+1) = 2n^2$

states corresponding to each value of n. Consequently, these hydrogenlike solutions have a $2n^2$-*fold degeneracy*, which means that there are $2n^2$ states that possess the same energy, $\mathscr{E}_n$, given by (3.10). This is illustrated in Fig. 3.1, where the energy levels available to the hydrogen's electron are presented. It can be seen that S-states have $l = 0$, P-states have $l = 1$, D-states have $l = 2$, and F-states have $l = 3$.

As we shall see later radiation is absorbed or emitted only when the electron makes certain *optically allowed* transitions between states. Thus the strong resonance line of hydrogen (the Lyman alpha line L_α) at 97.2 nm is emitted when the electron jumps from the $2p$ resonance level to the $1s$ ground level. The energy of the emitted photon is in general given by the Bohr relation

$$h\nu_{n_1n_2} = \mathscr{E}_{n_2} - \mathscr{E}_{n_1} \tag{3.12}$$

where $\nu_{n_1n_2}$ is the frequency of the photon and $\mathscr{E}_{n_2}$ and $\mathscr{E}_{n_1}$ represent the respective energies of the initial and final states. The quantum *selection rule* that determines which transitions are permissible can be stated in the simple form

$$\Delta l = \pm 1 \tag{3.13}$$

that is to say, the change in l is limited to ± 1.

In the case of atoms possessing more than one electron, the nuclear charge is partially screened and the outer electron can no longer be assumed to move

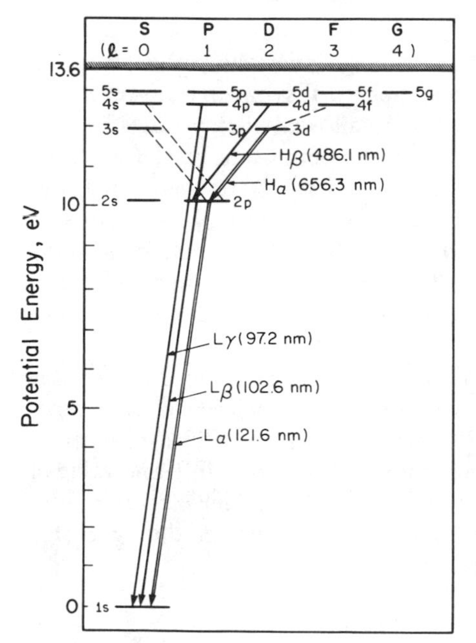

Fig. 3.1. A simplified Grotrian energy-level diagram for hydrogen.

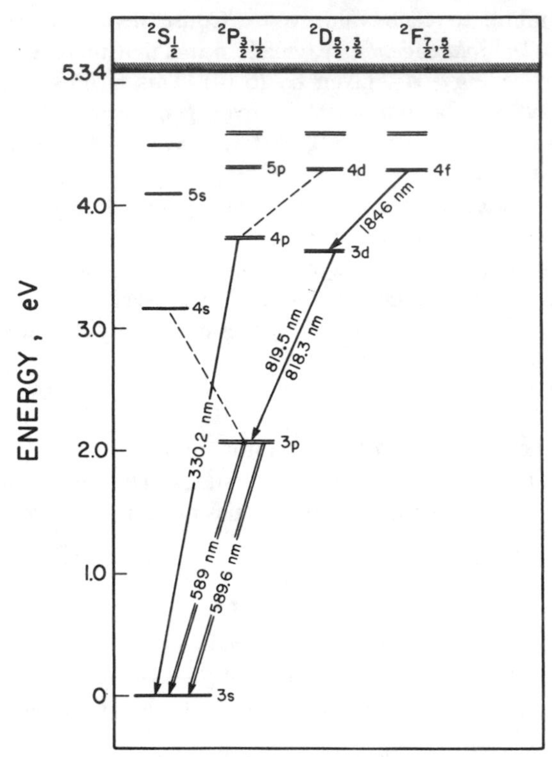

Fig. 3.2. Partial Grotrian energy-level diagram for sodium.

in a Coulomb potential field (i.e. one varying inversely with the radial separation of the electron from the nucleus). Spherical symmetry, however, can still be assumed. Under these circumstances the allowed energy states become dependent upon the quantum number l and the spin state of the electron; consequently there are no longer $2n^2$ degenerate energy states associated with the quantum number n. This leads to a more complex energy-level structure for the outermost electron, as illustrated in Fig. 3.2 for the case of sodium. Indeed, the removal of the so-called *l-fold degeneracy* is so complete in this instance that the resonance transition occurs between two states with the same n ($n = 3$ for sodium).

The importance of the electron spin is also apparent from Fig. 3.2, since the P and D states are now seen to be split into states with a *total angular momentum* quantum number j of $\frac{3}{2}$ and $\frac{1}{2}$ (the j-value is given as a subscript to the letter designating the l-value). By contrast the S-state has only one value of j, and that is $\frac{1}{2}$. This can be understood in terms of the *vector model* (Eisberg and Resnick, 1974) for angular momentum, which leads, in the case of a single-valence-electron atom, to the relation

$$j = l + \tfrac{1}{2} \quad \text{or} \quad j = l - \tfrac{1}{2} \tag{3.14}$$

where a quantum number of $\frac{1}{2}$ is associated with the spin state of the electron. The *multiplicity* of a state is in general given by $2S + 1$, where S represents the *total spin quantum number*. For sodium with one valence electron, $S = \frac{1}{2}$, and consequently the multiplicity is 2. This number is seen as a left superscript to the letter that designates the l-value of the state in Fig. 3.2. The quantum selection rules appropriate to transitions within this kind of atom are stated in terms of the permissible changes in the j and l values, Δj and Δl respectively, associated with the transition:

$$\Delta j = 0, \pm 1; \qquad \Delta l = \pm 1 \tag{3.15}$$

As an example of these selection rules, only three of the possible four transitions between the two pairs of states $(3^2D_{5/2}, 3^2D_{3/2})$ and $(3^2P_{3/2}, 3^2P_{1/2})$ are optically allowed. Only the two strongest lines are indicated in Fig. 3.2.

3.1.3. Expectation Values and Eigenvalue Equation

We have seen that the time-independent Schrödinger wave equation for a stationary (but not motionless) system can be expressed in the form

$$\mathscr{H}\psi(\mathbf{r}) = \mathscr{E}\psi(\mathbf{r}) \tag{3.16}$$

where $\mathscr{H}$ is the Hamiltonian operator and $\mathscr{E}$ the energy of the system. In a conservative system $\mathscr{H}$ comprises the sum of the kinetic- and potential-energy operators. Equation (3.16) is in the form of an *eigenvalue equation* in which the wave function $\psi(\mathbf{r})$ is the *eigenfunction* and $\mathscr{E}$ the *eigenvalue* of the operator $\mathscr{H}$. The eigenvalue is in fact the measurable value of the physical observable represented by the operator.

Multiplication of equation (3.16) by $\psi^*(\mathbf{r})$ (the complex conjugate of the wave function) gives

$$\psi^*\mathscr{H}\psi = \psi^*\mathscr{E}\psi = \mathscr{E}\psi^*\psi \tag{3.17}$$

In the case of an electron within an atom, $\psi^*(\mathbf{r})\psi(\mathbf{r})\,dV$ can be thought to represent the *probability* of finding the electron in the volume element dV centered about the point $\mathbf{r}$. Obviously,

$$\int \psi^*(\mathbf{r})\psi(\mathbf{r})\,dV = 1 \tag{3.18}$$

where the integration is extended over all space. It follows that

$$\int \psi^*(\mathbf{r})\mathscr{H}\psi(\mathbf{r})\,dV = \int \mathscr{E}\psi^*(\mathbf{r})\psi(\mathbf{r})\,dV = \mathscr{E} \tag{3.19}$$

and we see that the measurable entity, the energy $\mathscr{E}$, results from the integration indicated on the left-hand side of (3.19). Indeed, in general if a represents some classical variable (position, momentum, energy, etc.) that characterizes the system (particle) described by the wave function $\Psi(\mathbf{r}, t)$, then the *expectation* or average value (at instant t) is

$$\langle a \rangle \equiv \int \Psi^*(\mathbf{r}, t) \mathscr{A} \Psi(\mathbf{r}, t) \, dV \tag{3.20}$$

$\mathscr{A}$ being the *quantum operator* corresponding to a.

At this point we shall introduce an improved notation (having its origin with Dirac) such that an atom that is in a stationary quantum state $\psi(n, l, j)$, described by the set of quantum numbers n, l, and j, is written in the form of a *ket vector*, $|nlj\rangle$. This is often shortened for convenience to $|n\rangle$. In this notation the integral over all space of the product of two energy eigenstates $\psi(n_1 l_1 j_1)$ and $\psi(n_2 l_2 j_2)$ is

$$\int \psi^*(n_2, l_2, j_2) \psi(n_1, l_1, j_1) \, dV = \langle n_2 | n_1 \rangle \tag{3.21}$$

$\langle n_2|$ was named a *bra vector* by Dirac. The average or expectation value of the energy of an eigenstate $|n\rangle$ of the energy operator $\mathscr{H}$ can be written

$$\langle \mathscr{E} \rangle = \int \psi^*(n) \mathscr{H} \psi(n) \, dV = \langle n | \mathscr{H} | n \rangle = \mathscr{E}_n \tag{3.22}$$

using the Dirac form of (3.16), namely

$$\mathscr{H} |n\rangle = \mathscr{E}_n |n\rangle \tag{3.23}$$

If the system is known not to be in an energy eigenstate, then we can express the state function in terms of an infinite series of eigenvectors:

$$|\Psi\rangle = \sum_n c_n |n\rangle e^{i\mathscr{E}_n t/\hbar} \tag{3.24}$$

where $|c_n|^2$ represents the probability of finding the system in the eigenstate $|n\rangle$. It follows that

$$\sum_n |c_n|^2 = 1 \tag{3.25}$$

Under these circumstances the expectation (or average) energy

$$\begin{aligned}
\langle \mathscr{E} \rangle &\equiv \langle \Psi | \mathscr{H} | \Psi \rangle \\
&= \langle \Psi | \mathscr{H} \sum_n c_n |n\rangle e^{i\mathscr{E}_n t/\hbar} \\
&= \langle \Psi | \sum_n c_n e^{i\mathscr{E}_n t/\hbar} \mathscr{H} |n\rangle \\
&= \sum_m c_m^* e^{-i\mathscr{E}_m t/\hbar} \langle m | \Big(\sum_n c_n e^{i\mathscr{E}_n t/\hbar} \mathscr{E}_n |n\rangle \Big)
\end{aligned} \tag{3.26}$$

At this point we need to introduce the *orthonormal* properties of a set of eigenvectors, namely,

$$\langle m|n\rangle = \delta_{nm} \tag{3.27}$$

where $\delta_{nm} = 0$ for $n \neq m$ and 1 for $m = n$. If allowance for this is made in (3.26), then we can write

$$\langle \mathscr{E} \rangle = \sum_n |c_n|^2 \mathscr{E}_n \tag{3.28}$$

which states that the average energy is equal to the weighted sum of the eigenenergies where the weighting factors are just the probabilities of the system being in each of the eigenstates.

For the more general situation where we wish to evaluate the expectation value of some observable $\mathscr{A}$, then we can write

$$\langle a \rangle = \langle \Psi | \mathscr{A} | \Psi \rangle \tag{3.29}$$

and if we expand $|\Psi\rangle$ in terms of an infinite set of eigenfunctions of the operator $\mathscr{A}$,

$$|\Psi\rangle = \sum_p c_p |p\rangle \mathrm{e}^{i\mathscr{E}_p t/\hbar} \tag{3.30}$$

then (3.29) becomes

$$\langle a \rangle = \sum_q \sum_p c_q^* c_p \mathrm{e}^{i(\mathscr{E}_q - \mathscr{E}_p)t/\hbar} \langle q | \mathscr{A} | p \rangle \tag{3.31}$$

where $\mathscr{A}_{qp} \equiv \langle q | \mathscr{A} | p \rangle$ is called the *matrix element* of the operator $\mathscr{A}$.

3.2. QUANTIZED STATES OF MOLECULES

A complete description of any molecular state would involve a number of different degrees of freedom, each with its own coordinate. First there is translation of the molecular as a whole, then vibration and rotation of the nuclei with respect to each other, and then electron and nuclear spin. Finally there is the complex motion of the electrons with respect to the nuclei. The Schrödinger time-independent wave equation appropriate for a molecule can be expressed in the form

$$\left[-\frac{\hbar^2}{2m_e} \sum_i \nabla_i^2 - \sum_k \frac{\hbar^2}{2M_k} \nabla_k^2 + V(\mathbf{r}_i, \mathbf{R}_k) \right] \psi = \mathscr{E} \psi \tag{3.32}$$

where the first term within the brackets on the left-hand side represents the electron kinetic-energy operator, the subscript i allowing for a sum over several electrons; the second term represents the nuclear kinetic-energy operator, the subscript k allowing for a sum over several nuclei; and the third term represents the total potential-energy operator, which depends on the coordinates of all of the electrons and nuclei.

In order to make the analysis of a molecule more tractable the motion of its constituent parts is divided into (1) the relatively slow motion of the nuclei and (2) the much faster motion of the electrons about these nuclei. The physical basis for this decomposition stems from the smallness of the mass of the electron compared to that of any nucleus ($M_k/m_e \gtrsim 1800$). This *Born–Oppenheimer approximation*, as it is called, allows the Schrödinger equation for the molecule to be subdivided into two equations, one for the electrons and the other for the nuclei. To accomplish this the total wave function is expressed as a product of an electronic wave function $\psi_e(\ldots, \mathbf{r}_i, \ldots, \mathbf{R}_k, \ldots)$ and a nuclear wave function $\psi_N(\ldots, \mathbf{R}_k, \ldots)$, where $\mathbf{r}_i$ represents the position vector of the ith electron and $\mathbf{R}_k$ the position vector of the kth nucleus. Thus we may write

$$\psi = \psi_e(\mathbf{r}_i, \mathbf{R}_k)\psi_N(\mathbf{R}_k) \tag{3.33}$$

Now

$$\nabla_i^2 \psi_e \psi_N \approx \psi_N \nabla_i^2 \psi_e \tag{3.34}$$

as ψ_N is essentially independent of $\mathbf{r}_i$. However,

$$\nabla_k^2 \psi_e \psi_N = \psi_e \nabla_k^2 \psi_N + 2\nabla_k \psi_e \nabla_k \psi_N + \psi_N \nabla_k^2 \psi_e \tag{3.35}$$

Born and Oppenheimer showed that the second and third terms of (3.35) are in general negligible compared to the first. This is not unreasonable, as the electronic wave function would not be expected to be a sensitive function of the nuclear coordinates, since the electrons undergo many cycles during one nuclear motion cycle [Herzberg, 1967, Vol. I, p. 148]. Thus the first and second derivatives $\partial\psi_e/\partial x_k, \ldots$ and $\partial^2\psi_e/\partial x_k^2, \ldots$ can be neglected. Under these circumstances we may write

$$\sum_i \nabla_i^2 \psi_e + \frac{2m_e}{\hbar^2}\{\mathscr{E}_e - V_e\}\psi_e = 0 \tag{3.36}$$

and

$$\sum_k \frac{1}{M_k} \nabla_k^2 \psi_N + \frac{2}{\hbar^2}\{\mathscr{E} - \mathscr{E}_e - V_N\}\psi_N = 0 \tag{3.37}$$

where $\mathscr{E}_e$ represents the energy eigenvalue of the electronic system and in fact will depend upon $\mathbf{R}_k$. In splitting the Schrödinger equation for the molecule into one for the electrons and one for the nuclei, we have also assumed that the total potential energy V can be separated into an electronic component V_e (which includes the mutual potential energy of the electrons as well as the potential energy of the electrons with respect to the nuclei) and a nuclear component V_n.

The solutions of (3.36) are quite complex even for the simplest of molecules. In general the solutions lead to molecular orbitals that, just as with atoms, result in a series of electronic states $\Sigma, \Pi, \Delta, \ldots$ (similar to $S, P, D, \ldots$ for atoms) which are characterized by the quantization of the electron's angular momentum and the symmetry properties of the states. A more detailed discussion of this topic goes beyond the aims of this book. Many excellent texts on the subject of molecular structure and spectra can be found; for example, Steinfeld (1974), Mavrodineanu and Boiteux (1965), and Herzberg (1967).

For most molecules the energy difference between the ground electronic state and the first excited electronic state is generally more than 4 eV, which means that the wavelength of their resonance transition is too short to be of interest in laser remote sensing, due to attentuation within the atmosphere (or hydrosphere) and eye-safety considerations. The few exceptions of interest are SO_2, O_3, H_2O, and OH. It is possible that laser remote sensing from space platforms such as the Shuttle will not be limited in this manner.

If all the degrees of freedom are separable from one another, then the total energy of the molecule can be expressed as a simple sum of the individual contributions from each of the degrees of freedom:

$$\mathscr{E}_{\text{total}} = \mathscr{E}_{\text{trans}} + \mathscr{E}_{\text{electronic}} + \mathscr{E}_{\text{vib}} + \mathscr{E}_{\text{rot}} + \mathscr{E}_{\text{spin}} \tag{3.38}$$

Quantum-mechanically, this requires that the terms in the Hamiltonian operator corresponding to each mode commute so that the wave function for the total system can be written as a product of the form

$$\psi_{\text{total}} = \psi_{\text{trans}} \psi_{\text{electronic}} \psi_{\text{vib}} \psi_{\text{rot}} \psi_{\text{spin}} \tag{3.39}$$

With this in mind we can write for a diatomic molecule

$$\psi_N = \frac{1}{R} \psi_v(R) \psi_r(\theta, \phi) \tag{3.40}$$

where R is the internuclear separation and θ and ϕ represent the angular coordinates of a spherical polar coordinate system having its origin at the center of the diatomic molecule, (see Fig. 3.3). $\psi_v(R)$ represents the vibrational part of the wave function, and $\psi_r(\theta, \phi)$ its rotational part. If we substitute

(3.40) into the diatomic form of (3.37), we obtain

$$-\frac{\psi_r}{R^2}\frac{\partial}{\partial R}\left\{R^2\frac{\partial}{\partial R}\left(\frac{\psi_v}{R}\right)\right\} - \frac{\psi_v}{R^3}\left[\frac{1}{\sin\theta}\frac{\partial}{\partial\theta}\left(\sin\theta\,\frac{\partial\psi_r}{\partial\theta}\right) + \frac{1}{\sin^2\theta}\frac{\partial^2\psi_r}{\partial\phi^2}\right]$$
$$+\frac{2\mu}{\hbar^2}\{\mathscr{E}_N - V(R)\}\frac{\psi_v\psi_r}{R} = 0 \tag{3.41}$$

where μ is the reduced mass of the nuclei,

$$\mu = \frac{M_1 M_2}{M_1 + M_2} \tag{3.42}$$

M_1 and M_2 being the masses of the two nuclei. $\mathscr{E}_N$ is the difference in energy between the total internal energy (exclusive of spin) and the electronic excitation energy, that is, $\mathscr{E}_N = \mathscr{E} - \mathscr{E}_e$. Now the rotational component of the Hamiltonian operator

$$\mathscr{H}_{\text{rot}} = -\frac{\hbar^2}{2\mu R^2}\left[\frac{1}{\sin\theta}\frac{\partial}{\partial\theta}\left(\sin\theta\,\frac{\partial}{\partial\theta}\right) + \frac{1}{\sin^2\theta}\frac{\partial^2}{\partial\phi^2}\right] \tag{3.43}$$

and since $\psi_r(\theta, \phi)$ can be regarded as an eigenstate of this operator, it follows that we can write

$$\mathscr{H}_{\text{rot}}\psi_r = \mathscr{E}_J\psi_r \tag{3.44}$$

where according to the quantization of rotational energy (Eisberg and Resnick, 1974; Herzberg, 1967),

$$\mathscr{E}_J = \frac{\hbar^2}{2\mu R^2}J(J+1) \tag{3.45}$$

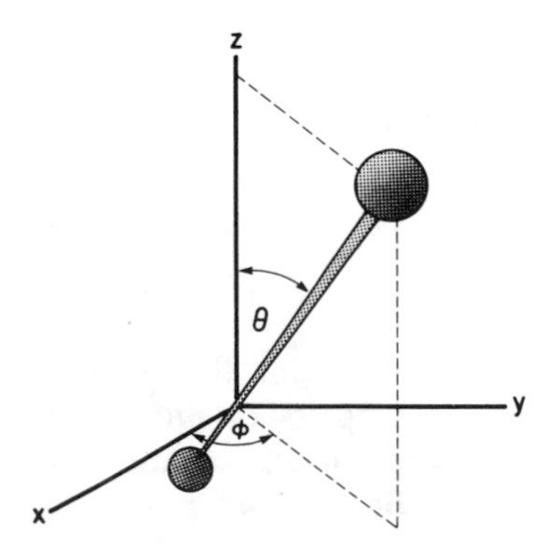

Fig. 3.3. Coordinate system for a diatomic molecule.

Here J is the rotational quantum number and can take on any integer value from 0 to ∞. Under these circumstances we can rewrite (3.41) in the form

$$\frac{\psi_r}{R}\left[-\frac{\partial^2\psi_v}{\partial R^2}+\frac{2\mu}{\hbar^2}\{\mathscr{E}_N-\mathscr{E}_J-V(R)\}\psi_v\right]=0 \tag{3.46}$$

Clearly this can be recast into a form which is equivalent to a one dimensional oscillator:

$$-\frac{\partial^2\psi_v}{\partial R^2}+\frac{2\mu}{\hbar^2}\{\mathscr{E}_v-V(R)\}\psi_v=0 \tag{3.47}$$

where $\mathscr{E}_v$ is the vibration energy and is given by

$$\mathscr{E}_v=\mathscr{E}_N-\mathscr{E}_J \tag{3.48}$$

Thus Schrödinger's equation in the case of a diatomic molecule can be split into two: one for the rotational mode, the other for vibration. We shall now look at each of these in more detail.

3.2.1. Vibrational States of a Diatomic Molecule

In the Born–Oppenheimer approximation, the potential energy $V(\mathbf{R})$ associated with a diatomic molecule possesses a single minimum at some equilibrium internuclear separation R_0 for each stable electronic state. This is illustrated in Fig. 3.4. If we expand $V(R)$ in a Taylor series about the equilibrium internuclear separation R_0, we obtain

$$V(R)=V_0+(R-R_0)\left(\frac{\partial V}{\partial R}\right)_{R=R_0}+\tfrac{1}{2}(R-R_0)^2\left(\frac{\partial^2 V}{\partial R^2}\right)_{R=R_0}+\cdots \tag{3.49}$$

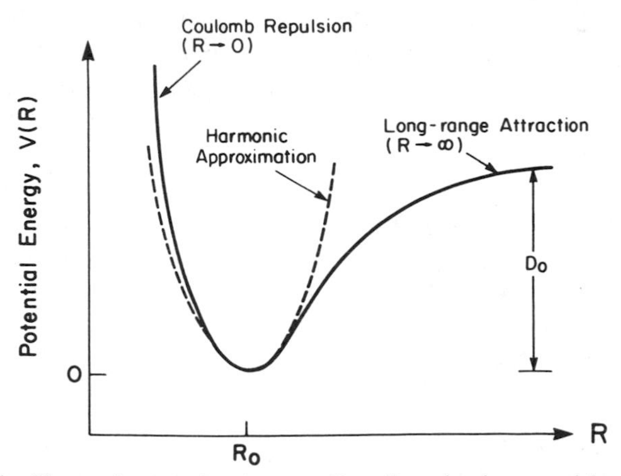

Fig. 3.4. Harmonic approximation to a diatomic molecular potential curve $V(R)$.

Since $\partial V/\partial R$ vanishes at $R = R_0$ (minimum potential), we can approximate the potential by that of a simple harmonic oscillator (Fig. 3.4),

$$V(R) \approx \tfrac{1}{2}(R - R_0)^2 K \tag{3.50}$$

where K $[= (\partial^2 V/\partial R^2)_{R=R_0}]$ is the effective restoring-force constant. The constant V_0 is neglected in (3.50), since only changes in potential are ultimately of interest.

If we substitute (3.50) into (3.47) we arrive at the Schrödinger equation for a simple harmonic oscillator,

$$-\frac{\partial^2}{\partial R^2}\psi_v(R) + \frac{2\mu}{\hbar^2}\{\mathscr{E}_v - \tfrac{1}{2}K(R - R_0)^2\}\psi_v(R) = 0 \tag{3.51}$$

Single-valued, finite, and continuous solutions to this equation are only possible for quantized energies (see Eisberg and Resnick, 1974), given by

$$\mathscr{E}_v = \hbar\omega(v + \tfrac{1}{2}) \tag{3.52}$$

where v is the vibrational quantum number, which is only permitted to take on integer values $0, 1, 2, \ldots,$ and

$$\omega = \left(\frac{K}{\mu}\right)^{1/2}$$

represents the angular frequency of the harmonic oscillator.

A diatomic molecule that is reasonably approximated by a simple harmonic oscillator is predicted (3.52) to possess an equally spaced energy level structure with a minimum vibrational energy of $\frac{1}{2}\hbar\omega$—the *zero-point energy*. The potential-energy curve $V(R)$ (shown broken) and several *vibrational eigenfunctions* $\psi_v(R)$ are schematically shown in Fig. 3.5. It might be noted that the vibrational quantum number v equals the number of nodes in $\psi_v(R)$. The ground vibrational state ($v = 0$) is of particular interest, as the corresponding probability distribution $|\psi_v|^2$ leads to a maximum at $R = R_0$, that is, at the minimum of the potential energy.

A diatomic molecule comprising two unlike atoms possesses a dipole moment and therefore can emit or absorb electromagnetic energy when it undergoes a transition from one vibrational state to another. The appropriate quantum selection rule for vibrational transitions in such a *heteronuclear* molecule takes the form

$$\Delta v = \pm 1 \tag{3.53}$$

That is to say, the change in the vibrational quantum number is restricted to values of ± 1. Clearly the frequency emitted (or absorbed) in a transition of

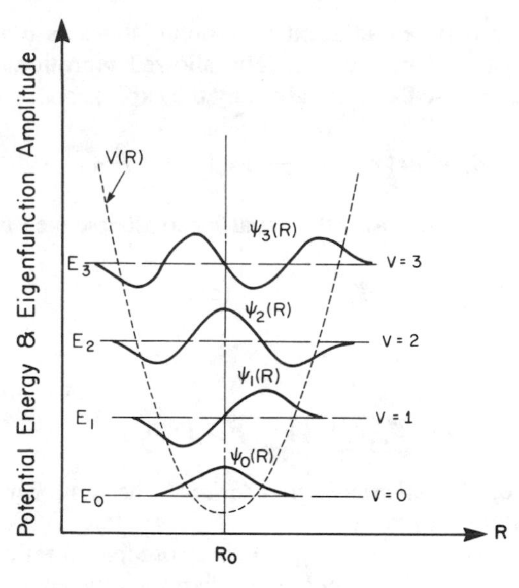

Fig. 3.5. Vibrational eigenfunctions $\psi_v(R)$ and potential-energy curve $V(R)$ (broken) for a diatomic molecule that is well represented by a simple harmonic oscillator.

this kind is given by

$$h\nu = \mathscr{E}_{v'} - \mathscr{E}_{v''} \tag{3.54}$$

where v' and v'' are the vibrational quantum numbers of the upper and the lower state, respectively. It must be pointed out that as the simple harmonic-oscillator approximation breaks down, so does the selection rule (3.53), to a small extent.

In a *homonuclear* molecule (O_2, H_2, N_2, etc.) there is no dipole moment at any value of internuclear separation, and consequently such molecules neither absorb nor emit infrared radiation, although they can be vibrationally excited or deexcited through collisions.

As Fig. 3.4 indicates, the parabolic potential approximation to the potential (3.50) is only valid for small displacements about the equilibrium internuclear separation (i.e. low-lying vibration states). For $R \to 0$ strong repulsion between the nuclei steepens $V(R)$ faster than predicted by (3.50), and for $R \to \infty$ the potential approaches a finite value corresponding to that of two separated atomic states. The finite energy between the minimum and its limiting ($R \to \infty$) value is the *dissociation energy* of the molecule, D_0. Note, that the actual energy required for dissociation is somewhat less than this, due to the zero-point energy of vibration, $\frac{1}{2}\hbar\omega$.

The departure of the molecular potential from that of the ideal harmonic oscillator—the *anharmonicity*, as it is called—leads primarily to a decrease in the energy spacing between adjacent vibrational states as v increases. This is schematically illustrated in Fig. 3.6. The allowed vibrational energies for a diatomic anharmonic oscillator are given by the expression

$$\mathscr{E}_v = h\nu(v + \tfrac{1}{2}) - h\nu x_e(v + \tfrac{1}{2})^2 + \cdots \tag{3.55}$$

or in terms of the *wave number*† (the usual spectroscopic measure of energy)

$$\mathscr{Y}_e = \frac{\nu}{c} \qquad (\mathrm{cm}^{-1}) \tag{3.56}$$

we can write

$$G(v) = \mathscr{Y}_e(v + \tfrac{1}{2}) - \mathscr{Y}_e x_e(v + \tfrac{1}{2})^2 + \cdots \tag{3.57}$$

where x_e is termed the *anharmonicity constant*. The term *kayser* is sometimes used in the literature for cm^{-1}.

The sign choice of this conventional spectroscopic notation is dictated by the fact that $\mathscr{Y}_e x_e > 0$. It is evident that anharmonicity leads to a considerable increase in the complexity of the molecular vibrational spectra, as many infrared frequencies are now possible where only one would be expected for an ideal harmonic oscillator. In Table 3.1 we present the vibrational parameters for a number of diatomic molecules. Clearly, even though $x_e \ll 1$, the infrared absorption spectrum often is found to contain overtones due to deviations from the simple harmonic representation. In reality this spectrum is further complicated by the presence of additional spectral lines that arise from the rotational energy structure—our next topic.

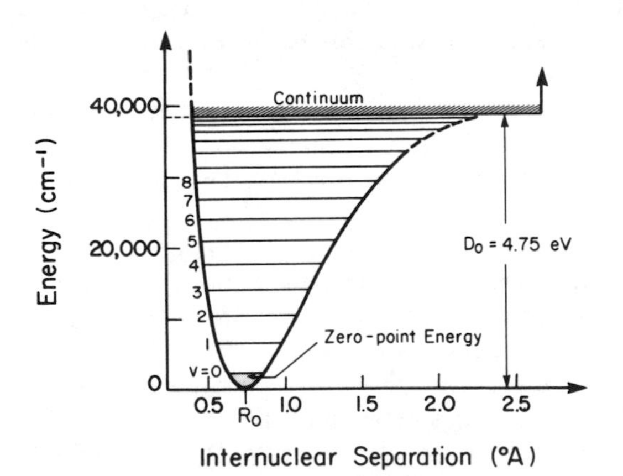

Fig. 3.6. Potential energy and vibrational levels for the electronic ground state of hydrogen (Herzberg, 1950).

† To avoid some of the confusion in the literature I have introduced $\mathscr{Y}$ to represent $1/\lambda$.

TABLE 3.1. VIBRATIONAL AND ROTATIONAL PARAMETERS FOR SELECTED DIATOMIC MOLECULES IN THEIR GROUND STATE[a]

Molecule	$\mathscr{Y}_e$ [b] (cm^{-1})	$\mathscr{Y}_e x_e$ (cm^{-1})	B_e (cm^{-1})	α_e (cm^{-1})	D_0 (eV)
CH	2861.6	64.3	14.46	0.534	3.47
Cl_2	564.9	4.0	0.244	0.0017	2.475
CN	2068.7	13.14	1.90	0.0174	(?)[c]
CO	2170.2	13.46	1.93	0.0175	11.11[c]
H_2	4395.2	113.9	60.80	2.99	4.48
HCl	2989.7	52.0	10.59	0.30	4.43
N_2	2359.6	14.46	2.01	0.019	7.37[c]
O_2	1580.4	12.07	1.45	0.016	5.08
OH	3735.2	82.81	18.87	0.714	4.35
NO	1904	13.97	1.71	0.0178	5.29

[a] From Herzberg (1967, Table 39).
[b] In much of the literature ω_e is used for the wave number value of the molecular vibration. $\mathscr{Y}_e$ is used here to avoid confusion with angular frequency.
[c] These values have some degrees of uncertainty, see Herzberg, *loc. cit.*

In the case of a *polyatomic* molecule, there are many different modes of vibration, and clearly this can lead to a complex band spectrum. Each nucleus of a free atom has three degrees of freedom, which is reflected in the total translational energy ($3kT/2$) of such an atom, since $kT/2$ represents the equilibrium mean thermal energy associated with each degree of freedom, according to statistical mechanics.

Thus an N-atom molecule also has $3N$ degrees of freedom. Three of these degrees of freedom are required for the total translational motion of the molecule, and three more (two in the case of a linear molecule) are involved in the free rotation of the molecule. This leaves $3N - 6$ (or $3N - 5$, for a linear molecule) possible modes of vibration. Consequently, simple molecules like H_2O and CO_2 have three modes of vibration, designated ν_1, ν_2, and ν_3. One of these can be thought of as a bending vibration, while the other two relate to symmetric and asymmetric vibration along the internuclear axis. More details on this subject are provided in texts on molecular spectra such as Herzberg (1967) and Steinfeld (1974).

3.2.2. Rotational States of a Diatomic Molecule

If we assume that the molecule is *rigid*, so that the internuclear separation is fixed at its equilibrium value R_0, then the rotational motion can be separated from the other degrees of freedom. Under these circumstances (3.44) and (3.45) yield

$$\mathscr{H}_{\text{rot}}\psi_r(\theta, \phi) = \frac{\hbar^2}{2\mu R_0^2} J(J + 1)\psi_r(\theta, \phi) \tag{3.58}$$

where $\mathscr{H}_{rot}$ represents the rotational component of the Hamiltonian (3.43) and J is the rotational quantum number. $\psi_r(\theta, \phi)$ is the rotational eigenstate for the molecule. The energy of the subsequent rotational states permitted for a simple rigid rotator,

$$\mathscr{E}_J = \frac{\hbar^2}{2\mu R_0} J(J+1) \tag{3.59}$$

clearly increases quadratically with J. The appropriate quantum selection rule for the rigid rotator molecule (Herzberg, 1967) takes the form

$$\Delta J = J' - J'' = \pm 1 \tag{3.60}$$

where J' and J'' represent the rotational quantum numbers for the upper and lower rotational states, respectively. The frequency of the photon emitted (or absorbed) in the transition between the eigenstates $\psi_{J'}$ and $\psi_{J''}$ is proportional to the difference in the rotational energies:

$$h\nu = \mathscr{E}_{J'} - \mathscr{E}_{J''} \tag{3.61}$$

In pure rotational transitions (i.e. those within one vibrational state of a given electronic state), we have $\Delta J = 1$ and so we can write

$$h\nu = 2B_e hc(J+1) \tag{3.62}$$

where for simplicity of notation J is taken to mean J'', the rotational quantum number of the lower level in (3.62), and

$$B_e \equiv \frac{h}{8\pi^2 \mu R_0^2 c} \tag{3.63}$$

is termed the *rotational constant*. The spectrum of a simple rigid rotator can be seen, according to (3.62), to comprise a series of equally spaced lines. This is illustrated by the energy-level structure in Fig. 3.7.

It should be quite apparent that our rigid-rotator model cannot be strictly valid, since we have already seen that the molecule undergoes vibration in the line joining the two nuclei. One approach to improving our model would be to assume that the two nuclei are connected by a massless spring. The quantized rotational energies appropriate to this *nonrigid rotator* are given by the expression

$$\mathscr{E}_J = B_e hcJ(J+1) - D_e hcJ^2(J+1)^2 \tag{3.64}$$

The *centrifugal distortion constant* D_e is always positive, and in the case of a harmonic oscillator

$$D_e = \frac{4B_e^3}{\mathscr{Y}_e^2} \tag{3.65}$$

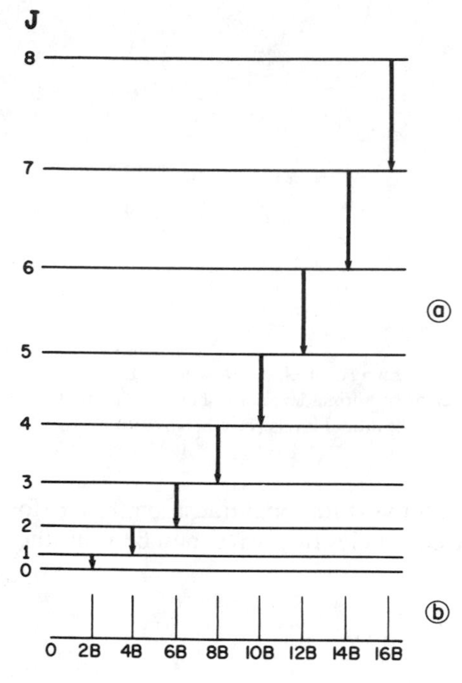

Fig. 3.7. (a) Rotational energy-level structure of a rigid diatomic molecule. (b) Schematic of the corresponding rotational spectrum.

Evaluation of D_e from Table 3.1 reveals that in most cases of interest (i.e. for $J \lesssim \mathscr{Y}_e/2B_e \approx 500$) the second term in (3.64) is negligible relative to the first.

A more important effect on the spectrum arises from the dependence of the rotational constant upon the vibrational state of the molecule. This dependence leads us to the *vibrating rotator* model and to the introduction of the *effective rotational constant* B_v. This is related to the equilibrium value B_e (Herzberg, 1967) by the expression

$$B_v = B_e - \alpha_e(v + \tfrac{1}{2}) + \cdots \tag{3.66}$$

where α_e is a positive constant that is small compared to B_e (see Table 3.1).

The total *term value* (energy expressed as the wave number in kaysers) for the vibrating-rotator model is

$$\begin{aligned} T(v, J) &\equiv \frac{\mathscr{E}_v + \mathscr{E}_J}{hc} \\ &\approx \mathscr{Y}_e(v + \tfrac{1}{2}) - \mathscr{Y}_e x_e(v + \tfrac{1}{2})^2 + B_v J(J + 1) \end{aligned} \tag{3.67}$$

The corresponding energy-level diagram is presented as Fig. 3.8. For the sake of clarity, the individual rotational levels are represented by shorter horizontal lines than the pure vibrational levels (where $J = 0$).

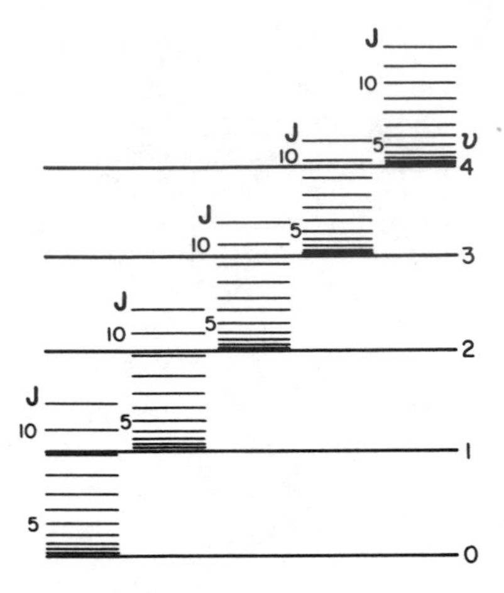

Fig. 3.8. Energy levels of a vibrating rotator. A number of rotational levels are shown for each of the first five vibrational levels (Herzberg, 1950).

The rotational spectrum associated with the vibrational transition from v' to v'' can be evaluated from (3.67). Accordingly the wave numbers of the resulting lines are given by

$$\mathscr{Y}(v', J' : v'', j'') = T(v', J') - T(v'', J'') \qquad (\text{cm}^{-1})$$

that is,

$$\mathscr{Y} = \mathscr{Y}_0 + B_{v'}J'(J' + 1) - B_{v''}J''(J'' + 1) \tag{3.68}$$

where from (3.57)

$$\mathscr{Y}_0 \equiv G(v') - G(v'') \tag{3.69}$$

are the wavenumbers of the pure vibrational transitions (i.e. $J' = J'' = 0$).

With $\Delta J = +1$ and $\Delta J = -1$, respectively, we obtain from (3.68)

$$\mathscr{Y}_R = \mathscr{Y}_0 + 2B_{v'} + (3B_{v'} - B_{v''})J + (B_{v'} - B_{v''})J^2, \qquad J = 0, 1, 2, \ldots \tag{3.70}$$

$$\mathscr{Y}_P = \mathscr{Y}_0 - (B_{v'} + B_{v''})J + (B_{v'} - B_{v''})J^2, \qquad J = 1, 2, 3, \ldots \tag{3.71}$$

Once again we have chosen to replace J'' (the lower level rotational quantum number) with J.

Equations (3.70) and (3.71) describe the R and P *branches* of the rotational spectrum. If we neglect, for the moment, the interaction between rotation and vibration, then $B_{v'} = B_{v''} = B_e$ and (3.70) and (3.71) simplify to

$$\begin{aligned} \mathscr{Y}_R &= \mathscr{Y}_0 + 2B_e + 2B_eJ \\ \mathscr{Y}_P &= \mathscr{Y}_0 \qquad\quad\; - 2B_eJ \end{aligned} \tag{3.72}$$

which produce two series of equidistant lines: the *R-branch* going from $\mathscr{Y}_0$ towards shorter wavelengths, and the *P-branch* going towards longer wavelengths. The rotational energy-level diagram and associated spectrum of lines are illustrated in Fig. 3.9. If this is compared with an observed rotational absorption spectrum, such as presented for HCl in Fig. 3.10, the distortion in the equispacing of the lines, arising from the rotation–vibration interaction, is quite apparent and can be accounted for by use of (3.70) and (3.71). The line corresponding to the band origin $\mathscr{Y}_0$ is missing as $\Delta J = 0$ is forbidden for molecules having zero electron angular momentum about the internuclear axis, that is, $\Lambda = 0$. The quantum number Λ, which reflects the electron's angular momentum about the internuclear axis, can in general take on any integer value (0, 1, 2, ...).

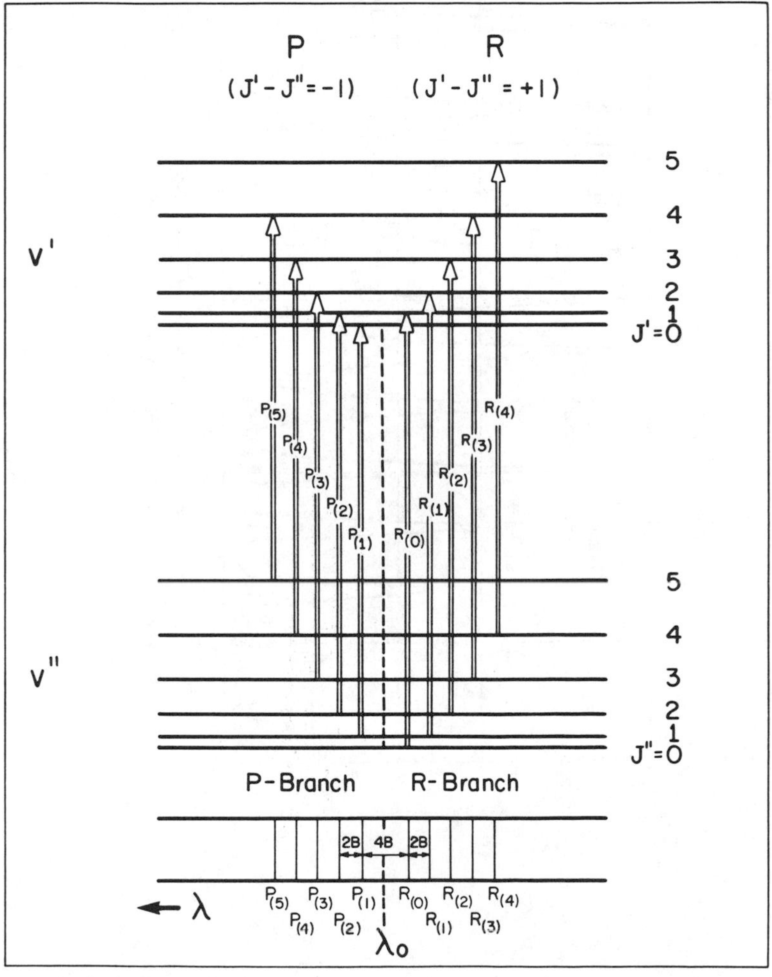

Fig. 3.9. Energy-level structure and rotational spectrum for a rigid rotator.

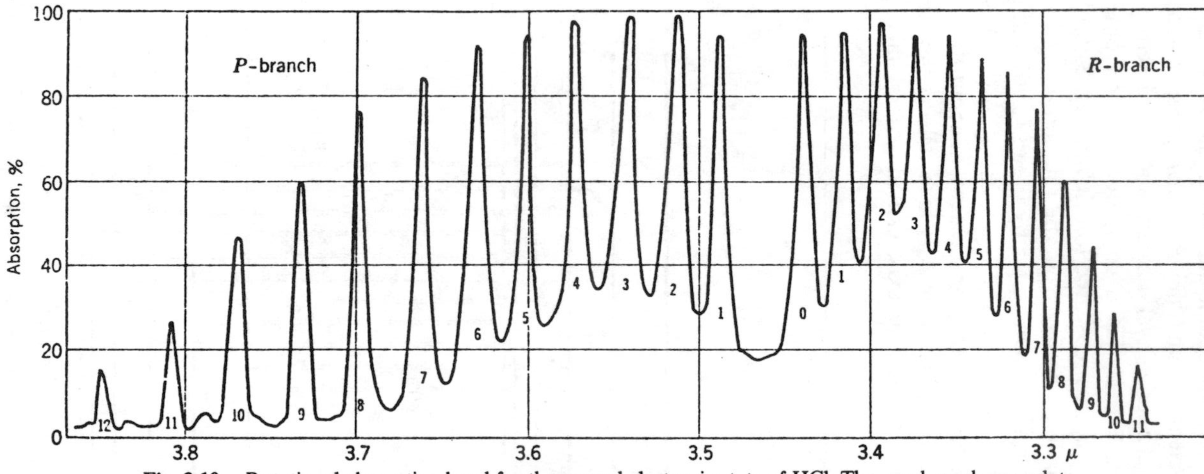

Fig. 3.10. Rotational absorption band for the ground electronic state of HCl. The numbers shown relate to the J-value of the lower level in each transition (Imes, 1919).

For values of $\Lambda \neq 0$, the selection rule pertaining to rotational transitions (3.60) is amended to the form

$$\Delta J = 0, \pm 1 \tag{3.73}$$

Transitions with $\Delta J = 0$ (note: $J = 0$ to $J = 0$ is still forbidden) give rise to an additional series of rotational lines, the *Q-branch*, for which

$$\mathscr{Y}_Q = \mathscr{Y}_0 + (B_{v''} - B_{v'})\Lambda^2 + (B_{v'} - B_{v''})J + (B_{v'} - B_{v''})J^2 \tag{3.74}$$

The term $(B_{v''} - B_{v'})\Lambda^2$ must also be added to the formulae for the R and P branches (3.70) and (3.71), respectively. All of the lines of the Q-branch practically sit on each other, since $B_{v'} \approx B_{v''}$ in the infrared spectra (i.e. for no change of the electronic state). This leads to a relatively intense line at about $\mathscr{Y}_0$ for the series of rotational transitions (corresponding to $\Delta J = 0$) that arise between two vibrational states within an electronic state with $\Lambda \neq 0$.

In the case of rotational transitions between two electronic states, the difference in the rotational constants $|B_{v'} - B_{v''}|$ can be sufficient to cause a *reversal* in either the R $(B_{v'} < B_{v''})$ or P $(B_{v'} > B_{v''})$ branch for high values of J. This reversal leads to the formation of *band heads* in the rotational fine structure of molecular bands. An example of a *red-edge band head* in CO is presented as Fig. 3.11.

$B_{v'}$ and $B_{v''}$ can differ appreciably in the case of transitions between two electronic states, because Δv is *not* restricted in such instances to values of ± 1.

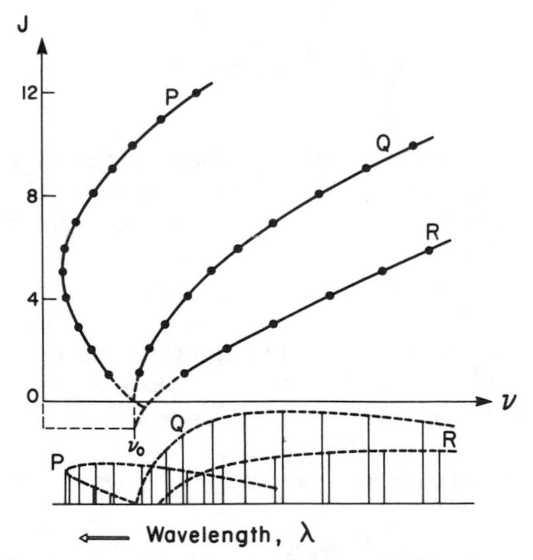

Fig. 3.11. Fortrat diagram of a vibronic band of CO. P, Q, and R values correspond to a change of $+1$, 0, and -1 in J. The sequence of transitions from $v = 0$ to $v = 0$ constitute the 0–0 band, while the sequence of transitions from $v = 1$ to $v = 0$ constitute the 1–0 band (Green and Wyatt, 1965).

As we shall see later, the change in the vibrational state is dictated by the Franck–Condon principle, and this quite often leads to a significant jump in v.

For a more detailed discussion of the terminology and spectral features of molecules, the reader should consult Mavrodineanu and Boiteux (1965), Steinfeld (1974), or the monumental series on molecular structure and spectroscopy by G. Herzberg (1967).

3.2.3. Thermal Distribution of Quantum States

As we shall see shortly, the strength of emission of a spectral line depends on the transition probability, the frequency, and the number density of molecules (or atoms) in the excited state. We shall now consider the distribution of molecules amongst their vibrational and rotational states under conditions of *thermal equilibrium*.

Statistical mechanics predicts that in equilibrium the velocity distribution of a classical ensemble of molecules is given by the *Maxwell–Boltzmann* formulae

$$f(u) = 4\pi u^2\left(\frac{m}{2\pi kT}\right)^{3/2} \mathrm{e}^{-mu^2/2kT} \tag{3.75}$$

where $f(u)\,du$ represents the probability of a molecule of mass m having a speed in the interval $(u, u + du)$, and T is the absolute temperature of the molecules. The corresponding equilibrium distribution of internal energy states is given by the *Boltzmann law*. In the case of a quantized system this takes the form

$$N_n = N_0 \frac{g_n}{g_0} \mathrm{e}^{-(\mathscr{E}_n - \mathscr{E}_0)/kT} \tag{3.76}$$

where N_n and N_0 represent the respective number density of molecules (or atoms) in the quantum states $|n\rangle$ and $|0\rangle$, the latter being taken as the ground state. $\mathscr{E}_n$ and $\mathscr{E}_0$ are the corresponding energies, and g_n and g_0 the respective degeneracies (or statistical weights) of the two states.

It might also be mentioned that in *true* equilibrium the radiation field would be described by Planck's distribution (3.2). In the lower region of the atmosphere, the density is sufficient to ensure (through the action of collisions) that both the velocity and internal distributions of molecular states are described by the equilibrium relations (3.75) and (3.76). This is not however the case for the radiation field.

The *vibrational distribution* for an ensemble of molecules can be written

$$N_v = N_0 \mathrm{e}^{-\mathscr{E}_{v0}/kT} \tag{3.77}$$

where N_v is now the density of molecules in the vibrational state $|v\rangle$ and $\mathscr{E}_{v0}$ is the difference in energy between $|v\rangle$ and $|0\rangle$. Also, since vibrational states are nondegenerate, the degeneracy factors are all unity. Summation of (3.77) leads to the more convenient expression

$$N_v = \frac{N}{Z_v(T)} e^{-hcG_{v0}/kT} \tag{3.78}$$

where N is the total number density of molecules of the given species, $G_{v0} \equiv G(v) - G(0)$ with $G(v)$ given by (3.57), and the *vibrational partition function* is

$$Z_v(T) \equiv \sum_{v \geq 0} e^{-hcG_{v0}/kT} \tag{3.79}$$

To first order this can be approximated by

$$Z_v(T) = [1 - e^{-hcG_{10}/kT}]^{-1} \tag{3.80}$$

due to the near-equal energy spacing of the vibrational levels.

The ratio of the number of molecules in the first few vibrational states at 300 K (room temperature) to the total number density of molecules for several diatomic molecules of interest in laser remote sensing is presented in Table 3.2. Clearly in most instances the degree of vibrational excitation at 300 K is negligible. If we take, for example, CO, we see that $N_1/N \approx 3.4 \times 10^{-5}$ at 300 K, so that for a concentration of 1 ppm of CO the number density of molecules in the first vibrational state at STP is

$$\begin{aligned} N_1 &\approx 3.4 \times 10^{-5} \times 2.69 \times 10^{19} \times 10^{-6}\ \text{cm}^{-3} \\ &\approx 9.15 \times 10^{8}\ \text{cm}^{-3} \end{aligned} \tag{3.81}$$

TABLE 3.2. VIBRATIONAL EQUILIBRIUM AT 300 K[a]

Molecule	G_{10} (cm^{-1})	N_1/N	N_2/N	N_3/N	N_4/N	N_5/N
CH	2733.0	2.00×10^{-6}	7.45×10^{-12}	5.13×10^{-17}	6.55×10^{-22}	1.55×10^{-26}
Cl_2	556.9	6.42×10^{-2}	4.61×10^{-3}	3.43×10^{-4}	2.66×10^{-5}	2.14×10^{-6}
CN	2042.4	5.52×10^{-5}	3.45×10^{-9}	2.45×10^{-13}	1.97×10^{-17}	1.80×10^{-21}
CO	2143.3	3.40×10^{-5}	1.32×10^{-9}	5.79×10^{-14}	2.90×10^{-18}	1.65×10^{-22}
H_2	4167.4	2.05×10^{-9}	1.25×10^{-17}	2.28×10^{-25}	1.24×10^{-32}	0
HCl	2885.7	9.63×10^{-7}	1.53×10^{-12}	3.99×10^{-18}	1.72×10^{-23}	1.22×10^{-28}
N_2	2330.7	1.38×10^{-5}	2.20×10^{-10}	4.01×10^{-15}	8.41×10^{-20}	2.03×10^{-24}
O_2	1556.3	5.69×10^{-4}	3.64×10^{-7}	2.61×10^{-10}	2.10×10^{-13}	1.90×10^{-16}
OH	3569.6	3.61×10^{-8}	2.89×10^{-15}	5.12×10^{-22}	2.01×10^{-28}	1.74×10^{-34}
NO	1876.1	1.23×10^{-4}	1.72×10^{-8}	2.76×10^{-12}	5.05×10^{-16}	1.06×10^{-19}

[a]Clearly in some instances, the ratio of population densities is so small that for all practical purposes no molecules reside in these vibrational states. In essence this can be taken to be the case when $N_m/N < 10^{-19}$

In the case of *rotational* states, quantum theory requires that each state having an angular-momentum quantum number J have a degeneracy of $2J + 1$ (which corresponds to the number of possible orientations of the angular-momentum vector) (Eisberg and Resnick, 1974, p. 463). The number density of molecules in rotational state $|J\rangle$ of vibrational state $|v\rangle$ at temperature T is given, according to (3.76), by

$$N_{J,v} = N_{0,v}(2J + 1)\mathrm{e}^{-(\mathscr{E}_J - \mathscr{E}_0)/kT} \tag{3.82}$$

or, in terms of the *total* number density N_v of molecules in the same vibrational state $|v\rangle$ (we also assume that all molecules are in the ground electronic state),

$$N_{J,v} = N_v \frac{(2J + 1)}{Z_r(T)} \mathrm{e}^{-hcB_eJ(J+1)/kT} \tag{3.83}$$

where the *rotational partition function* is

$$Z_r(T) = \sum_{J \geq 0} (2J + 1)\mathrm{e}^{-hcB_eJ(J+1)/kT} \tag{3.84}$$

For sufficiently large T (or small B_e) this sum can be approximated by an integral over J from $J = 0$ to ∞. This yields

$$Z_r(T) \approx \frac{kT}{hcB_e} \tag{3.85}$$

so that we can write

$$N_{J,v} = N_v \frac{hcB_e}{kT}(2J + 1)\mathrm{e}^{-hcB_eJ(J+1)/kT} \tag{3.86}$$

To obtain the relation between the number density of molecules in a particular rotational quantum state $|J\rangle$ and the total number density of molecules, we eliminate N_v in (3.86) through equation (3.78). In the case of HCl at 300 K, kT/hcB_e is 19.98, while the exact value of the rotational partition function is 20.39. In order to see the form of the rotational distribution curve we plot, in Fig. 3.12, the *normalized rotational population*

$$\frac{N_J}{N_r b} \equiv (2J + 1)\mathrm{e}^{-bJ(J+1)} \tag{3.87}$$

against J at 300 K for the ground state of CO, where $b \equiv hcB_e/kT$ and N_r represents the total rotational population. In effect, $N_J/N_r b$ represents the fraction of molecules in the state $|J\rangle$ multiplied by the rotational partition function $Z_r(T)$ (i.e. $1/b$).

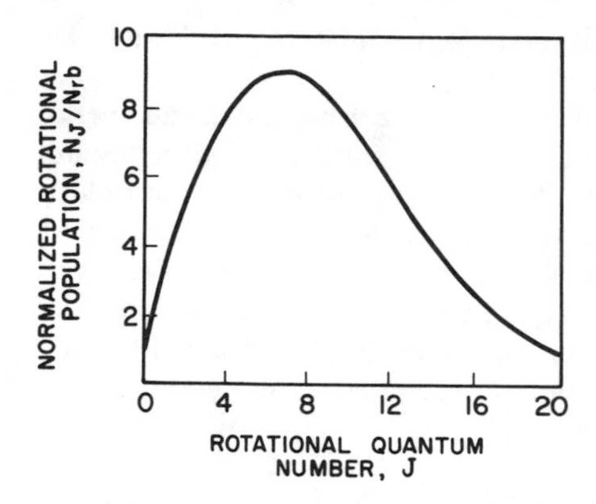

Fig. 3.12. Thermal distribution of rotational levels for the ground state of CO at 300 K. $N_J/N_r b = (2J + 1)e^{-bJ(J+1)}$.

3.3. RADIATIVE PROCESSES AND TRANSITION PROBABILITIES

Classical electromagnetic theory predicts [equation (2.78)] that the power radiated by an electron experiencing an acceleration $\dot{\mathbf{u}}$ is given by the relation

$$P = \frac{1}{6\pi\epsilon_0} \frac{e^2}{c^3} |\dot{\mathbf{u}}|^2$$

In the case of an electron within a *classical* atom or molecule, the electron equation of motion can be approximated by that of a simple harmonic oscillator. Under these circumstances the vector displacement $\boldsymbol{\xi}$ of the electron from its equilibrium position was shown earlier [equation (2.80)] to undergo sinusoidal oscillation at a resonant angular frequency ω_0:

$$\boldsymbol{\xi} = \boldsymbol{\xi}_0 e^{-i\omega_0 t}$$

We thus arrive at the power radiated,

$$P = \frac{1}{6\pi\epsilon_0} \frac{\omega_0^4}{c^3} |\mathbf{d}|^2 \tag{3.88}$$

in terms of the *dipole moment* of the molecule (or atom),

$$\mathbf{d} \equiv -e\boldsymbol{\xi} \tag{3.89}$$

3.3.1. Radiative Processes for Quantized States

In order to determine the actual power radiated from an atom or molecule we must evaluate $\mathbf{d}$ quantum-mechanically. As we have seen earlier [equation (3.29)], the expectation (or average) value of the dipole moment of an atom or molecule is

$$\langle \mathbf{d} \rangle \equiv \langle \Psi | \mathbf{d} | \Psi \rangle \tag{3.90}$$

where the quantum operator for the observable **d** is in fact just **d**, and $|\Psi\rangle$ represents the quantum state of the atom or molecule.

It can be shown (Liboff, 1980, p. 410) that if $|\Psi\rangle$ corresponds to an energy eigenstate of an atom or molecule, then $\langle \mathbf{d} \rangle = 0$ and so an atom (or molecule) in a stationary state does not radiate. At this point it is expedient to introduce the term *system* to mean either an atom or a molecule. We shall assume that such a system is in an excited eigenstate $|\Psi_n\rangle$ at time $t = 0$, where n is taken to represent the set of quantum numbers required to specify the eigenstate. The atom will subsequently decay to another eigenstate $|\Psi_m\rangle$ through either the emission of radiation or a collisional event. In the interim, the system can be regarded as being in a superposition state

$$|\Psi\rangle = |c_n\Psi_n + c_m\Psi_m\rangle \tag{3.91}$$

where $|c_n|^2$ represents the probability that the system is in the eigenstate $|\Psi_n\rangle$ and $|c_m|^2$ the probability that it is in $|\Psi_m\rangle$. Consequently it follows that even though these coefficients are time-dependent,

$$|c_n|^2 + |c_m|^2 = 1 \tag{3.92}$$

Furthermore, at $t = 0$, we have $|c_n|^2 = 1$ and $|c_m|^2 = 0$, while as $t \to \infty$, we have $|c_n|^2 \to 0$ and $|c_m|^2 \to 1$.

The expectation value of the dipole moment for the system described by (3.91) is

$$\begin{aligned}\langle \mathbf{d} \rangle &= \langle c_n\Psi_n + c_m\Psi_m|\mathbf{d}|c_n\Psi_n + c_m\Psi_m\rangle \\ &= |c_n|^2\langle \Psi_n|\mathbf{d}|\Psi_n\rangle + |c_m|^2\langle \Psi_m|\mathbf{d}|\Psi_m\rangle \\ &\quad + c_n^*c_m\langle \Psi_n|\mathbf{d}|\Psi_m\rangle + c_m^*c_n\langle \Psi_m|\mathbf{d}|\Psi_n\rangle\end{aligned} \tag{3.93}$$

As discussed earlier, eigenstates (or stationary states) of a system can be expressed in the form

$$|\Psi_n\rangle = |n\rangle \mathrm{e}^{-i\mathscr{E}_n t/\hbar} \tag{3.94}$$

When this is used in (3.93), we obtain

$$\begin{aligned}\langle \mathbf{d} \rangle &= |c_n|^2\langle n|\mathbf{d}|n\rangle + |c_m|^2\langle m|\mathbf{d}|m\rangle \\ &\quad + c_n^*c_m\langle n|\mathbf{d}|m\rangle \mathrm{e}^{i(\mathscr{E}_n - \mathscr{E}_m)t/\hbar} + c_m^*c_n\langle m|\mathbf{d}|n\rangle \mathrm{e}^{i(\mathscr{E}_m - \mathscr{E}_n)t/\hbar}\end{aligned} \tag{3.95}$$

As mentioned above, the first two terms are zero and the last two terms combine to yield

$$\begin{aligned}\langle \mathbf{d} \rangle &= 2\,\mathrm{Re}\{c_n^*c_m\langle n|\mathbf{d}|m\rangle \mathrm{e}^{i(\mathscr{E}_n - \mathscr{E}_m)t/\hbar}\} \\ &= 2|c_n^*c_m\langle n|\mathbf{d}|m\rangle|\cos(\omega_{nm}t)\end{aligned} \tag{3.96}$$

where

$$\omega_{nm} \equiv \frac{\mathscr{E}_n - \mathscr{E}_m}{\hbar} \tag{3.97}$$

is the *Bohr angular frequency*. If (3.96) is used for **d**, and ω_{nm} replaces ω_0 in (3.88), the power radiated from the system undergoing a transition from $|n\rangle$ to $|m\rangle$ is

$$P = \frac{2\omega_{nm}^4}{3\pi\epsilon_0 c^3}|\mathbf{d}_{nm}|^2\cos^2(\omega_{nm}t) \tag{3.98}$$

where

$$\mathbf{d}_{nm} \equiv \langle n|\mathbf{d}|m\rangle \tag{3.99}$$

is the *matrix element* of the dipole moment of the system and we have assumed that $|c_n^* c_m|^2$ is slowly varying and of order unity (Liboff, 1980, p. 411). Transitions between eigenstates of atoms or molecules typically occur over an interval that is of the order of 10^{-8} s, whereas the frequency of radiation emitted in such transitions is of the order of 10^{15} s^{-1}. Furthermore, most detectors of use in laser remote sensing have a response time that is not much better than 10^{-9} s. So we shall restrict our attention to the *time-averaged* (over one cycle) radiated power,

$$\langle P\rangle = \frac{\omega_{nm}^4}{3\pi\epsilon_0 c^3}|\mathbf{d}_{nm}|^2 \tag{3.100}$$

However, the time-averaged power radiated from a quantum system, P_{nm}, undergoing a spontaneous transition between two eigenstates $|n\rangle$ and $|m\rangle$, can be expressed in terms of the *Einstein transition probability* A_{nm} (s^{-1}), which gives the rate that a system in $|n\rangle$ will undergo spontaneous transitions to $|m\rangle$. Thus we may write

$$P_{nm} = \hbar\omega_{nm}A_{nm} \tag{3.101}$$

If we equate (3.100) with (3.101), we obtain

$$A_{nm} = \frac{\omega_{nm}^3}{3\pi\epsilon_0 \hbar c^3}|\mathbf{d}_{nm}|^2 \tag{3.102}$$

If at time $t = 0$, there are $N_n(0)$ systems per unit volume in the quantum state $|n\rangle$, then, in the absence of all processes other than the decay of $|n\rangle$ to $|m\rangle$ through spontaneous emission,

$$\frac{dN_n(t)}{dt} = -N_n(t)A_{nm} \tag{3.103}$$

where $N_n(t)$ represents the number density of quantum states $|n\rangle$ at time t. This has the solution

$$N_n(t) = N_n(0)e^{-A_{nm}t} \tag{3.104}$$

which enables us to introduce the *radiative lifetime* for the decay of $|n\rangle$ to $|m\rangle$:

$$\tau_{nm} \equiv \frac{1}{A_{nm}} \tag{3.105}$$

If there are several states into which $|n\rangle$ can radiatively decay, then the actual lifetime of this state is

$$\tau_n = \left\{ \sum_{m<n} A_{nm} \right\}^{-1} \tag{3.106}$$

If this quantum system is immersed in a radiation field of frequency ν that is close to ν_{nm} (how close will become apparent shortly), then two other radiative processes are possible. For systems in $|n\rangle$ the radiation field can *stimulate* (or induce) the transition from $|n\rangle$ to $|m\rangle$ with a probability given by $\mathsf{B}_{nm}\rho(\nu)$, where B_{nm} represents the *Einstein stimulated-emission coefficient* and $\rho(\nu)$ the spectral energy density of the radiation field. An important difference between spontaneous and stimulated emission is that in the latter case the emitted radiation reinforces the radiation field. That is to say, the stimulated emission has the *same frequency, same direction, and same phase* as the inducing radiation field. To put it another way, the stimulated photon is identical to the incident photon.

The total power radiated per unit volume into the frequency interval $(\nu, \nu + d\nu)$ can thus be expressed in the form

$$P^E_{nm}(\nu)\, d\nu = h\nu\left[A_{nm} + \mathsf{B}_{nm}\rho(\nu)\right] N_n \mathscr{L}(\nu)\, d\nu \tag{3.107}$$

where $N_n \mathscr{L}(\nu)\, d\nu$ represents the fraction of quantum systems in $|n\rangle$ at time t that can emit radiation into the interval $(\nu, \nu + d\nu)$. Here $\mathscr{L}(\nu)$ is termed the *line profile function* and will be discussed in detail shortly. It can be thought of as a probability of emission into $(\nu, \nu + d\nu)$, and consequently its integral over all frequencies must be unity.

For quantum systems in the lower $|m\rangle$ state, the radiation field can stimulate the transition from $|m\rangle$ to $|n\rangle$ through the absorption of a photon of energy $h\nu$. The probability of this event is $\mathsf{B}_{mn}\rho(\nu)$, where B_{mn} is the *Einstein absorption coefficient*. The power absorbed from the radiation field by this process per unit volume and in the frequency interval $(\nu, \nu + d\nu)$ is given by

$$P^A_{mn}(\nu)\, d\nu = h\nu \mathsf{B}_{mn}\rho(\nu) N_m \mathscr{L}(\nu)\, d\nu \tag{3.108}$$

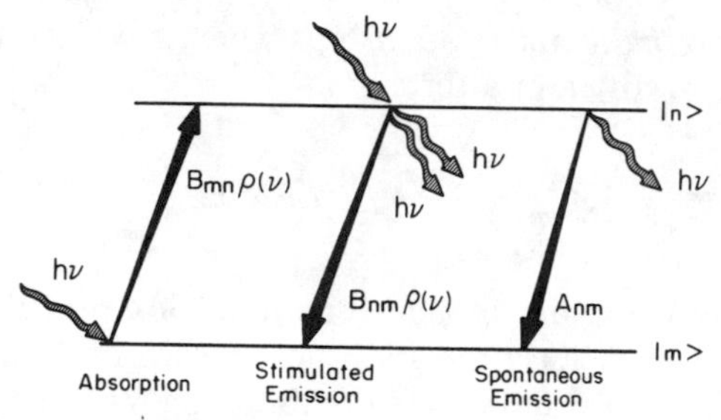

Fig. 3.13. Schematic illustration of three radiative processes.

These three radiative processes are illustrated in Fig. 3.13. If thermodynamic equilibrium exists, then detailed balance dictates that

$$P_{mn}^{A}(\nu) = P_{nm}^{E}(\nu) \tag{3.109}$$

from which we obtain

$$\rho(\nu) = \frac{A_{nm}}{\mathsf{B}_{nm}} \frac{1}{\dfrac{\mathsf{B}_{mn} N_m}{\mathsf{B}_{nm} N_n} - 1} \tag{3.110}$$

However, as we have seen, under equilibrium conditions the Boltzmann relation yields

$$\frac{N_m}{N_n} = \frac{g_m}{g_n} e^{h\nu/kT} \tag{3.111}$$

where g_m and g_n are the respective degeneracies of $|m\rangle$ and $|n\rangle$. We also know that $\rho(\nu)$ is given by Planck's formulae (3.2) if equilibrium holds. This leads us to conclude† that

$$\frac{A_{nm}}{\mathsf{B}_{nm}} = \frac{8\pi h \nu^3}{c^3} \tag{3.112}$$

and

$$g_n \mathsf{B}_{nm} = g_m \mathsf{B}_{mn} \tag{3.113}$$

†Note that

$$\frac{A_{nm}(\omega)}{\mathsf{B}_{nm}} = \frac{1}{2\pi} \frac{A_{nm}(\nu)}{\mathsf{B}_{nm}} = \frac{\hbar \omega_{nm}^3}{\pi^2 c^3} \tag{3.112a}$$

since $\rho(\omega)\, d\omega = \rho(\nu)\, d\nu$. The factor 2π arising in (3.112a) means that great care has to be taken when using the Einstein coefficients.

These represent two important relationships which have general applicability. The former can be rewritten in a form

$$A_{nm} = \frac{8\pi\nu^2}{c^3} \times h\nu \times \mathsf{B}_{nm}$$

suggestive of the fact that the spontaneous transition probability A_{nm} equals the stimulated probability for a radiation field having only *one photon* in each radiation mode, as $8\pi\nu^2/c^3$ represents the radiation-mode density.

The Einstein stimulated-emission and absorption coefficients can be expressed in terms of the *dipole-moment matrix element* $\mathbf{d}_{nm}$ for the $|n\rangle$-to-$|m\rangle$ transition by using (3.102), (3.112a), and (3.113). Thus we may write

$$\mathsf{B}_{mn} = \frac{\pi g_n}{3\hbar^2\epsilon_0 g_m}|\mathbf{d}_{nm}|^2 \tag{3.114}$$

Our discussion so far has been couched in general terms in order that the derived relations could be applied to either atoms or molecules. Although we have obtained the Einstein radiative coefficients in terms of the matrix element of the dipole moment d_{nm}, the literature in the spectroscopic field tends to use the *absorption oscillator strength* f_{mn} in preference to d_{nm}. We shall thus relate these two entities and provide the more convenient form of the relations.

The simplest way of doing this is to recall (2.109) for the classical absorption coefficient and recast it in terms of our present discussion:

$$\kappa(\nu) = \frac{N_m f_{mn} e^2}{2m_e\epsilon_0 c}\pi\mathscr{L}(\nu) = 2N_m f_{mn} r_e c\pi^2\mathscr{L}(\nu) \tag{3.115}$$

where the so-called *natural* line profile function (arising from radiation damping) is given by

$$\mathscr{L}(\nu) \equiv \frac{1}{\pi}\frac{\Gamma_N}{(\nu_{nm}-\nu)^2+\Gamma_N^2} \tag{3.116}$$

Here $\Gamma_N\,(\equiv \gamma/4\pi)$ represents the frequency half width at half maximum under conditions of *natural broadening*. We can also see from the Beer–Lambert attenuation law (2.108) that the appropriate radiative-transfer equation† is of the form

$$\frac{dI(\nu,z)}{dz} = -\kappa(\nu)I(\nu,z) \tag{3.117}$$

where $I(\nu, z)$ is the spectral irradiance of the radiation field.

† This will be considered in greater detail in Chapter 4.

We can obtain a similar relation in terms of the Einstein absorption coefficient B_{mn} by writing

$$\frac{dI(\nu, z)}{dz} = -N_m \mathsf{B}_{mn} \rho(\nu, z) h\nu \mathscr{L}(\nu) \tag{3.118}$$

If we express the right-hand side in terms of the spectral irradiance

$$I(\nu, z) = \rho(\nu, z) c \tag{3.119}$$

and equate the resulting right-hand side with that of (3.117), using (3.115), then we arrive at the relation

$$\mathsf{B}_{mn} = \frac{2 r_e c^2 \pi^2}{h\nu} f_{mn} \tag{3.120}$$

If we then use (3.112a) and (3.113), we obtain

$$A_{nm} = \frac{2\omega^2 r_e g_m}{c g_n} f_{mn} \tag{3.121}$$

or

$$g_m f_{mn} \approx 1.5 \times 10^{-8} \lambda^2 g_n A_{nm} \tag{3.122}$$

where λ is the wavelength in micrometers.

Tables of oscillator strengths and of radiative lifetimes are available in the literature, for example, Corliss and Bozman (1962) and Mavrodineanu and Boiteux (1965, p. 516).

3.3.2. Molecular Transition Probabilities and the Franck–Condon Principle

We shall now consider radiative transitions within molecules. To be more specific, let us treat the radiative decay of a molecule from an upper energy eigenstate $|bv'J'\rangle$ to a lower energy eigenstate $|av''J''\rangle$. The letters a and b denote the respective lower and upper electronic states, while v'' and v' represent the lower and upper *vibrational* quantum numbers. J'' and J' are the respective lower and upper *rotational* quantum numbers. The relevant matrix element of the dipole moment for this transition is

$$\langle \mathbf{d}(bv'J'; av''J'') \rangle \equiv \langle bv'J' | \mathbf{d} | av''J'' \rangle \tag{3.123}$$

Under the Born–Oppenheimer assumption the three internal energy modes are independent. This enables the spatial wave function for the lower state to be expressed as a product [see (3.33)] of an *electronic wave function* $\psi_a(\mathbf{r}_i, \mathbf{R}_k)$

which depends primarily upon the coordinates of the electron $\mathbf{r}_i$ (and weakly upon the separation coordinates of the nuclei $\mathbf{R}_k$) and a *nuclear wave function* $\psi_{v''}(\mathbf{R}_k)$ which only depends upon $\mathbf{R}_k$. The rotation state of the molecule does not play an important role in determining the strength of the transition (Herzberg, 1967, p. 203) and so can be neglected at this point. We also recognize that the dipole-moment operator can be written as the sum of two components

$$\mathbf{d} = \mathbf{d}_e + \mathbf{d}_N \tag{3.124}$$

where

$$\mathbf{d}_e \equiv -\sum_i e\mathbf{r}_i \quad \text{and} \quad \mathbf{d}_N \equiv \sum_k eZ_k\mathbf{R}_k \tag{3.125}$$

Then the matrix element of the dipole moment for the *band* $|bv'\rangle \to |av''\rangle$ is

$$\begin{aligned}\big\langle \mathbf{d}(bv'; av'')\big\rangle &= \langle \psi_b\psi_{v'}|\mathbf{d}_e + \mathbf{d}_N|\psi_a\psi_{v''}\rangle \\ &= \langle \psi_b|\mathbf{d}_e|\psi_a\rangle\langle\psi_{v'}|\psi_{v''}\rangle + \langle\psi_b|\psi_a\rangle\langle\psi_{v'}|\mathbf{d}_N|\psi_{v''}\rangle \end{aligned} \tag{3.126}$$

The first term expresses the fact that the electronic dipole moment operator $\mathbf{d}_e$ primarily operates on the electronic wave functions.† In a similar way the second term reflects the fact that the nuclear dipole moment operator $\mathbf{d}_N$ does not depend upon the electronic coordinates. Furthermore, since the electronic eigenfunctions form an orthonormal set, i.e., $\langle\psi_a|\psi_b\rangle = \delta_{ab}$, it follows that the second term is zero and we can write

$$\big\langle \mathbf{d}(bv'; av'')\big\rangle = \langle \mathbf{d}_{ba}\rangle\langle v'|v''\rangle \tag{3.127}$$

where $\langle \mathbf{d}_{ba}\rangle$ represents the electronic transition moment, and the dimensionless factor $\langle v'|v''\rangle$ determines the relative strength of the emission arising from the transition between the two vibrational states $|v'\rangle$ and $|v''\rangle$.

† In terms of overlap integrals,

$$\begin{aligned}\langle\psi_b\psi_{v'}|\mathbf{d}_e|\psi_a\psi_{v''}\rangle &= \int \psi_b^*\psi_{v'}^*\mathbf{d}_e\psi_a\psi_{v''}\,dV_e\,dV_N \\ &= \int\psi_{v'}^*\left[\int\psi_b^*\mathbf{d}_e\psi_a\,dV_e\right]\psi_{v''}\,dV_N \\ &= \int\psi_{v'}^*\langle\mathbf{d}_{ba}\rangle\psi_{v''}\,dV_N\end{aligned}$$

Now $\langle\mathbf{d}_{ba}\rangle$ has only a very weak dependence upon the internuclear coordinates $\mathbf{R}_k$, and so we take it out of the integral, recognizing that it is evaluated for $\mathbf{R}_k$ close to their equilibrium values. Thus we may write

$$\langle\psi_b\psi_{v'}|\mathbf{d}_e|\psi_a\psi_{v''}\rangle \approx \langle\mathbf{d}_{ba}\rangle\int\psi_{v'}^*\psi_{v''}\,dV_N = \langle\mathbf{d}_{ba}\rangle\langle\psi_{v'}|\psi_{v''}\rangle$$

The Einstein transition probability for the molecular band $|bv'\rangle \to |av''\rangle$ can be evaluated by combining (3.102) with (3.127) to yield

$$A(bv', av'') = \frac{\omega^3}{3\pi\epsilon_0 \hbar c^3} |\mathbf{d}_{ba}|^2 q_{v'v''} \tag{3.128}$$

where ω represents the average angular frequency over the band. We see that the relative probability of the vibrational transfer $|v'\rangle \to |v''\rangle$ during the electronic transition $|b\rangle \to |a\rangle$ is characterized by

$$q_{v'v''} \equiv |\langle v'|v''\rangle|^2 \tag{3.129}$$

and known as the *Franck–Condon factor*. It can easily be shown (Steinfeld, 1974, p. 119) that a sum rule for these factors exists and takes the form

$$\sum_{v''} q_{v'v''} = 1 \tag{3.130}$$

In essence the Franck–Condon factors determine the relative strengths of transition between the vibrational states of two electronic states; thus they take the place of a quantum selection rule for this purpose. Their physical basis can be traced to the Born–Oppenheimer separation of electronic and nuclear motion. Indeed, the *Franck–Condon principle* states that nuclear motion ($\approx 10^{-13}$ s) can be regarded as "frozen" on the time scale of electronic transitions ($\lesssim 10^{-15}$ s). As a consequence transitions (up or down) between electronic states of a molecule can be represented by vertical lines on the potential-energy diagram. Figure 3.14 illustrates this principle and in doing so explains why certain vibrational transitions are preferable to others. That is to say, the figure aids in understanding the Franck–Condon factors if we recall that the vibrational eigenstates with $v > 0$ give rise to maximum probabilities at the classical turning point (see Figure 3.5).

If a molecule initially resides in the $v'' = 0$ vibrational state of the ground electronic state, then application of the Franck–Condon principle makes clear that the most probable transition will lead to population of the $v' = 2$ vibrational state of the electronically excited state—that is, $q_{v'=2, v''=0}$ will be the largest Franck–Condon factor involving $v'' = 0$. Conversely, for transitions emanating from $v' = 0$, $q_{v'=0, v''=3}$ will be strongest. Reference to Figure 3.14 clearly indicates that for molecules with an electronically excited equilibrium separation R_0' that is greater than the value for the ground electronic state, R_0'', the maximum in the emission will be redshifted relative to the maximum in the absorption spectra. This spectral displacement is termed a *Stokes shift*.

If the rotational fine structure of a band is to be resolved and individual *ro-vibronic* (rotational, vibrational, and electronic) lines are to be studied, then an additional factor is involved. This dimensionless *rotational line-strength*

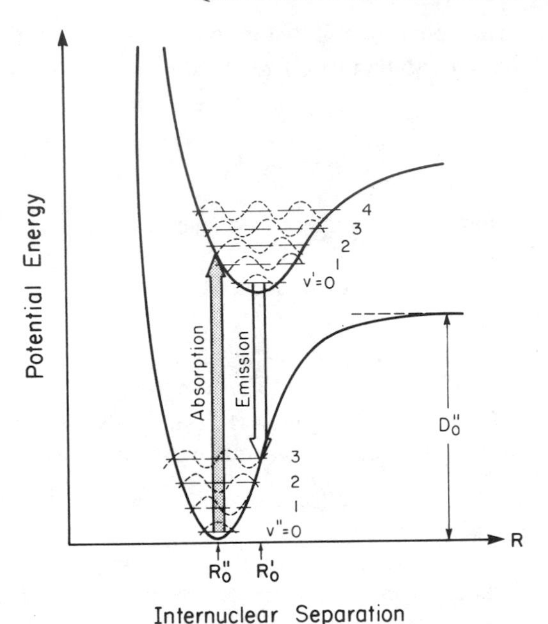

Fig. 3.14. Most probable transitions based on the Franck–Condon principle.

factor $S_{J'J''}$ is known as the Hönl–London factor (Herzberg, 1967, pp. 208, 382). A useful source of information on both Franck–Condon factors and Hönl–London factors is provided by Nicholls (1971).

3.4. SPECTRAL-LINE BROADENING

According to our view of the internal energy structure of atoms (or molecules), we might suspect that transitions between two quantum states would result in spectrally narrow emission or absorption lines. This is indeed the case in the limit of low temperature and density. However, even under these circumstances there is a finite spread of frequencies associated with the finite lifetime of the excited states. In general there are many causes of line broadening, some real and some apparent (instrumental).

The most important forms of spectral broadening, from our point of view, may be classified under the following headings:

Natural broadening, due to the finite lifetime of the energy states involved in the transition.

Doppler broadening, due to the Doppler shift associated with the thermal motion of the radiating (absorbing) species.

Collision broadening, due to elastic or inelastic encounters between the radiating (absorbing) species and neighboring atoms (or molecules).

Instrumental broadening, which arises when the spectral line is observed through an instrument with inadequate spectral resolution.

In addition to these, applied **E** or **H** fields can lead to broadening; in particular, if the radiating or absorbing species is immersed in a plasma, the electric microfields of the electrons and ions can give rise to *Stark broadening* (Griem, 1974). Still other forms of spectral distortion arise when *self-absorption* becomes important, this will be treated in Chapter 4.

3.4.1. Natural and Collision Broadening

We saw earlier that if we represent a molecule (or atom) by a simple harmonic oscillator with a resonant frequency ω_0, and imagine the oscillator to be given a sudden impulse, then the system would radiate electromagnetic energy at an angular frequency ω_0, for a length of time determined either by the radiation damping constant or by collisional decay. We shall show that this limited period of emission results in a finite bandwidth for the radiation field. In order to accomplish this we shall adopt a classical–quantum hybrid model. This *Lorentz approach* provides considerable physical insight into both natural and collision broadening with a minimum of mathematical complexity.

We shall assume that the radiated component of the electric field from an excited molecule (or atom) is

$$E(t) = \begin{cases} E_0 \mathbf{e}^{-i\omega_0 t}, & 0 < t < T \\ 0, & t \leq 0 \text{ or } t \geq T \end{cases} \tag{3.131}$$

where the *truncation period* T corresponds to the lifetime of the excited state. The corresponding Fourier component of the radiated field is

$$\begin{aligned} E(\omega, T) &\equiv \frac{1}{2\pi}\int_{-\infty}^{\infty} E(t)\mathbf{e}^{i\omega t}\, dt = \frac{E_0}{2\pi}\int_0^T \mathbf{e}^{i(\omega-\omega_0)t}\, dt \\ &= \frac{E_0}{\pi}\left[\frac{\sin\{(\omega-\omega_0)T/2\}}{(\omega-\omega_0)}\right]\mathbf{e}^{i(\omega-\omega_0)T/2} \end{aligned} \tag{3.132}$$

If τ represents the mean lifetime for a statistical ensemble of such radiating systems, then we can write

$$\frac{dN(t)}{dt} = -\frac{N(t)}{\tau}, \qquad \text{or} \quad N(t) = N(0)\mathbf{e}^{-t/\tau} \tag{3.133}$$

where $N(t)$ is the number density of the excited species at time t. Under these circumstances the probability of any given excited state undergoing a decay in the time interval $(T, T + dT)$ is given by

$$P(T)\,dT = -\frac{dN(T)}{N(0)} = e^{-T/\tau}\frac{dT}{\tau} \tag{3.134}$$

The observed radiation is a composite of the emission from all the excited species with this respective range of lifetimes. Under these circumstances the spectral irradiance is

$$I(\omega) = B\int_0^{\infty} |E(\omega, T)|^2 P(T)\,dT \tag{3.135}$$

where B is the appropriate proportionality constant. Substitution of (3.132) and (3.134) into (3.135) yields

$$I(\omega) = \frac{aA}{b^2}\int_0^{\infty} \sin^2\left(\frac{bT}{2}\right) e^{-aT}\,dT \tag{3.136}$$

where we have introduced $a \equiv 1/\tau$, $b \equiv \omega - \omega_0$, and the new constant of proportionality A. The integral in (3.136) is evaluated in any mathematical handbook and leads to the relation

$$I(\omega) = \frac{A}{(\omega_0 - \omega)^2 + (1/\tau)^2} \tag{3.137}$$

The constant of proportionality can be eliminated in terms of the total irradiance I_0 by spectrally integrating (3.137) over all frequencies:

$$I_0 \equiv \int_{-\infty}^{\infty} I(\omega)\,d\omega = \int_{-\infty}^{\infty} \frac{A\,d\omega}{(\omega_0 - \omega)^2 + (1/\tau)^2} = \frac{\pi A}{a} \tag{3.138}$$

Consequently, we may write

$$I(\omega) = I_0 \mathscr{L}(\omega) \tag{3.139}$$

where the line profile function

$$\mathscr{L}(\omega) = \frac{1}{\pi}\cdot\frac{(1/\tau)}{(\omega_0 - \omega)^2 + (1/\tau)^2} \tag{3.140}$$

is Lorentzian. We see that the spectral distribution of an emission line that is *lifetime-limited* is the same as that of a classical absorber (2.110) provided $1/\tau = \gamma/2$. The distinctive Lorentzian shape is illustrated in Fig. 3.15, and we can see that the angular-frequency half width at half maximum (HWHM) $\Delta\omega$

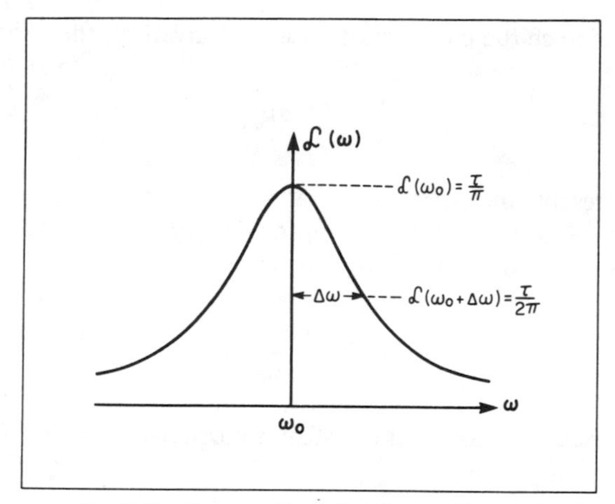

Fig. 3.15. Lorentz dispersion line profile function.

is given by

$$\mathcal{L}(\omega_0 + \Delta\omega) = \tfrac{1}{2}\mathcal{L}(\omega_0), \qquad \text{or} \quad \Delta\omega = \frac{1}{\tau} \tag{3.141}$$

In the case of natural broadening, τ_N is the radiative lifetime and typically has a value of about 10^{-8} s for atomic resonance lines. This leads to an angular-frequency HWHM

$$\Delta\omega_N = 10^8 \text{ rad/s}, \qquad \text{or} \quad \Delta\nu_N = \frac{10^8}{2\pi} = 16.7 \text{ MHz}$$

Furthermore, since $\lambda\nu = c$, it follows that the natural linewidth is

$$\Delta\lambda_N = \frac{\lambda^2}{c}\Delta\nu_N \tag{3.142}$$

Consequently, for the case cited above, $\Delta\lambda_N \approx 1.33 \times 10^{-5}$ nm if $\lambda = 500$ nm.

When collisions are effective in reducing the lifetime of the excited state or perturbing the energy separation between two quantum states of the emitting (absorbing) species, then additional broadening occurs. If the collisions are between *like* atoms (or molecules), we refer to *Holtsmark* broadening; if between *unlike* species, we refer to *van der Waals* broadening. If the lifetime of the excited state is reduced by collisions, the collision line profile takes the form

$$\mathcal{L}^c(\omega) = \frac{1}{\pi}\frac{1/\tau_c}{(\omega_0 - \omega)^2 + (1/\tau_c)^2} \tag{3.143}$$

where in this instance the collision lifetime τ_c is given by the expression

$$\frac{1}{\tau_c} = N\langle \sigma u \rangle_c \tag{3.144}$$

Here $\langle \sigma u \rangle_c$ represents the effective collision frequency per perturber and N the number density of perturbers. If the appropriate collision cross section is only a weak function of velocity, then we can write

$$\langle \sigma u \rangle_c \approx \sigma_c \left\{ \frac{8kT}{\pi\mu} \right\}^{1/2} \tag{3.145}$$

where μ is the reduced mass of the collision partners.

3.4.2. Doppler Broadening

If the radiating atom (or molecule) has a velocity component u_x along the line of sight, then the observed frequency ω is *Doppler-shifted* relative to the rest-frame frequency ω_0:

$$\omega = \omega_0\left(1 \pm \frac{u_x}{c}\right) \tag{3.146}$$

where c is the velocity of light. The positive sign applies if the emitter is moving towards the observer (see Fig. 3.16), and the negative sign is required when the emitter is receeding from the observer.

If the emitting species has a Maxwellian velocity distribution, then the probability that the radiating atom (or molecule) will have an x-component of velocity in the interval $(u_x, u_x + du_x)$ is given by

$$f(u_x)\,du_x = \left\{ \frac{m}{2\pi kT} \right\}^{1/2} e^{-mu_x^2/2kT}\,du_x \tag{3.147}$$

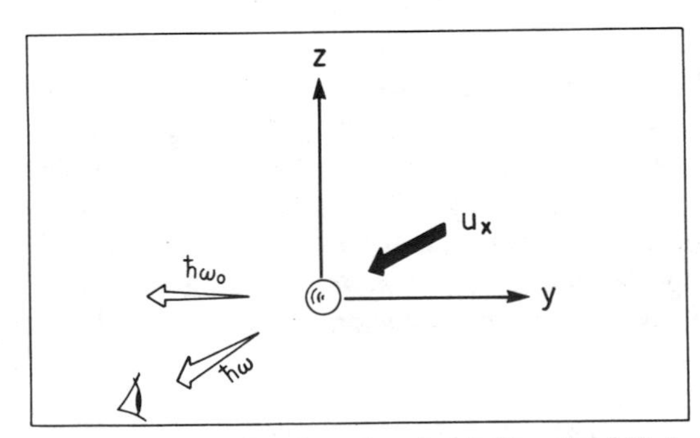

Fig. 3.16. Radiation emitted in the direction of motion is Doppler-shifted, so that the observed frequency is $\omega = \omega_0(1 + u_x/c)$.

where m is mass of the emitter and T its translational temperature. Combining (3.146) with (3.147), we see that the probability that a photon, with a rest angular frequency ω_0, will be observed with an angular frequency in the interval $(\omega, \omega + d\omega)$ is

$$\mathscr{L}_G(\omega) = \frac{1}{\beta\pi^{1/2}} e^{-(\omega-\omega_0)^2/\beta^2} \tag{3.148}$$

where we have introduced

$$\beta \equiv \left\{\frac{2kT\omega_0^2}{mc^2}\right\}^{1/2} \tag{3.149}$$

We see that the thermal motion (assumed Maxwellian) of the radiating species leads to a Gaussian line profile function (3.148); see Fig. 3.17. As we might expect, $\int \mathscr{L}(\omega)\,d\omega = 1$. The angular-frequency HWHM $\Delta\omega$ for the Doppler-broadened spectral line is given by the relation

$$\mathscr{L}_G(\omega_0 + \Delta\omega) = \tfrac{1}{2}\mathscr{L}_G(\omega_0)$$

or

$$\Delta\omega = \beta\{\ln 2\}^{1/2} \tag{3.150}$$

A comparison of the Lorentz and Gaussian line profile functions reveals that the spectral distribution associated with natural or collision broadening is quite different from that arising from thermal motion. The wing content of the Lorentz profile is much greater than that of the Gaussian profile if they have comparable spectral widths. Reference to (3.149) and (3.150) also indicates

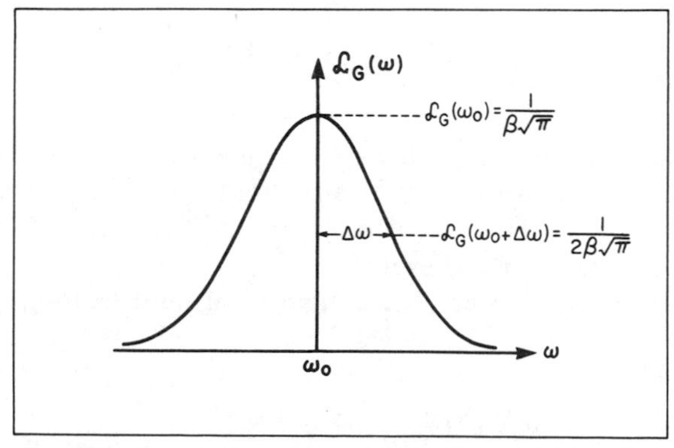

Fig. 3.17. Doppler-broadened Gaussian line profile function.

that the width of a Doppler broadened line is proportional to the square root of the temperature of the radiating species. It should be noted that although we have evaluated the emission line profile associated with the thermal motion of the atoms (or molecules), similar reasoning for the *absorption* of radiation would lead to an identical Gaussian line profile function.

3.4.3. Combined Natural, Collision, and Doppler Profiles

The rest-frame line profile of a radiating (or absorbing) species is most often Lorentzian (natural or collisional); however, invariably there is also relative motion between the emitter and the observer or between the source and the absorber. Where this motion arises from the thermal energy of the atoms (or molecules), we have seen that a Gaussian line profile results. Clearly, in most situations of interest the actual line profile will involve a convolution of these two types of broadening mechanisms. Although we shall consider emission, the analysis could just as easily be applied to the case of absorption, yielding the same profile function.

Thus the actual profile function can be written in the form of the convolution

$$\mathscr{L}(\omega) = \int_{-\infty}^{\infty} G(\omega^*) D(\omega, \omega^*)\, d\omega^* \tag{3.151}$$

where $D(\omega, \omega^*)$ represents the Lorentzian (dispersive) probability of emission at ω for an emitter having a line center frequency ω^*. $G(\omega^*)\, d\omega^*$ represents the Gaussian probability of an emitter having a line center frequency in the interval $(\omega^*, \omega^* + d\omega^*)$ due to its thermal motion. Thus in this instance we recognize that when there is relative motion between an emitter and an observer, the entire homogeneous line profile is Doppler shifted—which is the same as saying that the rest-frame line center frequency ω_0 is Doppler shifted to ω^*:

$$\omega^* = \omega_0\left(1 \pm \frac{u_x}{c}\right) \tag{3.152}$$

As before, $G(\omega^*)$ is derived from the appropriate Maxwellian velocity distribution and describes the probability of any emitter having a given velocity u_x relative to the observer. This is illustrated in Fig. 3.18, where $G(\omega^*)$ is seen to be the envelope of the $D(\omega^*, \omega^*)$-values.

From our earlier discussion [equations (3.140) and (3.148)] it is apparent that we can write

$$G(\omega^*) = \frac{1}{\beta\pi^{1/2}} e^{-(\omega_0 - \omega^*)^2/\beta^2} \tag{3.153}$$

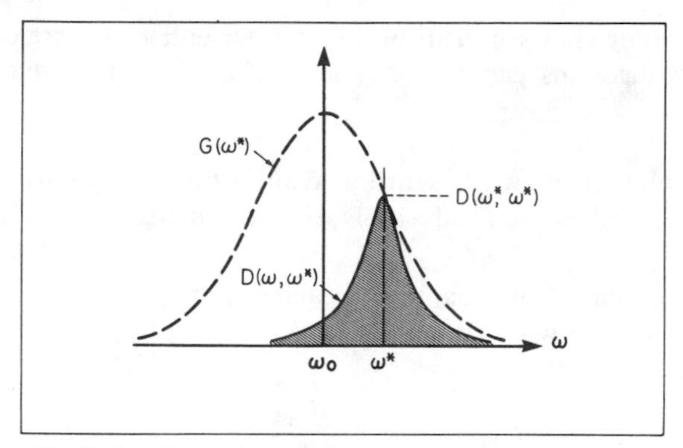

Fig. 3.18. Convolution of Gaussian and dispersive profiles.

and

$$D(\omega, \omega^*) = \frac{1}{\pi} \frac{1/\tau}{(\omega^* - \omega)^2 + (1/\tau)^2} \tag{3.154}$$

where β represents the Gaussian HWHM divided by $\{\ln 2\}^{1/2}$ and was defined by (3.149), and $1/\tau$ represents the Lorentzian HWHM. Consequently,

$$\mathscr{L}(\omega) = \frac{1}{\beta\tau\pi\sqrt{\pi}} \int_{-\infty}^{\infty} \frac{\mathrm{e}^{-(\omega_0 - \omega^*)^2/\beta^2}\, d\omega^*}{(\omega^* - \omega)^2 + (1/\tau)^2} \tag{3.155}$$

If we introduce

$$y \equiv \frac{\omega^* - \omega_0}{\beta}, \qquad u \equiv \frac{\omega - \omega_0}{\beta}, \quad \text{and} \quad a \equiv \frac{1}{\beta\tau} \tag{3.156}$$

then (3.155) becomes

$$\mathscr{L}(u) = \frac{a}{\beta\pi\sqrt{\pi}} \int_{-\infty}^{\infty} \frac{\mathrm{e}^{-y^2}\, dy}{(u - y)^2 + a^2} \tag{3.157}$$

the *Voigt line function*. In general (3.157) has to be evaluated numerically, although at core frequencies ($u \approx 0$), $\mathscr{L}(u)$ is close to being Gaussian, while in the wings of the profile ($u \gg 1$), $\mathscr{L}(u)$ is essentially Lorentzian. Useful tables for evaluating (3.157) are available in the literature (Aller, 1963, p. 325; Lochte-Holtgreven, 1968, pp. 128, 170).

If a spectral profile is being resolved by some kind of spectrometer, then some allowance may have to be made for the finite resolving power of the instrument. Indeed, it is often the case that the width of the instrumental profile is comparable to that of the spectral line under examination. Sometimes it is possible to assume that the instrumental profile is approximately described by a Lorentzian or Gaussian distribution. Under either of these circumstances the following theorems can be very useful for evaluating the *instrumental broadening*:

1. The convolution of two Lorentzian distributions whose widths are characterized by γ_1 and γ_2, respectively, is also a Lorentzian with a total width $\gamma = \gamma_1 + \gamma_2$.
2. The convolution of two Gaussian profiles characterized by widths β_1 and β_2 is also a Gaussian with a total width $\beta = (\beta_1^2 + \beta_2^2)^{1/2}$ (Corney, 1977, p. 256).

3.5. QUANTUM THEORY OF SCATTERING

The spontaneous emission of a photon of radiant energy by an excited atom is regarded as a first-order radiative process, whereas the scattering of electromagnetic energy by a molecule (or atom) is regarded as a second-order interaction, since two photons are involved. A photon $\hbar\omega$ of the incident radiation field is annihilated while a photon $\hbar\omega_s$ of scattered radiation is created. The scattering is said to be *elastic* if the scattered frequency ω_s is the same as the incident frequency ω. This form of scattering by quantized systems is termed *Rayleigh scattering*. If there is a change in the frequency, so that ω_s is not equal to ω, then we have *Raman scattering*, and the energy difference $\hbar(\omega - \omega_s)$ associated with this form of *inelastic* scattering is reflected in a change in the quantum state of the scatterer.

We have treated elastic scattering classically in the latter part of Chapter 2. In this chapter we shall briefly discuss the quantized view of Rayleigh scattering before going on to consider Raman scattering in some detail.

3.5.1. Elastic and Inelastic Scattering

A useful way of gaining some insight into the two forms of scattering can be obtained through considering the interaction of an electromagnetic wave, $E = E_0 \cos \omega t$, with an atomic system that has a polarizability $\wp = \wp_{x=0} + (\partial\wp/\partial x)_{x=0} x + \cdots$, where x is the charge displacement from equilibrium ($x = 0$), $\wp_{x=0}$ represents the equilibrium polarizability, and $(\partial\wp/\partial x)_{x=0}$ represents the field-induced polarizability. If the atomic system can be treated as a simple harmonic oscillator with angular frequency ω_1, then we can write $x = x_0 \cos \omega_1 t$, and the polarization of the atomic system becomes

$$\wp = \left\{ \wp_{x=0} + \left(\frac{\partial \wp}{\partial x} \right)_{x=0} x \right\} E_0 \cos \omega t$$

$$= \wp_{x=0} E_0 \cos \omega t + \tfrac{1}{2} E_0 x_0 \left(\frac{\partial \wp}{\partial x} \right)_{x=0} \{ \cos(\omega + \omega_1)t + \cos(\omega - \omega_1)t \}$$

We see that if $(\partial p/\partial x)_{x=0} = 0$, then there is no inelastic scattering. Consequently, it is the oscillating polarizability that can be thought of as modulating the scattered radiation and thereby leading to the appearance of the three frequencies ω, $\omega + \omega_1$, and $\omega - \omega_1$. The first corresponds to Rayleigh scattering, the second to *Raman anti-Stokes* scattering, and the third to *Raman Stokes* scattering. We see that although this classical picture makes it possible to understand the creation of $\omega + \omega_1$ and $\omega - \omega_1$ components of the scattered field, it is inadequate to account for the strength of the interaction.

A quantum-theory perturbation analysis (Loudon, 1973, p. 284) yields the basic differential scattering cross section for plane-polarized electromagnetic radiation of angular frequency ω incident upon an idealized nondegenerate molecule (or atom):

$$\frac{d\sigma}{d\Omega} = \sum_{f}^{\omega_f < \omega} \frac{\omega(\omega - \omega_f)^3}{(4\pi\epsilon_0 \hbar c^2)^2} \left| \sum_i \left\{ \frac{\langle f|\hat{\boldsymbol{\epsilon}}_s \cdot \mathbf{d}|i\rangle\langle i|\hat{\boldsymbol{\epsilon}} \cdot \mathbf{d}|0\rangle}{\omega_{i0} - \omega} + \frac{\langle f|\hat{\boldsymbol{\epsilon}} \cdot \mathbf{d}|i\rangle\langle i|\hat{\boldsymbol{\epsilon}}_s \cdot \mathbf{d}|0\rangle}{\omega_{if} + \omega} \right\} \right|^2 \tag{3.158}$$

In this *Kramers–Heisenberg* differential scattering cross section, $|0\rangle$ represents the initial quantum state, $|i\rangle$ some intermediate quantum state, and $|f\rangle$ the final quantum state. ω_{i0} represents the angular frequency corresponding to the 0-to-i virtual transition, and ω_{if} the angular frequency corresponding to the i-to-f virtual transition. The scattered angular frequency is $\omega_s \equiv \omega - \omega_f$. Finally, $\hat{\boldsymbol{\epsilon}}$ and $\hat{\boldsymbol{\epsilon}}_s$ represent the electric-field polarization unit vectors of the incident and scattered radiation, and $\mathbf{d}$ represents the electric-dipole-moment operator. A schematic representation of the interaction corresponding to each of the two kinds of terms is presented in Fig. 3.19. The sum over the

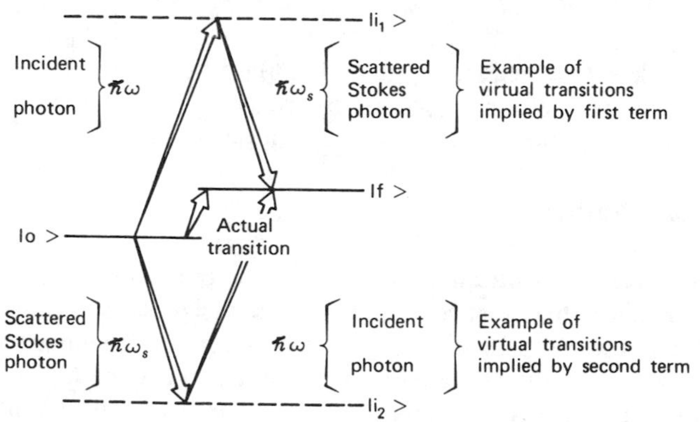

Fig. 3.19. Schematic representation of Raman (Stokes) scattering. The actual transition is from $|0\rangle$ to $|f\rangle$.

intermediate states $|i\rangle$ is intended to include an integral over the continuum of positive energy states (dissociated and ionized). The sum over the final states $|f\rangle$ is restricted to those states which contribute to the observed frequency.

It should be borne in mind that since (3.158) is derived from quantum perturbation theory, the summation arises as a mathematical consequence of expanding the perturbed wave function in terms of an infinite series of unperturbed wave functions. For this reason the transitions regarded as arising via the intermediate states are termed *virtual* transitions (Woodward, 1967). Although (3.158) has been used for H_2, its general application is made difficult by a lack of knowledge concerning the magnitude and sign of all the possible terms that could contribute and by a possible interference of terms that might have different phases (Woodward, 1967; Penney, 1974, pp. 191–217).

The differential cross section defined by (3.158) includes both elastic Rayleigh scattering, where $|f\rangle$ and $|0\rangle$ coincide (so $\omega_f = 0$), and inelastic Raman scattering, corresponding to the remaining terms in the summation over f. For elastic scattering the molecule (or atom) returns to its ground—or more generally, original— state $|0\rangle$ at the conclusion of the scattering event. In the case of elastic scattering we may write

$$\frac{d\sigma}{d\Omega} = \frac{\omega^4}{16\pi^2\epsilon_0^2\hbar^2c^4}\left|\sum_i\left\{\frac{\langle 0|\hat{\epsilon}_s\cdot\mathbf{d}|i\rangle\langle i|\hat{\epsilon}\cdot\mathbf{d}|0\rangle}{\omega_{i0}-\omega} + \frac{\langle 0|\hat{\epsilon}\cdot\mathbf{d}|i\rangle\langle i|\hat{\epsilon}_s\cdot\mathbf{d}|0\rangle}{\omega_{i0}+\omega}\right\}\right|^2 \tag{3.159}$$

Loudon (1973, pp. 285, 273) shows that in the limit $\omega \ll \omega_{i0}$, this equation can be reduced, in the case of hydrogen, to the form

$$\frac{d\sigma}{d\Omega} = \tfrac{81}{64}r_e^2\left(\frac{\omega}{\omega_H}\right)^4\{\cos^2\phi\cos^2\theta + \sin^2\phi\} \tag{3.160}$$

which is very close to the classical value (2.126) (in the limit $\omega \ll \omega_0$) if ω_H, the angular frequency corresponding to the hydrogen ground-state binding energy, is equated with the natural oscillating frequency of the harmonic oscillator, ω_0.

3.5.2. Raman Scattering

In Raman (inelastic) scattering the cross section is about three orders of magnitude smaller than the corresponding Rayleigh cross section, and the scattered signal consists of radiation that has suffered a frequency shift that is characteristic of the stationary energy states of the irradiated molecule. Raman spectroscopy represents a particularly powerful tool for laser remote sensing because it enables a trace constituent to be both identified and quantified relative to the major constituents of a mixture.

In the event that the molecule gains energy from the radiation field, the resulting lower-frequency scattered radiation is termed the *Stokes* component. If the molecule loses energy to the radiation field, the scattered radiation is referred to as the *anti-Stokes* component. These scattering processes are schematically illustrated in Fig. 3.20.

The structure of the Raman spectrum is often quite complex, even for diatomic molecules. In the special case of diatomic molecules possessing zero electron angular momentum around the internuclear axis ($\Lambda = 0$), the selection rules (Herzberg, 1967) allow vibrational–rotational transitions for which the change in the molecular rotational quantum number J can be only 0 or ± 2, and the change in the vibrational quantum number v can be only 0 or ± 1, that is,

$$\begin{aligned} \Delta v &= 0, \pm 1 \\ \Delta J &= 0, \pm 2 \end{aligned} \tag{3.161}$$

Under these circumstances the Raman spectrum consists of three branches: S ($\Delta J = +2$), Q ($\Delta J = 0$), and O ($\Delta J = -2$), plus a pure rotational structure centered about the exciting wavelength and corresponding to $\Delta v = 0$. Inaba and Kobayasi (1972) have computed the theoretical distribution of vibrational–rotational Raman lines corresponding to the $\Delta v = +1$ (Stokes shift) spectra of the N_2 molecule at 300 K. Their results are presented in Fig. 3.21. The ordinate gives the value of the differential scattering cross section for each of the Raman components corresponding to the vibrational transition $v = 0 \rightarrow 1$. As can be seen in Fig. 3.21, all lines in the Q-branch (for which $\Delta J = 0$) lie very close to each other and are not normally resolved. The S and O branches ($\Delta J = \pm 2$) are well separated and appear as side bands of the intense $\Delta J = 0$ line. It should be noted that although temperature variations do influence the intensity of the S and O branches, their effect on the Q-branch is often negligible.

Unfortunately, the S and O Raman side bands of a strongly scattering, high-concentration species may overlap and mask weaker Q-branch scattering from trace constituents in the atmosphere. This problem could arise in remote

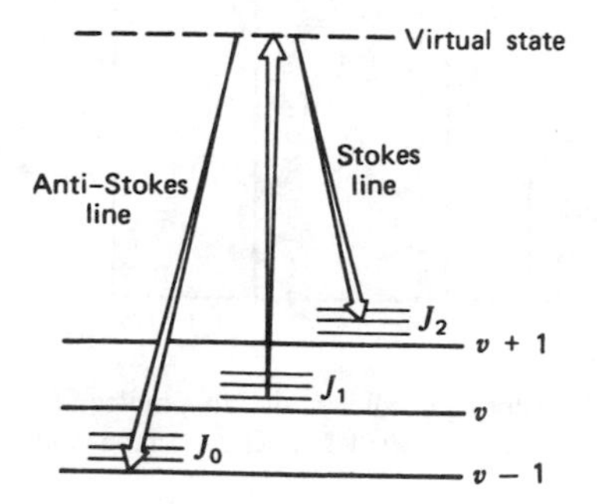

Fig. 3.20. Schematic representation of Raman vibrational ($\Delta v \pm 1$) Stokes and anti-Stokes scattering.

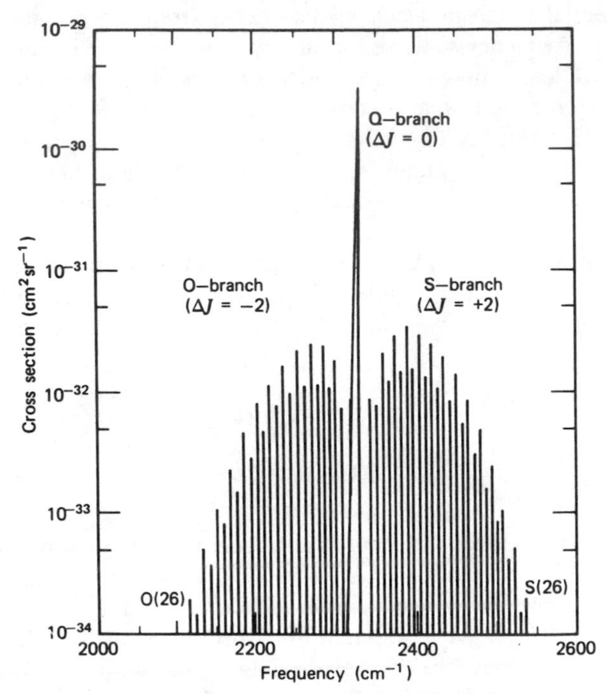

Fig. 3.21. Theoretical distribution of vibrational–rotational Raman spectrum ($v = 0 \rightarrow 1$ vibrational transition) at 300 K, showing the *O*-, *Q*-, and *S*-branch structures and the differential Raman-scattering cross section for N_2 molecules (Inaba and Kobayasi, 1972).

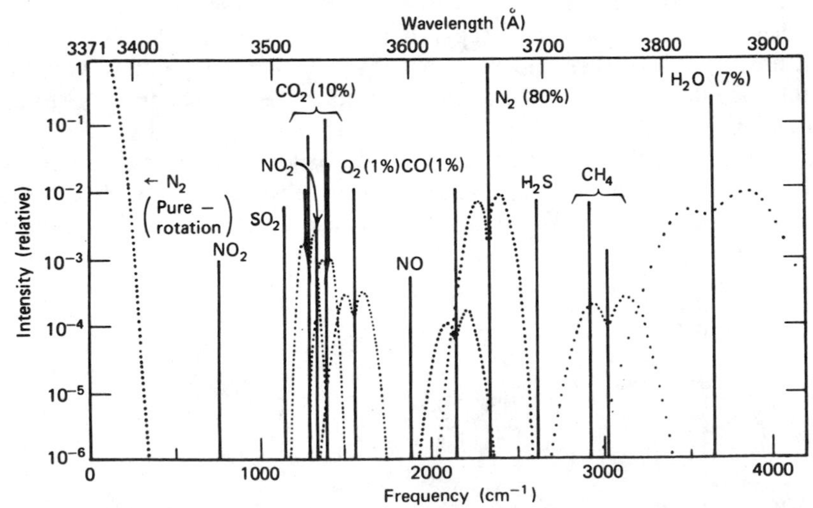

Fig. 3.22. Theoretical distribution of Raman volume backscattering coefficient due to a molecular mixture contained in a typical oil smoke as a function of Raman-shifted frequency (Inaba and Kobayasi, 1972).

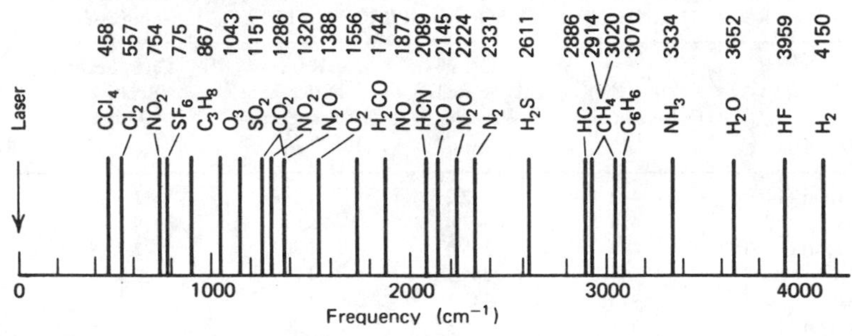

Fig. 3.23. Frequency shifts of the Q-branch of vibrational–rotational Raman spectra of typical molecular species present in polluted as well as ordinary atmosphere relative to the exciting laser frequency (Inaba and Kobayasi, 1972).

pollution monitoring. Inaba and Kobayasi (1972) have illustrated this problem with the theoretical calculation of the spectral distribution of the Raman scattering to be expected from a typical oil smoke plume (Fig. 3.22). They assumed that the partial pressures of the plume constituents were $[N_2] = 0.80$ atm, $[CO_2] = 0.10$ atm, $[H_2O] = 0.07$ atm, $[O_2] = [CO] = 10^{-2}$ atm, $[NO] = [SO_2] = [H_2S] = [CH_4] = 10^{-3}$ atm, and $[NO_2] = 10^{-4}$ atm. The solid line indicates the Q-branch, and the dotted lines the O and S branches, of the vibrational–rotational Raman band for each species excited by a N_2 laser operating at 337.1 nm.

The frequency shifts of the Q-branch of vibrational–rotational Raman spectra are summarized in Fig. 3.23 and Tables 3.3 and 3.4 for a large group of molecules of interest for air-pollution work involving laser remote sensing.

The Raman differential cross section is given by the inelastic term (i.e. $|f\rangle$ not the same as $|0\rangle$) of (3.158) provided the incident frequency ω is higher than the lowest molecular excitation frequency. In general the scattered radiation has as many angular-frequency components ω_s as there are energy states with

TABLE 3.3. RAMAN (Q, $O + S$ BRANCHES, AND TOTAL), PURE-ROTATION RAMAN, AND RAYLEIGH BACKSCATTERING CROSS SECTIONS[a]

Molecule	Raman Shift $\omega_j/2\pi c$ (cm^{-1})	$d\sigma/d\Omega$ (cm^2 sr^{-1}) Q-branch	$O + S$ Branches	Total	Rayleigh	Pure Rotation	Total
N_2	2329.66	2.9×10^{-30}	5.5×10^{-31}	3.5×10^{-30}	3.9×10^{-27}	1.1×10^{-28}	4.0×10^{-27}
O_2	1556.26	3.3×10^{-30}	1.3×10^{-30}	4.6×10^{-30}	3.3×10^{-27}	2.0×10^{-28}	3.5×10^{-27}
CO_2 (ν_1)	1388.15	3.4×10^{-30}	7.3×10^{-31}	4.2×10^{-30}	9.0×10^{-27}	8.3×10^{-28}	9.9×10^{-27}
CH_4 (ν_1)	2914.2	2.1×10^{-29}	0	2.1×10^{-29}	8.6×10^{-27}	0	8.6×10^{-27}

[a]Based on the polarizability tensor theory of Placzek (1934); reproduced from Inaba (1976).

TABLE 3.4. RAMAN WAVE-NUMBER SHIFTS AND MEASURED DIFFERENTIAL RAMAN BACKSCATTERING CROSS SECTIONS APPROPRIATE FOR 337.1-nm EXCITATION[a]

Molecule	Raman Shift (cm^{-1})	Raman-Shifted Wavelength (nm)	Raman Differential Cross Section (10^{-30} cm^2 sr^{-1})	Cross Section Relative to *Q*-branch of N_2	Ref.[b]
Freon 114[c]	442	342.2	4.2(*P*)	1.49(*P*)	1
CCl_4	459	342.4	26.0	9.3	2
Freon C-318[d]	699	345.2	7.8(*P*)	2.77(*P*)	1
NO_2 (ν_2)	754	345.7	24.0	8.6	3
SF_6	775	346.1	12.0	4.3	2
Freon 116[e]	807	346.5	7.3(*P*)	2.6(*P*)	1
Freon 114[f]	908	347.7	5.3(*P*)	1.9(*P*)	1
C_6H_6 (ν_2)	991	348.7	44.0	15.7	2, 3
O_3	1103.3	350.2	6.4	2.3	4
SO_2	1151.5	350.8	17.0	6.1	3, 5
CO_2 ($2\nu_2$)	1285	352.5	3.1	1.1	2, 3
NO_2 (ν_1)	1320	352.8	51.0	18.2	3
CO_2 (ν_1)	1388	353.7	4.2	1.5	2, 3
O_2	1556	355.9	4.6	1.6	2, 3
			3.3(*Q*)	1.2(*Q*)	2, 3
C_2H_4 (ν_2)	1623	356.6	5.4(*Q*)	1.9(*Q*)	2
NO	1877	360.0	1.5	0.54	3, 5
CO	2145	363.5	3.6	1.3	2, 3
N_2	2330.7	365.9	3.5	1.3	2, 6
			2.8(*Q*)	1.0(*Q*)	2, 6
H_2S	2611	369.7	19.0	6.8	2, 5
CH_3OH (ν_2)	2846	372.8	14.0	5.0	2, 3
C_5H_{12}	2885	373.4	124.0(*C*)	44.3(*C*)	1
C_3H_8	2886	373.4	81.8(*C*)	29.2(*C*)	1
C_6H_{14}	2886	373.4	134.0(*C*)	48.0(*C*)	1
C_4H_{10}	2890	373.5	93.5(*C*)	33.4(*C*)	1
CH_4	2914	373.8	32.2(*C*)	11.5(*C*)	1
—(ν_1)			21.0	7.5	2, 3, 6
C_5H_{10}	2941	374.2	102.5(*C*)	36.6(*C*)	1
C_3H_6	2942	374.2	63.6	22.7	1
C_2H_5OH	2943	374.2	19.0	6.8	2, 3
CH_3OH ($2\nu_6$)	2955	374.4	7.5	2.7	2, 3
C_4H_8	3010	375.2	89.6(*C*)	32.0	2
CH_4 (ν_3)	3017	375.3	14.0	5.0	2, 3
C_2H_4	3020	375.3	28.6	10.2	1
—(ν_1)			16(*Q*)	5.7(*Q*)	2
C_8H_{10}	3064	375.9	87.9(*C*)	31.4(*C*)	1
C_6H_6 (ν_1)	3070	376.0	30.0	10.7	2, 3
C_6H_6	3072	376.0	65.2	23.3	1
NH_3	3334	379.8	11.0	3.9	2
C_2H_2	3372	380.3	3.36	1.2	1
H_2O	3651.7	384.4	7.8(*Q*)	2.8(*Q*)	2, 6
H_2	4160.2	392.2	8.7	3.1	2, 3

[a]*Q* indicates the value of the *Q*-branch vibrational Raman backscattering cross section; *C* indicates a broad multipeaked structure associated with the C—H stretch mode; *P* indicates a cross section based on a ratio of peak intensities rather than spectrally integrated signals.

[b]*References*: 1, Stephenson (1974); 2, Murphy et al. (1969); 3, Inaba and Kobayasi (1972); 4, Schwiesow and Abshire (1973); 5, Fouche and Chang (1971); 6, Penney et al. (1974).

[c]1, 2-Dichlorotetrafluoroethane.

[d]Octafluorocyclobutane.

[e]Hexafluoroethane.

[f]Tetrafluoromethane.

frequency ω_f smaller than ω. Unfortunately, this quantum relation,

$$\frac{d\sigma}{d\Omega} = \sum_{f \neq 0}^{\omega_f < \omega} \frac{\omega(\omega - \omega_f)^3}{(4\pi\epsilon_0 \hbar c^2)^2} \left| \sum_i \left\{ \frac{(\hat{\epsilon}_s \cdot \mathbf{d}_{fi})(\hat{\epsilon} \cdot \mathbf{d}_{i0})}{\omega_{i0} - \omega} + \frac{(\hat{\epsilon} \cdot \mathbf{d}_{fi})(\hat{\epsilon}_s \cdot \mathbf{d}_{i0})}{\omega_{if} + \omega} \right\} \right|^2 \tag{3.162}$$

only provides accurate Raman scattering cross sections for H_2 and D_2. In equation (3.162), $\hat{\epsilon}_s \cdot \mathbf{d}_{fi}$ and $\hat{\epsilon} \cdot \mathbf{d}_{fi}$ represent the components of the dipole moment, for the $|f\rangle$-to-$|i\rangle$ transition (3.99), in the directions of polarization of the scattered and incident light ($\hat{\epsilon}_s$ and $\hat{\epsilon}$ respectively). In a similar manner, $\hat{\epsilon}_s \cdot \mathbf{d}_{i0}$ and $\hat{\epsilon} \cdot \mathbf{d}_{i0}$ represent the components of the dipole moment for the $|0\rangle$-to-$|i\rangle$ transition in the $\hat{\epsilon}_s$ and $\hat{\epsilon}$ directions. In the case of heavier molecules it has not been possible to calculate the Raman scattering cross sections from experimental measurements of the dipole-moment matrix elements (or oscillator strengths), due to insufficient knowledge and potential interference between the terms within the absolute square of (3.162) (Penney, 1974, p. 193). We shall see shortly that if the frequency of the incident radiation approaches that of an optically allowed transition, (3.162) can be simplified to yield an equation that makes quantitative calculations possible.

An alternative expression for the Raman differential cross section can be derived from a theoretical treatment of the polarizability of molecules exposed to a radiation field. According to Placzek (1934) and Inaba (1976), when the Raman scattering is observed in a direction perpendicular to the polarization direction of linearly polarized light, the vibrational Raman backscattering cross section for the Q-branch ($\Delta v = 1$, $\Delta J = 0$) is

$$\left[\frac{d\sigma_j}{d\Omega}\right]^Q = \frac{b_j^2(\omega - \omega_j)^4 g_j}{c^4[1 - e^{-\hbar\omega_j/kT}]} \{\mathring{\alpha}_j^2 + \tfrac{7}{180}\mathring{\gamma}_j^2\} \tag{3.163}$$

while the vibrational Raman backscattering cross section for the O and S branches ($\Delta v = 1$, $\Delta J = -2$ and $\Delta v = 1$, $\Delta J = 2$ respectively) is

$$\left[\frac{d\sigma_j}{d\Omega}\right]^{O+S} = \frac{b_j^2(\omega - \omega_j)^4 g_j}{c^4[1 - e^{-\hbar\omega_j/kT}]} \cdot \tfrac{7}{60}\mathring{\gamma}_j^2 \tag{3.164}$$

The total vibrational Raman backscattering cross section, obtained through adding (3.163) and (3.164), is

$$\frac{d\sigma_j}{d\Omega} = \frac{b_j^2(\omega - \omega_j)^4 g_j}{c^4[1 - e^{-\hbar\omega_j/kT}]} \{\mathring{\alpha}_j^2 + \tfrac{7}{45}\mathring{\gamma}_j^2\} \tag{3.165}$$

Here, ω and ω_j represent the respective angular frequencies of the incident radiation and that of the jth vibrational mode of the molecule; b_j [$=(\hbar/2\omega_j)^{1/2}$] represents the zero-point vibrational amplitude of this mode, and g_j its degree of degeneracy; $3\mathring{\alpha}_j$ and $\mathring{\gamma}_j^2$ correspond to the trace and anisotropy of the *derived polarizability tensor* associated with the normal coordinate q_j; T is the vibrational temperature of the molecules; and c is the velocity of light in vacuum. The factor

$$a_j = g_j\left(45\mathring{\alpha}_j^2 + 7\mathring{\gamma}_j^2\right) \tag{3.166}$$

has been called the *scattering activity* by Murphy, Holzer, and Bernstein (1969), who relate it to the *depolarization ratio*

$$\delta_{pj} = \frac{6\mathring{\gamma}_j^2 g_j}{a_j} \tag{3.167}$$

Experimental measurements of δ_{pj} and a_j have enabled both $\mathring{\alpha}_j^2$ and $\mathring{\gamma}_j^2$ to be determined for a large variety of molecules, a selection of which is provided in Table 3.3. An extensive list of prelaser data on vibrational Raman cross sections data has been prepared by Murphy, Holzer, and Bernstein (1969). Inaba and Kobayasi (1972) have summarized the measured values of the differential Raman backscattering cross section for a range of molecular constituents of the atmosphere. The recent interest in O_3 has prompted Schwiesow and Abshire (1973) to determine its Raman cross section, and Stephenson (1974) has evaluated the Raman cross section for a large number of hydrocarbon and Freon gases. In all instances these cross sections are experimentally evaluated relative to the 2331-cm^{-1} vibrational Q-branch Raman cross section of nitrogen. The absolute value of this cross section has been measured by several different methods, (Murphy et al., 1969; Hyatt et al., 1973; Penney, 1974), so that we can write with some degree of confidence

$$\frac{d\sigma}{d\Omega} \approx (4.3 \pm 0.2) \times 10^{-31}\ \text{cm}^2\ \text{sr}^{-1} \tag{3.168}$$

for the $0 \rightarrow 1$ vibrational Q-branch Raman transition in N_2 gas when excited at 514.5 nm. This corresponds to the value 2.8×10^{-30} cm^2 sr^{-1} given by Inaba and Kobayasi (1972) for 337.1-nm excitation when allowance is made for the ω_s^4 dependence of $d\sigma/d\Omega$.

In Table 3.4 we have prepared an updated list of Raman cross sections, corresponding wave-number shifts, and wavelengths appropriate to excitation by the nitrogen laser operating at 337.1 nm. Where the symbol Q appears in the table, the value of the Q-branch vibrational Raman backscattering cross section is given instead of the total cross section.

Placzek (1934) has also used a quantum treatment of the polarizability theory to derive the differential scattering cross section for the rotational Raman effect in diatomic molecules for the case of plane-polarized incident light. The pure rotational Raman and Rayleigh backscattering cross sections are also given by (3.163) and (3.164) if the factor $b_j^2 g_j/(1 - e^{-\hbar\omega_j/kT})$ is set equal to unity and $\mathring{\alpha}_j^2$ and $\mathring{\gamma}_j^2$ are interpreted as the isotropic and anisotropic parts of the polarizability tensor. Again the above is strictly valid only if ω is much less than any allowed transition frequency. An example of a pure rotational Raman spectrum is presented as Fig. 3.24.

Several advantages could accrue from use of the pure rotational Raman spectra (Barrett, 1974; Smith, 1972). In the first place, the rotational scattering cross section of a molecule is usually larger (sometimes by two orders of magnitude) than its vibrational–rotational Raman cross section (Fenner et al., 1973). Second, through the use of a Fabry–Perot interferometer as a comb filter, it has been possible to utilize all of the rotational Raman lines as a single signal (Barrett, 1974, 1976, 1977). Third, the luminosity of a Fabry–Perot interferometer is greater than that of a grating spectrometer for the same resolution (Jacquinot, 1954). The overall gain in signal could be three orders of magnitude, and such a technique has great potential for monitoring gas properties remotely.

The difficulty with using rotational Raman scattering for pollution detection lies in the overlapping of the rotational transitions of the major atmospheric

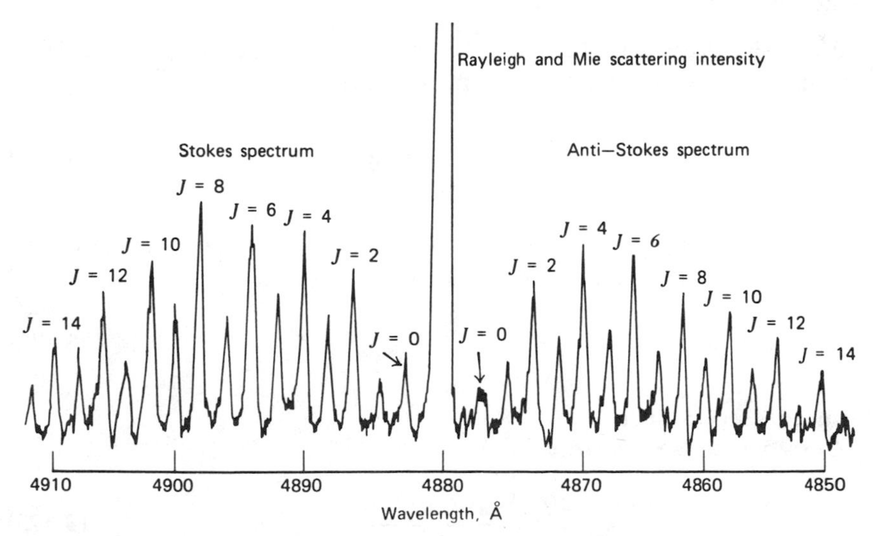

Fig. 3.24. Experimentally measured pure rotational Raman spectrum of nitrogen. Laser excitation wavelength 488.0 nm (Salzman, 1974).

constituents and the large Rayleigh–Mie backscattered signal. Both Barrett (1976, 1977) and Smith (1972) have indicated that these problems could be overcome with careful design of the Fabry–Perot interferometer.

3.5.3. Resonance Raman Scattering

When the exciting frequency approaches an isolated absorption line, one term, corresponding to the transition from one particular state $|g\rangle$ to an intermediate state $|r\rangle$, can dominate the sum in (3.162). Under these circumstances the gr transition corresponds to an optically allowed transition, so that the transition moments become large and allowance has to be made for the finite lifetime Γ^{-1} of $|r\rangle$. In the absence of collisions, radiation damping limits the bandwidth of the gr transition, thereby avoiding the mathematical singularity associated with ω becoming equal to the center angular frequency of the resonant gr transition, ω_0. Under these circumstances we can write

$$\frac{d\sigma}{d\Omega} = \frac{\omega\omega_s^3}{\left(4\pi\epsilon_0\hbar c^2\right)^2} \frac{\left|(\hat{\boldsymbol{\epsilon}}_s \cdot \mathbf{d}_{fr})(\hat{\boldsymbol{\epsilon}} \cdot \mathbf{d}_{rg})\right|^2}{(\omega_0 - \omega)^2 + \Gamma^2} \tag{3.169}$$

The total (near resonance) scattering cross section integrated over all scattering angles and light polarizations (Loudon, 1973, p. 287) can be expressed in the form

$$\sigma_{gf} = \frac{16\pi}{9} \cdot \frac{\omega\omega_s^3}{\left(4\pi\epsilon_0\hbar c^2\right)^2} \cdot \frac{|\mathbf{d}_{fr}|^2|\mathbf{d}_{rg}|^2}{(\omega_0 - \omega)^2 + \Gamma^2} \tag{3.170}$$

The radiative decay rate A_{rf} of the rf transition can be related [see (3.102)] to the square of the appropriate dipole matrix element:

$$A_{rf} = \frac{\omega_s^3}{3\pi\epsilon_0\hbar c^3}|\mathbf{d}_{fr}|^2 \tag{3.171}$$

as $\omega_s = \omega_{fr}$. The Einstein absorption coefficient B_{gr} of the gr transition can be related [see (3.114)] to the square of its transition moment:

$$\mathsf{B}_{gr} = \frac{\pi}{3\hbar^2\epsilon_0}|\mathbf{d}_{gr}|^2 \tag{3.172}$$

for nondegenerate states.

Consequently we can write

$$\sigma_{gf} = \frac{\hbar\omega\mathsf{B}_{gr}}{c\pi} \cdot \frac{A_{rf}}{(\omega_0 - \omega)^2 + \Gamma^2} \tag{3.173}$$

Now we can express equation (3.108) in the form

$$P^A(\omega) = \frac{\hbar\omega \mathrm{B}_{gr}\mathscr{L}(\omega) N_g I(\omega)}{c}$$

which allows us to introduce an absorption cross section σ_{gr}^A for a molecule (or atom) in $|g\rangle$ in the form

$$\sigma_{gr}^A = \frac{\hbar\omega \mathrm{B}_{gr}\mathscr{L}(\omega)}{c} \tag{3.174}$$

In the absence of Doppler broadening, the absorption-line profile

$$\mathscr{L}(\omega) = \frac{1}{\pi}\frac{\Gamma}{(\omega_0 - \omega)^2 + \Gamma^2} \tag{3.175}$$

is essentially in the same form as given earlier [equation (3.140)]. The *quantum yield* ϕ_{rf}^F for emission (fluorescence) into the *rf* transition is defined by the relation

$$\phi_{rf}^F \equiv \frac{A_{rf}}{\Gamma} \tag{3.176}$$

In essence ϕ_{rf}^F expresses the probability that a molecule (or atom) excited to state $|r\rangle$ will undergo a radiative decay through the spontaneous emission of a photon of energy $\hbar\omega_s$. Combining (3.176) with (3.174) enables us to write

$$\sigma_{gf} = \sigma_{gr}^A \phi_{rf}^F \tag{3.177}$$

and we see that the cross section for inelastic scattering from $|g\rangle$ to $|f\rangle$ is equal to the product of the cross section for absorption from $|g\rangle$ to $|r\rangle$ and the quantum yield for radiative decay from $|r\rangle$ to $|f\rangle$.

There is an interesting controversy associated with this analysis, for it appears that under resonance conditions there is little distinction between the scattering cross section σ_{gf} and the fluorescence cross section σ_{gf}^F. The term fluorescence is normally used to refer to that process for which the decay time is in the nanosecond (or longer) range at low pressures and decreases due to collisional quenching at pressures greater than a few torr. On the other hand, Raman scattering is generally taken to refer to the two-photon process that is relatively instantaneous and suffers no quenching (i.e., its intensity per molecule is insensitive to gas composition or pressure, at least up to several atmospheres). Equation (3.170) appears to describe scattering at frequency separations that are large compared to the total linewidth, yet also can be seen to describe fluorescence close to an allowed transition. At high pressures the influence of homogeneous and inhomogeneous broadening and quenching confuses the issue. The pressure sensitivity and time-dependent nature of the cross sections have only been partially resolved by theory and experiment (Placzek, 1934; Fouche et al., 1972).

We see that for atoms (3.170) predicts large cross sections near resonance for strong isolated atomic lines. In the case of aluminum atoms, where the Raman shift is about 2 nm, an enhancement of 10^5 has been predicted for a spectral misalignment of about 5 nm (Fouche et al., 1972). In the case of vibrational Raman scattering within molecules, (3.170) can be rewritten in the form

$$\sigma_{gf} = \frac{16\pi\omega\omega_s^3}{\left(12\pi\epsilon_0\hbar c^2\right)^2}\cdot\frac{|\mathbf{d}_{m,v_g''}^{n,v_r'}|^2|\mathbf{d}_{m,v_f''}^{n,v_r'}|^2}{(\omega_0-\omega)^2+\Gamma^2} \tag{3.178}$$

where $\mathbf{d}_{m,v_g''}^{n,v_r'}$ and $\mathbf{d}_{m,v_f''}^{n,v_r'}$ represent the dipole matrix elements for the $m, v_g'' \to n, v_r'$ and $n, v_r' \to m, v_f''$ transitions; n and m refer to the intermediate and lower electronic state; and v_g'', v_r', and v_f'' represent the vibrational quantum numbers of the initial, intermediate, and final states, respectively. If we express these dipole matrix elements in terms of the mn electronic absorption oscillator strength f_{mn} and the appropriate Franck–Condon factors $q_{v_r', v_g''}$ and $q_{v_r', v_f''}$, by combining (3.102), (3.121), and (3.128) we obtain

$$|d_{m,v_g''}^{n,v_r'}|^2 = \frac{6\pi\epsilon_0\hbar c^2 r_e}{\omega_0} f_{mn} q_{v_r', v_g''} \tag{3.179}$$

and

$$|d_{m,v_f''}^{n,v_r'}|^2 = \frac{6\pi\epsilon_0\hbar c^2 r_e}{\omega_s} f_{mn} q_{v_r', v_f''} \tag{3.180}$$

Then we may write

$$\sigma_{gf} = 4\pi r_e^2\left(\frac{\omega\omega_s^2}{\omega_0^3}\right) f_{mn}^2 q_{v_r', v_g''} q_{v_r', v_f''} \frac{\omega_0^2}{(\omega_0-\omega)^2+\Gamma^2} \tag{3.181}$$

where r_e represents the classical electron radius (2.82).

With molecules the variety of vibrational–rotational transitions possible ensures that the potential enhancement is much less. The reduction in the enhancement can be attributed to four factors (Penney, 1974):

1. The fraction of molecules capable of near resonance may be much less than unity due to the rotational level spread of the vibrational ground state's population.

2. The average molecular transition moment can be several orders of magnitude smaller than the corresponding atomic transition moment, due to the large number of nondegenerate transitions to the manifold of rotational–vibrational states associated with the excited electronic state.
3. There are a large number of alternative radiative decay modes associated with the intermediate level at resonance.
4. The closer the exciting frequency approaches the resonance condition, the greater is the possibility of collisions that can introduce quenching effects, as observed in fluorescence.

Nevertheless, Fouche and Chang (1972), St. Peters and Silverstein (1973), and St. Peters et al. (1973) have reported strong enhancement when I_2 is excited within 0.01 nm of individual molecular lines. Although there is some disagreement between these authors as to the degree of quenching observed in their experiments and the subsequent interpretation in terms of resonance Raman scattering or resonance fluorescence, in general this emission appears to have a weaker quenching dependence upon background gas pressure than normally observed for fluorescence. This observation has been explained, however, by a near-balance between the opposing effects of collisional broadening (which increases absorption in the wings of the line) and quenching (which reduces the probability of remission following absorption) (Fouche et al., 1972; St. Peters and Silverstein, 1973).

It is clear that the ambiguity inherent in quenching data tends to confuse the issue. Indeed, St. Peters and Silverstein (1973) point out that quenching data can only be used to determine scattering time if elastic collisional broadening is unimportant. In order to avoid this difficulty, Williams et al. (1973) measured directly the scattering time associated with detuning a single-mode argon laser from an I_2 resonance line. They observed a marked shortening of this scattering time when the laser was detuned off resonance by as little as 0.002 nm.

3.5.4. Fluorescence and Absorption

Although there may be some uncertainty as to the operative process (resonance Raman scattering or fluorescence) in cases of near-resonance, there is little doubt that fluorescence is the relevant process when the exciting wavelength λ_l approaches to within the line width of the absorption feature. Under these circumstances the fluorescence cross section per unit wavelength interval is given by an equation that is practically identical to (3.177), namely,

$$\sigma^F(\lambda, \lambda_l) = \sigma^A(\lambda_l)F(\lambda, \lambda_l) \tag{3.182}$$

where $F(\lambda, \lambda_l)$ represents the *quantum yield factor* and is in turn given by the product of the *quenching factor* Q^F and the *fluorescence profile function* $\mathscr{L}^F(\lambda)$:

$$F(\lambda, \lambda_l) = Q^F \mathscr{L}^F(\lambda) \tag{3.183}$$

In the case of an excited molecule in $|r\rangle$ many radiative and collisional decay modes could be available. However, if the only radiative process to yield emission at λ corresponds to the $|r\rangle$-to-$|f\rangle$ transition, then we can write

$$F(\lambda, \lambda_l) = \frac{A_{rf}}{\Gamma_R + \Gamma_Q} = \frac{\Gamma_R}{\Gamma_R + \Gamma_Q} \cdot \frac{A_{rf}}{\Gamma_R} \tag{3.184}$$

where clearly

$$Q^F \equiv \frac{\Gamma_R}{\Gamma_R + \Gamma_Q}$$

and

$$\mathscr{L}^F(\lambda) \equiv \frac{A_{rf}}{\Gamma_R}$$

represents the fraction of total emission in the band corresponding to the *rf* transition, from which it follows that

$$\Gamma_R \equiv \sum_{m<r} A_{rm} \tag{3.185}$$

and Γ_Q represents the collision quenching rate.

The quantum yield factor $F(\lambda, \lambda_l)$ can be seen to represent the probability that a photon of wavelength λ is emitted for each photon absorbed, and the fluorescence profile function $\mathscr{L}^F(\lambda)$ represents the fraction of the total fluorescence emitted into unit wavelength interval centered about λ. Thus $\int \mathscr{L}^F(\lambda)\, d\lambda = 1$, and the total fluorescence cross section can be expressed in the form

$$\sigma^F(\lambda_l) = \sigma^A(\lambda_l) Q^F \tag{3.186}$$

Furthermore since fluorescence is usually isotropic, the appropriate differential cross section per unit wavelength interval is

$$\frac{d\sigma^F(\lambda, \lambda_l)}{d\Omega} = \frac{1}{4\pi} \sigma^A(\lambda_l) Q^F \mathscr{L}^F(\lambda) \tag{3.187}$$

and the total fluorescence differential cross section is

$$\frac{d\sigma^F(\lambda_l)}{d\Omega} = \frac{1}{4\pi} \sigma^A(\lambda_l) Q^F \tag{3.188}$$

The quenching factor Q^F is given by the *Stern–Volmer relation*

$$Q^F = \frac{1}{1 + \Gamma_Q/\Gamma_R} = \frac{1}{1 + \sum_s a_s p_s} \tag{3.189}$$

where a_s represents the *quenching coefficients* for collisions between the radiating molecule and species s. p_s denotes the *partial pressure* of this species. Inaba (1976) has evaluated the total fluorescence differential cross section, the total absorption cross section, and the quenching factor for a few transitions of interest in several typical molecules at atmospheric pressure. These data are presented in Table 3.5.

It is evident that Raman cross sections are in general much smaller than the fluorescence cross sections (compare the values in Table 3.4 with those in Table 3.5). Furthermore, $\Gamma_Q \gg \Gamma_R$ and so $Q^F \ll 1$ under atmospheric conditions, so that from (3.186) it follows that invariably the spectrally integrated fluorescence cross section is much less than the corresponding absorption cross section:

$$\sigma^F(\lambda_l) \ll \sigma^A(\lambda_l)$$

Consequently, it might be thought that absorption would constitute the most sensitive technique for the detection of trace pollutants within the atmosphere. We shall see later that although this is true in principle, other considerations, such as absorption from other more abundant species or attenuation through aeroticulate scattering, complicate the situation.

TABLE 3.5. CALCULATED VALUES[a] OF TOTAL FLUORESCENCE DIFFERENTIAL CROSS SECTION $d\sigma^F(\lambda_l)/d\Omega$, TOTAL ABSORPTION CROSS SECTION $\sigma^A(\lambda_l)$, AND QUENCHING FACTOR Q^F

Molecule	Excitation Wavelength (nm)	Q^F	$\sigma^A(\lambda_l)$ (cm^2)	$d\sigma^F(\lambda_l)/d\Omega$ ($cm^2\ sr^{-1}$)
SO_2	290	1.2×10^{-5}	3.5×10^{-19}	3.2×10^{-25}
	300.1	4.9×10^{-5}	5.0×10^{-19}[b]	2.0×10^{-24}
NO_2	400	2.5×10^{-5}	2.8×10^{-19}	5.6×10^{-25}
	435.8	3.0×10^{-5}	3.0×10^{-19}	7.2×10^{-25}
I_2	589.5	1.5×10^{-3}	4.6×10^{-18}	6.1×10^{-22}
NO	226.5	$\approx 3 \times 10^{-3}$	1.3×10^{-18}	3×10^{-22}
OH	282.6	10^{-2}–10^{-3}	1.2×10^{-17}	10^{-20}–10^{-21}

[a]Under Atmospheric pressure (Inaba, 1976, Table 5.10).
[b]Recent measurements by Brassington (1981) indicate $\sigma^A \approx 10^{-18}\ cm^2$. For SO_2 at 300.1 nm, see Fig. 3.25.

TABLE 3.6. MEASURED ABSORPTION CROSS SECTIONS FOR GASEOUS SPECIES[a]

Molecule	$\mathscr{Y}$ (cm^{-1})	λ	$\sigma^A(\lambda_l)$ (10^{-18} cm^2)	$\kappa_A(\lambda)$ at STP [(ppm cm)$^{-1}$]
Acetylene, C_2H_2	719.9	13.89 μm	9.2	2.48×10^{-4}
Ammonia, NH_3	1084.6	9.220 μm	3.6	9.68×10^{-5}
Benzene, C_6H_6	1037.5	9.639 μm	0.09	2.42×10^{-6}
1,3-Butadiene, C_4H_6	1609.0	6.215 μm	0.27	7.26×10^{-6}
1-Butene, C_4H_8	927.0	10.787 μm	0.13	3.50×10^{-6}
Carbon monoxide, CO	2123.7	4.709 μm	2.8	7.53×10^{-5}
Carbon tetrachloride, CCl_4	793.0	12.610 μm	4.8	1.29×10^{-4}
Ethylene, C_2H_4	949.5	10.531 μm	1.34	3.60×10^{-5}
	950	10.526 μm	1.70	4.57×10^{-5}
Fluorocarbon-11,	847	11.806 μm	4.4	1.18×10^{-4}
CCl_3F (Freon-11)	1084.6	9.220 μm	1.24	3.34×10^{-5}
Fluorocarbon-12,	920.8	10.860 μm	11.0	2.96×10^{-4}
CCl_2F_2 (Freon-12)	923.0	10.834 μm	3.68	9.90×10^{-5}
Fluorocarbon-113, $C_2Cl_3F_3$	1041.2	9.604 μm	0.77	2.07×10^{-5}
Methane, CH_4	2948.7	3.391 μm	0.6	1.61×10^{-5}
	3057.7	3.270 μm	2.0	5.38×10^{-5}
Nitric oxide, NO	1900.1	5.265 μm	0.6	1.61×10^{-5}
	1917.5	5.215 μm	0.67	1.80×10^{-5}
Nitrogen dioxide,	1605.4	6.229 μm	2.68	7.21×10^{-5}
NO_2	22311.0	448.2 nm	0.2	5.38×10^{-6}
Ozone, O_3	1051.8	9.508 μm	0.9	2.42×10^{-5}
	1052.2	9.504 μm	0.56	1.51×10^{-5}
	39425.0	253.6 nm	12.0	3.23×10^{-4}
Perchloroethylene, C_2Cl_4	923.0	10.834 μm	1.14	3.07×10^{-5}
Propane, C_3H_8	2948.7	3.391 μm	0.8	2.15×10^{-5}
Propylene, C_3H_6	1647.7	6.069 μm	0.09	2.42×10^{-6}
Sulfur dioxide,	1108.2	9.024 μm	0.25	6.73×10^{-6}
SO_2	1126.0	8.880 μm	0.2	5.38×10^{-6}
	2499.1	4.001 μm	0.02[b]	5.38×10^{-7}
	33330.0	300.1 nm	1.0	2.69×10^{-5}
Trichloroethylene, C_2HCl_3	944.2	10.591 μm	0.56	1.51×10^{-5}
Vinyl chloride, C_2H_3Cl	940.0	10.638 μm	0.4	1.08×10^{-5}

[a] Based on the data presented by Hinkley et al. (1976).

[b] Recent measurements (Altmann and Pokrowsky, 1980) indicate an absorption cross section of 0.416×10^{-18} cm^2 for SO_2 at 3.9843 μm, corresponding to the $P_4(6)$ line of a DF laser.

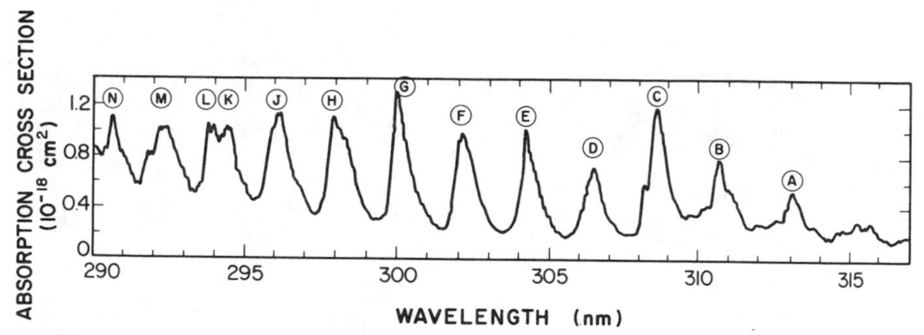

Fig. 3.25. SO_2 absorption cross section as a function of wavelength (Brassington, 1981).

An overview of the techniques and their relevant differential cross sections will be provided in Chapter 6. A selection of measured absorption cross sections and corresponding absorption coefficients are provided in Table 3.6. Note that the absorption coefficient in (ppm cm)$^{-1}$ under STP conditions is given by the expression

$$\kappa_A(\lambda) = N_L \sigma^A(\lambda) \times 10^{-6} \tag{3.190}$$

where $N_L = 2.69 \times 10^{19}$ cm^{-3} (Loschmidt's number) represents the number density of molecules in the atmosphere under STP conditions.

In differential-absorption work accurate measurements of the absorption cross section as a function of wavelength are required. Brassington (1981) has made such measurements in the case of SO_2 using a frequency-doubled dye-laser source with a 0.05-nm linewidth. His measurements extend from 290 to 317 nm and are reproduced as Fig. 3.25. Takeuchi et al. (1978) have provided the absorption cross section for NO_2 over the interval 430 to 455 nm, and their results are presented as Fig. 3.26. Browell et al. (1979) have used a

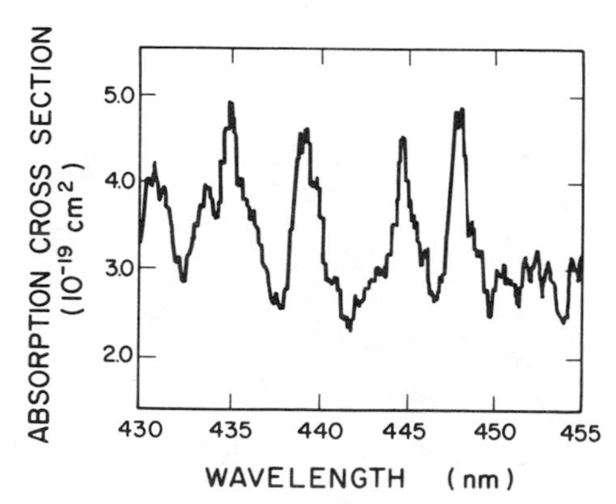

Fig. 3.26. NO_2 absorption cross section as a function of wavelength (Takeuchi et al., 1978).

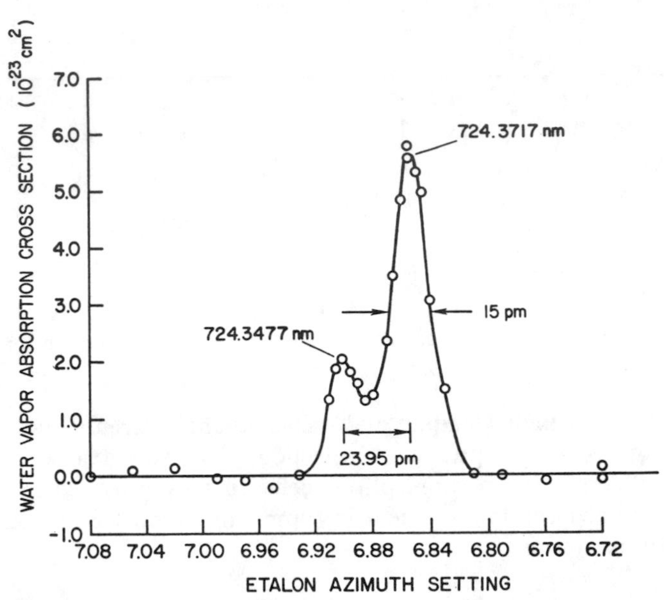

Fig. 3.27. H_2O absorption spectrum obtained with 300-m absorption cell and ruby-pumped dye laser (Browell et al., 1979).

0.008-nm-linewidth ruby-laser-pumped dye laser to study the 724.348- and 724.372-nm absorption lines of H_2O. Their results are included as Fig. 3.27. Additional absorption spectra for water vapor have been reported by Zuev (1976), and the Freon infrared absorption bands of concern with regard to the thermal balance in the atmosphere have been measured by Varansai and Ko (1977). Finally, the U.S. Air Force Geophysical Laboratory (AFGL) compilation of atmospheric absorption lines represents one of the most comprehensive sources of molecular absorption (Rothman, 1981).

4

Interaction and Propagation of Radiation

In the field of laser remote sensing the propagation of electromagnetic radiation and the attenuation characteristics of the intervening and target media are obviously important subjects. In this chapter we shall develop the basic radiative-transfer equation that can be used for describing the propagation of a collimated beam of radiation. Although initially we shall consider only the effects of processes such as absorption and spontaneous and stimulated emission, nevertheless, this will enable us to appreciate the significance of optical depth in regard to both the transmitting and target media. Besides being of interest to laser remote sensing, this work will also lay the foundations for our discussion of lasers.

Later in this chapter we shall extend our deliberations of radiative transfer to include the effects of scattering. The complexity of scattering is such that in the present context we shall only provide an introduction to the subject and indicate its relevance to laser remote sensing. Although, in general, scattering represents a serious impediment to observation, there are many situations where it provides important information. In probing the atmosphere single scattering often dominates; however, in the vicinity of clouds or when studying natural bodies of water such as the oceans, multiple scattering can be important and indeed limit the penetration depth of a probing laser beam.

4.1. RADIATIVE-TRANSFER EQUATION

We consider the interaction of an ensemble of molecules (or atoms) with electromagnetic radiation. We shall assume that the frequencies of this band of radiation lies close to the center frequency ν_0 of the transition between states

$|n\rangle$ and $|m\rangle$ of the molecules (or atoms). Continuity of energy requires

$$\frac{\partial}{\partial t}[\rho(\nu)\,d\nu] + \nabla \cdot [\rho(\nu)\mathbf{v}\,d\nu] \\ = h\nu N_n A_{nm} \mathcal{L}(\nu)\,d\nu + h\nu[N_n \mathsf{B}_{nm} - N_m \mathsf{B}_{mn}]\rho(\nu)\mathcal{L}(\nu)\,d\nu \tag{4.1}$$

where $\rho(\nu)\,d\nu$ represents the energy density of the radiation field in the interval $(\nu, \nu + d\nu)$, $h\nu$ the photon energy, and N_n and N_m the population densities in states $|n\rangle$ and $|m\rangle$ respectively. $\mathcal{L}(\nu)$ represents the line profile function, and A_{nm}, B_{nm}, and B_{mn} are the Einstein coefficients for spontaneous emission, stimulated emission, and absorption respectively.

The first term on the left-hand side of (4.1) takes account of the time rate of change of the energy density of the radiation field in the frequency interval $(\nu, \nu + d\nu)$, the second term allows for the flow of energy, and the three *source* and *sink* terms on the right-hand side of (4.1) take account of spontaneous emission, stimulated emission, and absorption respectively. For most of the work of interest in the field of laser remote sensing we can restrict our attention to the steady state. In addition it is often possible to confine ourselves to the propagation of radiation within a small solid angle $\Delta\Omega$ (chosen to be centered about the z-axis). Under these circumstances we can simplify equation (4.1) significantly to obtain

$$\frac{d}{dz}[c\,\Delta\rho(\nu)] = h\nu N_n A_{nm} \mathcal{L}(\nu)\frac{\Delta\Omega}{4\pi} + h\nu[N_n \mathsf{B}_{nm} - N_m \mathsf{B}_{mn}]\mathcal{L}(\nu)\,\Delta\rho(\nu) \tag{4.2}$$

where $\Delta\rho(\nu)$ represents the fraction of the spectral energy density that is confined to propagate within $\Delta\Omega$. The fraction of spontaneous emission that falls within the solid angle of the beam is accounted for by the factor $\Delta\Omega/4\pi$. If we introduce the spectral radiance [see (2.48)], then we can rewrite the *steady-state one-dimensional radiative-transfer equation* in the form

$$\frac{dJ(\nu)}{dz} = \frac{h\nu}{4\pi} N_n A_{nm} \mathcal{L}(\nu) + \frac{h\nu}{4\pi}[N_n B_{nm} - N_m B_{mn}] J(\nu) \mathcal{L}(\nu) \tag{4.3}$$

where we introduce the *Milne coefficients* for absorption and stimulated emission:

$$B_{nm} \equiv \frac{4\pi}{c}\mathsf{B}_{nm} \quad \text{and} \quad B_{mn} \equiv \frac{4\pi}{c}\mathsf{B}_{mn} \tag{4.4}$$

We shall also introduce the *volume emission coefficient* per steradian,

$$\epsilon(\nu, z) \equiv \frac{h\nu}{4\pi} N_n A_{nm} \mathcal{L}(\nu) \qquad (\mathrm{W\ cm^{-3}\ Hz^{-1}\ sr^{-1}}) \tag{4.5}$$

and the *volume absorption coefficient*

$$\kappa(\nu, z) \equiv \frac{h\nu}{4\pi}[N_m B_{mn} - N_n B_{nm}]\mathscr{L}(\nu) \qquad (\text{cm}^{-1}) \tag{4.6}$$

The latter is sometimes expressed in terms of the *absorption* cross section $\sigma^A(\nu) = (h\nu/4\pi)B_{mn}\mathscr{L}(\nu)$ and the *stimulated-emission* cross section $\sigma_S(\nu) = (h\nu/4\pi)B_{nm}\mathscr{L}(\nu)$.

The one-dimensional radiative-transfer equation can then be written

$$\frac{dJ(\nu, z)}{dz} = \epsilon(\nu, z) - \kappa(\nu, z)J(\nu, z) \tag{4.7}$$

Under most conditions of interest, including equilibrium, the *upper-level population density* N_n (viz., the population in the higher excited state) will be much less than *the lower-level population density* N_m. Since B_{nm} is usually of the same order of magnitude as B_{mn} [see equations (3.113) and (4.4)], it follows that $\kappa(\nu, z) > 0$ and we have net absorption of radiation. As we shall see later, it is possible to create a situation in which $N_n B_{nm} > N_m B_{mn}$ and consequently $\kappa(\nu, z) < 0$. Under these circumstances stimulated emission dominates absorption, and we shall see in Chapter 5 that the medium provides energy to the radiation field, thereby amplifying the beam rather than attenuating it.

4.2. OPTICAL-DEPTH CONSIDERATIONS

The solution of (4.7) is aided by the introduction of the *source function*

$$S(\nu, z) \equiv \frac{\epsilon(\nu, z)}{\kappa(\nu, z)} \tag{4.8}$$

and the *optical depth* $\tau(\nu, l)$ corresponding to a *path length* l and defined by the relation

$$\tau(\nu, l) \equiv \int_0^l \kappa(\nu, z)\, dz \tag{4.9a}$$

or

$$d\tau(\nu, z) = \kappa(\nu, z)\, dz \tag{4.9b}$$

Substitution of these quantities into the one-dimensional radiative transfer equation yields

$$\frac{dJ}{d\tau} = S - J \tag{4.10}$$

The solution of this equation is quite straightforward:

$$\frac{dJ}{d\tau} = S - J$$

$$\frac{dJ}{d\tau} + J = S$$

$$\frac{d}{d\tau}\{J\mathrm{e}^{\tau}\} = \mathrm{e}^{\tau}S$$

$$[J\mathrm{e}^{\tau}]_0^{\tau} = \int_0^{\tau} \mathrm{e}^{\tau}S\,d\tau$$

$$J(\tau)\mathrm{e}^{\tau} - J(0) = \int_0^{\tau} S(\tau)\mathrm{e}^{\tau}\,d\tau$$

and for a semiinfinite medium, occupying the $z \geq 0$ region of space, the spectral radiance at a depth ($z = l$) can be written

$$J(\nu, l) = J(\nu, 0)\mathrm{e}^{-\tau(\nu, l)} + \mathrm{e}^{-\tau(\nu, l)}\int_{\tau(\nu, 0)}^{\tau(\nu, l)} S(\nu, z)\mathrm{e}^{\tau(\nu, z)}\,d\tau(\nu, z) \quad (4.11)$$

where $J(\nu, 0)$ represents the spectral radiance incident upon the $z = 0$ interface.

This integral relation cannot always be solved in closed form, but several simplified situations are worth discussing, as they provide considerable physical insight into the nature of the solution and the influence of optical depth (thickness).

The most obvious simplification that can be made, and one that often approximates the real situation, stems from assuming a homogeneous medium. Under these conditions, the volume emission coefficient, the volume absorption coefficient, and therefore the source function all become independent of position:

$$S(\nu) = \frac{\epsilon(\nu)}{\kappa(\nu)} \quad (4.12)$$

and the optical depth becomes proportional to the actual depth of the medium:

$$\tau(\nu, l) = \kappa(\nu)l \quad (4.13)$$

Substitution of (4.12) and (4.13) into (4.11) leads to a closed-form solution of the form

$$J(\nu, l) = J(\nu, 0)\mathrm{e}^{-\kappa(\nu)l} + S(\nu)[1 - \mathrm{e}^{-\kappa(\nu)l}] \quad (4.14)$$

Basically the medium can either be *cold*, that is to say unexcited, so that $\epsilon(\nu) = 0$ and therefore $S(\nu) = 0$, or it can be *hot*, so that $\epsilon(\nu) \neq 0$. It is evident that the optical depth $\kappa(\nu)l$ represents the key parameter in determining the behavior of $J(\nu, l)$. If $\kappa(\nu)l \ll 1$, we say that the medium is *optically thin*, while if $\kappa(\nu)l \gtrsim 1$, we say it is *optically thick*.

For a cold medium that is irradiated we can set $S(\nu) = 0$ and $J(\nu, 0) \neq 0$. In this case we can write

$$J(\nu, z) = J(\nu, 0)e^{-\kappa(\nu)z} \tag{4.15}$$

which of course is just the Beer–Lambert law of exponential attenuation. For an optically thin absorbing medium and an incident spectral radiance that is independent of frequency over some bandwidth ($-\nu^*$ to ν^*), the absorption profile of the medium is clearly imprinted on the transmitted beam (see Fig. 4.1). As the penetration depth increases, the absorption profile burned into the propagating beam deviates from that of the medium. This spectral distortion associated with increasing optical depth leads to a flattening of the hole created in the spectral profile of the propagating beam. Clearly in Fig. 4.1 the medium is optically thick at a depth z_2 for the frequency range ($-\nu_1$ to ν_1).

In the case of a hot (self-luminous) medium, with no external illumination, $J(\nu, 0) = 0$ and $S(\nu) \neq 0$. Consequently, equation (4.14) becomes

$$J(\nu, z) = S(\nu)[1 - e^{-\kappa(\nu)z}] \tag{4.16}$$

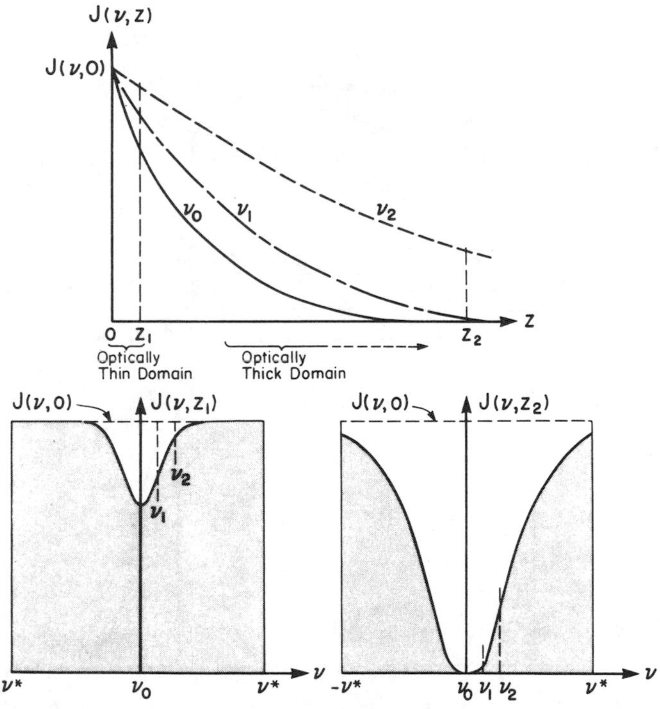

Fig. 4.1. Propagation and attenuation of radiation through a cold absorbing medium. The incident radiation is taken to be spectrally flat from $-\nu^*$ to ν^*.

If the medium can be regarded as optically thin over the frequency range of interest, then $\kappa(\nu)z \ll 1$ and expansion of the exponential factor leads to a relation that constitutes the foundation of analytical emission spectroscopy:

$$J(\nu, z) \approx S(\nu)[1 - \{1 - \kappa(\nu)z\}] \approx \epsilon(\nu)z \tag{4.17}$$

Using (4.5), we see that we can write

$$J(\nu, z) \approx \frac{h\nu}{4\pi} N_n A_{nm} \mathscr{L}(\nu)z \tag{4.18}$$

which indicates that the observed spectral radiance has the same spectral distribution as the radiating species [i.e., $J(\nu, z)$ is proportional to $\mathscr{L}(\nu)$; see Fig. 4.2]. In addition $J(\nu, z)$ is proportional to N_n.

As the optical depth of the hot (self-luminous) medium increases, the spectral distribution of the observed radiation starts to depart from that of the radiating species and a form of *saturation* sets in: the spectral radiance at the line center frequency ν_0 approaches the source-function value. With increasing optical thickness, $J(\nu, z)$ approaches $S(\nu)$ over a progressively wider band of frequencies. We have attempted to illustrate these features in Fig. 4.2.

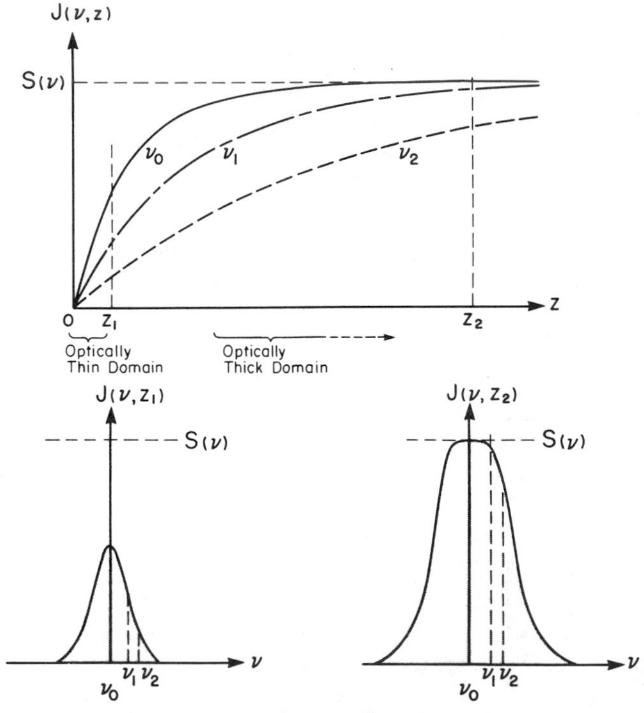

Fig. 4.2. Variation of the observed spectral radiance with the optical depth of the source. The two lower spectra reveal the distortion in the emitted spectral profile with increasing optical depth.

It might be noted from (4.5), (4.6), and (4.8) that the source function can be written in the form

$$S(\nu) \equiv \frac{A_{nm}}{B_{nm}} \cdot \frac{1}{\dfrac{N_m B_{mn}}{N_n B_{nm}} - 1} \tag{4.19}$$

If we take account of (3.112), (3.113), and (4.4), and in addition assume that the hot medium is in thermal equilibrium so that (3.111) applies, then we find that

$$S(\nu) = \frac{2h\nu^3}{c^2} \frac{1}{e^{h\nu/kT} - 1} \equiv B_\nu(T) \tag{4.20}$$

the *Planck blackbody function*. Note that $B_\nu(T) = (c/4\pi)\rho(\nu)_{\text{Planck}}$ and so represents the blackbody spectral radiance, where $\rho(\nu)_{\text{Planck}}$ is the blackbody spectral energy density and is given by (3.2).

Most sources of luminosity, including the sun, are nonuniform, with a temperature that decreases towards the boundary. Consequently, as the radiation from the hot interior passes through the cooler outer regions it can suffer some degree of absorption. This can lead to an effect known as *self-reversal*, in which a dip appears in the center of the line profile of the emitted radiation. In the case of the sun, absorption lines from certain elements located in the outer zones show up in the continuous solar spectrum as dark (*Fraunhofer*) lines.

4.3. LASER-INDUCED FLUORESCENCE AND SATURATION EFFECTS

The development of the radiative-transfer equation to this point has not taken into account the influence of the radiation field upon the population densities of the atomic (or molecular) states of the medium through which it passes. Clearly, if the radiation is suitably tuned, absorption will occur, and this will lead to both fluorescence and attenuation of the beam. Those atoms (or molecules) which are excited can no longer absorb the radiation, but can in fact contribute energy back to the radiation field, either randomly through spontaneous emission or coherently through stimulated emission. The attenuation of the medium will consequently be affected by the radiation field. In this section we shall examine this coupling of the radiation field with the medium through which it passes. We shall see that as the strength of the radiation field increases, it will affect both the transmission and fluorescence properties of the medium.

If we assume that we can represent the atoms (or molecules) of the medium by a simple two-level system and that initially they all reside in the lower level (the ground state), then the temporal variation of the excited-state population density $N_2(t)$ can be described by a rate equation of the form

$$\frac{dN_2(t)}{dt} = N_1(t)R_{12}(t) - N_2(t)\left[R_{21}(t) + \frac{1}{\tau_{21}}\right] \tag{4.21}$$

where $N_1(t)$ represents the population density of the lower level and τ_{21} represents the lifetime of the excited state in the absence of the radiation field. This can be less than the radiative lifetime τ_{21}^R ($\equiv A_{21}^{-1}$) if collisional effects are significant. The rate of absorption per atom (or molecule),

$$R_{12}(t) \equiv \mathsf{B}_{12}\int_{-\infty}^{\infty} \Delta\rho^L(\nu, t)\mathscr{L}(\nu)\, d\nu \tag{4.22}$$

follows from the equations (3.108) if we assume that the medium is exposed to a collimated beam of laser radiation, $\mathscr{L}(\nu)$ represents the line profile function of the constituents of the medium, and $\Delta\rho^L(\nu, t)$ represents the spectral energy density in the beam. If we introduce the laser's spectral irradiance

$$I^L(\nu, t) \equiv c\,\Delta\rho^L(\nu, t) \tag{4.23}$$

and use (4.4), then we can write

$$R_{12}(t) = \frac{B_{12}}{4\pi}\int_{-\infty}^{\infty} I^L(\nu, t)\mathscr{L}(\nu)\, d\nu \tag{4.24}$$

In a similar manner the rate of stimulated emission per atom (or molecule) can be written

$$R_{21}(t) = \frac{B_{21}}{4\pi}\int_{-\infty}^{\infty} I^L(\nu, t)\mathscr{L}(\nu)\, d\nu \tag{4.25}$$

and from (3.113), we see that

$$R_{12} = gR_{21} \tag{4.26}$$

where $g \equiv g_2/g_1$, the ratio of the degeneracies for the upper and lower levels, respectively. If we adhere strictly to a two-level system, continuity requires

$$N_1(t) + N_2(t) = N_0 \tag{4.27}$$

where N_0 is the number density of the species (essentially all in the ground state) prior to irradiation. If we eliminate $N_1(t)$ from (4.21) by means of (4.27), we can write

$$\frac{dN_2(t)}{dt} = N_0 R_{12}(t) - \frac{N_2(t)}{\tau} \tag{4.28}$$

where

$$\frac{1}{\tau} \equiv (1+g)R_{21}(t) + \frac{1}{\tau_{21}} \tag{4.29}$$

and can be interpreted as the effective lifetime of the excited state in the presence of the radiation field.

In any real situation the laser pulse will be attenuated as it propagates through the medium. Consequently, equation (4.28) will have to be applied at each position z along the path of the laser beam. The relevant pair of coupled equations needed to fully describe the one-dimensional propagation of a laser pulse through an absorbing medium are

$$\frac{\partial N_2(z,t)}{\partial t} = N_0 \frac{B_{12}}{4\pi} \int_{-\infty}^{\infty} I^L(z,t,\nu)\mathscr{L}(\nu)\,d\nu - N_2(z,t)\left[(1+g)\frac{B_{21}}{4\pi}\int_{-\infty}^{\infty} I^L(z,t,\nu)\mathscr{L}(\nu)\,d\nu + \frac{1}{\tau_{21}}\right] \tag{4.30}$$

and

$$\frac{\partial I^L(z,t,\nu)}{\partial z} = \frac{1}{4\pi}\left[N_2(z,t) - gN_1(z,t)\right]B_{21}I^L(z,t,\nu)\mathscr{L}(\nu)\,h\nu \tag{4.31}$$

This pair of equations have been studied by Measures and Herchen (1983). In reality, however, they only apply strictly along the axis of the laser beam, since lateral variations have been neglected. If we approximate the temporal variation of the laser pulse by a rectangular function and also assume for the moment that its shape is independent of location, then we can write

$$I^L(z,t,\nu) = \begin{cases} I^L(\nu) & \text{for} \quad 0 < t \le \tau_L \\ 0 & \text{for} \quad t > \tau_L \end{cases} \tag{4.32}$$

Under these circumstances (4.30) simplifies to a form somewhat like (4.28) and we can write

$$\frac{dN_2(t)}{dt} = N_0 R_{12} - \frac{N_2(t)}{\tau} \tag{4.33}$$

This has the simple solution

$$N_2(t) = N_2(0)e^{-t/\tau} + N_0 R_{12}\tau\{1 - e^{-t/\tau}\} \tag{4.34}$$

If there were no excitation within the medium prior to the arrival of the laser pulse, then $N_2(0) = 0$, and

$$N_2(t) = \frac{N_0 R_{12}\tau_{21}}{1 + S_I}\{1 - e^{-t(1+S_I)/\tau_{21}}\} \tag{4.35}$$

where we have introduced the *saturation parameter*

$$S_I = (1 + g)R_{21}\tau_{21} \tag{4.36}$$

into equation (4.29).

In remote laser sensing there are two limiting situations that are often useful. In the case of differential absorption the laser bandwidth is usually much less than that of the absorption line. For this case we approximate the laser spectral distribution by a delta function centered at the absorption-line center frequency ν_0:

$$I^L(\nu) = I^L\delta(\nu - \nu_0) \tag{4.37}$$

Here I^L represents the laser irradiance. Under these circumstances†

$$R_{21} = \frac{B_{21} I^L \mathscr{L}(\nu_0)}{4\pi} \tag{4.38}$$

and the saturation parameter is

$$S_I = \frac{(1 + g)B_{21}\mathscr{L}(\nu_0)\tau_{21} I^L}{4\pi} \tag{4.39}$$

which can be expressed in the form of a ratio of the laser irradiance to the *saturated irradiance* I^S defined by the relation

$$I^S = \frac{4\pi}{(1 + g)B_{21}\mathscr{L}(\nu_0)\tau_{21}} = I^S(\nu)\,\delta\nu \tag{4.40}$$

where $\delta\nu \equiv \mathscr{L}^{-1}(\nu_0)$ represents the effective bandwidth of the absorption line, and $I^S(\nu)$, *saturated spectral irradiance*, is given by

$$I^S(\nu) = \frac{8\pi h\nu^3}{(1 + g)c^2}\left(\frac{\tau_{21}^R}{\tau_{21}}\right) \tag{4.41}$$

Thus for a narrowband laser

$$S_I = \frac{I^L}{I^S} \tag{4.42}$$

† $\int_{-\infty}^{\infty}\delta(\nu - \nu_0)\mathscr{L}(\nu)\,d\nu = \mathscr{L}(\nu_0)$.

The other limiting situation often arises in the case of laser-induced fluorescence, where the laser bandwidth can be greater than that of the absorption line. In this instance we assume that the laser is so broadband that we can use the approximation

$$R_{21} = \frac{B_{21} I^L(\nu)}{4\pi} \tag{4.43}$$

and under these conditions†

$$S_I = \frac{I^L(\nu)}{I^S(\nu)} \tag{4.44}$$

It is also apparent from (4.26) that

$$R_{12}\tau_{21} = GS_I \tag{4.45}$$

where

$$G = \frac{g}{1+g} \tag{4.46}$$

It follows that for either of these two limiting situations

$$N_2(t) = GN_0\left(\frac{S_I}{1+S_I}\right)\{1 - e^{-t(1+S_I)/\tau_{21}}\} \tag{4.47}$$

At the end of the period of illumination the excited states decay with the original lifetime τ_{21}, and the excited-state rate equation simplifies to the form

$$\frac{dN_2(t)}{dt} = -\frac{N_2(t)}{\tau_{21}} \tag{4.48}$$

The solution for $t > \tau_L$ takes the form

$$N_2(t) = N_2(\tau_L)e^{-(t-\tau_L)/\tau_{21}} \tag{4.49}$$

and is subject to the condition that the solutions (4.47) and (4.49) match at the end of the laser pulse, namely,

$$N_2(\tau_L) = GN_0\left(\frac{S_I}{1+S_I}\right)\{1 - e^{-\tau_L(1+S_I)/\tau_{21}}\} \tag{4.50}$$

The temporal variation of the excited-state density, normalized by GN_0, for several values of the saturation parameter S_I and for $\tau_L = \tau_{21}$, are presented as

†Here we have assumed $\int_{-\infty}^{\infty} I^L(\nu)\mathscr{L}(\nu)\,d\nu \approx I^L(\nu)\int_{-\infty}^{\infty}\mathscr{L}(\nu)\,d\nu \approx I^L(\nu)$.

Fig. 4.3. It is quite apparent that for large values of S_I, the excited-state density attains a maximum value that remains constant for the residual period of laser excitation. It is also evident that this maximum value of N_2 tends to the value GN_0 as S_I increases. This behavior is even more apparent from the top right insert of Fig. 4.3, where the normalized peak excited-state density is plotted against the saturation parameter for $\tau_L = \tau_{21}$.

Measures (1968) was the first to show that under conditions of saturation $(S_I \gg 1) N_2^{\max} \rightarrow GN_0$, and that this occurs in a time

$$\tau_s = \frac{1}{(1+g)R_{21}} \tag{4.51}$$

which from the definition of τ [equation (4.29)] amounts to stating that

$$\frac{1}{\tau} = \frac{1}{\tau_s} + \frac{1}{\tau_{21}} \tag{4.52}$$

The saturation condition, $N_2^{\max} = GN_0$, corresponds to *radiative balance*,

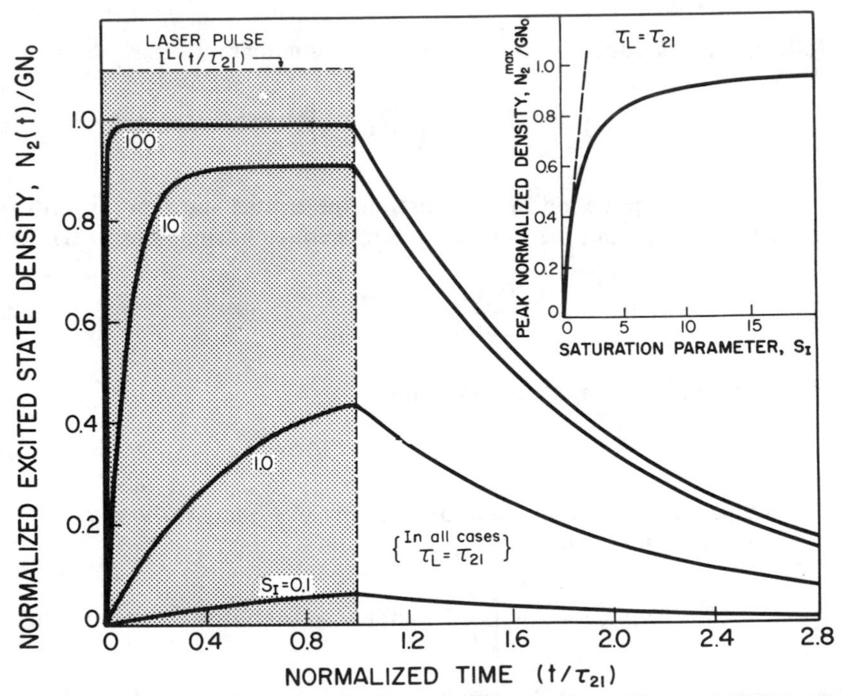

Fig. 4.3. Variation with time of the normalized excited-state density for atoms irradiated by a suitably tuned rectangular-shaped laser pulse of duration τ_L for four values of the saturation parameter S_I. The insert indicates the variation of the peak normalized excited-state density with the saturation parameter.

wherein the volume rate of stimulated emission balances the volume rate of absorption and the excited-state and ground-state populations are locked in the ratio of their degeneracies:

$$\frac{N_2}{N_1} = g \tag{4.53}$$

This could be interpreted as an infinite-temperature distribution if we assumed that Boltzmann statistics applied.

The total spontaneous emission (fluorescence) per cm³ per steradian from the excited atoms (or molecules) can be obtained from (4.47) and (4.49):

$$\begin{aligned}\int_0^\infty \frac{h\nu N_2(t)}{4\pi\tau_{21}^R} dt &= \frac{h\nu G N_0 S_I}{4\pi\tau_{21}^R(1 + S_I)} \\ &\quad \times \left[\int_0^{\tau_L} (1 - e^{-t/\tau})\, dt + e^{\tau_L/\tau_{21}}(1 - e^{-\tau_L/\tau}) \int_{\tau_L}^\infty e^{-t/\tau_{21}}\, dt\right] \\ &= \frac{h\nu G N_0 S_I \tau_{21}}{4\pi\tau_{21}^R(1 + S_I)} \left[\frac{\tau_L}{\tau_{21}} + \left(\frac{S_I}{1 + S_I}\right)\{1 - e^{-\tau_L(1+S_I)/\tau_{21}}\}\right]\end{aligned} \tag{4.54}$$

If we divide this fluorescence by its low-intensity (nonsaturating) limit, we obtain the normalized fluorescence coefficient

$$\mathscr{I}\left(\frac{\tau_L}{\tau_{21}}, S_I\right) = \frac{\dfrac{\tau_L}{\tau_{21}} + \left(\dfrac{S_I}{1 + S_I}\right)\{1 - e^{-(\tau_L/\tau_{21})(1+S_I)}\}}{(\tau_L/\tau_{21})(1 + S_I)} \tag{4.55}$$

This might better be termed the *saturation correction factor*, for it can be used to allow for the reduction in the growth of fluorescence with increasing laser irradiance. The variation of this saturation correction factor with increasing saturation parameter for three values of the ratio τ_L/τ_{21} is shown in Fig. 4.4. The decrease of $\mathscr{I}(\tau_L/\tau_{21}, S_I)$ with increasing S_I (or laser irradiance) can be understood in terms of stimulated emission limiting (or saturating) the excited-state population density. It is evident from this figure that the longer the duration of the laser pulse (in terms of the excited-state lifetime), the smaller the value of the saturation correction factor.

We return now to consider the propagation of the laser pulse through an absorbing medium. We derived earlier the one-dimensional (i.e., plane-wave approximation) radiative-transfer equation (4.31) appropriate to this situation. If we spectrally integrate this equation and take account of continuity, then we can write

$$\frac{\partial I^L(z,t)}{\partial z} = [N_2(z,t) - GN_0]h\nu(1 + g)R_{21} \tag{4.56}$$

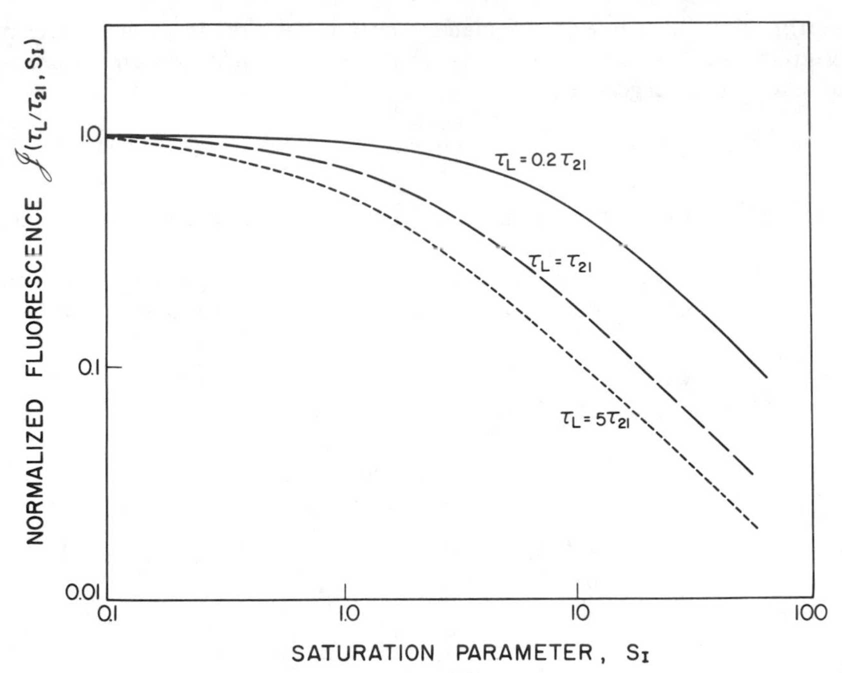

Fig. 4.4. Variation of the normalized fluorescence (sometimes termed saturation correction factor) $\mathscr{J}(\tau_L/\tau_{21}, S_I)$ with the saturation parameter S_I, for three ratios of τ_L/τ_{21} (0.2, 1, 5).

For $\tau_L > t > \tau_{21}/(1 + S_I)$, we can use the steady-state solution

$$N_2(z) = \frac{GN_0 S_I}{1 + S_I} \tag{4.57}$$

to yield

$$\frac{dS_I}{dz} = -\frac{GN_0 h\nu}{I^S \tau_{21}}\left(\frac{S_I}{1 + S_I}\right) \tag{4.58}$$

The solution of this equation was shown by Gires and Combaud (1965) to take the form

$$S_I(z) + \ln S_I(z) = S_I(0) + \ln S_I(0) - GN_0\sigma_{21}^T z \tag{4.59}$$

where we have introduced the total radiative cross section at the line center frequency [see (3.174)],

$$\sigma_{21}^T = \frac{h\nu \mathscr{L}(\nu_0)[B_{12} + B_{21}]}{4\pi} \tag{4.60}$$

There are two limiting forms of (4.59). In the weak-field limit, where $S_I \ll 1$, equation (4.59) tends to the Beer–Lambert attenuation equation

$$I^L(z) = I^L(0)\mathrm{e}^{-GN_0\sigma_{21}^T z} \tag{4.61}$$

while in the strong-field (saturation, $S_I \gg 1$) limit it tends to the linear attenuation form

$$I^L(z) = I^L(0) - GN_0\sigma_{21}^T I^S z \tag{4.62}$$

first proposed by the author (Measures, 1970).

In summary, we see that coupling the radiation field to the population rate equation leads to fluorescence and attenuation. If the laser radiation field is weak, the resulting laser-induced fluorescence is proportional to the laser irradiance and the beam is attenuated exponentially. On the other hand, a strong laser beam can saturate the fluorescence and suffer only linear attenuation as a consequence of *bleaching* the absorbing species. A more detailed discussion of this bleaching phenomenon in the case of a molecular gas has been provided by Douglas-Hamilton (1978).

4.4. RADIATIVE-TRANSFER EQUATION WITH SCATTERING

To this point we have considered the propagation of a collimated beam of electromagnetic radiation through an absorbing but nonscattering medium. However, in any real medium attenuation can result from scattering in addition to absorption. Scattering considerably complicates the radiative transfer problem (makes it hazy) by introducing unwanted radiation into the propagating beam. The theory of radiative transfer has been treated in many texts, for example Chandrasekhar (1960), Pomraning (1973), and Zuev (1982).

In a scattering interaction a photon with energy $h\nu$ and momentum $\hbar\mathbf{k}$, described by $(\nu, \mathbf{k})$, is replaced by one described by $(\nu^*, \mathbf{k}^*)$. We can think of this as a change in the characteristics of the photon (rather than its annihilation and the creation of another), namely,

$$\nu \to \nu^* \quad \text{and/or} \quad \mathbf{k} \to \mathbf{k}^*$$

where $\mathbf{k}^* = (2\pi/\lambda^*)\hat{\mathbf{k}}^*$. If $\nu^* = \nu$, then we speak of *elastic* scattering and there is only a change in the direction of propagation.

In order to cope with this interaction we introduce the *differential scattering coefficient* $\beta_s(\nu, \mathbf{k} : \nu^*, \mathbf{k}^*)$, defined as the probability per unit distance traveled that a photon $(\nu, \mathbf{k})$ is scattered from ν into the band of frequencies $d\nu^*$, centered about ν^*, and from $\mathbf{k}$ into $d\Omega^*$, centered about $\mathbf{k}^*$. If the z-axis of a Cartesian coordinate system is chosen to coincide with $\mathbf{k}$, and θ is the scattering angle, then

$$\hat{\mathbf{k}}^* \cdot \hat{\mathbf{k}} = \cos\theta$$

and

$$d\Omega^* = \sin\theta \, d\theta \, d\phi \tag{4.63}$$

(see Fig. 4.5).

The *total* scattering coefficient

$$\beta_s(\nu) \equiv \int_0^\infty d\nu^* \int_{4\pi} d\Omega^* \, \beta_s(\nu, \mathbf{k} : \nu^*, \mathbf{k}^*) \tag{4.64}$$

represents the probability per unit distance traveled that a photon $(\nu, \mathbf{k})$ will suffer a change of frequency and direction. Then the rate of scattering out of the beam for photons of frequency ν, per unit volume, is

$$S_0(\nu) \equiv \int_{\Delta\Omega} d\Omega \int_0^\infty d\nu^* \int_{4\pi} d\Omega^* \, \beta_s(\nu, \mathbf{k} : \nu^*, \mathbf{k}^*) f(\nu, \mathbf{k}) c \tag{4.65}$$

where $f(\nu, \mathbf{k})$ represents the photon distribution function—that is to say, the number density of photons of frequency ν, per unit frequency interval, and propagating within unit solid angle centered about the **k**-direction (z-axis). Then the total number density of photons within the frequency interval $(\nu, \nu + d\nu)$ within the beam of solid angle $\Delta\Omega$ is

$$\Phi(\nu) \, d\nu = \int_{\Delta\Omega} f(\nu, \mathbf{k}) \, d\Omega \, d\nu = \frac{\Delta\rho(\nu) \, d\nu}{h\nu} \tag{4.66}$$

Consequently, the spectral radiance within the beam is given by

$$J(\nu, \mathbf{k}) = ch\nu f(\nu, \mathbf{k}) \tag{4.67}$$

and the spectral irradiance of the beam by

$$I(\nu) = \int_{\Delta\Omega} J(\nu, \mathbf{k}) \, d\Omega = ch\nu \int_{\Delta\Omega} f(\nu, \mathbf{k}) \, d\Omega \tag{4.68}$$

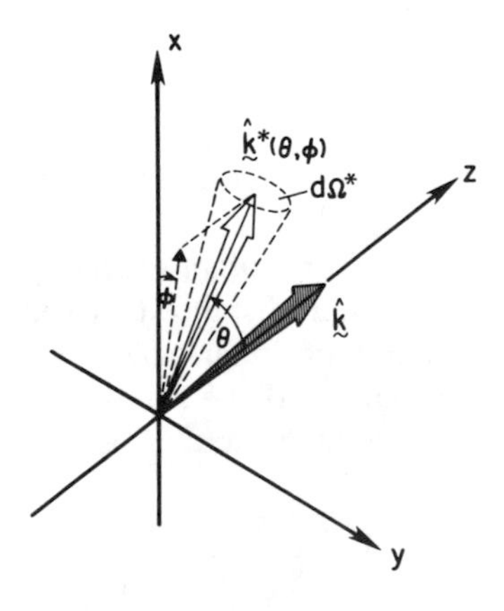

Fig. 4.5. Scattering diagrams for $\mathbf{k} \rightarrow \mathbf{k}^*$.

Under these circumstances the *outscattering rate* per unit volume can be written in the form

$$S_0(\nu) = \frac{1}{h\nu} \int_{\Delta\Omega} d\Omega \int_0^\infty d\nu^* \int_{4\pi} d\Omega^* \, \beta_s(\nu, \mathbf{k} : \nu^*, \mathbf{k}^*) J(\nu, \mathbf{k}) \tag{4.69}$$

In a similar manner we can write

$$S_i(\nu) = \frac{1}{h\nu} \int_{\Delta\Omega} d\Omega \int_0^\infty d\nu^* \int_{4\pi} d\Omega^* \, \beta_s(\nu^*, \mathbf{k}^* : \nu, \mathbf{k}) J(\nu^*, \mathbf{k}^*) \frac{\nu}{\nu^*} \tag{4.70}$$

for the *inscattering rate* per unit volume, where $\beta_s(\nu^*, \mathbf{k}^* : \nu, \mathbf{k})$ represents the probability per unit distance traveled that a $(\nu^*, \mathbf{k}^*)$ photon will suffer a scattering event and be replaced with a $(\nu, \mathbf{k})$ photon.

If we now take account of these additional scattering sources and sinks for photons propagating in the beam (i.e., confined within $\Delta\Omega$), we can rewrite the energy conservation equation (4.2) in the form

$$\frac{d}{dz}\{c \, \Delta\rho(\nu)\} = \epsilon(\nu) \, \Delta\Omega - \beta_A(\nu) \, \Delta\rho(\nu) c - h\nu S_0(\nu) + h\nu S_i(\nu) \tag{4.71}$$

where $\Delta\rho(\nu) \, d\nu$ represents the fraction of the radiant energy density that is propagating within $\Delta\Omega$ and lies in the frequency interval $(\nu, \nu + d\nu)$. Also

$$\beta_A(\nu) = \frac{h\nu}{c} [N_m \mathrm{B}_{mn} - N_n \mathrm{B}_{nm}] \mathscr{L}(\nu) = \kappa_A(\nu) \tag{4.72}$$

where $\kappa_A(\nu)$ is the volume absorption coefficient as defined in (4.6).

Since $J(\nu, \mathbf{k})$ is independent of both ν^* and $\mathbf{k}^*$, equation (4.69) can be rewritten in the form

$$S_0(\nu) = \frac{1}{h\nu} \int_0^\infty d\nu^* \int_{4\pi} d\Omega^* \, \beta_s(\nu, \mathbf{k} : \nu^*, \mathbf{k}^*) \int_{\Delta\Omega} J(\nu, \mathbf{k}) \, d\Omega \tag{4.73}$$

From the definition (4.64) of the total outscattering coefficient $\beta_s(\nu)$ we see that

$$h\nu S_0(\nu) = \beta_s(\nu) \int_{\Delta\Omega} J(\nu, \mathbf{k}) \, d\Omega = \beta_s(\nu) \, \Delta\rho(\nu) c \tag{4.74}$$

Substituting (4.74) into (4.71) yields

$$\frac{d}{dz}\{c \, \Delta\rho(\nu)\} = \epsilon(\nu) \, \Delta\Omega - c\kappa_\epsilon(\nu) \, \Delta\rho(\nu) + h\nu S_i(\nu) \tag{4.75}$$

where we have introduced the *total attenuation* (or *extinction*) coefficient

$$\kappa_\epsilon(\nu) \equiv \beta_A(\nu) + \beta_s(\nu) \tag{4.76}$$

The integrodifferential equation (4.75) is very difficult to solve even for highly simplified situations and requires spatial and boundary conditions. For a well-confined beam propagating in the z-direction, the steady-state radiative transfer equation, with scattering, takes the form†

$$\frac{dI(\nu)}{dz} = \epsilon(\nu)\,\Delta\Omega - \kappa_\epsilon(\nu)I(\nu) + \int_{\Delta\Omega} d\Omega \int_0^\infty d\nu^* \int_{4\pi} d\Omega^*\, \beta_s(\nu^*, \mathbf{k}^* : \nu, \mathbf{k}) J(\nu^*, \mathbf{k}^*) \frac{\nu}{\nu^*} \quad (4.77)$$

where

$$I(\nu) \equiv \int_{\Delta\Omega} J(\nu, \mathbf{k})\, d\Omega = c\,\Delta\rho(\nu) \quad (4.78)$$

represents the spectral irradiance of the beam.

This form of the radiative-transfer equation will be relevant when we consider the propagation of laser beams through the atmosphere under low-visibility (hazy) conditions or in natural bodies of water. Under these circumstances attenuation due to scattering can be quite severe. Indeed, in the case of laser bathymetry (water depth measurements) scattering limits the depth that can be monitored in most situations to about 10 to 15 monopath (single-scattering) attenuation lengths, (Duntley, 1963; Hickman, 1973; Levis et al., 1973). Multiple scattering further complicates the radiative-transfer equation to an extent that is beyond the scope of the present treatment (Pomraning, 1973). It has also been shown to lead to enhanced rotational Raman scattering near the base of clouds (Cohen et al., 1978). Tam and Zardecki (1980, 1982) have considered multiple-scattering corrections to the Beer–Lambert law by means of a rigorous small-angle solution of the radiative-transfer equation. They present numerical algorithms and results relating to the multiple-scattering effects on laser propagation in fog, cloud, and rain.

For a reasonably well-collimated beam under clear weather conditions, the contributions of spontaneous emission and inscattering are often negligible and so we can write

$$\frac{dI(\nu)}{dz} = -\kappa_\epsilon(\nu)I(\nu) \quad (4.79)$$

We see that scattering contributes to the attenuation of a propagating beam in the same way as absorption, leading again to the Beer–Lambert law

$$I(\nu, l) = I(\nu, 0)\mathrm{e}^{-\int_0^l \kappa_\epsilon(\nu, z)\, dz} \quad (4.80)$$

4.5. PROPAGATION THROUGH THE ATMOSPHERE

In the case of a collimated beam of electromagnetic radiation propagating through the atmosphere, equation (4.80) applies, where in this instance

$$\kappa_\epsilon(\nu) = \sum_i \{\kappa_E^i(\nu) + \kappa_R^i(\nu) + \kappa_A^i(\nu)\} + \kappa_M(\nu) \quad (4.81)$$

† Note that the explicit z-dependence of the variables in this equation has been omitted for the sake of brevity.

The sum extends over each of the atmospheric constituents, and $\kappa_E(\nu)$, $\kappa_R(\nu)$, $\kappa_A(\nu)$, and $\kappa_M(\nu)$ represents the elastic (Rayleigh), Raman, absorption, and Mie volume attenuation coefficients, respectively. In regard to attenuation, the contribution arising from inelastic (Raman) scattering is negligible. When the wavelength of the radiation coincides with that of a relatively strong absorption line or band of even a minor constituent of the atmosphere, $\kappa_A(\nu)$ can dominate and appreciable attenuation result.

For wavelengths less than about 200 nm, the atmosphere is totally opaque as a result of the Schumann–Runge bands of molecular oxygen (O_2). The absorption due to O_2 decreases with increasing wavelength so that beyond 250 nm it is unimportant and likely to be exceeded by the effect of small quantities of ozone (O_3) (Green, 1966). In the infrared part of the spectrum many atmospheric constituents contribute to the absorption, leaving only a few spectral windows through which optical probing is possible (Smith et al., 1968; Gebbie et al., 1951). Reference to Fig. 4.6 reveals that water vapor (H_2O) and carbon dioxide (CO_2) are the principal absorbers in the unpolluted atmosphere. Between 300 nm and 1 μm there are few absorption bands, and under clear sky conditions it is Rayleigh–Mie scattering that determines the attenuation characteristics of the atmosphere in this portion of the spectrum.

The electronic absorption bands of most molecules, other than O_3, SO_2 and NO_2, lie in the far ultraviolet at wavelength below 185 nm. The most intense and broad vibration–rotation band of *water vapor* is centered at about 6.27 μm and completely absorbs electromagnetic radiation in an interval that extends from 5.5 μm to around 7.5 μm. Other vibration–rotation band centers are at 2.73, 2.66, 1.87, 1.38, 1.10, 0.94, 0.81, and 0.72 μm. The large dipole moments of the water molecule and its isotopes give rise to an intense rotational spectrum that runs from about 8 μm through to the far infrared.

The weak far-red absorption bands of water vapor overlap the range of frequencies attainable from a ruby laser and are particularly useful for remote meteorological studies. The fine structure of part of this band is shown in Fig. 4.7.

The *carbon dioxide* molecule has two main infrared absorption bands centered at about 4.3 and 15 μm (the third is optically inactive). In addition to the main bands, CO_2 has overtone combination bands and hot bands with centers near 10.4, 9.4, 5.2, 4.3, 2.7, 2.0, 1.6, and 1.4 μm. The strong absorption observed beyond 14 μm is also primarily due to CO_2.

Electronic transitions in the *ozone* molecule produce Hartley and Huggins bands located in the ultraviolet at wavelengths shorter than 340 nm. Weaker Chappius bands are found between 450 and 740 nm. The three main vibration–rotation bands of O_3 have centers at 9.0, 14.1, and 9.6 μm, with additional weaker bands at 5.75, 4.75, 3.59, 3.27, and 2.7 μm.

In the case of a polyatomic molecule, there are many different modes of vibration, and clearly this can lead to a complex band spectrum. Each nucleus of a free atom has three degrees of freedom, which is reflected in the total translational energy, $3kT/2$ of such an atom, since $kT/2$ represents the

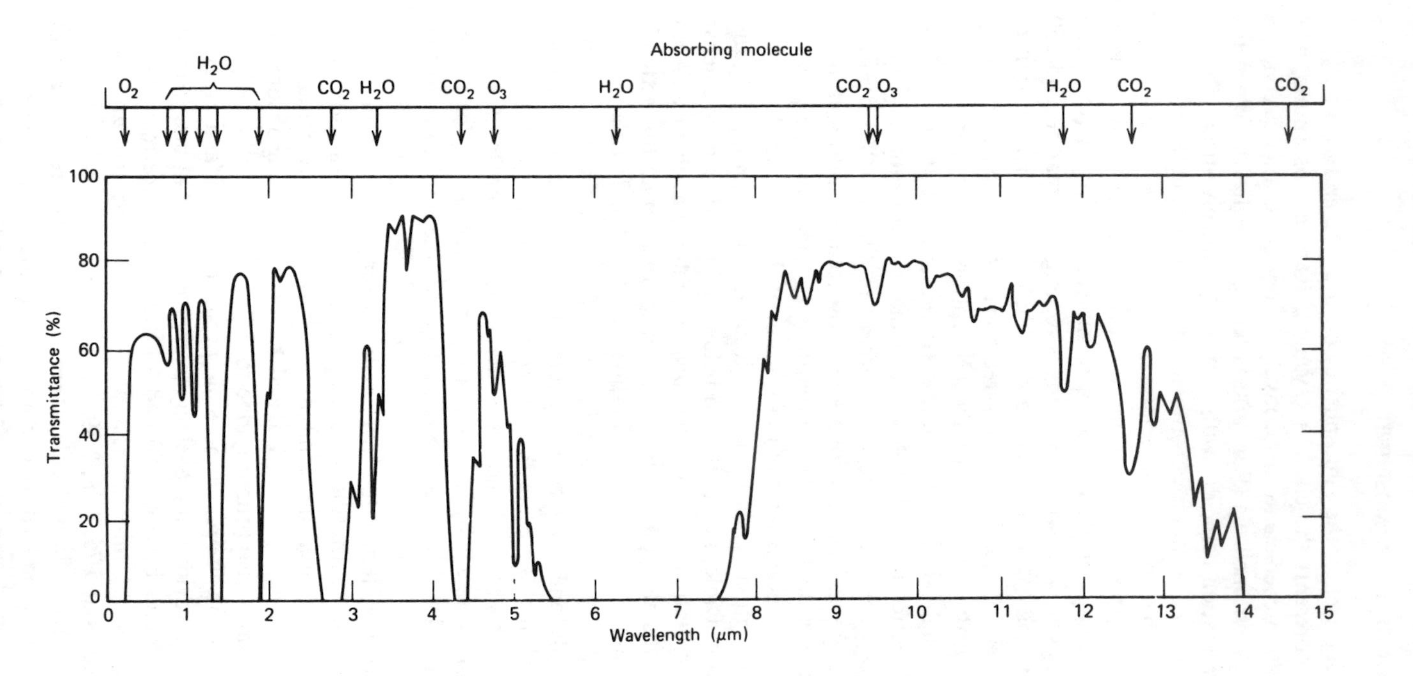

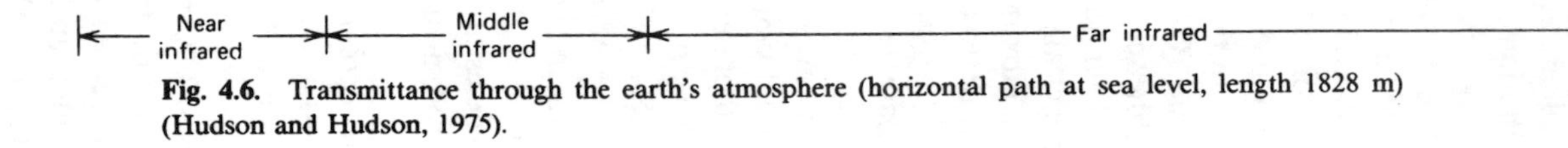

Fig. 4.6. Transmittance through the earth's atmosphere (horizontal path at sea level, length 1828 m) (Hudson and Hudson, 1975).

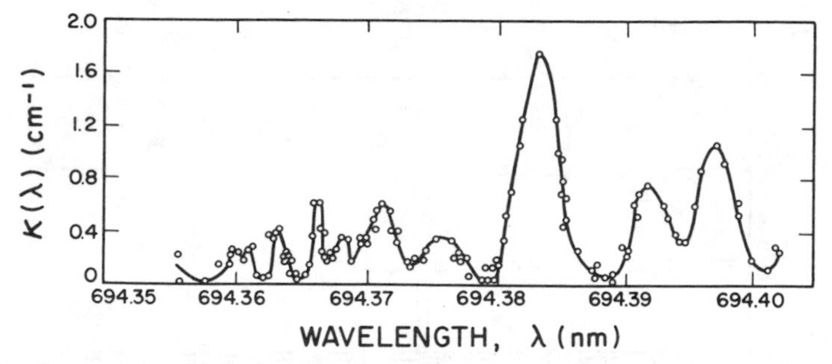

Fig. 4.7. Fine structure of the water-vapor absorption spectrum over a 0.7-cm^{-1}-wide region, obtained with a laser spectrometer having very high resolution (Zuev, 1976).

equilibrium mean thermal energy associated with each degree of freedom according to statistical mechanics.

Thus an N-atom molecule also has $3N$ degrees of freedom. Three of these degrees of freedom are required for the total kinetic energy of the molecule; three more (two in the case of a linear molecule) are involved in the free rotation of the molecule. This leaves $3N - 6$ (or $3N - 5$, for a linear molecule) possible modes of vibration. Consequently, simple molecules like H_2O and CO_2 have three modes of vibration, designated ν_1, ν_2, and ν_3. One of these can be thought of as a bending vibration, while the other two relate to symmetric and asymmetric vibration along the internuclear axis. More details on this subject are provided in texts on molecular spectra such as Herzberg (1967) and Steinfeld (1974). Additional information on the infrared absorption characteristics of these and other molecules is provided by Zuev (1976) and by Pressley's *Handbook of Lasers* (1971), and detailed calculations of atmospheric transmittance have been undertaken by McClatchey et al. (1971, 1973), LaRocca (1975), Selby and McClatchey (1975), and Roberts et al. (1976). Specific atmospheric-transmittance calculations for a number of laser lines emitted by a few of the more common gas lasers (CO_2, CO, HF, and DF) have been reported by Kelley et al. (1976). A list of the more important absorption bands of a number of atmospheric pollutants has been provided in Table 3.6. The most prominent infrared absorption bands of several atmospheric constituents are shown in relation to the overall atmospheric absorption spectra in Fig. 4.8. Also shown are the spectral locations of the output wavelengths of the HF, DF, CO, and CO_2 gas lasers.

As indicated earlier, the transmission characteristics of the atmosphere in the visible portion of the spectrum (400 to 700 nm approximately) is determined under clear sky conditions by Rayleigh–Mie scattering of particulates and aerosols. The attenuation of electromagnetic radiation over a wide range of wavelengths becomes progressively more severe with the formation of haze (see Fig. 4.9).

With increasing buildup of water droplets or dust, molecular Rayleigh scattering becomes negligible, and elastic (Rayleigh–Mie) scattering from these

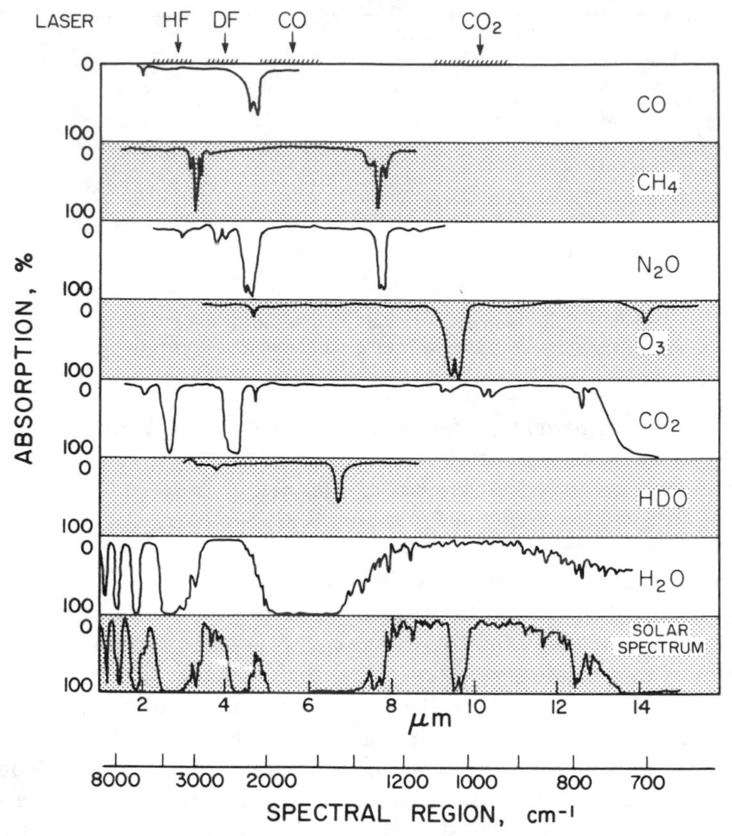

Fig. 4.8. The most prominent absorption bands of several atmospheric constituents relative to the overall atmospheric absorption spectra. Also shown are the wavelengths available from several of the more powerful infrared gas lasers (Vergez-Deloncle, 1964).

aeroticulates dominates the attenuation. The relevant attenuation coefficient under these circumstances can be expressed in the form of a sum

$$\kappa_M(\lambda) = \sum_{n=n_1, n_2, \ldots} \beta_s(n, \lambda) \tag{4.82}$$

over the atmospheric constituents of different refractive index, $n = n_1, n_2, \ldots,$ etc. The attenuation (extinction) coefficient for a component of refractive index n was given earlier [equation (2.161)] and can be written

$$\beta_s(n, \lambda) = \int_{a_1}^{a_2} \pi a^2 Q_s(a, n, \lambda) N(a, n)\, da \tag{4.83}$$

where $N(a, n)\, da$ represents the number density of the aeroticulates (of refrac-

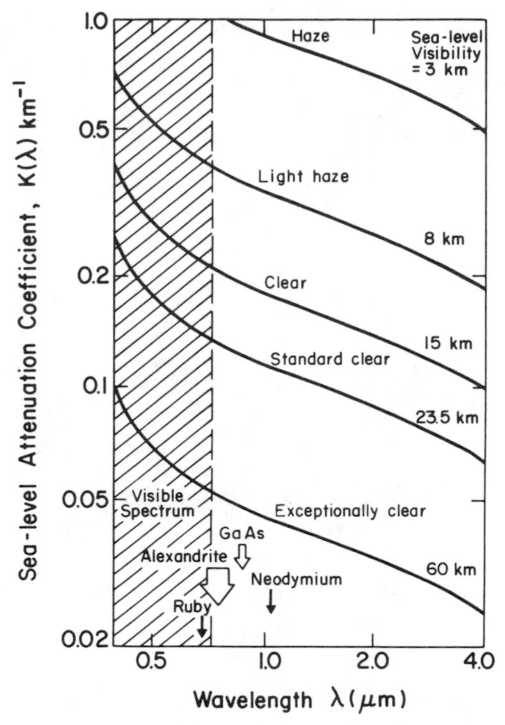

Fig. 4.9. Variation of sea-level attenuation coefficient $\kappa(\lambda)$ with wavelength for various atmospheric conditions. Also shown are wavelengths of some relevant lasers (Pressley, 1971).

tive index n) having a mean radius in the interval $(a, a + da)$, and $Q_s(a, n, \lambda)$ corresponds to their total scattering efficiency, defined by (2.159). Unfortunately, the complexity of $Q_s(a, n, \lambda)$ and the local fluctuations in $N(a, n)$ makes $\beta_s(n, \lambda)$ both highly variable and difficult to evaluate. Nevertheless, Nilsson (1979) has attempted to calculate the atmospheric extinction associated with aerosols for a variety of weather conditions over the 0.2 to 40-μm wavelength range.

A considerable simplification has been found possible (Kruse et al., 1963) in certain situations by relating the mean value of this atmospheric extinction coefficient $\kappa_M(\lambda)$, to the visibility through the empirical formula

$$\kappa_M(\lambda) \approx \frac{3.91}{R_v}\left\{\frac{550}{\lambda}\right\}^q \text{ km}^{-1} \tag{4.84}$$

where

$$\begin{aligned} q &= 0.585 R_v^{1/3} \qquad \text{for} \quad R_v \lesssim 6 \text{ km} \\ &\approx 1.3 \qquad \text{for average seeing conditions} \end{aligned} \tag{4.85}$$

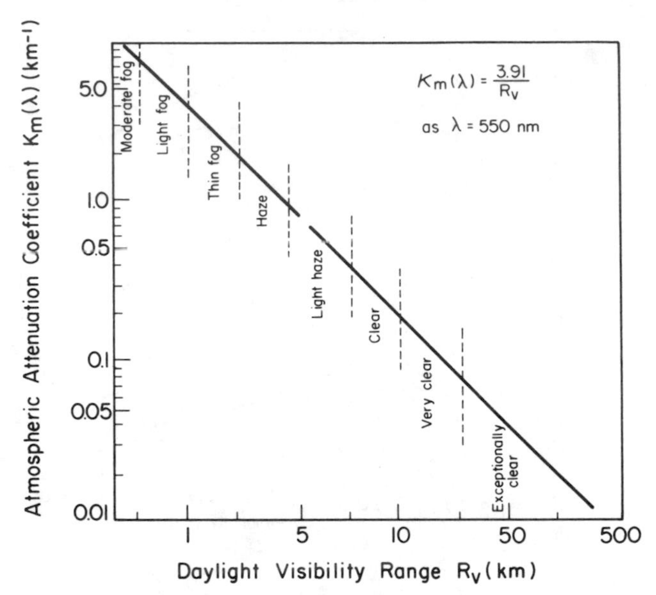

Fig. 4.10. Variation of atmospheric attenuation coefficient $\kappa_m(\lambda)$ with visibility range R_v, at a wavelength of 550 nm (Pressley, 1971).

In (4.84) λ is the wavelength of the radiation in nm and R_v is the meteorological visual range in km (defined as the horizontal range at which the transmission at 550 nm is 2%). The variation in the atmospheric attenuation (extinction) coefficient $\kappa_M(\lambda)$ with the daylight visibility range R_v for visible radiation at 550 nm is indicated in Fig. 4.10.

Woodman (1974) has discussed the reliability of (4.84) and concluded that although its accuracy is questionable in the infrared ($\lambda > 2$ μm), it can probably be used to provide a rough estimate of $\kappa_M(\lambda)$ from R_v in the visible wavelength range. However, Twomey and Howell (1965) and Fenn (1966) have also questioned the reliability of (4.84) in the case of laser radiation. Clay and Lenham (1981) have measured the attenuation coefficient $\kappa_M(\lambda)$ for fogs at several wavelengths.

Example. Calculate the total attenuation coefficient at a wavelength of 300.1 nm for a given location where the visibility range $R_v = 5$ km and the sulfur dioxide concentration is 10 ppm. STP conditions may be assumed. If a laser beam propagates across a 20-m plume of SO_2 at 10 ppm, calculate the percentage attenuation of the beam.

Answer. For SO_2 at 300.1 nm, $\kappa_A(\lambda) = 2.69 \times 10^{-5}$ (ppm cm)$^{-1}$ according to Table 3.6. Thus if the concentration of SO_2 is 10 ppm and STP conditions apply, then the effective absorption coefficient is $\kappa_A(\lambda) \approx 2.69 \times 10^{-5}$ (ppm cm)$^{-1}$ $\times$ 10 ppm $\simeq 2.69 \times 10^{-4}$ cm^{-1}. Since the visibility range is

5 km, the attenuation coefficient due to scattering is given by (4.84):

$$\kappa_M(\lambda) = \frac{3.91}{5}\left(\frac{550}{300}\right)^{0.585\times 5^{1/3}} \approx 1.43\ \text{km}^{-1} \approx 1.43 \times 10^{-5}\ \text{cm}^{-1}$$

and so the total attenuation (extinction) coefficient at 300.1 nm is

$$\kappa_\epsilon(\lambda) = \kappa_A(\lambda) + \kappa_M(\lambda) = 2.69 \times 10^{-4} + 1.43 \times 10^{-5}$$

$$\approx 2.83 \times 10^{-4}\ \text{cm}^{-1}$$

If the SO_2 plume extends over 20 m, the attenuation of a laser beam at 300.1 nm can be evaluated from (4.80):

$$\frac{I(300.1\ \text{nm}, 2 \times 10^3\ \text{cm})}{I(300.1\ \text{nm}, 0)} = \mathrm{e}^{-2.83\times 10^{-4}\times 2\times 10^3} = 0.57$$

Thus the beam suffers 43% attenuation in propagating across the 20-m plume of SO_2.

5

Laser Fundamentals

Spontaneous emission represents the primary mechanism of electromagnetic radiation in conventional sources at wavelengths of less than a few tens of micrometers. In the case of an incandescent source, the temperature is high enough to provide the excitation directly, while in a discharge lamp the current-carrying free electrons are responsible for the excitation. In general such conventional sources emit radiation that is spectrally broadband and incoherent (both spatially and temporally).

By contrast, lasers rely on stimulated emission (usually within an optical resonator) for the generation of their radiation. This leads to a significant compression of electromagnetic energy relative to that produced conventionally. This compression can occur in frequency space, solid angle, and time.

Lasers are either *continuous wave* (CW) or *pulsed*. Pulsed lasers can be ideal for remote probing of the environment if their radiant output forms a well-collimated beam that is close to being monochromatic and of very short duration.

Another important feature of a laser is its very high energy density relative to that in a conventional source. The latter is limited to blackbody temperatures of only a few thousand degrees, while in the case of a laser beam, the radiant energy density can be equivalent to that of a blackbody source at a temperature in excess of 10^{20} K. Indeed as we shall see shortly, the excited-state distribution within a laser is such that if we attempt to describe it in terms of a thermal distribution we are faced with the concept of a *negative temperature*. The reader should not be too alarmed by this comment, however, as it stems from our attempt to describe a nonequilibrium situation in terms of an equilibrium parameter (somewhat like trying to describe fire to an intelligent fish). We see that the laser is not just a different kind of light (infrared to ultraviolet), but is qualitatively different from other light sources.

The word *laser* is an acronym standing for "light amplification by stimulated emission of radiation." The laser represents an extraordinary versatile tool with applications that span the entire spectrum of man's endeavors, ranging from genetic engineering that could improve the human race to weapons that could aid in our annihilation.

In this chapter we shall briefly review the fundamentals of lasers and discuss the types and properties of lasers with special emphasis on those aspects that are suitable for remote probing of the environment. Our discussion of lasers must obviously be limited, and we refer the interested reader to several excellent texts on the subject: Maitland and Dunn (1969), Siegman (1971), Schäfer (1973), Yariv (1976), and Verdeyen (1981).

5.1. POPULATION INVERSION AND OPTICAL GAIN

In Section 4.1 we saw that electromagnetic radiation of frequency coincident with that of an optically allowed transition within an ensemble of molecules (or atoms) undergoes both absorption and stimulated emission. Under normal circumstances (and in particular if equilibrium holds) the number density of molecules in the upper energy state $|n\rangle$ will be much less than the number density in the lower energy state $|m\rangle$ ($N_n \ll N_m$). This leads to a positive value for the volume absorption coefficient $\kappa(\nu)$ as defined by (4.6), and consequent attenuation of the radiation with penetration.

If on the other hand we create a highly nonequilibrium situation in which

$$\frac{N_n}{N_m} > \frac{g_n}{g_m} \tag{5.1}$$

where g_n/g_m is the degeneracy ratio for $|n\rangle$ and $|m\rangle$, then $\kappa(\nu)$ becomes negative and we are led to introduce the *volume gain coefficient*

$$\alpha(\nu) \equiv \frac{h\nu}{4\pi}[N_n B_{nm} - N_m B_{mn}]\mathscr{L}(\nu) \tag{5.2}$$

which is invariably positive if (5.1) holds, due to the relationship between the Milne coefficients,

$$g_n B_{nm} = g_m B_{mn} \tag{5.3}$$

which can be seen from (3.113) and (4.4).

This kind of nonequilibrium is termed a *population inversion* and can be created in a number of ways, ranging from intense optical pumping to an electrical discharge. The change in the energy-state population density brought about by the creation of a population inversion is illustrated in Fig. 5.1.

If we introduce the *inverted population density*

$$\mathscr{N} \equiv N_n - \frac{g_n N_m}{g_m} \tag{5.4}$$

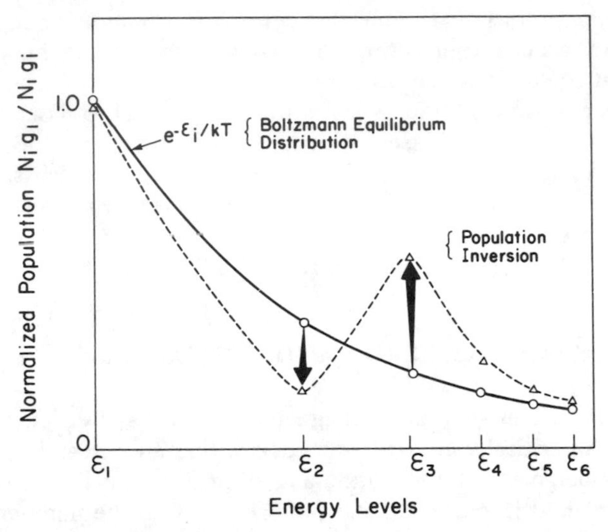

Fig. 5.1. Representative modification of an equilibrium distribution associated with the creation of a population inversion between levels 3 and 2. —O—, equilibrium values; ---Δ---, nonequilibrium values.

and the *stimulated-emission cross section*

$$\sigma(\nu) \equiv \frac{h\nu}{4\pi} B_{nm} \mathscr{L}(\nu) \tag{5.5}$$

then we can redefine the medium's gain coefficient† in terms of these quantities:

$$\alpha(\nu) = \sigma(\nu)\mathscr{N} \tag{5.6}$$

provided there are no other loss mechanisms.

An *active* or *gain* medium is one in which a population inversion has been created, that is, $\mathscr{N} > 0$. In such a medium radiation of appropriate frequency will be amplified rather than attenuated in the absence of other loss mechanisms (see Fig. 5.2). This amplification is possible because a population inversion represents a source of potential energy and stimulated emission is the mechanism by which the radiation field is reinforced through the conversion of excited-state energy into electromagnetic energy. If other loss mechanisms, such as scattering or absorption in some other constituent of the medium, are

†An alternate and often cited, form for the gain coefficient,

$$\alpha(\lambda) = \frac{\lambda^4 \mathscr{N} \mathscr{L}(\lambda)}{8\pi c \tau_{nm}} \tag{5.6a}$$

is easily derived from (5.6). Remember to use $\mathscr{L}(\lambda)\, d\lambda = \mathscr{L}(\nu)\, d\nu$.

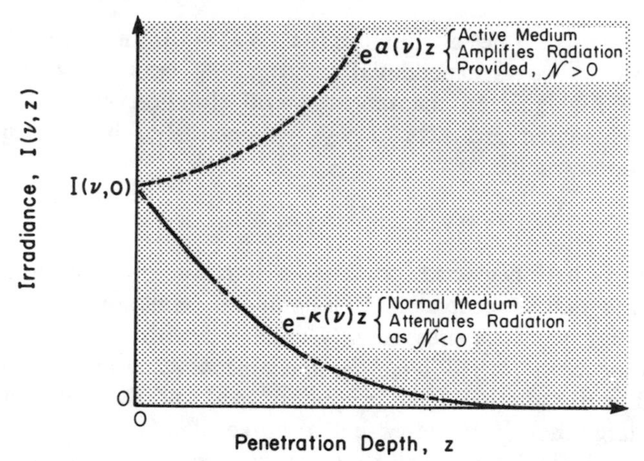

Fig. 5.2. Variation in radiation field with penetration of a medium. Amplification occurs for $\mathcal{N} > 0$; attenuation, for $\mathcal{N} < 0$.

present, then the volume gain coefficient $\alpha(\nu)$ has to be redefined in the following manner:

$$\alpha(\nu) \equiv \sigma(\nu)\mathcal{N} - \tilde{\kappa}(\nu) \tag{5.7}$$

where $\tilde{\kappa}(\nu)$ represents the *net attenuation coefficient*† and comprises the sum of the absorption and scattering coefficients for the medium. $\alpha(\nu)$ can be thought of as the net gain per unit length in the medium.

For radiation propagating essentially in the z-direction, the steady-state radiative-transfer equation was given by (4.79). In the case of an active medium this equation can be recast in the form

$$\frac{dI(\nu, z)}{dz} = \alpha(\nu, z)I(\nu, z) \tag{5.8}$$

where $I(\nu, z)$ is the spectral irradiance at position z within the medium. For a semiinfinite medium (existing for $z > 0$) with a spectral irradiance $I(\nu, 0)$ incident on the $z = 0$ surface, the solution of (5.8) takes the form

$$I(\nu, l) = I(\nu, 0)\mathrm{e}^{\int_0^l \alpha(\nu, z)\, dz} \tag{5.9}$$

or, for a homogeneous medium,

$$I(\nu, z) = I(\nu, 0)\mathrm{e}^{\alpha(\nu)z} \tag{5.10}$$

† This does not include absorption associated with the *mn* transition, which is included in the first term.

Clearly, if there is no net attenuation [i.e., $\tilde{\kappa}(\nu) = 0$], then the radiation field either grows or decays exponentially, depending upon whether $\mathcal{N} > 0$ or $\mathcal{N} < 0$ (see Fig. 5.2). In the more general situation where $\tilde{\kappa}(\nu) \neq 0$, the radiation field again grows or decays exponentially, but in this instance depending upon whether

$$\alpha(\nu) \equiv \sigma(\nu)\mathcal{N} - \tilde{\kappa}(\nu) > 0 \quad \text{or} \quad < 0 \tag{5.11}$$

Although we have neglected spontaneous emission up to this point, it should be recognized that in general it is spontaneous emission that provides the initial radiation which is subsequently amplified within a medium having a positive gain coefficient. This *amplified spontaneous emission* (ASE) can lead to bursts of intense radiation if the product of $\alpha(\nu)$ and the scale length for the medium is large enough. However such a source, albeit powerful, would still be incoherent and would not really constitute a laser, due to the very large number of radiation modes possible.

A *mode* of the radiation field can be thought of as a stationary electromagnetic-field configuration which satisfies Maxwell's equations and the boundary conditions. The electric-field component for such a mode takes the form

$$E(\mathbf{r}, t) = E_0 u(\mathbf{r}) \mathbf{e}^{-i\omega t} \tag{5.12}$$

where $u(\mathbf{r})$ is purely a spatial function. It can be shown (Loudon, 1973) that the number of radiation modes per unit frequency interval per unit volume—the *mode density*— is given by

$$p_0(\nu) = \frac{8\pi\nu^2}{c^3} \tag{5.13}$$

At optical frequencies and typical bandwidths $p(\nu)$ is very large ($\gtrsim 10^{10}$ cm^{-3}).

Thus in order to produce a powerful coherent source of radiation from an active medium it is necessary to severely restrict the number of radiation modes for which there is appreciable gain. This is done in most instances by shaping the active medium (making it much longer in the desired direction of propagation) and placing it within a suitable resonator. One of the simplest forms of resonator (a Fabry–Perot resonator), and one that was used in the first laser (Schawlow and Townes, 1958; Maiman 1960), comprises two flat mirrors that are aligned parallel to each other and perpendicular to the axis of the system (essentially the direction of desired output).

5.2. THRESHOLD FOR LASER ACTION

A schematic diagram of an optically pumped (for example, ruby, neodymium, or dye) laser is presented as Fig. 5.3. In this simplified design a flashlamp is used to pump optical energy into a suitable medium within a plane-mirror

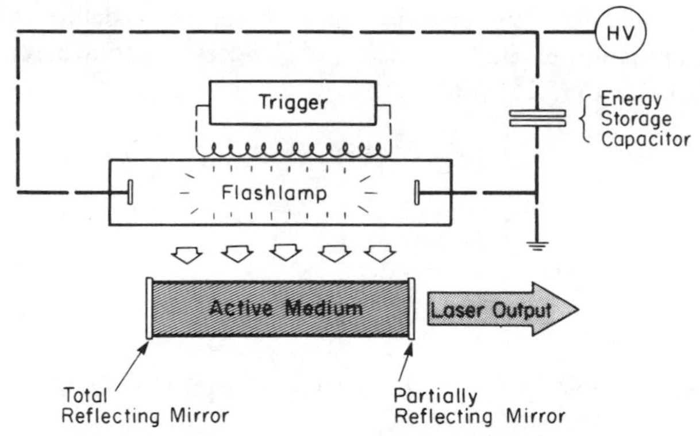

Fig. 5.3. Schematic of a flashlamp-pumped laser.

(Fabry–Perot) resonator. Laser action arises as soon as the population inversion is sufficient to overcome the optical losses of the system. For an active medium within a resonator these optical losses include the radiation transmitted through the mirrors in addition to any internal loss within the active medium. If the mirror reflection coefficients (with respect to irradiance) are R_1 and R_2, then the *round-trip gain* (i.e., double-transit gain) $\mathscr{G}(\nu)$ in irradiance is defined according to the relation

$$\mathscr{G}(\nu) \equiv R_1 R_2 e^{2\alpha(\nu)l} \tag{5.14}$$

where l is the length of the active medium. We assume that the mirrors are placed at the ends of the active medium. Laser action will arise when there is an increase in the irradiance after a round trip through the system, that is to say, when

$$R_1 R_2 e^{2\alpha(\nu)l} > 1, \qquad \text{or} \quad \alpha(\nu)l > \gamma_R \tag{5.15}$$

where

$$\gamma_R \equiv \ln \frac{1}{R} \tag{5.16}$$

represents the *resonator loss coefficient*, and

$$R \equiv (R_1 R_2)^{1/2}$$

represents the geometrical-mean reflection coefficient. In principle this requirement (that the net medium gain be greater than the resonator loss per round trip) would lead to a singularity† for the irradiance without the intervention of

† The amplitude would grow to an infinite value after an infinite number of round trips.

some limiting process. This criterion can be recast to define the *critical* (or threshold) population inversion density, $\mathcal{N}_c$, necessary to achieve laser action. Thus using (5.7) with (5.15) yields

$$e^{[\sigma(\nu)\mathcal{N} - \tilde{\kappa}(\nu) - \gamma_R/l]l} > 1 \tag{5.17}$$

which can be rewritten

$$\mathcal{N} > \frac{\gamma}{\sigma(\nu)l} \tag{5.18}$$

where $\gamma = \gamma_R + \tilde{\kappa}(\nu)l$ represents the net loss per pass. It follows from this that

$$\mathcal{N}_c \equiv \frac{\gamma}{\sigma_{\max}(\nu)l} \tag{5.19}$$

In general the most important loss mechanisms, over and above the resonator loss, are:

1. Scattering from optical inhomogeneities (primarily in solid-state lasers).
2. Absorption within the gain medium (primarily triplet losses in dye lasers).
3. Diffraction by mirror aperture (primarily in gas lasers).

If these losses can be kept small compared to the resonator loss, then we can write

$$\mathcal{N}_c \approx \frac{\gamma_R}{\sigma_{\max}(\nu)l} \tag{5.20}$$

This expression can be rewritten in a form that more clearly reveals the important parameters that determine $\mathcal{N}_c$. If we assume a high-quality resonator so that $R \approx 1$, or $\gamma_R \approx 1 - R$, and introduce the *cavity residence time*

$$\tau_c \equiv \frac{l}{c(1 - R)} \tag{5.21}$$

then for a Lorentzian-broadened laser transition

$$\mathcal{N}_c \approx p\frac{\tau_{nm}}{\tau_c} \tag{5.22}$$

Here, τ_{nm} is the radiative lifetime (A_{nm}^{-1}) of the laser transition, and

$$p = \frac{8\pi^2 \Delta\lambda}{\lambda^4} \tag{5.23}$$

represents the radiation mode density over the effective bandwidth of the laser transition. In deriving (5.22) we have invoked (3.112), (3.140), (4.4), and (5.5). It is evident from (5.22) and (5.23) that if we wish to keep $\mathcal{N}_c$ small we must chose a long-wavelength transition that is spectrally narrow and has a short radiative lifetime.

5.3. MODE STRUCTURE AND SPECTRAL NARROWING

We shall now derive the criterion for laser action in a more rigorous fashion. We start by constructing a one-dimensional model of the laser (see Fig. 5.4). This is seen to comprise basically an active medium that is positioned between two plane and parallel mirrors (a Fabry–Perot resonator). The amplitude reflection coefficient of the totally reflecting mirror is r_2, while that of the partially reflecting (output) mirror is r_1. The amplitude transmission coefficient of the latter is t_1. We assume that spontaneous emission gives rise to a burst of radiation that originates adjacent to the totally reflecting mirror. As this radiation propagates back and forth between the two mirrors it is amplified (if the net gain coefficient α is positive) and suffers a phase shift k $(= 2\pi/\lambda)$ per unit distance traveled.

If a steady-state condition is assumed, then a wave of amplitude E_0 at the $z = 0$ plane becomes $E_0 e^{(\alpha/2+ik)l}$ at the output mirror. Note that α is divided by 2 in this expression because the gain coefficient as defined by (5.7) relates to the amplification of irradiance. A fraction $t_1 E_0 e^{(\alpha/2+ik)l}$ is transmitted through this mirror, and the remainder $r_1 E_0 e^{(\alpha/2+ik)l}$ (assuming no absorption in the mirror) is reflected back into the active medium. After a further double pass a

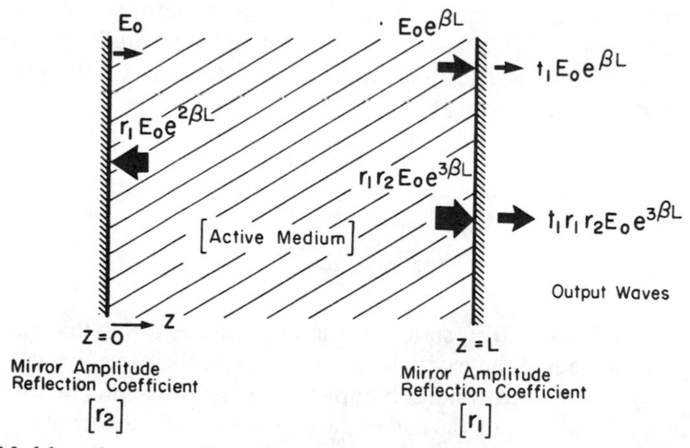

Fig. 5.4. Model used to derive the laser-action criterion for an active medium within a plane Fabry–Perot cavity. E_0 is taken to represent the amplitude of the initial spontaneous-emission wave. $\beta \equiv \alpha/2 + ik$ represents the complex amplitude gain coefficient.

wave of amplitude $r_1 r_2 t_1 E_0 e^{(\alpha/2+ik)3l}$ is transmitted through the output mirror. The total output amplitude having its origin in E_0 is thus

$$E = t_1 E_0 e^{(\alpha/2+ik)l} + r_1 r_2 t_1 E_0 e^{(\alpha/2+ik)3l} + r_1^2 r_2^2 t_1 E_0 e^{(\alpha/2+ik)5l} + \cdots \tag{5.24}$$

This represents an infinite geometrical progression with the sum

$$E = \frac{t_1 E_0 e^{(\alpha/2+ik)l}}{1 - r_1 r_2 e^{2(\alpha/2+ik)l}} \tag{5.25}$$

Thus the transmitted spectral irradiance is

$$I = \tfrac{1}{2}\epsilon_0 c|E|^2 = I_0 t_1^2 \frac{e^{(\alpha/2+ik)l}}{1 - Re^{2(\alpha/2+ik)l}} \cdot \frac{e^{(\alpha/2-ik)l}}{1 - Re^{2(\alpha/2-ik)l}} \tag{5.26}$$

where

$$R \equiv r_1 r_2 \tag{5.27}$$

is equivalent to the geometrical mean reflection coefficient of the resonator introduced earlier. I_0 represents the initial irradiance of the wave. Equation (5.26) can be simplified and rewritten in the form

$$I = \frac{I_0 t_1^2 e^{\alpha l}}{\{1 - Re^{\alpha l}\}^2 + 4Re^{\alpha l}\sin^2 kl} \tag{5.28}$$

Clearly, I suffers a singularity when (1) $Re^{\alpha(\nu)l} = 1$ and (2) $kl = n\pi$ simultaneously, where n is an integer between one and infinity. This singularity corresponds to laser oscillation, and we see that two conditions have to be simultaneously fulfilled. The first condition is identical to the criterion we obtained earlier, that is from (1),

$$\alpha(\nu)l = \gamma_R \tag{5.29}$$

where, as before,

$$\gamma_R \equiv \ln\frac{1}{R}$$

Equation (5.29), in effect, states that laser action requires that the medium net gain per pass equal the cavity loss, and leads to the same critical value for $\mathcal{N}_c$ as (5.19). The second condition implies that laser action is restricted to the resonance frequencies ν_n of the cavity, i.e., from (2),

$$\nu_n = \frac{nc}{2l} \tag{5.30}$$

In terms of wavelength, $n\lambda = 2l$, and we see that n represents the number of *antinodal* planes (perpendicular to the optical, or z, axis) between the cavity boundaries, $z = 0$ and $z = l$.

Thus we see that laser oscillation is possible only at those frequencies which satisfy the two criteria $\alpha(\nu)l \geq \gamma_R$ and $\nu_n = nc/2l$. This can be better understood by reference to Fig. 5.5, where we present the medium-gain and cavity-loss curves as a function of frequency. According to this figure, laser action is expected to occur only on three of the cavity modes: $n - 1$, n, and $n + 1$; for in this illustration only the frequencies ν_{n-1}, ν_n, and ν_{n+1} satisfy the two criteria (5.29) and (5.30).

The frequency spacing between adjacent cavity modes (sometimes referred to as *temporal* or *axial* modes),

$$\Delta\nu_{n,\,n+1} = \frac{c}{2l} \tag{5.31}$$

is often smaller than the bandwidth of the laser transition, and so the output of such a laser can comprise many temporal modes.

A more complete derivation of the output irradiance of a gain medium within a cavity would take account of the amplification of spontaneous emission generated throughout the active medium. Such an analysis leads to a pre-lasing output spectral irradiance

$$I(\nu) = I_{sp}\frac{\mathscr{L}(\nu)}{[1 - R\mathscr{g}(\nu)]^2} \cdot \frac{1}{1 + \dfrac{4R\mathscr{g}(\nu)\sin^2(2\pi\nu l/c)}{[1 - R\mathscr{g}(\nu)]^2}} \tag{5.32}$$

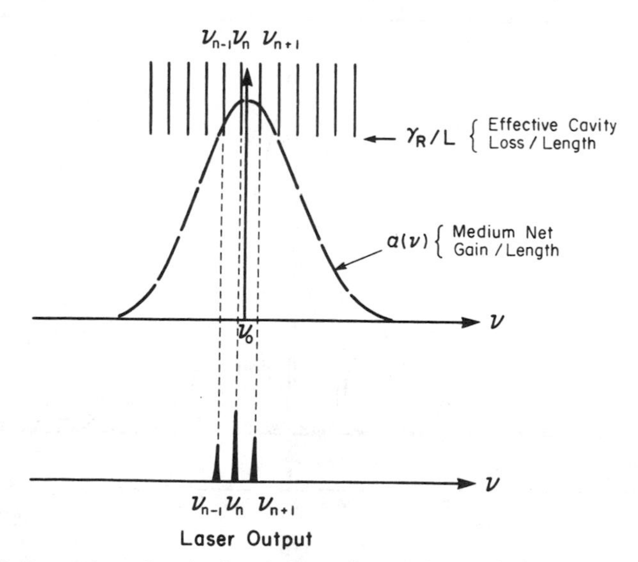

Fig. 5.5. Medium gain and cavity loss indicate frequencies at which laser action is possible, namely, $\alpha(\nu) > \gamma_R/l$.

where I_{sp} is related to the power density of spontaneous emission from the active medium and

$$\mathscr{g}(\nu) = e^{\alpha(\nu)l} \tag{5.33}$$

represents the medium gain per pass.

Although equation (5.32) is based on a simplified one-dimensional model of a laser, it nevertheless indicates several of the important features observed in the output of lasers:

1. It leads to the two criteria for laser action [(5.29) and (5.30)] and predicts a mode frequency spacing $\Delta\nu_{n,\,n+1} = c/2l$.

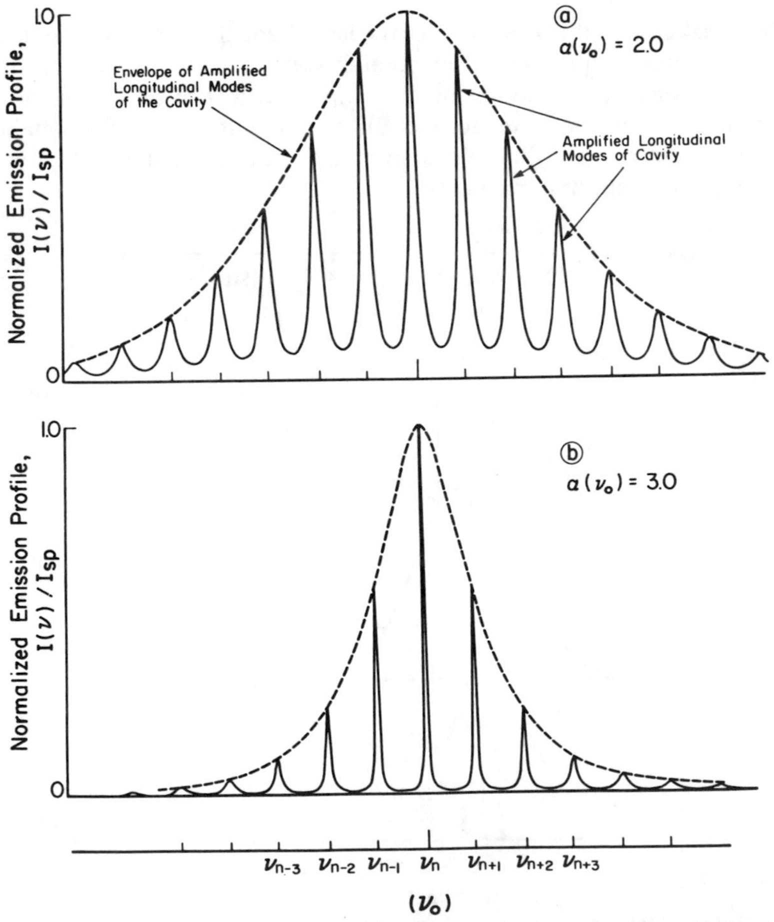

Fig. 5.6. Spectral distribution of amplified spontaneous emission within a Fabry–Perot cavity for two values of center-line-frequency gain coefficient: (a) $\alpha(\nu_0) = 2.0$ and (b) $\alpha(\nu_0) = 3.0$.

2. It reveals that the gain-modified transition profile function [the second factor in (5.32)] is the envelope of the amplified temporal modes [described by the third factor of (5.32)— a gain-modified Airy function].
3. It also indicates that the temporal-mode envelope narrows with increasing gain and a progressive spectral narrowing of modes as they approach the line center frequency.

These features are illustrated in Fig. 5.6, where the only difference between (a) and (b) is a 50% increase in the population inversion density. Note that since $I(\nu)/I_{sp}$ is plotted as a function of frequency, both curves have the same peak amplitude.

5.4. GAIN SATURATION

We have seen that the conditions $Re^{\alpha(\nu)l} = 1$ and $kl = n\pi$ lead to a mathematical singularity in equation (5.32) whereby the output irradiance tends to infinity. Although these conditions serve as the criteria for laser action, in reality a self-limiting process called *gain saturation* prevents the actual occurrence of this singularity. In essence, depletion of the population inversion through stimulated emission limits the growth of the radiation field within the cavity. Insight into this process of gain saturation can be obtained through considering the somewhat simplified three-energy-level model of a laser presented in Fig. 5.7.

In order to appreciate the significance of the energy-level system portrayed in Fig. 5.7, we need to digress for a moment. Lasers, in general, are classified as three- or four-level systems depending upon the position of the terminal level for the laser transition—and not the number of levels, as one might suppose. A *four-level laser* is one in which the terminal level is energetically well above the ground level (i.e., $E_{mg} \gg kT$ in Fig. 5.7), so that its equilibrium population is small. In a *three-level laser* the ground level serves as the terminal level. Clearly, a given population inversion is created within a four-level laser with much less energy than in a comparable three-level laser.

As it happens most four-level lasers are in fact well represented by four energy levels (see the small insert in Fig. 5.7). However, in many instances a very fast nonradiative relaxation process is responsible for transferring the bulk of the excited-state population to the laser level, thereby justifying the three-energy level model presented as Fig. 5.7.

We return to the subject of gain saturation. The rate of populating the laser level can be expressed in the form

$$\frac{dN_n}{dt} = S_n - N_n A_{nm} - (N_n B_{nm} - N_m B_{mn}) \int \frac{I^L(\nu) \mathscr{L}(\nu)\, d\nu}{4\pi} \tag{5.34}$$

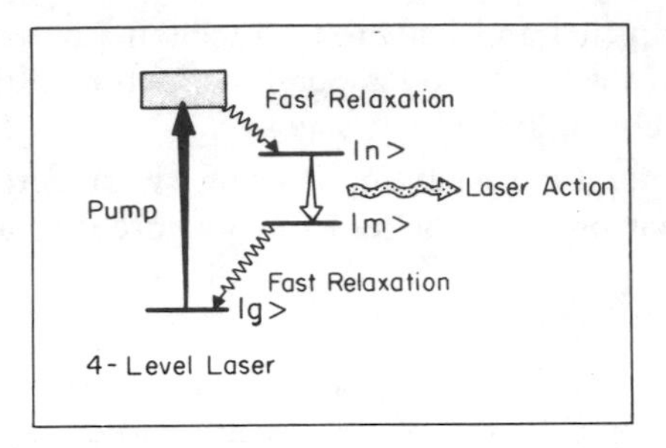

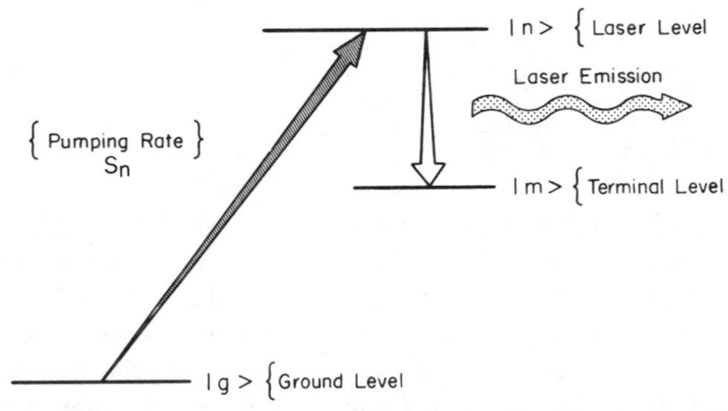

Fig. 5.7. Simplified three-energy-level representation of a four-level (see top insert) laser.

where S_n represents the rate of pumping the laser level. $\mathscr{L}(\nu)$ represents the line profile function of the lasing transition, and $I^L(\nu)$ the laser spectral irradiance within the active (gain) medium. N_n and N_m represent the population densities of the states $|n\rangle$ and $|m\rangle$, respectively.

We shall assume that the radiation field is confined to one cavity mode with frequency coincident with the line center frequency ν_0 of the lasing transition. With this assumption we can write

$$I^L(\nu) = I^L\delta(\nu - \nu_0) \tag{5.35}$$

where $\delta(\nu - \nu_0)$ represents the Dirac delta function and I^L the laser irradiance of that mode. Under these circumstances

$$\int I^L(\nu)\mathscr{L}(\nu)\,d\nu = \int I^L\delta(\nu - \nu_0)\mathscr{L}(\nu)\,d\nu = I^L\mathscr{L}(\nu_0) \tag{5.36}$$

and if we introduce the *line-center stimulated-emission cross section*

$$\sigma_0 \equiv \frac{h\nu}{4\pi}B_{nm}\mathscr{L}(\nu_0) \tag{5.37}$$

we can write equation (5.34) in the form

$$\frac{dN_n}{dt} = S_n - \mathcal{N}\left[\frac{\sigma_0 I^L}{h\nu} + \frac{1}{\tau}\right] - \frac{g_n N_m}{g_m \tau} \tag{5.38}$$

We have written τ^{-1} for A_{nm} in (5.38), but if the lifetime of $|n\rangle$ is in fact shorter than the radiative lifetime, then we must use the actual lifetime. The last term in (5.38) can be neglected in an efficient laser, and so the steady-state population inversion density is

$$\mathcal{N} \approx \frac{\tau S_n}{1 + I^L/I^s} \tag{5.39}$$

where

$$I^s \equiv \frac{h\nu}{\sigma_0 \tau} \tag{5.40}$$

is termed the *saturated irradiance*.†

The actual gain per unit length of the active medium at frequency ν is

$$\sigma(\nu)\mathcal{N} = \frac{\sigma(\nu)\tau S_n}{1 + I^L/I^s} \tag{5.41}$$

To be consistent with steady-state conditions, this gain per unit length must just equal the effective loss per unit length, i.e.,

$$\sigma(\nu)\mathcal{N} = \frac{\gamma}{l} \tag{5.42}$$

and so $\mathcal{N}$ is *clamped* to its critical value $\mathcal{N}_c$ of (5.19). If this were not the case, the field energy in the cavity would run away and invalidate our steady-state assumption. We are thus forced to conclude that

$$\frac{\tau S_n}{1 + I^L/I^s} = \mathcal{N}_c \tag{5.43}$$

or put another way,

$$I^L = I^s\left\{\frac{\tau S_n}{\mathcal{N}_c} - 1\right\} \tag{5.44}$$

†It should be noted that the saturated irradiance I^s in this instance is a factor $1 + g$ times the saturated irradiance I^S defined by equation (4.40) in the context of fluorescence saturation.

The laser irradiance within the cavity can be seen to be proportional to the saturated irradiance of the medium. It should be noted that τS_n represents the steady-state population density of the laser level that would exist in the absence of the radiation field. Furthermore, since I^s is fixed for the medium, it is clear that I^L increases linearly with pumping rate S_n. If we introduce the *critical* (or *threshold*) pumping rate S_n^c defined by the relation

$$S_n^c \equiv \frac{\mathcal{N}_c}{\tau} \tag{5.45}$$

then we can write

$$I^L = I^s \left\{ \frac{S_n}{S_n^c} - 1 \right\} \tag{5.46}$$

At this juncture we shall reexamine the saturated irradiance I^s by drawing upon (3.112), (4.4), and (5.37). These enable us to write

$$I^s = \rho_s c \tag{5.47}$$

where the *saturated energy density* of the radiation field is

$$\rho_s \equiv \frac{8\pi\nu^2}{c^3} \delta\nu \, h\nu \tag{5.48}$$

Here, $\delta\nu$ represents the *effective bandwidth* of the lasing transition and is defined by the expression

$$\delta\nu \equiv \frac{1}{\mathcal{L}(\nu_0)} \tag{5.49}$$

Bearing in mind that $8\pi\nu^2/c^3$ represents the number density of radiation modes per unit frequency interval, it is clear from (5.48) that the saturated energy density corresponds to the situation where there is one photon in each radiation mode (the $8\pi\nu^2/c^3$ takes account of the two planes of polarization). The growth of the radiation field with pumping rate is illustrated in Fig. 5.8. We see that for $S_n = 2S_n^c$, the coherent laser output in the single oscillating mode equals the spontaneous emission into all modes.

The radiative-transfer equation appropriate to the laser radiation within the cavity takes the form

$$\frac{dI^L(\nu)}{dz} = \sigma(\nu)\mathcal{N} I^L(\nu) \tag{5.50}$$

where we assume that for a reasonably efficient laser the medium loss coefficient

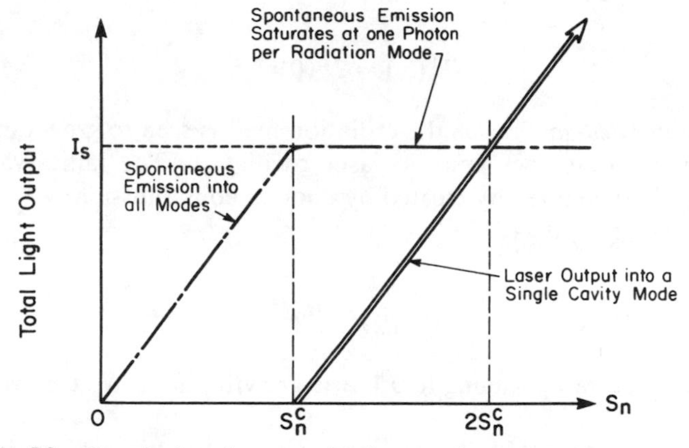

Fig. 5.8. Growth of total output with the rate of pumping of laser medium.

$\tilde{\kappa}(\nu)$ is negligible. In the case of laser oscillation on a single cavity mode at the line center frequency, integrating (5.50) over all frequencies yields

$$\frac{dI^L}{dz} = \sigma_0 \mathcal{N} I^L \tag{5.51}$$

As we have seen above, if steady-state conditions prevail, then the population-inversion density is clamped to its critical value, and so combining (5.43) with (5.51) leads to the growth equation for the laser irradiance as it propagates through the active medium:

$$\frac{dI^L}{dz} = \left\{ \frac{\sigma_0 \tau S_n}{1 + I^L/I^s} \right\} I^L \tag{5.52}$$

If we introduce the *small (unsaturated) gain coefficient*

$$\alpha_0 \equiv \sigma_0 \tau S_n \tag{5.53}$$

then we can write

$$\frac{dI^L}{dz} = \frac{\alpha_0 I^L}{1 + I^L/I^s} \tag{5.54}$$

Clearly there are two limiting situations:

1. If $I^L \ll I^s$, then

$$\frac{dI^L}{dz} \approx \alpha_0 I^L \tag{5.55}$$

and

$$I^L(z) = I^L(0)e^{\alpha_0 z} \tag{5.56}$$

In this *unsaturated-gain region* the radiation field is seen to grow exponentially with distance, as it does prior to laser oscillation. The initial value of the irradiance $I^L(0)$ is in reality created by spontaneous emission.

2. When $I^L \gg I^s$, then

$$\frac{dI^L}{dz} \approx \alpha_0 I^s \tag{5.57}$$

and the gain saturates, leading to a linear growth phase. We can write

$$I^L(z) \approx I^L(z_s) + (z - z_s)\alpha_0 I^s \tag{5.58}$$

for $z > z_s$, where z_s is the value of z where $I^L = I^s$, that is, at the location where the rate of the depopulation of $|n\rangle$ by stimulated emission just equals that of spontaneous emission.

The growth of the laser field with propagation distance over the entire length of an active medium is presented in Fig. 5.9. Since $\alpha_0 = \sigma_0 \tau S_n$, we see that (5.58) can be rewritten in the form

$$I^L(z) = I^L(z_s) + (z - z_s)h\nu S_n \tag{5.59}$$

which indicates that under saturated-gain conditions, the incremental growth of the radiation field within the active medium just corresponds to the volume rate of pumping times the photon energy times the length of the gain region.

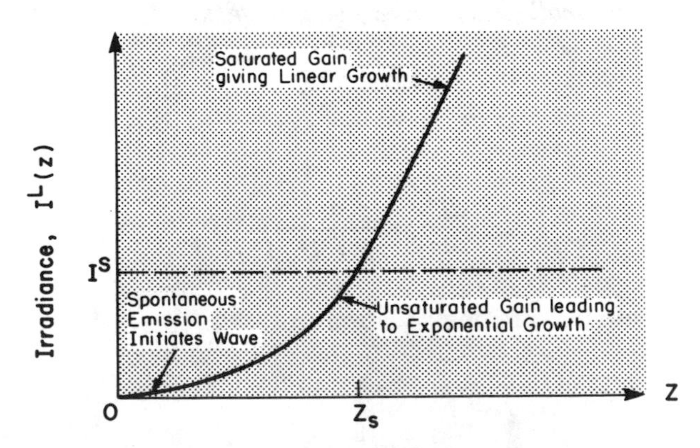

Fig. 5.9. Growth of laser irradiance with propagation distance into an active (gain) medium.

This can be interpreted to mean that under ideal conditions each inverted molecule (or atom) contributes one photon to the laser field.

In summary we see that (for a steady-state laser) the gain coefficient of an active medium will increase with the pumping rate (through increasing $\mathscr{N}$), until the gain/length equals the loss/length, at which point the gain will saturate and will no longer increase with an increase in the pumping rate. This can be understood in terms of the inverted population density being clamped to the threshold value.

The question arises, how does saturation affect the gain curve? In particular, is the saturation localized in frequency space, or does the entire gain curve saturate? In order to answer this question we have to know which kind of line-broadening mechanism dominates the laser transition.

Line broadening is termed *homogeneous* when each molecule (or atom) is affected in the same way. Consequently, in the case of *homogeneous broadening* every molecule (or atom) has the same spectral profile and bandwidth and therefore interacts equally with radiation at any frequency. Both natural and collision broadening give rise to homogeneously broadened lines. Under these

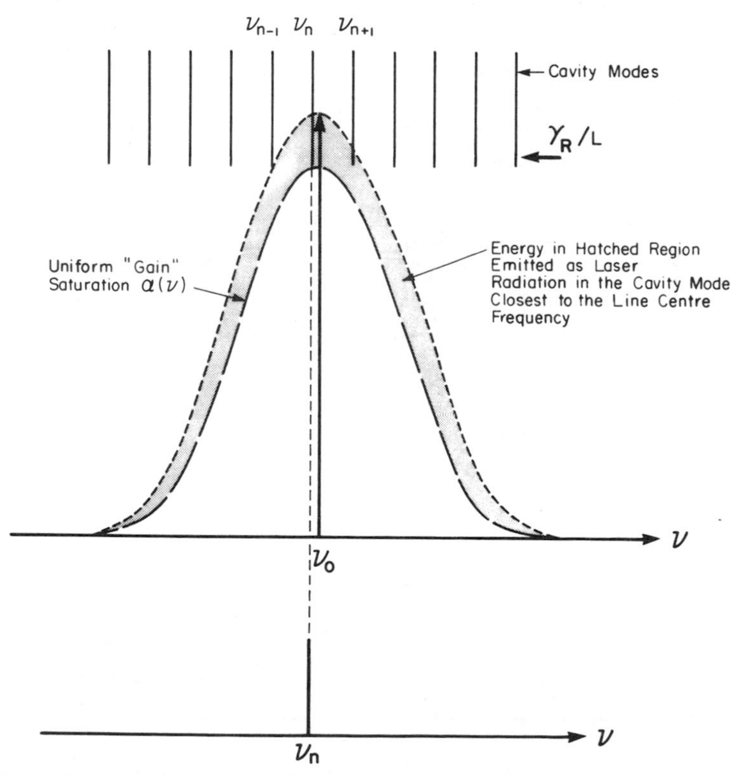

Fig. 5.10. Gain saturation and subsequent laser oscillation on only one mode for a homogeneously broadened line.

circumstances the gain saturates uniformly, and in principle laser oscillation occurs at the frequency of the first cavity mode to saturate. This situation is illustrated in Fig. 5.10. The broken curve represents the potential gain curve in the absence of laser oscillation, while the lower curve displays the actual gain curve arising from saturation and subsequent oscillation at ν_n.

In the case of *inhomogeneous broadening* only a small group of molecules (or atoms) interact with radiation at a given frequency. Consequently, the gain saturates only over a small portion of the lasing-transition bandwidth, and so laser oscillation can occur on many cavity modes simultaneously. This results in what is termed *hole burning* of the gain curve, as seen in Fig. 5.11. This can be understood if we recognize that *inverted* molecules (or atoms), having their homogeneous center line frequency displaced from that of the radiation field by more than the homogeneous width, have little chance of feeding their energy to the radiation field. Doppler broadening that arises from the thermal motion of the radiating species is a good example of an inhomogeneous broadening mechanism. In this instance radiation at some frequency ν_n can only interact with those inverted molecules (or atoms) whose thermal motion is such that their homogeneous line center frequency is displaced sufficiently from ν_0 that ν_n falls within their homogeneous linewidth. This is illustrated in Fig. 5.12.

The spectral output of real lasers is often further complicated by other effects, such as *spatial hole burning*—in which the cavity standing waves at different frequencies feed on different parts of the spatial distribution of inverted molecules (or atoms). In this way even a laser operating on a

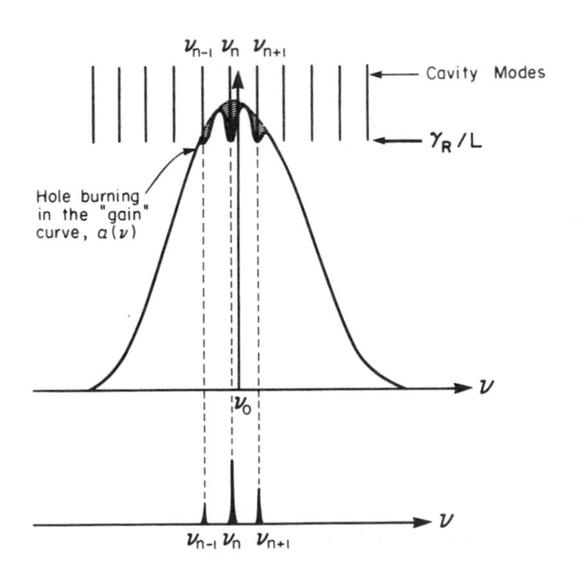

Fig. 5.11. Gain saturation, lasing, and hole burning for an inhomogeneously broadened line.

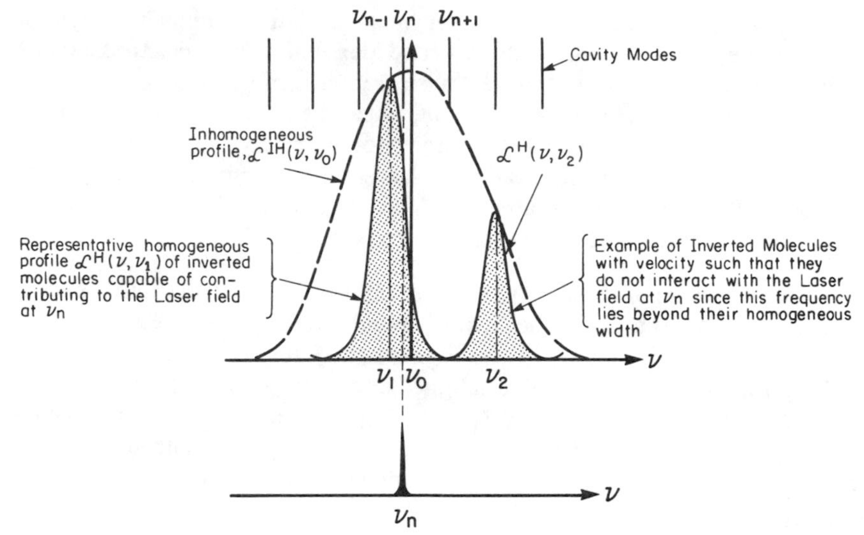

Fig. 5.12. Illustration of how gain saturation is confined to the homogeneous width of the inverted molecules (or atoms).

homogeneously broadened transition can oscillate on several cavity modes. Another point to bear in mind is that in a laser anomalous dispersion can lead to shifts in the frequency spacing of the cavity modes—so-called *mode pulling* (Maitland and Dunn, 1969, p. 210).

5.5. TEMPORAL MODULATION OF LASERS

The output of the earliest optically pumped solid-state lasers typically comprised a burst of sharp spikes that lasted for many microseconds. The duration of each spike was in the tens-of-nanoseconds range and their peak power was rarely more than ten kilowatts. A laser output of this nature would be of limited value for laser remote sensing. *Q-switching* is a technique that enables much of the laser energy to be compressed into a single, very intense pulse with peak powers that can exceed 10^8 W and a duration of about 20 ns.

Q-switching can be applied to most high-energy lasers provided the lifetime of the lasing level is comparable to the pumping period, for in this technique the energy is stored in the inverted population. This is accomplished by making the cavity loss initially very high. This prevents laser action from commencing and enables an appreciable fraction of the pumping energy to be stored in the inverted population. Then at an appropriate moment (ideally when the inversion density is saturating), the cavity loss is suddenly reduced to be commensurate with the lasing configuration (that is to say, the *quality*, or *Q*, of the cavity is suddenly increased). The gain per pass of the system is now well above the threshold value, and consequently oscillation builds up rapidly, converting the energy stored in the inverted population into electromagnetic energy. This leads to an output power pulse of intense magnitude. A schematic illustration of this is presented as Fig. 5.13, where the temporal variation in the population inversion density $\mathcal{N}(t)$ and the photon density $\Phi(t)$ are shown.

Q-switching works well with solid-state lasers, such as ruby and neodymium glass (or YAG), but is not suitable for liquid organic-dye lasers or most gas lasers. There are three basic approaches to Q-switching:

(a) *Misalignment of the Cavity.* In this approach one of the mirrors (or a retroreflecting prism) forming the cavity is rapidly rotated. During most of the rotation the cavity is obviously misaligned; then suddenly at some preset angular position the cavity becomes aligned and laser action can occur.

(b) *Deflection Out of the Cavity.* In this approach the radiation is deflected out of the cavity by some device, such as an acoustooptical deflector, then at the appropriate moment this deflection is suddenly switched off, allowing laser oscillation to start.

(c) *Absorption Within the Cavity.* Another method of ensuring that the gain per pass is kept very low during the pumping period is to include an absorber within the cavity. The problem then becomes one of suddenly eliminating its absorption. One of the most common and least expensive techniques is to use a *saturable absorber*. In this approach a thin layer of dye solution (such as cryptocyanine) becomes photobleached when the radiation level exceeds some suitable value. At that moment the attenuation of the radiation within the cavity falls dramatically and laser action commences.

One of the most reliable methods of rapidly reducing the attenuation within the cavity is to use a *Pockels cell* to rotate the plane of polarization of the radiation field so that it is no longer blocked by a polarizer placed within the

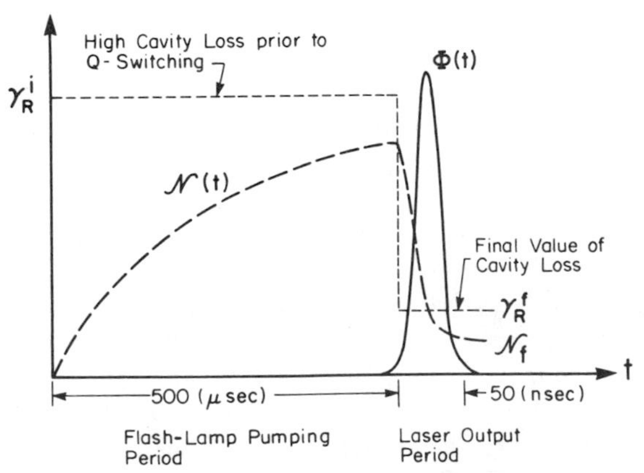

Fig. 5.13. Variation of population-inversion density $\mathcal{N}(t)$ and laser photon density $\Phi(t)$ during Q-switching of a flashlamp-pumped laser.

cavity. This technique provided much better control over the amplitude and timing of the resulting *giant laser pulse* (Röss, 1969).

It is easy to show that the increase in peak power attained through Q-switching can be quite substantial. Suppose that $\mathcal{N}_i$ represents the inverted population density just prior to Q-switching, and $\mathcal{N}_f$ the value at the end of the emission. The energy emitted during the pulse is

$$\mathcal{E}_Q \approx \tfrac{1}{2}(\mathcal{N}_i - \mathcal{N}_f)Vh\nu \tag{5.60}$$

where V is the volume of the active medium and $h\nu$ the laser photon energy. The factor $\frac{1}{2}$ appears because the population-inversion density changes by 2 for each transition between the laser and terminal levels (i.e., N_m increases by one and N_n decreases by one for each laser photon emitted). If we imagine that the emission takes place in roughly twice the final value of the cavity residence time τ_c^f, then the peak power in the pulse is

$$P_Q \approx \frac{(\mathcal{N}_i - \mathcal{N}_f)Vh\nu}{4\tau_c^f} \tag{5.61}$$

In order to obtain a guide to the degree of enhancement of the peak power achieved through Q-switching, we shall assume that $\mathcal{N}_i$ and $\mathcal{N}_f$ are approximately equal to the critical values (5.20) associated with the two states of cavity loss:

$$\mathcal{N}_i \approx \frac{1}{\sigma_0 l}\ln\frac{1}{R_i} \tag{5.62}$$

and

$$\mathcal{N}_f \approx \frac{1}{\sigma_0 l}\ln\frac{1}{R_f} \tag{5.63}$$

where σ_0 represents the peak-stimulated emission cross section and l the length of the cavity. R_i and R_f represents the initial and final values of the cavity geometrical mean reflection coefficients, respectively. If the laser was operated exclusively at the lower value of the cavity loss, then the energy emitted in the same period would be

$$\mathcal{E}_N \approx \tfrac{1}{2}\mathcal{N}_f Vh\nu \tag{5.64}$$

and the corresponding peak power

$$P_N \approx \frac{\mathcal{N}_f Vh\nu}{4\tau_c^f} \tag{5.65}$$

Using (5.62) and (5.63) with (5.61) and (5.65), we may write the peak power ratio

$$\frac{P_Q}{P_N} \approx \frac{\ln 1/R_i}{\ln 1/R_f} - 1 \tag{5.66}$$

Typical values for R_f and R_i are 95% and 0.1%, which leads to a peak power enhancement of about 134 according to (5.66). In the case of a ruby laser $h\nu \approx 1.8$ eV, $\sigma_0 = 10^{-19}$ cm^2, and typically $l = 50$ cm and $V = 10$ cm^3. Consequently, from (5.21)

$$\tau_c^f = \frac{l}{(1 - R_f)c} \approx 33 \text{ ns}$$

and using this in (5.61) we obtain

$$P_Q \approx 30 \text{ MW}$$

Q-switching is inappropriate for short-lived laser levels, and in this case an alternative method of producing a short-duration high-power pulse, termed *cavity dumping*, is employed. This technique is virtually the opposite of Q-switching in that two totally (or as nearly as can be achieved) reflecting mirrors are used to permit the energy of the radiation field to build up within the cavity. Then at an appropriate moment, when the radiation field energy has almost peaked, the radiation is suddenly coupled out of the cavity by either an electrooptical switch such as Pockels cell and Glan prism polarizer (see Fig. 5.14) or an acoustooptical deflector. Some degree of pulse energy compression from a flashlamp pumped organic dye laser has been achieved recently through cavity dumping.

Ultrashort laser pulses are generated through a process known as *mode locking*. Although this technique is not as efficient as Q-switching or cavity

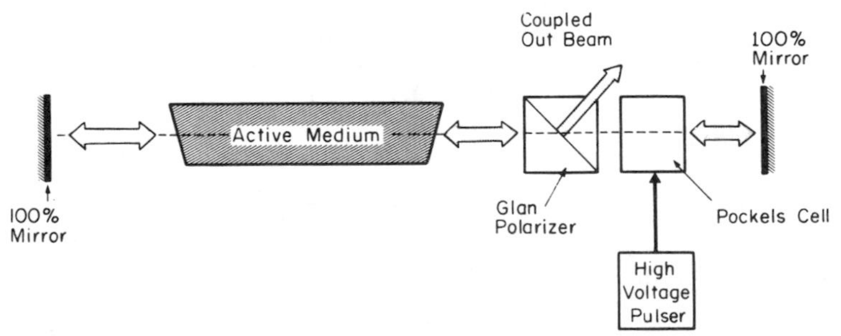

Fig. 5.14. Cavity-dumping configuration employing a Pockels cell to suddenly rotate the plane of polarization of the radiation field within the cavity, thereby causing it to be deflected out within one transit time.

dumping (viz., the fraction of the pump energy utilized is much less) it does lead to the production of extremely short-duration pulses. These are often about a thousandth the duration of pulses created by Q-switching or cavity dumping and can yield subpicosecond-length pulses. The instantaneous peak power in these pulses can be quite high, but their energy content is generally very low (microjoules).

The electric field of radiation oscillating on N temporal (cavity) modes can be written in the form

$$E(t) = E_0 \sum_{n=1}^{N} e^{i(\omega_n t + \delta_n)} \tag{5.67}$$

where ω_n represents the angular frequency of the nth mode and δ_n its relative phase. For simplicity we are assuming that all of the modes have the same maximum amplitude E_0. As we have seen earlier (5.31), the modes of a Fabry–Perot cavity differ in angular frequency by

$$\Delta\omega = |\omega_n - \omega_{n\pm 1}| = 2\pi \frac{c}{2l} \tag{5.68}$$

In general, the modes are not related, so that the relative phase δ_n has random values. Under these circumstances the modes are incoherent with each other and the total irradiance is found by adding the intensities of the modes:

$$I = \tfrac{1}{2} N \epsilon_0 c E_0^2 = N I_0 \tag{5.69}$$

In a *mode-locked* laser, the modes are forced to all have the same phase, in which case the total irradiance is found by adding the electric fields. The total electric field is

$$E(t) = E_0 e^{i\delta} \sum_{n=1}^{N} e^{i\omega_n t} \tag{5.70}$$

or, using (5.68),

$$E(t) = E_0 e^{i(\omega_N t + \delta)} \left[1 + e^{-i\phi} + e^{-2i\phi} + \cdots + e^{-Ni\phi}\right] \tag{5.71}$$

where ω_N is the angular frequency of the highest-frequency mode, and

$$\phi \equiv t\,\Delta\omega = \frac{\pi c t}{l} \tag{5.72}$$

The terms in the brackets of (5.71) form a geometric series, and consequently we can write the total irradiance within the cavity as

$$I(t) = \tfrac{1}{2}\epsilon_0 c |E|^2 = \tfrac{1}{2}\epsilon_0 c E_0^2 \left[\frac{\sin^2(N\phi/2)}{\sin^2(\phi/2)}\right] \tag{5.73}$$

Now for $\phi/2 \to 0$, $\sin(N\phi/2)/\sin(\phi/2) \to N$, and so we see that the maximum irradiance

$$I_{\max} \approx N^2 \tfrac{1}{2}\epsilon_0 c E_0^2 = N^2 I_0 \tag{5.74}$$

so that the peak irradiance for the mode locked laser is N times the value obtained for the case of uncorrelated modes [equation (5.69)]. Furthermore, in general, the quantity within the square brackets in (5.73) equals N^2 for $\phi/2 = m\pi$, where m is an integer.

The temporal variation of the irradiance is thus a series of narrow pulses with a separation between consecutive pulses of $2l/c$. This can be seen by reference to Fig. 5.15, where $(\sin^2 N\pi\tau)/(N^2 \sin^2 \pi\tau)$ is plotted as a function of the *normalized time* $\tau \equiv ct/2l$ for the two cases of 4 and 8 modes. Fourier-transform theory predicts that the duration (FWHM) of the pulses τ_p will be inversely proportional to the total bandwidth:

$$\tau_p = \frac{2l}{Nc} \tag{5.75}$$

This kind of temporal output can be understood in terms of an ultrashort pulse that propagates back and forth within the cavity.

In the case of a ruby or neodymium laser N can be a thousand or more, and so substantial narrowing can be achieved. For dye lasers that can lase over a much broader bandwidth, subpicosecond pulses have been attained. For most

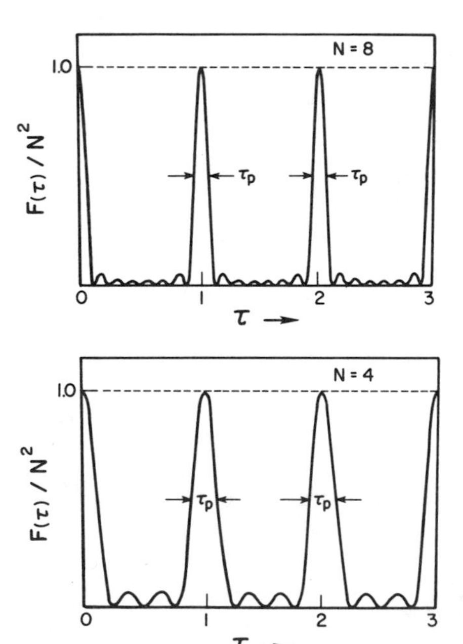

Fig. 5.15. Variation of $F(\tau)/N^2$ with τ, showing the pulse sequence from an idealized mode-locked laser, where $F(\tau) = (\sin^2 N\pi\tau)/(\sin^2 \pi\tau)$ and $\tau = ct/2l$.

applications that require a single pulse, one of the sequence of mode-locked pulses is gated out of the system. Where more energy or higher peak power is required, this pulse is then enhanced by passing it through a suitable traveling-wave amplifier.

At present there is no immediate remote-sensing application for such ultrashort pulses, and so we shall not discuss this topic further. The interested reader is directed to the text edited by Shapiro (1977) and the review article by Bradley and New (1974) for additional information, including methods of generating mode-locked laser pulses.

5.6. OPTICAL PROPERTIES OF LASER BEAMS

In order to obtain a better knowledge of the optical modes within a cavity and to determine the far-field divergence of a laser beam, we have to consider the field distribution within the cavity. We shall use the *scalar approximation*, which assumes that the electromagnetic field is nearly transverse and uniformly (linearly) polarized. Also scale sizes are assumed very much larger than the wavelength of the radiation.

Consider an electromagnetic wave propagating more or less in the z-direction. If $E(\mathbf{r}_1)$ represents the amplitude of the electric field at position $\mathbf{r}_1 = (x_1, y_1, z_1)$ on some input plane at z_1 (this could be, for example, the surface of mirror 1 of a plane-parallel Fabry–Perot cavity) then the magnitude of the electric field as position $\mathbf{r}_2$ on the output plane at z_2 is given by the Fresnel–Kirchhoff diffraction integral (essentially using Huygens' principle):

$$E(\mathbf{r}_2) = \frac{ik}{4\pi}\int_{S_1} E(\mathbf{r}_1)\frac{1 + \cos\theta}{r}\mathrm{e}^{-ikr}\,dS_1 \tag{5.76}$$

where dS_1 represents the element of surface on the input plane and $\mathbf{r} \equiv \mathbf{r}_2 - \mathbf{r}_1$. The z-separation of the two planes is taken to be ℓ, and both S_1 and $S_2 = 4a^2$, where a is half the length of the side of each plane (or mirror). θ is the angle between the propagation direction (line joining dS_1 and dS_2) and the normal to the input plane (or the axis of the system); see Fig. 5.16. As before, $k = 2\pi/\lambda$. The integral in (5.76) is evaluated across the entire surface of the input plane.

Since we are concerned only with relatively narrow well collimated beams we make the following simplifying assumptions:

1. The *obliquity factor* $(1 + \cos\theta)/2$ is ≈ 1.
2. $r \equiv |\mathbf{r}_2 - \mathbf{r}_1| \approx z_2 - z_1 = \ell$ in the denominator of (5.76).
3. In the exponential factor we use the expansion

$$kr \approx k\left\{\ell + \frac{(x_2 - x_1)^2}{2\ell} + \frac{(y_2 - y_1)^2}{2\ell} + \cdots\right\} \tag{5.77}$$

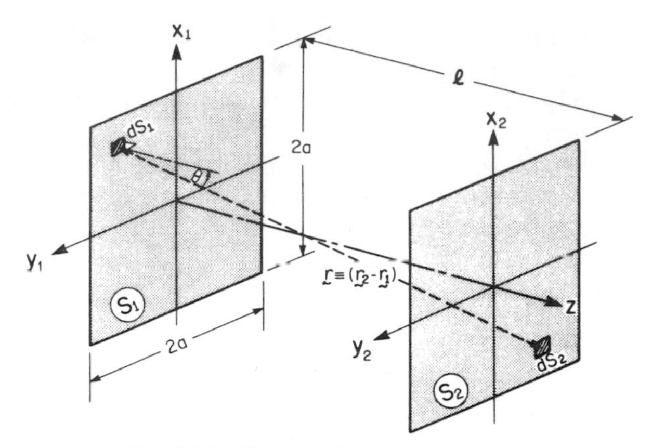

Fig. 5.16. Plane-parallel resonator geometry.

With these approximations, the simplified Huygens integral becomes

$$E(x_2, y_2) = \frac{i\mathbf{e}^{-ik\ell}}{\ell\lambda} \int_{-a}^{a} \int_{-a}^{a} dx_1\, dy_1\, E(x_1, y_1)\mathbf{e}^{-(ik/2\ell)[(x_2-x_1)^2+(y_2-y_1)^2]} \tag{5.78}$$

Fox and Li (1961) showed that after a sufficient number of passes between two plane and parallel mirrors a field distribution is reached that no longer changes from pass to pass, regardless of the initial field distribution on mirror 1. This final distribution can be thought of as an eigensolution of (5.78) and as such represents a transverse mode of the cavity. The shape of this *self-reproducing* field distribution is found to depend on an important parameter of the system known as the *Fresnel number*,

$$N_F \equiv \frac{a^2}{\ell\lambda} \tag{5.79}$$

According to Siegman (1971, p. 306), spherical waves with a Gaussian amplitude distribution form the basic elements in the analysis of the propagating beams within a laser cavity. In the case of such Gaussian spherical waves the magnitude of the electric field is given by the expression (also Maitland and Dunn, 1969, p. 157)

$$E(x, y) = \underbrace{E_0\left\{\frac{2}{\pi}\right\}^{1/2} \frac{1}{w}}_{\substack{\text{normalizing}\\ \text{factor}}} \underbrace{\mathbf{e}^{-(ik/2R)(x^2+y^2)}}_{\substack{\text{spherical wave}\\ \text{of radius } R}} \underbrace{\mathbf{e}^{-(x^2+y^2)/w^2}}_{\substack{\text{Gaussian}\\ \text{distribution}\\ \text{with spot size } w}} \tag{5.80}$$

where R represents the radius of curvature of the wave and w its spot size (or

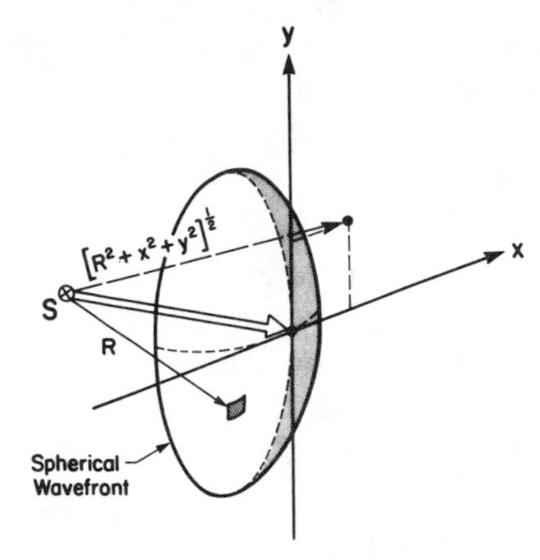

Fig. 5.17. The phase distribution over a transverse plane at a distance R from a point source S is given by

$$e^{-ik(R^2+x^2+y^2)^{1/2}} \approx e^{-ikR-ik(x^2+y^2)/2R}$$

for $R^2 \gg x^2 + y^2$.

exponential radius). The total power in the wave is

$$P = \iint \tfrac{1}{2}\epsilon_0 c |E(x, y)|^2 \, dx\, dy = \tfrac{1}{2}\epsilon_0 c E_0^2 \tag{5.81}$$

which enables us to write

$$E_0 = \left\{\frac{2P}{\epsilon_0 c}\right\}^{1/2} \tag{5.82}$$

The spherical nature of the wave is revealed by the factor

$$e^{-(ik/2R)(x^2+y^2)}$$

for, as seen in Fig. 5.17, the phase distribution over a transverse plane at a distance R from a point source is given by the factor

$$e^{-ik[R^2+x^2+y^2]^{1/2}} \approx e^{-ikR}e^{-(ik/2R)(x^2+y^2)}$$

provided $R^2 \gg x^2 + y^2$.

Siegman (1971) shows that a beam starting as a Gaussian plane wave at the $z = 0$ plane, that is to say,

$$E(x_0, y_0) = \left\{\frac{4P}{\pi\epsilon_0 c}\right\}^{1/2} \frac{1}{\omega_0} e^{-(x_0^2+y_0^2)/\omega_0^2} \tag{5.83}$$

evolves through diffraction into a Gaussian spherical wave with a diverging radius of curvature

$$R(z) = z\left[1 + \left(\frac{\pi\omega_0^2}{z\lambda}\right)^2\right] \tag{5.84}$$

where ω_0 is the spot size at the *waist* (i.e., the $z = 0$ plane). This is illustrated in Fig. 5.18. It is apparent that at sufficiently large distances a spherical wavefront develops with its center of curvature located essentially at the waist. Furthermore, the spot size at some axial position z is given by

$$w(z) = w_0 \left[1 + \left(\frac{z\lambda}{\pi w_0^2} \right)^2 \right]^{1/2} \tag{5.85}$$

This suggests that at large distances from the waist, i.e., $z \gg \pi w_0^2/\lambda$ (the Rayleigh range), the beam diverges linearly with z, and the corresponding diffraction-limited far-field Gaussian-beam divergence half angle is

$$\theta_d^G \equiv \frac{w(z)}{z} \approx \frac{\lambda}{\pi w_0} \tag{5.86}$$

which is in fact the minimum that can be achieved.

It is instructive to compare this result with that for an ideal plane wave (i.e., having perfect spatial coherence and uniform irradiance). Application of the Fresnel–Huygens principle to such a plane wave yields a divergence half angle

$$\theta_d = 1.22 \frac{\lambda}{D} \tag{5.87}$$

where D is the beam diameter. This divergence, which also arises from diffraction, can be compared directly with that obtained in the case of a Gaussian beam (5.86) by setting $D = 2w_0$. Clearly for comparable distances the divergence of the Gaussian beam is about half that for the plane wave.

A diffraction-limited Gaussian laser beam of radius 0.5 cm and wavelength 500 nm would be expected to spread to only 0.707-cm radius at the Rayleigh range of 157 m. Beyond that distance it would spread linearly and have a

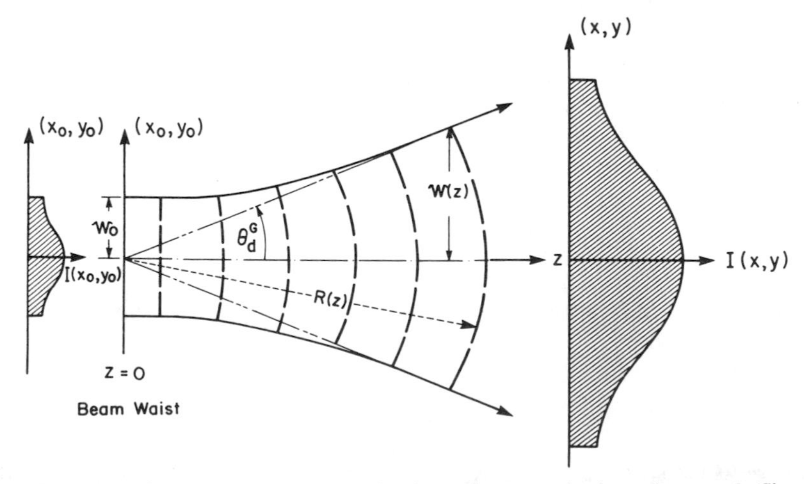

Fig. 5.18. Outward propagation of a Gaussian beam from a waist located at $z = 0$ (Siegman, 1971).

radius of roughly 159 cm at a range of 5 km. The divergence of the beam in the far field is of obvious importance to the subject of laser remote sensing, for it is clear from the above discussion that if a laser beam is to be kept as well confined as possible it is important that its divergence angle be as small as possible. This can be accomplished for a given wavelength by attempting to make the beam diffraction-limited and the output aperture as large as possible, commensurate with other considerations.

5.7. TRANSVERSE MODES AND DIFFRACTION LOSS

An important property of a Gaussian spherical waveform is that it represents the simplest waveform that can reproduce itself upon being processed by the Huygens integral (5.78). As such it represents the simplest or lowest-order transverse mode of a laser cavity (that is to say, only the amplitude and not the shape of the field distribution is changed by successive bounces within the cavity).

There are, however, more complicated waveforms that are also self-reproducing and as such represent possible transverse modes of a laser cavity. The existence of transverse modes is necessitated by the finite lateral dimensions of any real cavity, since diffraction losses mean that the plane-wave solutions considered earlier in the Fabry–Perot analysis (Section 5.3, where the criteria for laser action were developed) are only approximations that are valid close to the optical (z) axis of the cavity.

These more general waveforms comprise a doubly infinite sequence of higher-order *Hermite–Gaussian modes*. Each retains the same transverse field distribution at every longitudinal position z, except for a change in the amplitude. Boyd and Gordon (1961) have shown that in the case of a *confocal resonator* (formed by two curved mirrors whose focal points coincide) the magnitude of the electric field in a $\mathrm{TEM}_{l,m}$ (transverse electric and magnetic) mode can be expressed in the form

$$E_{l,m}(x, y, z) = E_{l,m}^{0} H_l\left\{\frac{x\sqrt{2}}{w(z)}\right\} H_m\left\{\frac{y\sqrt{2}}{w(z)}\right\} \mathrm{e}^{-(x^2+y^2)/w^2(z)} \mathrm{e}^{-ik_{l,m}z} \tag{5.88}$$

where $H_l\{x\sqrt{2}/w(z)\}$ is a Hermite polynomial of order l and argument $x\sqrt{2}/w(z)$. In a similar manner $H_m\{y\sqrt{2}/w(z)\}$ is a Hermite polynomial of order m and argument $y\sqrt{2}/w(z)$. Finally, $E_{l,m}^0$ is the appropriate amplitude factor and $k_{l,m}$ is the relevant propagation constant. The transverse variation of the electric field along x (or y) is seen to be of the form $H_l(\xi)\mathrm{e}^{-\xi^2/2}$, where $\xi = x(\text{or } y)\sqrt{2}/w(z)$ and $l = 0, 1, 2, \ldots$.

A few of the lowest-order x (or y) distributions are presented in Fig. 5.19. It is apparent that the lth-order waveform contains l zeros or null points. These

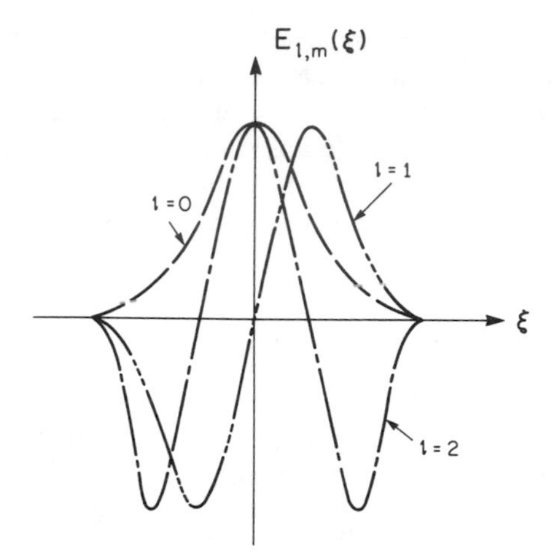

Fig. 5.19. Hermite–Gaussian functions $C_l H_l(\xi) e^{-\xi^2/2}$ corresponding to higher-order transverse modes within a confocal cavity. Modes are normalized to same peak amplitude (Yariv, 1976).

null points become translated into nodal planes with respect to the field distribution given by (5.88). Moreover, since $H_0(\xi) = 1$, it is clear that the TEM_{00} mode corresponds to a simple Gaussian (or uniphase) distribution. The electric field orientation and relevant nodal planes associated with the first few transverse modes (for both rectangular and circular configurations) are indicated in Fig. 5.20.

The standing-wave condition used previously [equation (5.30)] to determine the ideal longitudinal (or temporal) mode frequencies can also be applied to the more general situation. Basically the resonance condition states that the axial phase shift per pass must equal $q\pi$, where q is an integer. Siegman (1971, p. 333) indicates that the resonance frequency of the qth axial mode in the

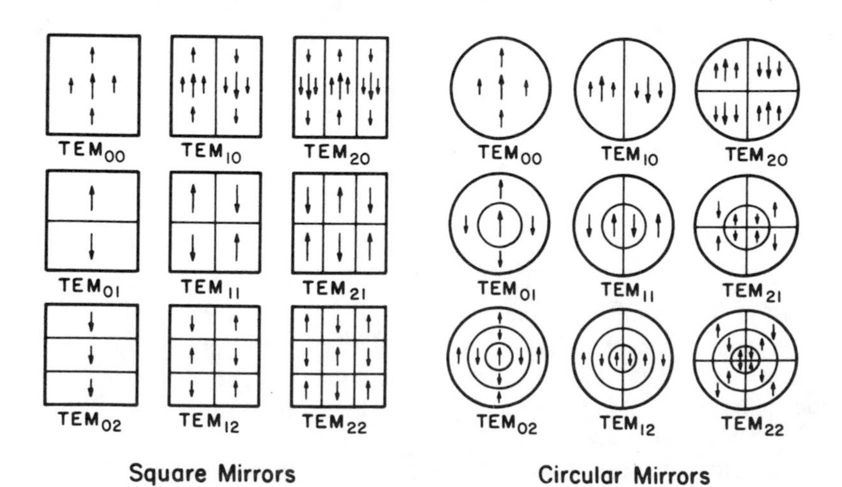

Fig. 5.20. Electric-field orientations and nodal planes associated with the $TEM_{l,m}$ modes for l, $m = 0, 1, 2$.

*lm*th transverse-mode pattern for a confocal arrangement in given by the relation

$$\nu_{qlm} = \frac{c}{2\ell}\left[q + \tfrac{1}{2}(l + m + 1)\right] \tag{5.89}$$

In general l and m are very much smaller than q, which means that the simpler expression (5.30) is quite often adequate. Put another way, the frequency separation of consecutive transverse modes is very much less than the frequency spacing between longitudinal (or temporal) modes. The small differences in frequency between the different transverse modes could, however, be important where heterodyning is employed as a means of detecting weak signals. Care must obviously be used under these circumstances to reduce the transverse-mode content of the laser beam.

The work of Fox and Li (1961) also revealed that the diffraction loss per transit for a given longitudinal (or temporal) mode depended upon the Fresnel number N_F [equation (5.79)] of the resonator and the order of the transverse mode, the lowest-order transverse modes suffering less diffraction loss than the higher order ones. This is illustrated in Fig. 5.21, where the percentage of diffraction loss per transit is plotted as a function of N_F for the TEM_{00} and TEM_{10} modes of a plane (Fabry–Perot) resonator.

It is quite evident that for small values of $a^2/\lambda\ell$, the diffraction loss associated with plane resonators is too large for them to be employed, and other low-loss configurations have to be used. One of the most popular is the confocal resonator; as reference to Fig. 5.21 indicates, the diffraction losses associated with this cavity are very much less than that of the plane resonators.

A *symmetric confocal resonator* comprises two spherical mirrors separated by their radius of curvature, and—as with all confocal systems—the focal points of the two mirrors coincide. The waist spot size w_0 for a cavity of this kind is

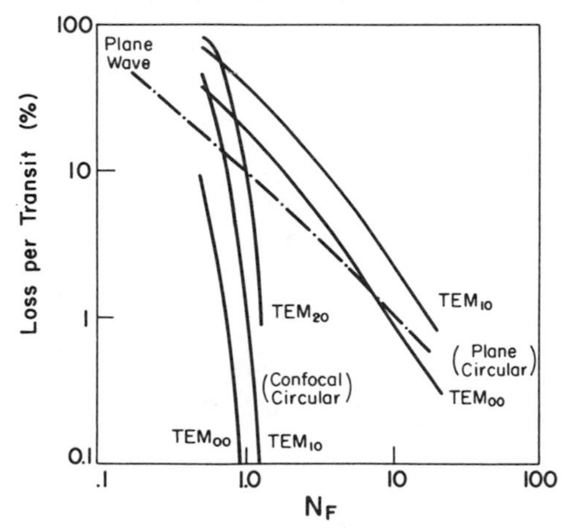

Fig. 5.21. Loss curves as a function of the Fresnel number N_F (from Fox and Li, 1961) for circular plane-parallel and confocal-cavity modes. The diffraction loss for a plane wave is also presented.

given by the simple relation

$$w_0 = \left\{ \frac{\ell\lambda}{2\pi} \right\}^{1/2} \tag{5.90}$$

where ℓ is the axial separation of the mirrors (see Fig. 5.22). The mirror spot size in this instance is $w_0\sqrt{2}$ or $(\ell\lambda/\pi)^{1/2}$ and corresponds to the smallest attainable with any stable symmetric configuration. The far-field divergence half angle for a Gaussian (TEM_{00} mode) output beam for a symmetric confocal resonator is given by

$$\theta_d^G = \frac{\lambda}{\pi w_0} = \left\{ \frac{2\lambda}{\pi\ell} \right\}^{1/2} \tag{5.91}$$

Higher-order modes would have a greater divergence. The influence of optical components, such as lenses, on Gaussian beams can be handled by the concept of *ray-transfer matrix* (e.g., Maitland and Dunn, 1969, p. 158). For certain remote-sensing applications, for example in heterodyne detection, it is preferable if the laser beam has a flat-top field distribution in the far field. Veldkamp and Kastner (1982) have proposed a method of beam-profile shaping based on the use of a diffraction grating and an anamorphic prism beam compressor.

5.8. TYPES OF LASERS AND THEIR PROPERTIES

There are many ways to classify lasers: we can separate them into pulsed and continuous; infrared, visible, and ultraviolet; high-power and low-power; and so on. Probably the most important classification is into *solid-state* (i.e.,

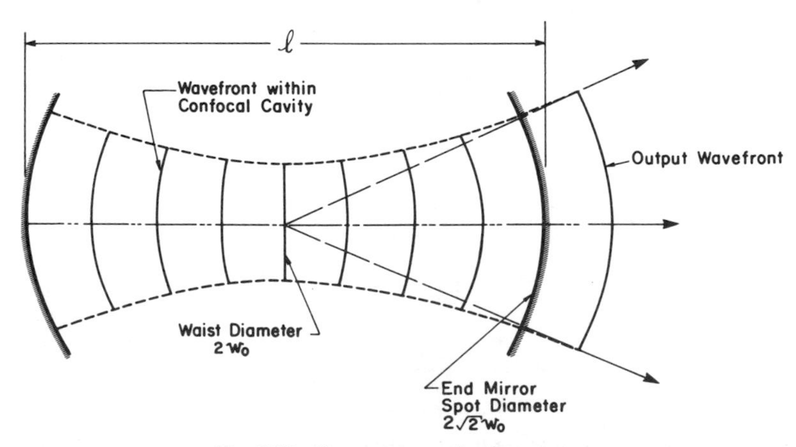

Fig. 5.22. Symmetric confocal resonator.

TABLE 5.1. SOME PULSED LASERS COMMERCIALLY AVAILABLE AND SUITABLE FOR LASER ENVIRONMENTAL SENSING[a]

Type; Wavelength (μm)	Output (J)		Pulse Width (μs)	Pulse Rate (pps)	Beam size (mm) @ Divergence (mrad)	Manufacturer
	TEM	Multimode				
				Gas Lasers, Pulsed		
Carbon dioxide						
9.1–11	3	15	0.1	1–5	30 @ 0.6	Lumonics
Carbon monoxide						
5–7	0.01	—	10–1000	1–1000	5 @ 3	Advanced Kinetics
Copper vapor						
0.5106, 0.5782	0.0002	0.0005	0.02	2000	35 @ 1	Laser Consultants
Deuterium fluoride						
3.5–4	0.5	0.7	0.1–1	0.5	30 @ 0.5	Lumonics
Hydrogen fluoride						
2.64–3.01	0.9	1.3	0.5	0.1–1	30 @ 0.2	Laser Applications
Krypton chloride						
0.222	—	0.06	0.015	1–10	12 × 32 @ 1 × 2	Lambda Physik
Krypton fluoride						
0.248	—	0.02	0.006	1–100	4 × 10 @ 1 × 2	Sopra
0.248	—	1	0.025	1–10	12 × 32 @ 1 × 2	Lambda Physik
Nitrogen						
0.3371		0.003	0.003	1–200	6 × 17 @ 2 × 4	Lambda Physik
0.3371		0.009	0.01	0–50	6 × 22 @ 1 × 7	Molectron
Nitrous oxide						
10.5–11	0.3	1	0.1	1–5	30 @ 0.6	Lumonics
Xenon bromide						
0.282	—	0.025	0.025	0.3	8 × 20 @ 2 × 6	Tachisto
Xenon chloride						
0.308	—	0.5	0.02	1–10	12 × 32 @ 1 × 2	Lambda Physik
Xenon fluoride						
0.351	0.1	0.1	0.014	1–5	9 × 25 @ 2 × 5	Lambda Physik
0.351	—	0.15	0.04	0–3	7 × 19 @ 2 × 6	Tachisto

TABLE 5.1. (*Continued*)

Type; Wavelength (μm)	Output (J) TEM	Output (J) Multimode	Pulse Width (μs)	Pulse Rate (pps)	Beam size (mm) @ Divergence (mrad)	Manufacturer
			Solid-State Lasers, Pulsed			
Alexandrite						
0.73 → 0.78[b]		0.1–0.7	0.07–0.1	20	6 @ 6	Allied Chemical
Erbium-doped YLF						
0.85		0.05	0.15	10	7 @ 2	Sanders Assoc.
1.73		0.005				
Neodymium glass						
0.266 (4th harm.)	0.15		0.003	$\frac{1}{50}$	25 @ 0.3	Quantel Intl.
0.355 (3rd harm.)	0.1		$0.25–2 \times 10^{-4}$			
0.53 (2nd harm.)	—	2	0.03	$\frac{1}{30}$	30 @ 3	Raytheon
1.06		1	0.015	$\frac{4}{60}$	9 @ 3–5	Holobeam Laser
1.06		4	0.02	$\frac{1}{30}$	15 @ 3	Apollo Lasers
Neodymium YAG						
0.265 (4th harm.)		0.02	0.015	20	6 @ 3–5	Holobeam Laser
0.355 (3rd harm.)	0.1		0.005	2–22	6 @ 0.08	Quanta-Ray
0.532 (2nd harm.)	0.3		0.015	20	7 @ 0.6	Quantel
1.06		1.2–1.5	0.015	≤ 30	45 @ 0.8	Intl. Laser Systems
Ruby						
0.6943		2	0.015	$\frac{6}{60}$	9 @ 3–5	Holobeam Laser
0.6943	5		0.003–0.02	$\frac{1}{10}$	13 @ 0.6	Quantel Intl.
0.6943	0.05–1	3–30	0.02	1	15 @ 2–5	Apollo Lasers
0.6943	10	10	0.025	$\frac{1}{10}$	16 @ 0.6	J K Lasers

Semiconductor Lasers

Wavelength (μm)	Peak Power, (W)	Diode or Array	Rep. Rate (Hz)	Pulse Width, (ns)	Beam Divergence (mrad)	Manufacturer
0.85	3	Diode	3000	200	250 × 140	RCA, Electro-Optics
0.904	10	Diode	5000	200	250 × 140	
0.85	75–1000	Array	1000	100	125	Laser Diode Labs
0.85	2025	Diode	5000	100	125	
0.86	3000	Array	10^4	1000	300	Optelecom
0.904	50–300	Array	500	200	157 × 187	ITT Comp. Group

Tunable Dye Lasers

Tunable Spectral Range (nm)	Power (W)	Rep. Rate (Hz)	Line Width (nm)	Pulse Width ns	Pumping Method	Manufacturer
200–3000	10^7	20	0.001	4–6	Nd–YAG	Quanta-Ray
320–1000	1.5×10^6	100	0.001	10–12	Excimer	Lambda Physik
320–1000	6×10^6	50	0.02	6–10	Nd–YAG	
217–760	10^7	10	0.01/0.001	4	Nd–YAG	Molectron
217–950	2.5×10^5	100	0.01	6	N_2 laser	
340–800	10^6	0.1	0.1	300	Coax. flashlamp	Candela

[a]Source: Laser Focus Buyers' Guide
[b]Tunable

insulating—crystalline or glass), *gas*, *liquid*, and *semiconductor*. A detailed study of the many different types of lasers is well beyond the present text; instead we shall concentrate on those lasers that are capable of emitting high-power, short-duration, narrow-bandwidth pulses of radiant energy with a low degree of divergence. Lasers of this kind are required for probing the environment. A high repetition rate is also needed where airborne mapping is undertaken or the return signal is very weak.

In the early days of laser remote sensing the availability of only fixed-frequency lasers operating in the red or infrared by and large restricted this activity to studying Rayleigh and Mie backscattering signals from the atmosphere. The subsequent explosive development of laser technology has considerably expanded the repertoire of laser interactions that can be exploited for the purpose of probing the environment remotely. The range of lasers available for remote sensing can be gauged from the selection of commercial pulsed lasers presented in Table 5.1 and the overview of lasers and their properties displayed in Table 5.2. An even wider spectrum of possibilities exist if use is made of second (or third) harmonic generation, parametric conversion, or Raman shifting.

5.8.1. Solid-State Lasers

Impurity ions in a crystalline lattice (or glassy material) constitute the active ingredients in this class of laser. The very first laser was a *ruby* laser, in which Cr^{3+} ions were embedded in an Al_2O_3 (sapphire) crystal. Nd^{3+} is the impurity ion in the $Y_3Al_2O_{15}$ (yttrium aluminum garnet, or YAG) laser. Recently, a new class of *tunable* solid state lasers has been developed (Walling et al., 1980). These *alexandrite* lasers comprise Cr^{3+} ions in a $BeAl_2O_4$ (chrysoberyl) crystal host.

Each of these lasers is optically pumped over a fairly broad band by means of a flashlamp (see Fig. 5.23) and operates as a high-energy system. *Q*-switching transforms them into high-power, short-pulse lasers that are ideal for certain remote-sensing applications. Although the output wavelength of a ruby laser can be varied over a very limited range (± 0.4 nm) through careful control of the ruby's temperature, by and large ruby and Nd–YAG lasers can be thought of as operating at a fixed frequency or wavelength: 694.3 nm for ruby and 1.06 μm for Nd–YAG.

Although the alexandrite lasers can operate in a three-level mode that is similar to that of the ruby laser (but at 680.4 nm), they can also lase on a band of *vibronic* transitions which endow them with four-level tunability from 701 to 818 nm. This tuning can be accomplished through the use of birefringent filters, giving linewidths of about 0.1 nm. An order-of-magnitude reduction in the laser linewidth can be attained through the combination of a birefringent filter and an intracavity etalon. The peak stimulated-emission cross section for the *R*-line at 680.4 nm is 3×10^{-19} cm^2, which is about a factor of three larger than for ruby. The peak cross section in the tunable region of these alexandrite

TABLE 5.2. TYPES OF LASERS RELEVANT TO REMOTE SENSING

	Solid State	Gas	Liquid	Semiconductor
Representative examples	Ruby Neodymium (YAG) Alexandrite	XeCl (rare-gas halide) N_2 (transient) $HgBr_2/HgBr$ (dissociation) CO_2 (molecular)	Organic dyes such as: Rhodamine 6G Coumarin Cresyl violet	GaAs GaAsP InAs $Pb_{1-x}Sn_xSe$
Primary pumping technique	Flashlamp	Intense electrical discharge in gas	Flashlamp or laser	High current injection leading to n, p radiative annihilation at an n–p junction
Range of wavelengths and tuning	Ruby (694.3 nm)—thermal tuning ± 0.4 nm Nd–YAG (1.06 μm) Alexandrite—tunable (701–818 nm) Second (or third) harmonic generation possible with all three kinds	H_2 (116, 160 nm) Xe_2 (170 nm) KrF (249 nm) XeCl (308 nm) N_2 (337 nm) $HgBr_2/HgBr$ (502–504 nm) DF or HF (2.7–4.0 μm) CO (5.0–5.7 μm) CO_2 (9.0–11 μm) HCN (337 μm)	Large range of dyes provide wavelengths from 340 nm to 1.1 μm Typical tuning range per dye $\approx$ 40 nm with widths of 0.1–0.01 nm possible with grating or prism (+etalon) arrangement	GaAsP–$Pb_{1-x}Sn_xSe$ (550 nm to 32 μm) Tuning possible by changing current, applying pressure or magnetic field
Modes of operation and pulse duration	Q-switching leads to 10–100-ns pulses Mode-locking can yield 10-ps pulses	Fast discharges lead to pulses that typically range from 1 ns to 1 μs Q-switching possible with certain molecular gas lasers, cavity dumping with others	When N_2 laser pumped pulses are ~ 5–10-ns When flashlamp pumped 0.3–1-μs pulses Cavity dumping of latter can yield 30-ns pulses	Current pulsed but requires cooling and efficient heat sink 10 ns to 1μs possible
Peak power and energy/pulse attainable	For ruby and Nd–YAG 10^6–10^8 W and 1–10 J when Q-switched; for Alexandrite lasers 10^7 W and 500 mJ	10^4–10^7 W and 1 mJ to 1 J	10^4–10^6 W in narrow, tunable bandwidth; 0.1– 3 J	100 W possible from laser diode arrays

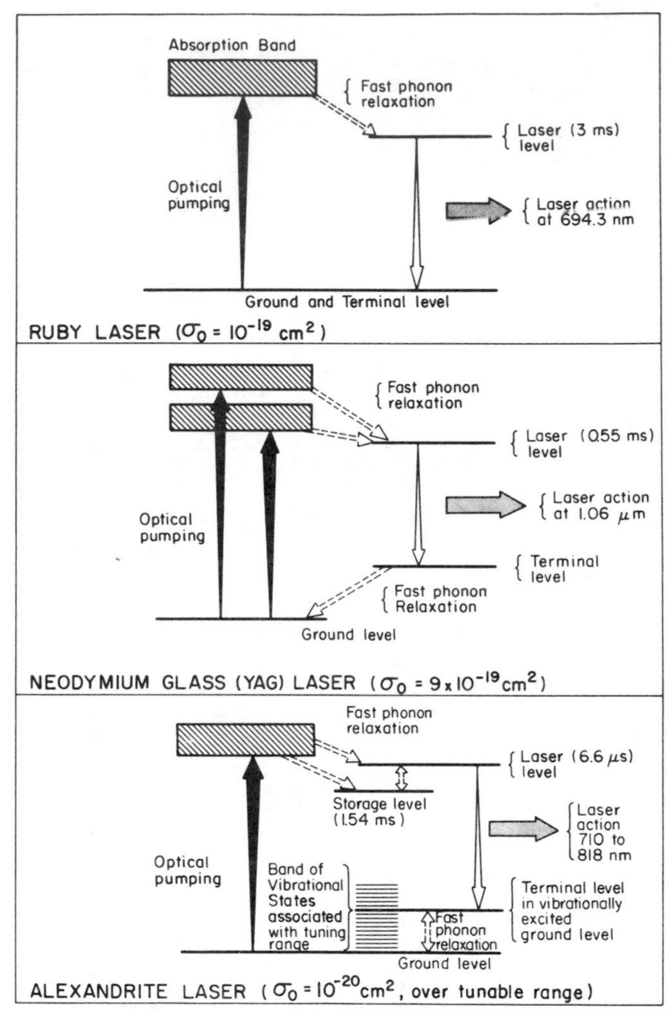

Fig. 5.23. Comparison of laser schemes for solid-state lasers [Walling et al. (1980), "Tunable Alexandrite Lasers," *IEEE J. Quant. Electron.*, 1309, **QE-16**, © 1980 IEEE].

lasers is closer to 10^{-20} cm^2 at room temperature. The three basic energy level configurations associated with ruby, Nd–YAG, and alexandrite lasers are presented as Fig. 5.23.

In each case an intense flash of optical energy pumps a broad absorption band of the medium. A sizable fraction of the resulting excited-state population rapidly decays through a fast nonradiative (lattice vibration) relaxation to a much sharper level. This level has a sufficiently long lifetime that a population inversion can develop between it and a terminal level (in a four-level system) or the ground level (in a three-level system).

It is apparent from Fig. 5.23 that when alexandrite is operated as a four-level laser, its threshold for laser action should be much less than ruby and its efficiency somewhat higher. This is in fact borne out experimentally and illustrated in Fig. 5.24.

A typical solid state laser (ruby or Nd–YAG) used in remote sensing applications has an output pulse energy in the range 100 mJ to 1 J. When Q-switched this translates into a peak power of 1 to 50 MW with a pulse duration of between 10 and 35 ns. The beam divergence of about 1 mrad is usually acceptable, and a repetition rate of up to 10 Hz is possible. The penalty for requiring a high repetition rate is usually water cooling of the laser.

Second- and third-harmonic conversion is often required, particularly with Nd–YAG, where the primary wavelength of 1.06 μm is of little value except for Rayleigh and Mie scattering. Frequency doubling or tripling is accomplished through the use of suitable nonlinear crystals such as KDP (potassium dihydrogen phosphate). The efficiency of this conversion depends upon the quality of the crystal, wavelength, irradiance, and coherence properties of the laser beam. A value of 10% is quite normal, 50% representing an upper level of performance.

Recent technological improvements in the design of Nd–glass lasers have led to the development of a compact version that is suitable for airborne operations (NASA SP-433, 1979). This has renewed interest in this kind of laser, particularly that it can be tuned across its spontaneous linewidth ($\approx$ 200 cm^{-1}).

5.8.2. Gas Lasers

Gas lasers are the most versatile class of laser: within this classification fall the shortest-wavelength laser, the longest-wavelength laser, the highest-efficiency laser, and the most powerful (cw) laser. Gas lasers can be pumped by a wide variety of techniques that include the use of electrical discharges, electron beams, chemical-flow combinations, thermo-gasdynamic expansions, and nuclear-fission fragmentation. The last technique raises the possibility of direct nuclear-pumped lasers. The mechanisms for creating population inversions in gas lasers include electron excitation or ionization, molecular dissociation,

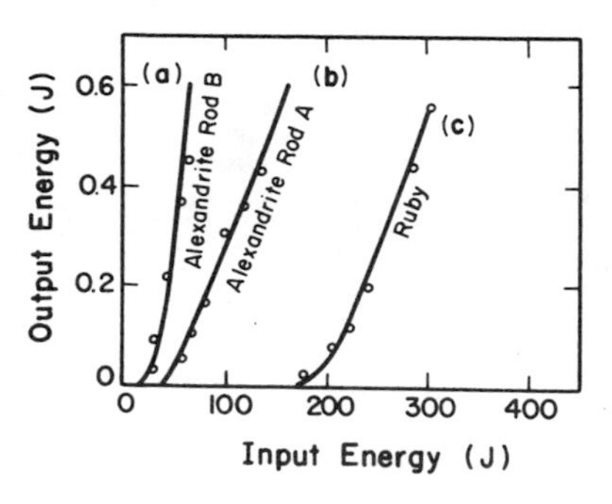

Fig. 5.24. Comparative performance of alexandrite and ruby in the same apparatus with optimized output coupling. The total Cr concentrations and slope efficiencies are: (a) alexandrite rod B, 4.2 $\times$ 10^{19} ions/cm^3, 1.5 atm %; (b) alexandrite rod A, 1.5 $\times 10^{19}$ ions/cm^3, 0.50 atm %; (c) ruby 1.6 $\times$ 10^{19} ions/cm^3, 0.57 atm %. In alexandrite, only 53% of the Cr ions are on the active mirror sites [Walling et al. (1980), "Tunable Alexandrite Lasers," *IEEE J. Quant. Electron.*, 1313, **QE-16**, © 1980 IEEE].

chemical reactions, the formation of excited-state complexes, Penning ionization, and resonant energy transfer.

In the case of gas lasers that are likely to be of interest in the field of laser remote sensing, an intense electrical discharge will probably constitute the basic energy source. We shall limit our discussion of gas lasers to a brief review of the operation and properties of a few representative examples. In each case we shall attempt to select the laser which is likely to be of most interest. We shall commence with the best-known infrared laser and progress through to an exciting new laser that operates in the near ultraviolet. Although we shall not discuss in detail the inert-gas ion lasers, we should mention in passing that this class of laser provides continuous, stable, spectrally very narrow output at quite modest power levels (tens of watts) for several wavelengths in the near ultraviolet, blue, and red regions of the spectrum. The most intense lines from the argon ion laser are at 488 and 514.5 nm (Gordon et al., 1964; Bridges et al., 1971).

A representative gas laser is schematically illustrated in Fig. 5.25. In most instances the gain per unit length of a gas laser is so small that a long cavity is required to ensure sufficient gain per pass. Unfortunately, the diffraction loss from such a low-Fresnel-number [equation (5.79)] cavity is excessive unless a confocal resonator is employed (see Fig. 5.21). In addition, reflection losses from the discharge tube windows can be eliminated for radiation with its electric field in the plane of the page provided the windows are tilted at the Brewster angle [equation (2.72)] with respect to the optical axis of the system.

The Carbon Dioxide Laser. Laser action in the carbon dioxide molecule arises between pairs of vibrational levels of the ground electronic state (see Fig. 5.26). This triatomic linear molecule possesses three nondegenerate vibrational modes: (1) the *symmetric mode* ($\nu_1 00$), (2) the *bending mode* ($0\nu_2 0$), and (3) the *asymmetric mode* ($00\nu_3$), where ν_1, ν_2, and ν_3 represent the appropriate vibrational quanta in each mode. The most prominent laser emission occurs at about 10.6 μm on a series of rotational lines between the (001) and (100) vibrational levels shown in Fig. 5.26. The (001) laser level is pumped by direct electron excitation of the ground state (000) or, more efficiently, by resonant

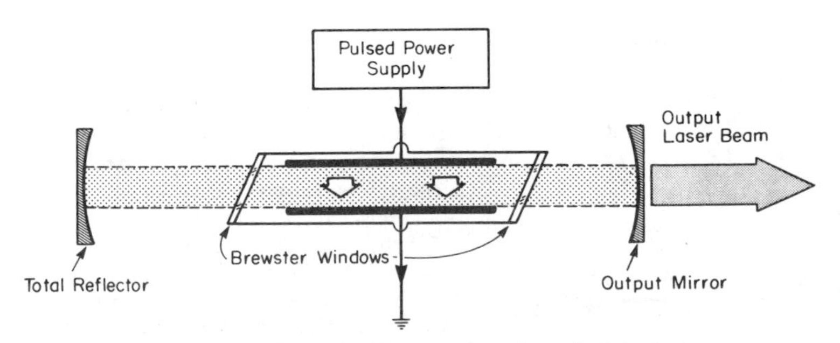

Fig. 5.25. Schematic representation of a pulsed gas laser.

energy transfer with vibrationally excited nitrogen molecules that are also present in the discharge tube. In fact optimum performance of a CO_2 laser is obtained when both N_2 and He gases are present, usually in the ratio $CO_2 : N_2 : He \approx 1 : 1 : 10$. The nitrogen aids in populating the (001) laser level:

$$N_2(v = 1) + CO_2(000) \rightarrow N_2(v = 0) + CO_2(001)$$

while the helium atoms are quite efficient at depopulating the terminal level, thereby preventing a bottleneck (buildup of population in the terminal level due to a low rate of radiative decay) and maintaining an appropriate electron energy distribution.

Some degree of tuning is possible by using a diffraction grating in place of one of the cavity mirrors. Under the normal operating conditions (i.e., pressures of less than 1 atm), rotation of the grating causes the laser emission to jump from one rotational line to another. At high pressures (> 15 atm), continuous tuning is possible. Commercial pulsed CO_2 lasers that emit pulses of 1 to 10 J with a repetition rate of 1 to 10 Hz are readily available (see Table 5.1). Wood (1974) has written a fairly comprehensive review of high-pressure pulsed molecular lasers.

In a low-pressure CO_2 laser the Doppler-broadened gain profile is only about 50 MHz wide, while the mode separation of a typical 1-m-long cavity is 150 MHz. Consequently, this kind of CO_2 laser operates in a single longitudinal mode and can work with a *conversion efficiency* (electrical energy in to coherent infrared radiant energy out) of up to 20%. Many of the pulsed CO_2 lasers that are useful in remote sensing are classified as TEA (transverse electric discharge at atmospheric pressure) lasers. Recently, TEA lasers using other gases such as HF, DF, and CO have been finding application in remote sensing.

The Mercurous Bromide Laser. A relatively new laser that is being viewed as one of the prime candidates for oceanographic work because of its blue–green (502- and 504-nm) output is the mercurous bromide (HgBr) laser. This laser

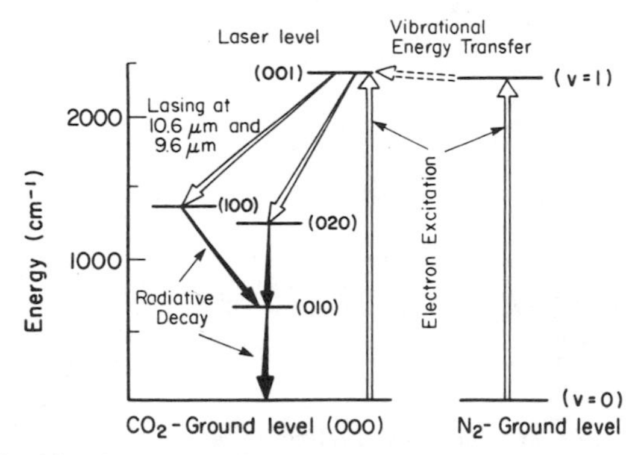

Fig. 5.26. Low-lying vibrational energy-level structure and lasing transitions in a CO_2–N_2 laser.

operates on vibronic transitions between the first excited state and the ground state of the HgBr radical. The electronic structure of HgBr is quite well suited for laser action because the lasing state is strongly (ionically) bound, while the ground state is weakly (covalently) bound. This tends to result in a large Franck–Condon shift so that the lasing transition terminates on a high vibrational state of the ground electronic state. These terminal states are rapidly deactivated by collisions with a buffer gas, leading to efficient laser action. HgBr is created either by photodissociation of mercuric bromide ($HgBr_2$) using an ArF laser at 193 nm (White, 1977) or by collisional dissociation in a discharge (Burnham and Schimitschek, 1981); see Fig. 5.27.

$HgBr_2$–HgBr lasers with an output energy of about 0.5 J have been demonstrated with an overall efficiency of close to 1%. There has also been some indication that operation at high repetition rates ($\approx$ 100 Hz) might be possible. The free-running output of this laser is centered around two peaks at 502 and 504 nm. There is, however, evidence that single-line operation, with a 0.05-nm width, tunable between 495 and 505 nm, should be attainable, (Burnham and Schimitschek, 1981).

The Nitrogen Laser. The nitrogen laser is representative of a class of *self-terminating*, or *transient*, lasers in which laser action ceases as a result of a buildup of the population density in the terminal level. Several metal vapor lasers (Cu at 510.5 and 570 nm, Pb at 722.9 and 363.9 nm, and Tl at 535.0 nm) and the neon laser at 540.1 nm are also self-terminating lasers. All of these

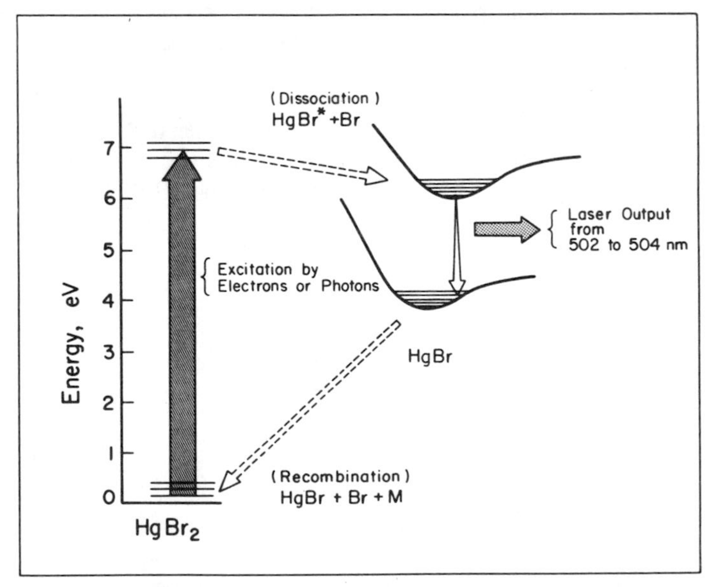

Fig. 5.27. Energy levels and processes involved in an $HgBr_2$–HgBr laser, (Burnham and Schimitschek, 1981).

lasers have very high gain coefficients when pumped adequately (i.e., fast enough), and can operate with only one mirror. Indeed, the nitrogen laser often operates without any mirrors and in effect represents a single-pass stimulated-emission device.

Laser action in the nitrogen molecule occurs principally in a 0.1-nm band centered about 337.1 nm and arises between two vibronic states. The laser level is one of the low-lying vibrational states of the $C\,^3\Pi_u$ electronic state, while the terminal level is one of the low-lying vibrational states of the $B\,^3\Pi_g$ electronic state. This is illustrated in Fig. 5.28. The creation of a population inversion between these states is possible because the probability of electron-impact excitation is greater for the $X \rightarrow C$ transition than it is for $X \rightarrow B$. The radiative lifetime of the laser level is about 40 ns, which necessitates very fast pumping. This is generally accomplished by means of a rapid transverse (to the direction of propagation) discharge in a long channel of nitrogen gas (Leonard, 1965; Basting et al., 1972).

The output pulse duration of the nitrogen laser can range from subnanosecond for a high-pressure system to 10 ns when operated at around 20 torr. Although the peak power of these lasers can exceed 10^6 W, their divergence is inherently quite large (10 mrad typically) because of their single-pass nature. Consequently, some form of output collimating optics is often necessary. Nevertheless, the low cost, simple and rugged construction, and high repetition rate (1000 Hz) possible with this kind of laser makes it attractive for certain roles in remote sensing (Measures et al., 1973).

The Rare-Gas Monohalide Laser. The new class of lasers termed *excimer* lasers possess characteristics that make them potentially interesting for remote sensing. Rare-gas halide lasers constitute one family of such lasers that operate primarily in the ultraviolet (Sze, 1979). The most important members of this family and their salient properties are presented in Table 5.3. In general these lasers emit pulses of 10- to 20-ns duration, although pulse lengths of 100 ns have been achieved in large systems operating at above atmospheric pressure (Ewing, 1978; Sze, 1979).

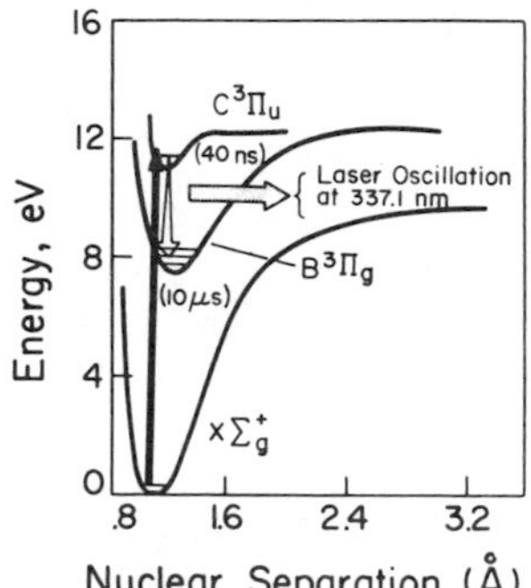

Fig. 5.28. Laser action between two electronic states of the nitrogen molecule (Leonard, 1967).

TABLE 5.3. RARE-GAS HALIDE LASERS[a]

Lasing Molecule	Wavelength (nm)	Pulse Energy (mJ)
ArF	193	160
KrCl	222	160
KrF	249	380
XeBr	282	100
XeCl	308	400
XeF	352	160

[a]Sze (1979).

The rare-gas monohalides belong to a large class of molecules broadly known as excimers (or more correctly *exciplexes*, as the molecule is heteronuclear). These molecules have a ground state that is unstable and tends to dissociate in about 10^{-12} s, whereas the excited states in which they are formed through collisions are strongly bound transfer states (for example $Kr^{+}F^{-}$) which have radiative lifetimes of 10^{-9} to 10^{-8} s. As such, they are ideal candidates for laser action and can be made to operate with an efficiency in excess of 1%. The excited state of the excimer can radiate into a broad band, and as a result of this the stimulated-emission cross section (10^{-17} to 5×10^{-16} cm^2) is somewhat smaller than that associated with similar atomic or molecular transitions (10^{-15} to 10^{-12} cm^2) of comparable oscillator strengths in the

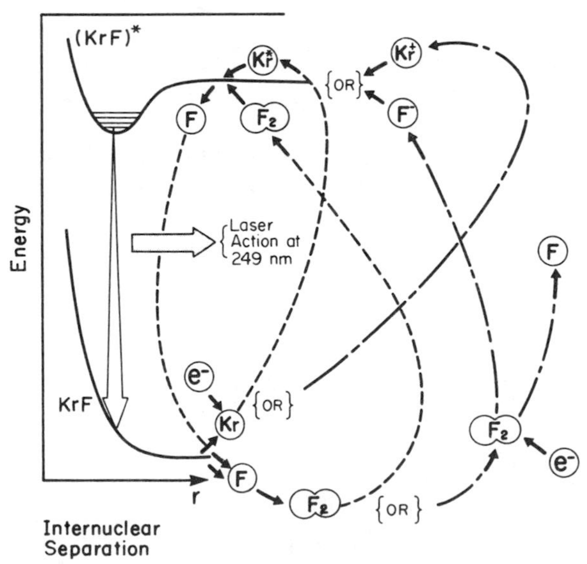

Fig. 5.29. Schematic illustration of the two collision mechanisms leading to the formation of the excimer KrF* and its subsequent laser action at 249 nm.

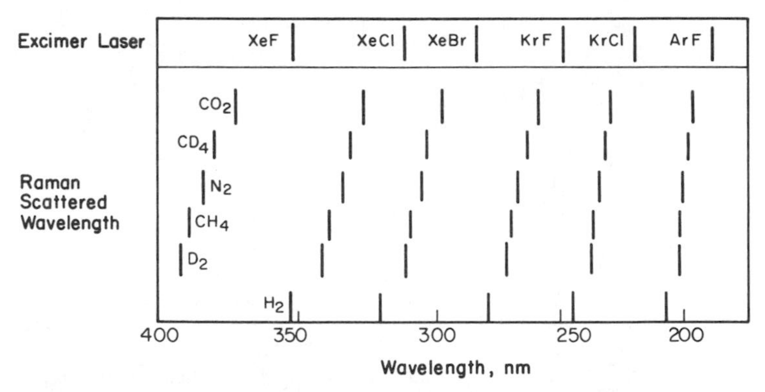

Fig. 5.30. Excimer-laser frequencies and Stokes-shifted frequencies (Ewing 1978).

ultraviolet. As a consequence the gain coefficient of these excimers is quite small, and intense pumping is required within the short radiative lifetime of the excited state. The two primary creation paths of the KrF excimer are illustrated with the relevant part of the energy-level diagram for the krypton fluoride laser in Fig. 5.29.

Although these lasers can be excited by electron beams, the commercial systems of interest for remote sensing utilize fast transverse electrical discharges. At high pressures of operation the formation of triatomic species, such as Kr_2F^*, becomes both a blessing and a curse. The latter results from their absorption at laser wavelengths, the former lies in their promise of leading to the development of a new class of broadband tunable ultraviolet lasers (Tittel et al., 1980). Another possibility in regard to attaining laser radiation over a wide range of wavelengths is to Raman-shift the outputs from the rare-gas halide lasers (Ewing, 1978). An idea of the spectral coverage attainable by this means can be gauged by reference to Fig. 5.30.

5.8.3 Liquid Organic-Dye Lasers

The class of organic molecules which absorb strongly and fluoresce intensely are termed *dyes*. The electronic energy-level structure of such molecules primarily arises from two valence electrons and consists of singlet ($S_0, S_1, S_2, \ldots$) and triplet ($T_1, T_2, \ldots$) levels where S_0 is the ground level. In the singlet states the spins of the two electrons are antiparallel, while in the triplet states they are parallel. In general, *spin-flip* transitions ($S \to T$ or $T \to S$) are optically much less probable than $S \to S$ or $T \to T$ transitions. The absorption and emission spectra of these molecules are fairly broad and continuous (Fig. 5.31), due to the substructure of vibrational and rotational levels.

Laser action for such a medium involves excitation ($S_0 \to S_1$) through intense optical pumping—this is accomplished through a flashlamp or another laser—followed by rapid (10^{-12} s) internal conversion to the lowest vibra-

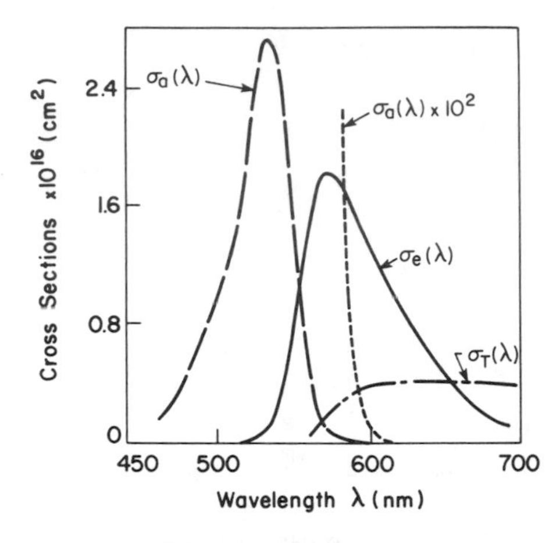

Fig. 5.31. Wavelength variation of the singlet–singlet absorption and emission cross sections, $\sigma_a(\lambda)$ and $\sigma_e(\lambda)$ respectively, for the dye rhodamine 6G. Also indicated is the wavelength variation of the triplet–triplet absorption cross section $\sigma_T(\lambda)$ (Schafer, 1973).

tional–rotational level of the S_1 electronic state. This leads to the development of a population inversion and subsequent laser emission between this low-lying S_1 level and a high vibrational level of the S_0 ground electronic state (see Fig. 5.32). The gain coefficient for a dye laser can be much greater than that of any other medium. Values of 10^3 cm^{-1} have been achieved and can be compared with a value of 0.1 cm^{-1} for a ruby laser. Only semiconductor lasers have gains ($\approx 10^2$ cm^{-1}) approaching that of a dye laser.

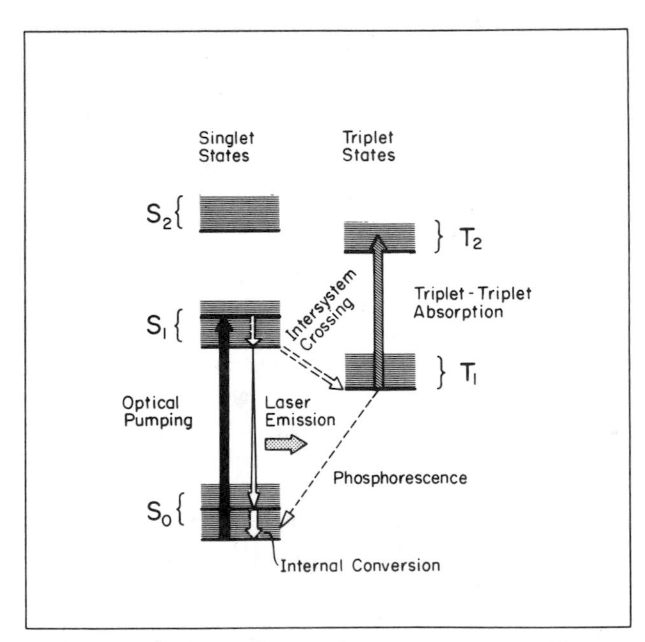

Fig. 5.32. Schematic representation of transitions and energy-level structure associated with an organic-dye laser.

Although the emission spectral width for these dyes can be quite extensive (typically 20 to 50 nm FWHM), the range of wavelengths over which laser action can occur is appreciably less. Laser oscillation on the short-wavelength side of the emission profile is prevented by singlet absorption ($S_0 \rightarrow S_1$ transitions) while on the long-wavelength side it is the drop in the probability of the emission that restricts the range of wavelengths available for laser action (see Fig. 5.33).

Under normal (fluorescence) conditions triplet absorption ($T_1 \rightarrow T_2$ transitions) is negligible because of the low T_1-level populations. When the dye is subjected to the intense optical pumping required for lasing, *intersystem crossing* can lead to the development of an appreciable T_1 population and a corresponding triplet absorption term. Indeed, it was just this triplet absorption that prevented laser action from being accomplished in the very early work on dye lasers. Since the intersystem-crossing rate is small compared to the radiative lifetime of the lasing level, the key to attaining laser action in these dyes was to pump the dye in a time that is short compared to some critical time that permits an appreciable buildup of the triplet-state population.

Basically there are two kinds of pulsed dye laser: those pumped by very fast flashlamps and those pumped by short-pulsed lasers. In the former case, the flashlamps are either collinear or coaxial and are designed to have as small an inductance as possible. They are usually triggered with a thyratron or a spark gap and have risetimes of about 100 ns or less. A schematic representation of this kind of laser is shown in Fig. 5.34. The dye has to be circulated through some form of heat exchanger to avoid the development of thermal (and thereby refractive-index) inhomogeneities. A filter is often used to reduce the scattering loss from bubbles.

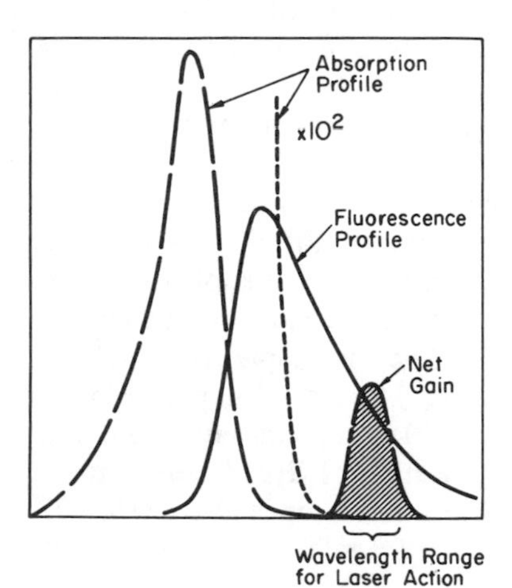

Fig. 5.33. Schematic illustration of why only a small portion of the fluorescence profile is amenable to laser action.

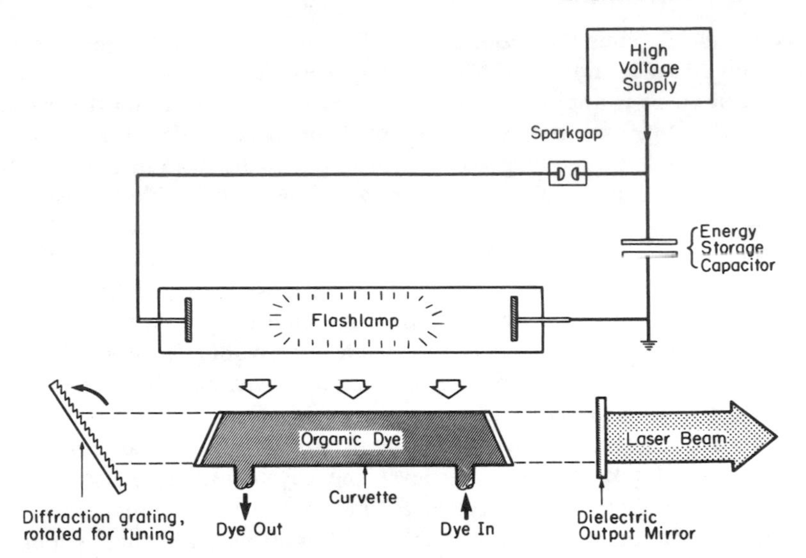

Fig. 5.34. Schematic arrangement of a grating-tuned flashlamp-pumped dye laser.

Spectral tuning and narrowing is often accomplished through the use of a diffraction-grating reflector. The orientation θ of this grating, with respect to the optical axis of the system, determines which band of wavelengths is reflected back through the active medium (see Fig. 5.34). This so-called *Littrow arrangement* can be used to tune the laser across its gain curve, laser action being permitted only at a wavelength that satisfies the relation

$$\lambda = \frac{2d}{m} \cos \theta$$

where d is the *ruling distance* of the grating, and m the order of diffraction. Alternatively, combinations of prisms and etalons can be used for the same purpose. Indeed, for spectral narrowing below about 0.02 nm some kind of *intracavity etalon* is required even if the tuning and initial narrowing are achieved through the use of a grating.

Flashlamp-pumped dye lasers can generate 300-ns to 1-μs pulses of tunable (340–700 nm) radiation with peak powers in excess of 10^6 W in a beam of about 1-cm^2 cross section. Recently, cavity dumping has been applied to these lasers to yield pulses of about 30-ns duration and even higher peak powers.

Morrow and Price (1974) used a simple coaxial flashlamp-pumped dye laser to compare the properties of a number of dyes. Their results, presented as Table 5.4, provide us with a feel for the wavelength, concentration, and associated output energy expected from this selection of dyes when pumped coaxially with a 100 J of electrical energy. Their flashlamp had a 20-to-90% risetime of 150 ns and a FWHM of 500 ns. Experimental tuning curves‡ for

‡A comprehensive set of tuning curves is available from the Exciton Chemical Co. Inc., Dayton, Ohio.

TABLE 5.4. LASER PARAMETERS OF DYES TESTED WITH 100-J ELECTRICAL INPUT ENERGY TO A COAXIAL FLASHLAMP[a]

Dye	Wavelength (mm)	Solvent	Concentration (10^{-4} M)	Cavity Reflectivity	Output Energy (mJ)
DTDCI[b]	760	DMSO[c]	1.0	0.55	40
Nile blue perchlorate[d]	690	Ethanol	2.5	0.55	180
DODCI[e]	665	DMSO[c]	2.5	0.55	95
Rhodamine 6G	595	Ethanol	5.0	0.30	400
Coumarin 6[d]	540	Ethanol	1.0	0.30	60
Esculin monohydrate	460	Ethanol	1.0	0.30	300
Carbostyril 165[d]	425	Ethanol	3.0	0.55	150
Dimethyl POPOP[f]	420	Ethanol	1.0	0.88	—
p-Quaterphenyl[d]	375	DMF[g]	0.7	0.55	12
PBD[h]	362.5	Ethanol	2.5	0.92	—
p-Terphenyl	347.5	*p*-Dioxane	1.0	0.84	10

[a] Morrow and Price (1974).
[b] 3,3′-diethyl-2,2′-thiadicarbocyanine iodide.
[c] Dimethylsulfoxide.
[d] Dyes supplied by Eastman Kodak.
[e] 3,3′-diethyloxadicarbocyanine iodide.
[f] 1,4-bis-2-(4-methyl-5-phenyloxazole) benzene.
[g] Dimethylformamide.
[h] 2-phenyl-5-(4-biphenyl)-1,3,4-oxadiazole.

another selection of dyes also pumped by a coaxial flashlamp was provided by Marling et al. (1974) and is reproduced here as Fig. 5.35. In this work the energy in the flashlamp varied between 73 and 86 J, and the dye concentration ranged between 10^{-5} and 10^{-3} mole/liter. The output reflectivity and grating efficiency also varied from dye to dye so Fig. 5.35, can only be used as a guide in selecting dyes. As we shall see later, it is often desirable, in laser remote

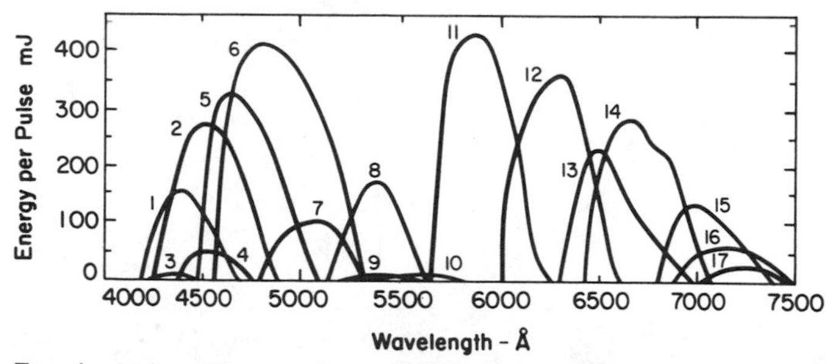

Fig. 5.35. Experimental tuning curves for a selection of dyes used in a coaxial flashlamp-pumped dye laser (Marling et al., 1974). Optimum dyes: (curve 1) Coumarin 120; (curve 2) 4,6-Dimethyl-7-ethylaminocoumarin; (curve 5) Esculin monohydrate; (curve 6) Coumarin 102; (curve 8) Coumarin 6; (curve 11) Rhodamine 6G; (curve 12) Kiton Red S; (curve 14) Cresyl violet nitrate + R6G; (curve 15) Carbazine with triethylamine + R6G.

sensing, to be able to generate two independently tunable laser pulses simultaneously. Kittrell and Bernheim (1976) have proposed several methods of attaining this goal.

For applications where short-duration (1 to 20 ns) tunable laser pulses are required, the dye is optically pumped by another laser. This can be the fundamental or harmonic of a Q-switched solid-state laser, a rare-gas halide laser, or a nitrogen laser. The latter provides somewhat less energy than the other lasers, but represents a relatively low-cost method of generating short tunable laser pulses in the visible.

Tuning and spectral narrowing for a laser-pumped dye laser is generally achieved through the use of a diffraction grating and a *beam expander* (Hänsch, 1972; see Fig. 5.36). The beam expander is necessary in this instance because the active zone within the dye is very narrow, which means that in the absence of a beam expander only a small fraction of the grating would be illuminated, resulting in rather poor spectral narrowing. Although a linewidth of about 0.01 nm can be obtained with the grating–beam-expander arrangement shown in Fig. 5.36, an additional factor-of-ten reduction in the linewidth can be achieved through the use of a suitable intracavity etalon.

Simultaneous tunable two-wavelength oscillation of a nitrogen-laser-pumped dye laser was demonstrated by Inomata and Carswell (1977). These two outputs were then frequency-doubled using a KDP crystal. A particularly attractive feature of their arrangement (especially in regard to laser remote sensing) is the collinearity of the four wavelengths and the fact that it can also be used with a more powerful Q-switched solid-state-laser-pumped dye laser. Another, even more powerful, source of tunable ultraviolet laser radiation was developed by Bücher and Chow (1977), who used an 8×10^6-W, 15-ns

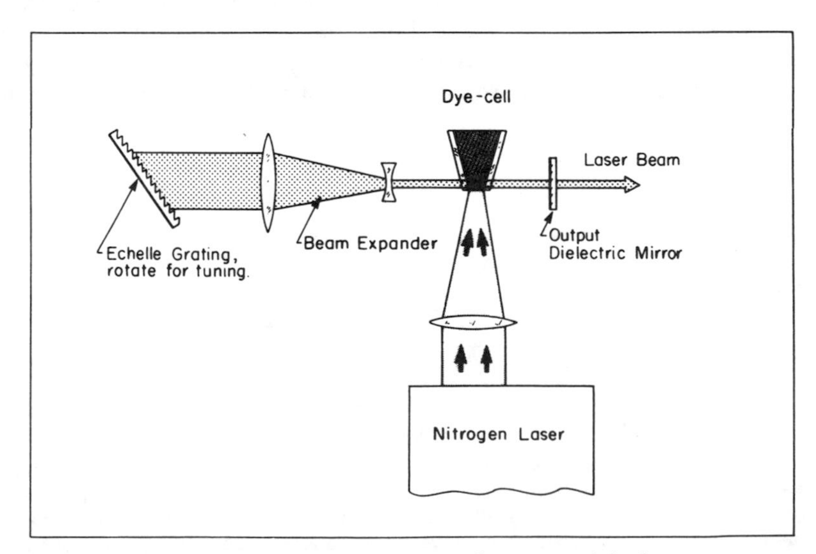

Fig. 5.36. A schematic of a nitrogen-laser-pumped dye laser.

KrF-laser pulse to pump a *p*-terphenyl dye laser. They demonstrated that laser oscillation could be obtained between 321.8 and 365.3 nm with a peak output of about 10^6 W.

The lasing efficiency and photochemical stability of infrared (710 nm to 1.08 μm) dyes pumped by a *Q*-switched ruby laser were studied by Oettinger and Dewey (1976). In addition to tuning with gratings, etalons, and prisms, dye lasers can be rapidly tuned through acousto-optical deflection (Hutcheson and Hughes, 1974) or the use of Lyot (birefringent) filters (Hodgkinson and Vukusic, 1978). Peak powers of 5×10^4 W in the visible with a linewidth of 6×10^{-5} nm have been generated by a two-stage dye-laser amplifier pumped by a 10^6-W nitrogen laser (Wallenstein and Hänsch, 1975).

5.8.4. Semiconductor Lasers

The physics of semiconductor lasers is very different from that of the other kinds of laser. This stems principally from the nonlocalized wave function needed to describe an electron within the semiconductor. In this type of material electrons are constrained to move through the crystal in either of two energy bands. Each band essentially represents a continuum of energy levels, the number of energy levels (or states) being equal to the number of atoms within the crystal. According to the *Pauli exclusion principle*, only one electron can occupy each energy state.

In an intrinsic (undoped) semiconductor at very low temperatures, all of the states in the lower, or *valence*, energy band are filled with electrons, whereas those in the upper, or *conduction*, energy band are empty. Within a semiconductor the probability $f(\mathscr{E})$ that an electron state at energy $\mathscr{E}$ is occupied by an electron is given by Fermi–Dirac statistics rather than Maxwell–Boltzmann statistics, i.e.,

$$f(\mathscr{E}) = \frac{1}{1 + e^{(\mathscr{E} - \mathscr{E}_F)/kT}} \tag{5.92}$$

where $\mathscr{E}_F$ is the so-called *Fermi energy* and T the temperature. At 0 K this Fermi energy represents, in essence, the boundary between the filled and occupied states. In an undoped crystal it lies halfway between the bottom of the conduction band and the top of the valence band. Free electrons (that is to say, those described by nonlocalized wave functions) can never reside within the *energy gap* that lies between the conduction and valence bands [see Fig. 5.37(a)].

In a semiconductor that is heavily doped with *donor* impurity atoms the Fermi level lies in the conduction band, due to the extra electrons contributed to the conduction band from the impurity atoms. Such a material is called an *n-type* semiconductor [see Fig. 5.37(b)]. In contrast, the Fermi level is found to lie in the valence band of a semiconductor that is heavily doped with *acceptor* impurity atoms which remove electrons from the valence band [see Fig.

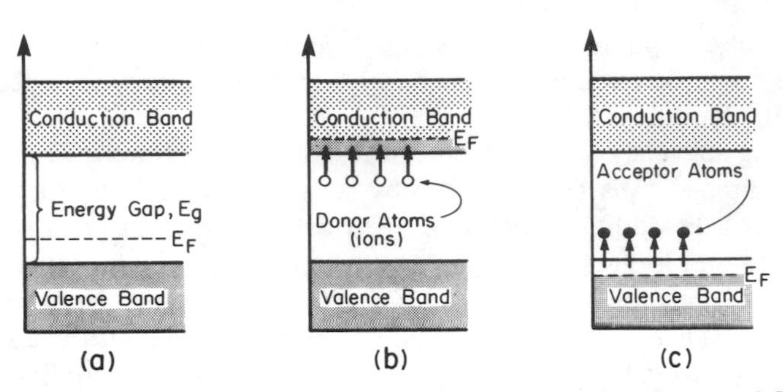

Fig. 5.37. (a) Undoped semiconductor. (b) Degenerate *n*-type semiconductor at 0 K. Impurity atoms donate electrons to the conduction band. (c) Degenerate *p*-type semiconductor at 0 K. Impurity atoms accept electrons from the valence band.

5.37(c)]. The electron vacancies, or *holes*, at the top of the valence band act like positive charges, leading to the name *p-type* semiconductor. The energy levels of the acceptor and donor impurity atoms can merge with the valence and conduction bands, respectively, due to the high level of doping ($\gtrsim 10^{18}$ cm^{-3}). At temperatures above 0 K, there is no longer a sharp boundary between the occupied and unoccupied levels, and in general the Fermi level lies at an energy for which the probability of occupation falls to 0.5 (see Fig. 5.38).

In a *p–n* junction the *p* and *n* regions are both degenerate, and in equilibrium the Fermi level has the same value across the junction, as shown in

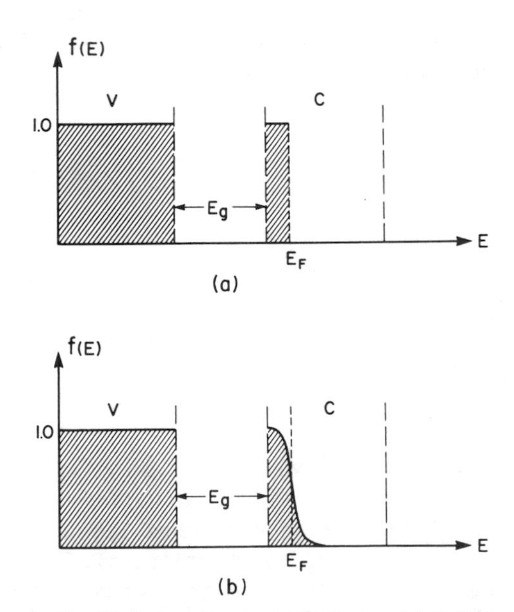

Fig. 5.38. Electron distribution for an *n*-type semiconductor: (a) at 0 K and (b) above 0 K.

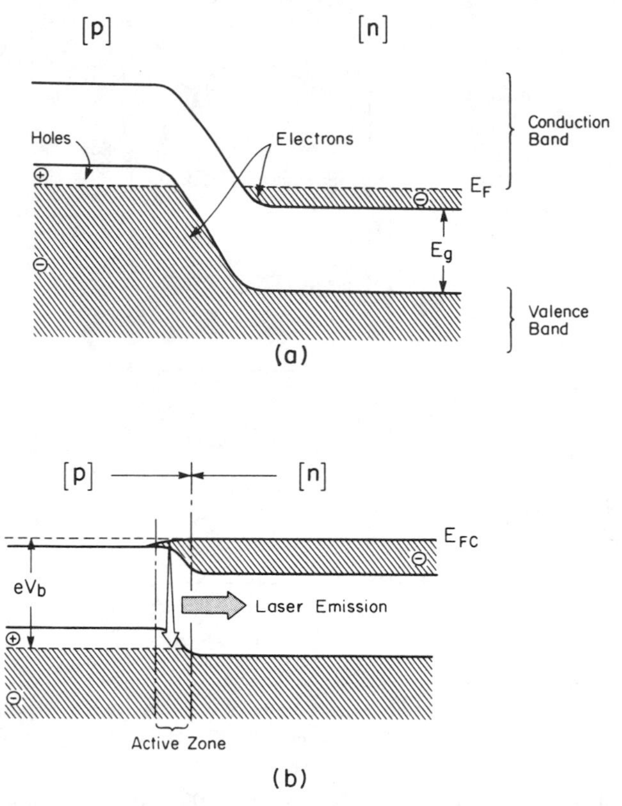

Fig. 5.39. Degenerate p–n junction with (a) zero bias voltage and (b) forward bias voltage $V_b = E_g/e$. Within the active zone spontaneous or stimulated recombination can occur as an electron from the conduction band falls into a hole within the valence band.

Fig. 5.39(a). If a forward bias voltage V_b that is nearly equal to the energy-gap voltage $(\mathscr{E}_g/e)$ is applied, the Fermi level in the n-region is raised by eV_b with respect to that in the p-region. Under these conditions there exists a narrow zone, called the *active region*, that contains both free electrons and holes and is doubly degenerate. In essence the equivalent of a population inversion has been created in this zone through the application of a forward bias voltage. Electromagnetic radiation of frequency ν, satisfying the condition $\mathscr{E}_g/h < \nu < (\mathscr{E}_{Fc} - \mathscr{E}_{Fv})/h$ and propagating through this active or depletion zone, will be amplified through stimulated recombination. That is to say, an electron from the conduction band is stimulated by the action of the radiation field to recombine with a hole from the valence band [see Fig. 5.39(b)]. $\mathscr{E}_{Fc}$ and $\mathscr{E}_{Fv}$ represent the quasi-Fermi energy levels for the conduction and valence bands respectively. This double degeneracy represents a nonequilibrium condition that is maintained in a p–n junction laser by the application of the forward bias voltage and driving a large enough current through the junction.

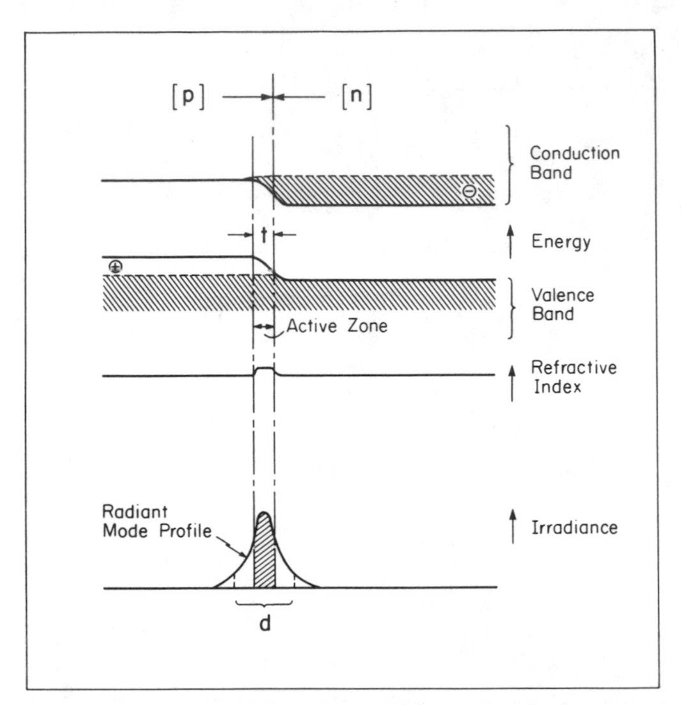

Fig. 5.40. Schematic representation of the active zone of a p–n junction homostructure laser, the associated small increase in the refractive index and the radiant mode distribution.

The thickness of the active (or depletion) zone, t, is very small, and in the case of junction lasers is usually less than the radiation mode width d, as indicated schematically in Fig. 5.40. The laser beam therefore extends into the absorbing p and n regions that are adjacent to the active zone. For example, in a GaAs laser ($\lambda \approx 0.84\ \mu$m), t $\lesssim 1\ \mu$m whereas $d \approx 2$–$5\ \mu$m.

An estimate for the lasing threshold current density i_{th} can be obtained by equating the net gain per double pass of the semiconductor to the reflection losses. To account for the transverse spread of the radiation relative to the active zone, we use n_e/V for the inverted-electron density, where n_e is the total number of inverted electrons in the active volume and V is the mode volume. Under these circumstances the gain coefficient (5.6) becomes

$$\alpha(\nu) = \frac{c^2 n_e \mathscr{L}(\nu)}{8\pi\nu^2 V \tau_R^e} \tag{5.93}$$

where $\mathscr{L}(\nu)$ is the line profile function for spontaneous-recombination emission in the junction, and τ_R^e is the lifetime of a conduction electron in the p-region before recombining with a hole in the valence band.

At low temperatures n_e can approximately be related to the injection current $\mathfrak{I}$ through the steady-state relation

$$\frac{n_e}{\tau_R^e} \approx \frac{\mathfrak{I}\eta_R}{e} \tag{5.94}$$

where η_R, the *radiative quantum efficiency*, represents the fraction of injected carriers (electrons or holes) that recombine radiatively, and e is the electron charge. If $\kappa(\nu)$ is the loss per unit length associated with that portion of the radiation mode that extends laterally beyond the active zone, then from (5.15) the laser threshold requires

$$l\{\alpha(\nu) - \kappa(\nu)\} \approx 1 - R \tag{5.95}$$

where l is assumed to be the length of the active zone and R the mean reflectivity of the end surfaces of the semiconductor crystal. These are cleaved or polished to act as mirrors for the laser (see Fig. 5.41). Combining (5.93), (5.94), and (5.95) yields

$$i_{\text{th}} \approx \frac{8\pi\nu^2 ed}{c^3\mathscr{L}(\nu_0)\eta_R\tau_c}\{1 + c\tau_c\kappa(\nu)\} \tag{5.96}$$

where we have written $V = Ad$ and assumed $i = \mathfrak{I}/A$. We have also introduced the cavity residence time τ_c as defined by (5.21). If we can assume a Lorentzian line profile function with a spectral linewidth $\Delta\lambda$, then (5.96) becomes

$$i_{\text{th}} \approx \frac{8\pi^2 ed\,\Delta\lambda}{\lambda^4\eta_R\tau_c}\{1 + c\tau_c\kappa(\nu)\} \tag{5.97}$$

We can see from (5.97) that the spread of radiation beyond the active region increases i_{th} through the use of d (the electromagnetic mode confinement

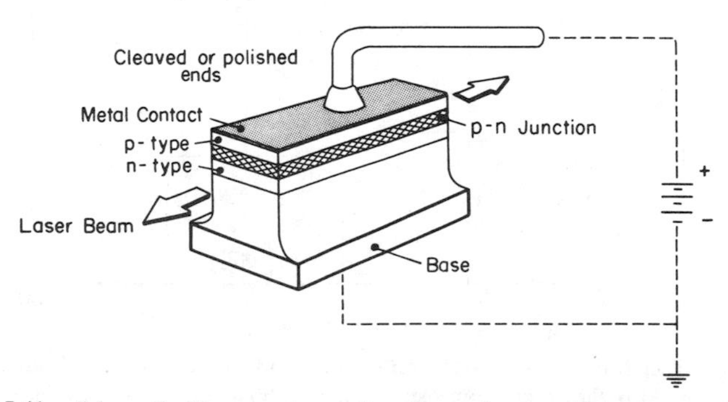

Fig. 5.41. Schematic diagram of a semiconductor homostructure junction laser.

distance) in place of t (the width of the active zone), and through the introduction of $\kappa(\nu)$. This threshold current density can be reduced by more complex designs of the junction. A discussion of these *heterojunction lasers* is beyond the scope of this text. Kressel (1972) provides a detailed discussion of this topic. Another problem associated with the narrow gain region is the large output divergence from these lasers (roughly λ/d). This necessitates the use of optics if a reasonably well collimated beam is required.

The wavelength of emission from semiconductor lasers can be varied by control of the energy gap, which has been made possible by the development of ternary semiconductor compounds of adjustable chemical composition. A selection of the materials from which semiconductor lasers can be treated, with the wavelength range associated with each, is presented in Fig. 5.42. It is apparent that semiconductor lasers can operate from about 600 nm to 34 μm, a range that encompasses the absorption bands of most molecules, some of which are included in Fig. 5.42. There are several methods of tuning: varying the current, changing the temperature, subjecting the crystal to great pressure, or applying a magnetic field.

In summary, the primary advantages of semiconductor lasers are:

1. Small rugged design with no need for external mirrors.
2. No gases or liquids required.
3. Direct electrical excitation with fast current tuning possible.
4. High efficiency ($\approx$ 50%).
5. Broad wavelength range of operation (0.6 to 34 μm).
6. Mass production resulting in low price.
7. Pulsed or cw operation.

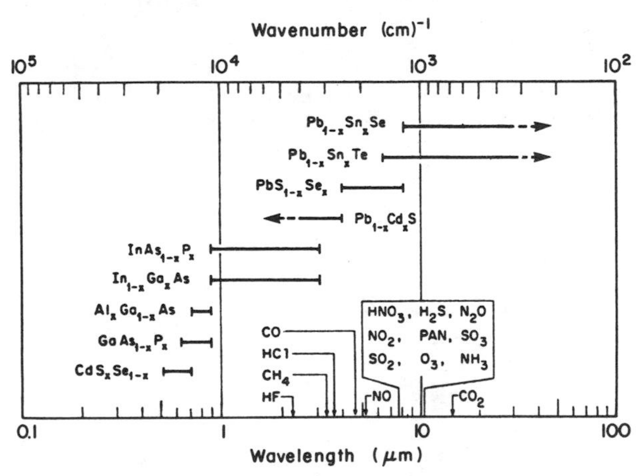

Fig. 5.42. Wavelength ranges for a selection of semiconductor lasers made from different alloys and composition. Also shown are some strongly absorbing regions for several common atmospheric pollutants (Hinkley, 1972).

8. Tunability.

The main disadvantages of these lasers are:

1. Low peak power ($\approx$ 100 W pulsed).
2. Cryogenic cooling (20 to 77 K) required.
3. Large divergence of output beam.
4. Restriction to long wavelengths ($\lambda > 0.6\ \mu m$).

5.9. EYE-SAFETY CONSIDERATIONS

A general limitation that will eventually encompass all operational laser remote sensors is that of safety. This may be more relevant to earth-oriented, airborne systems than to their sky-facing counterparts, but invariably both will have to comply with fairly strict regulations due to the high energy or peak power transmitted by their probing laser beams. The American National Standards Institute (ANSI) (1973) has recommended eye-safety maximum permissible exposure (MPE) levels for various regions of the spectrum. However, these values have been contested as not being sufficiently stringent (Zuchlich and Connolly, 1976; Laser Focus, 1976).

Studies have shown that there are at least three principal mechanisms of injury: thermal–mechanical (acoustic transient), thermal, and photochemical. Threshold retinal lesions from short-duration exposure may result from an acoustic transient which accompanies localized heating in the vicinity of the highly absorbing pigment granules, (Wolbarsht and Sliney, 1974). For longer exposure times the damage is more likely to arise from thermal denaturation of complex organic molecules (Vassiliadis, 1974) or photochemical activity. Of particular relevance to laser environmental sensing has been the recent work of Zuchlich and Connolly (1976), who have found evidence that the threshold for damage in the near ultraviolet may, as a result of photochemical injury, be much lower than the previously accepted value. Table 5.5 provides a guide as

TABLE 5.5. MAXIMUM PERMISSIBLE EXPOSURE[a]

Wavelength	Exposure time (s)	MPE (J cm^{-2})
310 to 400 nm	—	10^{-3}
400 nm to 1.4 μm	10^{-9} to 10^{-5}	5×10^{-7}
	10^{-5} to 10	$1.8 \times 10^{-3} t^{3/4}$
1.4 to 13 μm	10^{-9} to 10^{-7}	10^{-2}
	10^{-7} to 10	$0.56 t^{1/4}$

[a]Based on the ANSI standard with some allowance for the work of Zuchlich and Connolly (1976).

to the MPE for the eye and is based on the ANSI standards, making some allowance for the work of Zuchlich and Connolly.

The severity of the constraint imposed on any laser remote sensor by an eye-safety standard will depend to a large extent upon its mode of operation. Systems that operate at night and are directed skyward will be restricted far less than airborne surveillance systems that operate during the day. An atmospheric lidar system, operating at night, can expand its laser beam to meet the eye-safety standard, then expand its field of view to avoid loss of signal return without incurring any appreciable extra noise (there would of course be some loss of spatial resolution). For daytime operation any expansion of the field of view will invariably lead to an increase in the background radiation accepted by the optical system, and so this option is not as acceptable. In the case of an airborne system the increased risk of exposure and the patchiness of the target will impose counterdemands with regard to operational ceiling and beam expansion, particularly in the light of the findings by Zuchlich and Connolly (1976).

Heaps (1980) has considered the problem of eye safety in connection with the proposed Shuttle-mounted lidar measurements of the concentration of hydroxyl radicals in the upper atmosphere. At first sight this appears to represent an extreme example of potential eye hazard, for it is expected that such measurements would require a laser pulse of about 1 J, at a wavelength of either 282 or 308 nm, to be projected towards the earth from the orbiting lidar. Unfortunately, although the ozone layer in the stratosphere leads to complete attenuation of the 282-nm radiation, some fraction of the 308-nm laser pulse will reach the earth's surface. Even though this wavelength is strongly absorbed by glass windows, eyeglasses, and most plastics that appear transparent in the visible, to the unshielded eye it represents a possible hazard.

The severity of this hazard can be assessed in terms of the exposure of the unaided eye to the sun. The solar spectral irradiance between 310 and 320 nm is, according to Heaps (1980), 6.35×10^{-1} W m^{-2} nm^{-1} (see also Fig. 1.1). If there were no attenuation, an observer on the ground would be exposed to an irradiance of 6.35 W m^{-2} within the 310- to 320-nm spectral interval. By comparison the 1-J laser pulse (at 308 nm) emitted from the Shuttle lidar would illuminate an area of about 314 m^2 on the ground, assuming an output divergence of 0.1 mrad, no atmospheric attenuation, and an orbit altitude of 200 km. This amounts to an irradiance of 3.18×10^{-3} W m^{-2} and corresponds to staring at the sun for 0.5 ms. Such exposures are commonplace and appear to have little long-term harmful effect, and on this basis Heaps (1980) states that the Shuttle-borne lidar poses no eye-safety hazard.

6

Laser Systems as Remote Sensors

Optical probing of the atmosphere through elastic scattering actually predates the invention of the laser. Nevertheless, the superior qualities in regard to power and collimation of even the early ruby lasers made them obvious replacements for the conventional searchlights previously used. It was, however, the development of *Q*-switching by McClung and Hellwarth (1962) that made possible the generation of very short, single pulses of laser energy and thereby range-resolved measurements—in a manner somewhat analogous to radar. In this mode of operation the time between the transmission of the laser pulse and the arrival of the scattered return signal can be directly related (through the velocity of light) to the range at which the scattering occurred. From this we derive the acronym *lidar* (standing for "light detection and ranging").

The energy compression afforded by *Q*-switching also gave rise to laser power densities of such magnitude that remote measurements based on inelastic scattering from specific molecules could be considered for certain applications. Additional developments in laser technology, such as second-harmonic generation and the invention of the nitrogen and the tunable dye laser, have led to new prospects for laser remote sensing by means of absorption and fluorescence. In the case of these techniques *lidar* might more appropriately stand for "light identification, detection, and ranging."

6.1. REMOTE-SENSING TECHNIQUES USING LASERS

Today the range of processes amenable to laser remote sensing includes Rayleigh scattering, Mie scattering, Raman scattering, resonance scattering, fluorescence, absorption, and differential absorption and scattering (DAS). A brief description of each is provided in Table 6.1, and the range of cross

TABLE 6.1. OPTICAL INTERACTIONS OF RELEVANCE TO LASER ENVIRONMENTAL SENSING

Technique	Physical description
Rayleigh scattering	laser radiation elastically scattered from atoms or molecules is observed with no change of frequency VIRTUAL LEVEL hν hν GROUND LEVEL
Mie scattering	laser radiation elastically scattered from small particulates or aerosols (of size comparable to wavelength of radiation) is observed with no change in frequency hν hν
Raman scattering	laser radiation inelastically scattered from molecules is observed with a frequency shift characteristic of the molecule ($h\nu - h\nu^* = E$) VIRTUAL LEVEL hν hν* VIBRATIONALLY EXCITED LEVEL GROUND LEVEL
Resonance scattering	laser radiation matched in frequency to that of a specific atomic transition is scattered by a large cross section and observed with no change in frequency EXCITED LEVEL hν hν GROUND LEVEL
Fluorescence	laser radiation matched to a specific electronic transition of atom or molecule suffers absorption and subsequent emission at lower frequency; collision quenching can reduce effective cross section of this process; broadband emission is observed with molecules 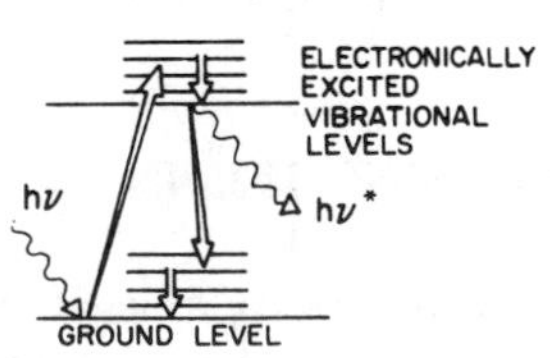

TABLE 6.1 (*Continued*)

Technique	Physical description
Absorption	observe attenuation of laser beam when frequency matched to the absorption band of given molecule [diagram: hν; EXCITED LEVEL; GROUND LEVEL]
Differential absorption and scattering (DAS)	the differential attenuation of two laser beams is evaluated from their backscattered signals when the frequency of one beam is closely matched to a given molecular transition while the other's frequency is somewhat detuned from the transition [diagram: $h\nu_1$; $h\nu_2$; EXCITED LEVEL; GROUND LEVEL; $h\nu_2$]

sections observed for each process is schematically presented in Fig. 6.1. It is evident from this figure that the cross section for Mie scattering can be so large that just a few appropriate-size scatters could give rise to a scattered signal that would completely swamp any Rayleigh- or Raman-scattered component. This implies that quite low concentrations (or changes in concentration) of dust or aerosols can be detected.

Although resonance scattering, sometimes referred to as atomic or resonance fluorescence, also has an inherently large cross section, collision quenching with the more abundant atmospheric species generally ensures that the detected signal is small; consequently this technique is used to best effect in studies of the trace constituents in the upper atmosphere (Hake et al., 1972; Gibson and Sandford, 1972, 1971; Aruga et al., 1974; Felix et al., 1973). The influence of collision quenching on molecular fluorescence can be equally detrimental, particularly where long-lived states are involved (Measures and Pilon, 1972). The broadband nature of molecular fluorescence invariably leads to a low value for the signal-to-noise ratio where background radiation forms the major component of noise (Wang, 1974; Byer, 1975). In the event that the fluorescence is long-lived, the spatial resolution can be degraded.

As we have seen earlier (Section 3.5.2), Raman scattering is an inelastic scattering process wherein the laser radiation may be thought of as raising the molecule to a virtual level from which it immediately decays (in $< 10^{-14}$ s), with the subsequent emission of radiation having a different wavelength. The difference in energy between the incident and emitted photons is a characteris-

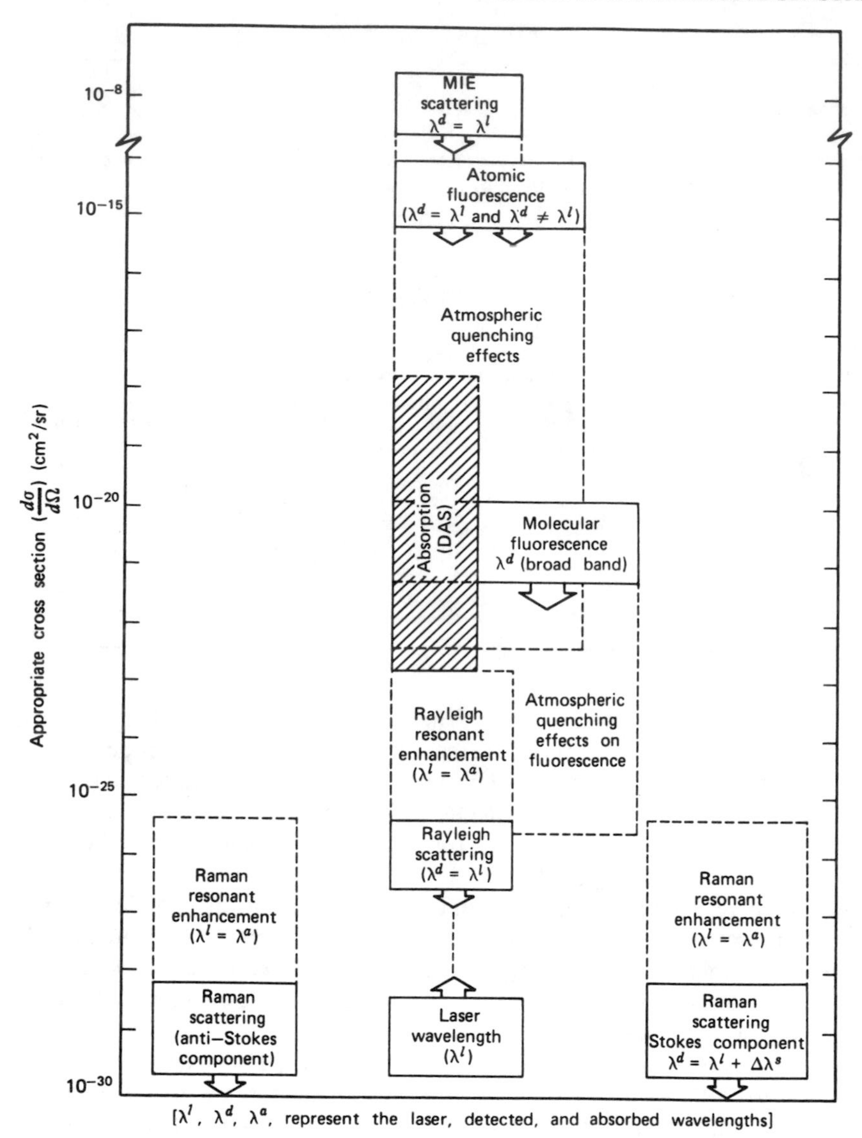

Fig. 6.1. Optical interactions of relevance to laser environmental sensing.

tic of the irradiated molecule and usually corresponds to a change of one vibrational quantum.

The frequency shifts of the *Q*-branch of vibrational–rotational Raman spectra were summarized for a large group of molecules in Fig. 3.23. It is evident that most of the molecules likely to be of interest in any air-pollution study are included. Unfortunately, the Raman cross sections are so small that the range and sensitivity of this technique are rather limited. Raman scattering is consequently most likely to be employed for remote monitoring of effluent plumes where the concentrations can be quite high—tens to hundreds of ppm, as opposed to the few ppm or fractions of ppm more typical of dispersed contaminants.

As seen from Fig. 6.1, some degree of enhancement (at least a factor of 10^3) in cross section can, in principle, be achieved if the excitation frequency is made to closely coincide with an allowed transition. Unfortunately, Hochebleicher et al. (1976) have shown that the improvement in sensitivity expected of the resonance Raman scattering may not materialize, due to absorption of both the exciting and the scattered light.

A particularly attractive feature of Raman scattering relates to the ease with which the concentration of any species relative to some reference species, such as nitrogen, can be evaluated from the ratio of the respective Raman signals, provided the relevant cross section ratio is known, (Fenner et al., 1973; Stephenson, 1974; Fouche and Chang, 1971, 1972).

The cross section for absorption of radiation is in general much greater than either the effective (quenched) fluorescence cross section or the cross section for Raman scattering. Consequently, the attenuation of a beam of suitably tuned laser radiation is a sensitive method of evaluating the mean density of a given constituent. In order to separate absorption by the molecule of interest from other causes of attenuation, a differential approach is usually adopted. In this instance two frequencies are employed, one centered on a line within the absorption band of interest, the other detuned into the wing of the line. With a few notable exceptions, most of the absorption bands of interest lie in the infrared and correspond to vibrational–rotational transitions (Smith et al., 1968). Although in principle this approach involves a *bistatic* arrangement,† clever use of retroreflectors or topographical scatterers allows the more convenient monostatic configuration to be employed (Hodgeson et al., 1973; Asai and Igarashi, 1975; Henningsen et al., 1974; Hinkley and Kelly, 1971; Byer and Garbuny, 1973; Schnell and Fischer, 1975; Schewchun et al., 1976; Murray and van der Laan, 1978; Altmann et al., 1980a, b). Lack of spatial resolution and poor infrared-detector sensitivity represent the major drawbacks of this approach.

High sensitivity with good spatial resolution can be achieved by the combination of differential absorption and scattering (DAS). This technique was first suggested by Schotland (1966) for the purpose of remotely evaluating the water-vapor content of the atmosphere. In this approach a comparison is made between the atmospheric backscattered laser radiation monitored when the frequency of the laser is tuned to closely match that of an absorption line (within the molecule of interest) and when it is detuned to lie in the wing of the line. In this way, the large Mie scattering cross section is employed to provide spatial resolution and to ensure a strong return signal at both frequencies, while the ratio of the signals yields the required degree of specificity due to

†A configuration involving a considerable separation between the laser transmitter and the receiver optics.

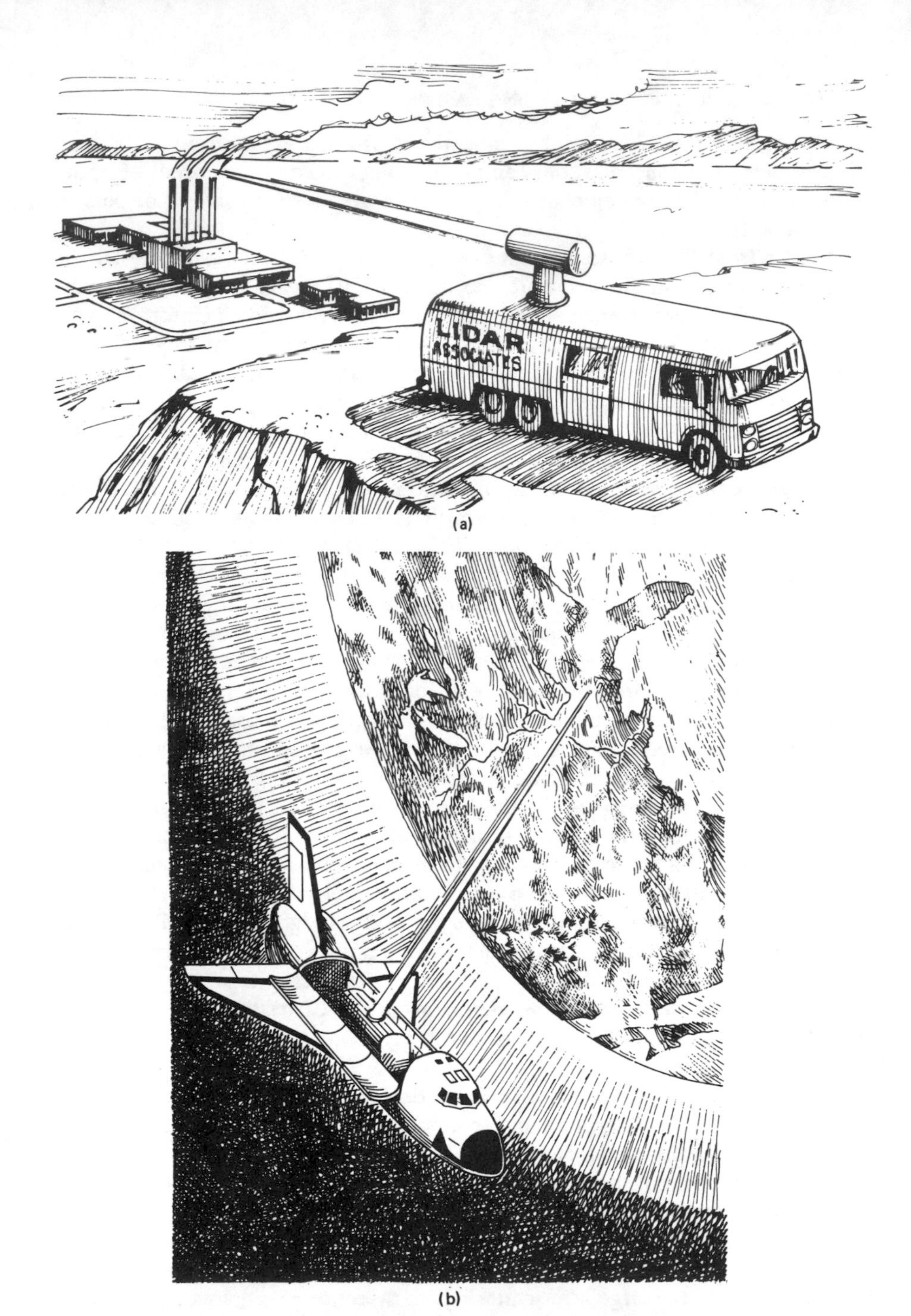

Fig. 6.2. (a) Lidar monitoring of atmospheric pollution sources will enable remote surveillance of industrial emission. (b) Laser sensors mounted aboard the NASA Shuttle will make remote measurements of the upper atmosphere. (c) An airborne laser fluorosensor might be used to study the dispersion of water effluents from industrial and municipal treatment plants. (d) Laser bathymetry of coastal regions could be undertaken from helicopters.

Fig. 6.2. (*Continued*)

differential absorption. These advantages appear to bestow upon the DAS technique the greatest sensitivity for long-range monitoring of specific molecular constituents (Measures and Pilon, 1972; Byer, 1975; Grant et al., 1974; O'Shea and Dodge, 1974; Ahmed, 1973; Rothe et al., 1974a, b; Hoell et al., 1975; Fredriksson et al., 1979; Baumgartner et al., 1979; Fredriksson et al., 1981).

Although detector sensitivity makes both fluorescence and the DAS techniques more amenable to those molecules which possess an absorption band in the visible or near-ultraviolet part of the spectrum, recent improvements in infrared detector sensitivity has given the DAS approach more universal appeal (Murray et al., 1976; Weitkamp, 1981). The acronym DIAL, standing for "differential-absorption lidar," has gained considerable popularity recently for all of the laser remote-sensing techniques that rely on differential absorption.

Toward the end of the laser's first decade some consideration was given to the development of earth-oriented laser sensors that could be used from mobile platforms such as aircraft and helicopters. In the initial applications, these downward-pointing laser systems were operated in a mode somewhat analogous to radar, where surface scattering and reflection represented the dominant form of interaction. Surface-wave studies and bathymetric measurements in coastal waters were the first topics to be given serious consideration (Hickman and Hogg, 1969; Prettyman and Cermak, 1969; Kim et al., 1975). The possibility of undertaking studies of water turbidity grew naturally from the latter series of experiments (Hickman et al., 1974).

An important advance was made with the realization that use of short-wavelength lasers could broaden the spectrum of applications as a result of laser-induced fluorescence. This led to the development of a new form of remote sensor termed a *laser fluorosensor* (Measures and Bristow, 1971; Fantasia et al., 1971; Measures et al., 1973; Bristow et al., 1973; O'Neil et al., 1973; Kim and Hickman, 1973; Fantasia and Ingrao, 1974). Detailed spectroscopic studies (Fantasia et al., 1971; Fantasia and Ingrao, 1974; Houston et al., 1973; Measures et al., 1974) of both crude oils and petroleum products have indicated that an airborne laser fluorosensor with high spectral resolution may indeed be capable of classifying an oil slick with sufficient precision for this information to be entered as evidence in a court of law. The discovery of fluorescent decay spectra (Houston et al., 1973; Measures et al., 1974) should allow an improvement in the design of laser fluorosensors by providing adequate identification capability with somewhat lower spectral resolution.

Early in these studies laser-induced fluorescence from natural bodies of water was considered to constitute a source of background emission that could interfere with the oil fluorescence signal (Fantasia et al., 1971; Measures and Bristow, 1971; Measures et al., 1973; Bristow et al., 1973; Hoge and Swift, 1980; O'Neil et al., 1980, 1981). Further studies not only have diminished this concern but have discovered that this *apparent* water fluorescence signal might serve to indicate the presence of high organic contamination and thereby enable the dispersion of such effluent plumes to be mapped remotely (Measures et al., 1975). The fluorescence of chlorophyll has long been known, and the possibility of employing a laser fluorosensor to remotely map the chlorophyll concentration of natural bodies of water has also been studied (Hickman and Moore, 1970; Mumola et al., 1973; Kim, 1973; Browell, 1977; Bristow et al., 1981; Hoge and Swift, 1981).

It is clear that the scope of lasers in environmental sensing is extensive. They can be used to undertake: (1) concentration measurements of both major and minor constituents and are consequently well suited for pollution monitoring; (2) evaluation of thermal, structural, and dynamic properties of both the atmosphere and the hydrosphere; (3) threshold detection of specific constituents, as might be required by some pollution-alert schemes; (4) mapping of effluent plume dispersal; and (5) spectral fingerprinting of a specific target such as an oil slick.

Furthermore, these observations can be made remotely with both spatial and temporal resolution. As indicated in Fig. 6.2, laser sensors can be operated from the ground or mobile platforms such as boats, helicopters, aircraft, or satellites. Satellite observations are likely to be limited to studies of the outer atmosphere (Yeh and Browell, 1982), except where reflection from the earth or a cloud is employed. In the latter case it is proposed that the atmospheric burden of specific constituents, integrated between the satellite and the earth, would be determined by differential absorption using the earth or a cloud top as a diffuse reflector for the two probe wavelengths (Singer, 1968; Seals and Bair, 1973; Guagliardo and Bundy, 1975).

6.2. BASIC LASER REMOTE-SENSOR SYSTEM

The invention of the laser has led to the development of a wide array of laser-based environmental sensors and a corresponding rich selection of names: laser radar, lidar, laser fluorosensor, laser bathymeter, and so on. In order to discuss the common elements of this new class of remote active sensor we shall use the acronym *lidar* to encompass all such instruments. The functional elements and manner of operation of most remote environmental lidars are schematically illustrated in Fig. 6.3. An intense pulse of optical energy emitted by a laser is directed through some appropriate output optics toward the target of interest. The function of the output optics can be threefold—to improve the beam collimation, provide spatial filtering, and block the transmission of any unwanted broadband radiation, including the emission that arises from some lasers. Often a small fraction of this pulse is sampled to provide a zero-time marker (a reference signal with which the return signal can be normalized in the event that the laser's output reproducibility is inadequate) and a check on the laser wavelength where this is important.

The radiation gathered by the receiver optics is passed through some form of spectrum analyzer on its way to the photodetection system. The spectrum analyzer serves to select the observation wavelength interval and thereby

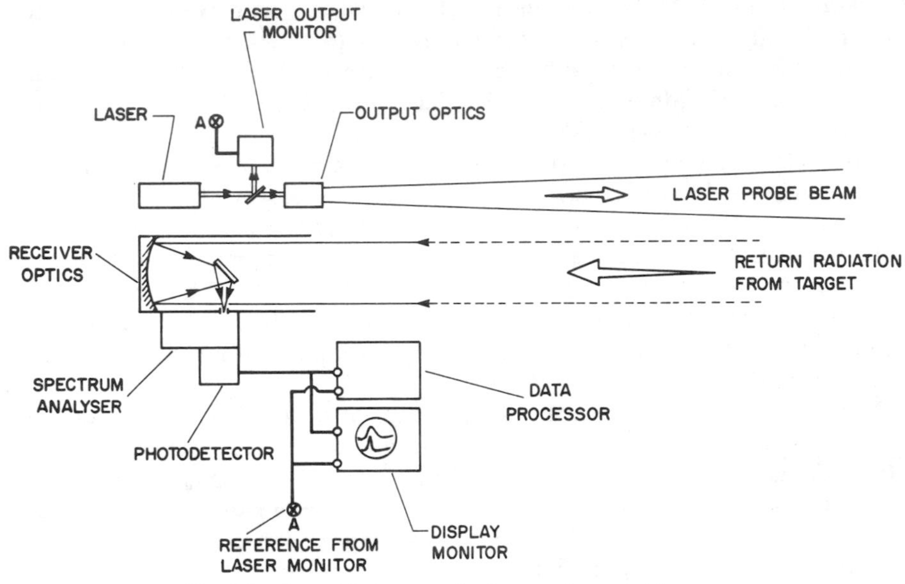

Fig. 6.3. Essential elements of a laser environmental sensor.

discriminate against background radiation at other wavelengths. It can take the form of a monochromator, a polychromator, or a set of narrowband spectral filters together with a laser-wavelength-blocking filter (unless elastically scattered light is of interest). The choice of photodetector is often dictated by the spectral region of interest, which in turn is determined by the kind of application and the type of laser employed.

In principle, there are two basic configurations for laser remote sensors. The *bistatic* arrangement involves a considerable separation of the transmitter and receiver to achieve spatial resolution in optical probing studies. Today, this arrangement is rarely used, as nanosecond lasers are capable of providing spatial resolution of a few feet, and so in most instances a *monostatic* configuration is employed. Under these circumstances the transmitter and receiver are at the same location, so that in effect one has a single-ended system. A monostatic lidar can either be coaxial or biaxial. In a *coaxial* system the axis of the laser beam is coincident with the axis of the receiver optics, while in the *biaxial* arrangement the laser beam only enters the field of view of the receiver optics beyond some predetermined range. This configuration avoids the problem of near-field backscattered radiation saturating the photodetectors, but is optically not quite as efficient as the coaxial approach. The near-field backscattering problem in a coaxial system can be overcome by either gating of the photodetector or use of a fast shutter (Poultney, 1972a, b). A relatively simple method of aligning the transmitter and receiver optics is described by Oppenheim and Menzies (1982).

Newtonian and Cassegrainian reflecting telescopes form the mainstay of the receiver optics to date and are illustrated in Fig. 6.4. A biaxial Newtonian system is portrayed in Fig. 6.3. The combined virtues of compact design and long focal length have given the Cassegrainian system growing popularity. Telescopes based on large plastic Fresnel lenses may offer some advantages with regard to cost, weight, and size (Grams and Wyman, 1972), and thereby be of particular interest in the development of operational airborne lidars. The size of the receiver's aperture depends to a large extent on the technique and range involved. Observations based on Raman scattering appear to be the most demanding, with 30- or 40-in. collecting telescopes being not uncommon. Hirschfeld et al. (1973) used a 36-in. telescope to detect CO_2, H_2O, and SO_2 at a range of 200 m.

The signal from the photodetector may be processed via analogue or digital techniques. The early work invariably involved the former (A-scope) approach where the backscattered signal intensity was displayed as a function of elapsed time (proportional to range) on a wide-bandwidth oscilloscope and photographed. The set of representative photographs presented in Fig. 6.5 is taken from Melfi et al. (1969) and represents the Raman returns from water vapor and from nitrogen. From such photographs the data could be processed manually or the curves digitized and a computer employed. The development of very fast dual-waveform digitizers, such as the Biomation 4500 and Tektronix 7612D, make real-time data processing possible. Uthe and Allen (1975) provided a brief review of the data-handling techniques employed in atmospheric

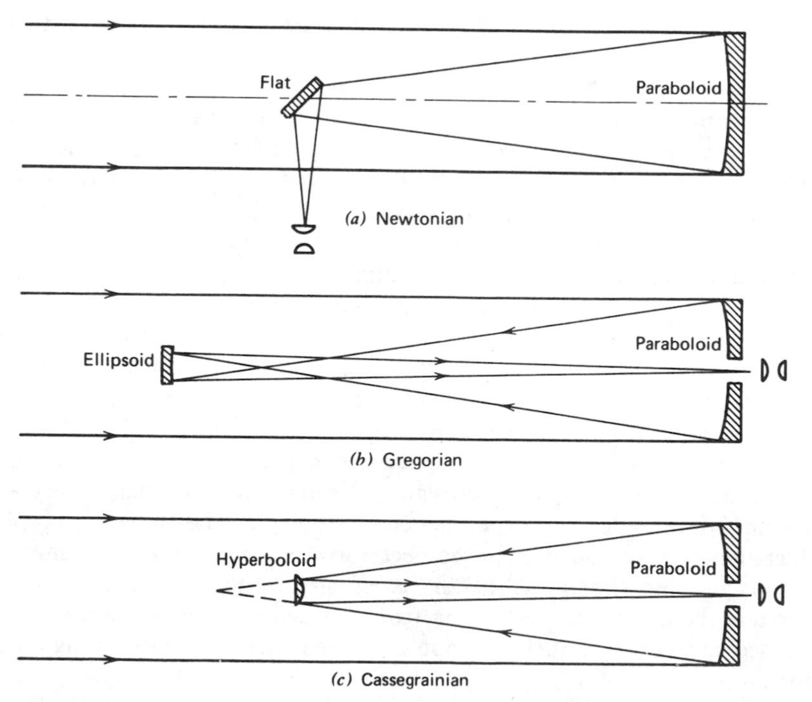

Fig. 6.4. Telescope configurations: (a) Newtonian; (b) Gregorian; (c) Cassegrainian (Ross, 1966).

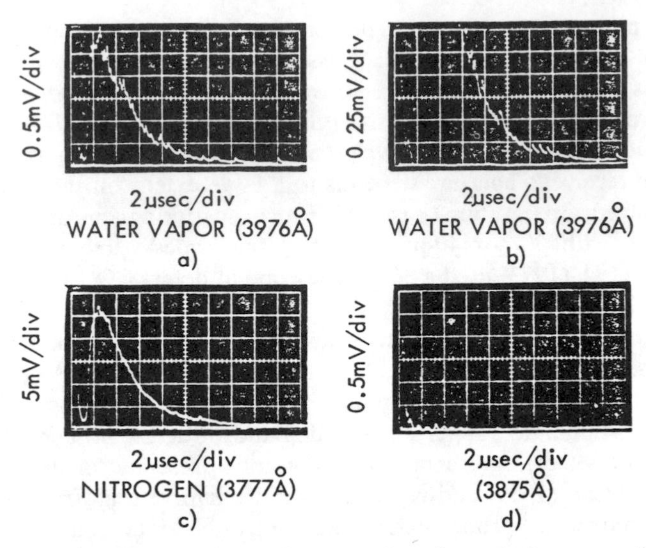

Fig. 6.5. (a, b) Typical Raman backscattering returns from the atmosphere versus time after laser emission; oscillograms taken with monochromator set at 397.6 nm (3654.5-cm^{-1} band of water molecule). (c) Oscillogram taken with monochromator set at 377.7 nm (2330.7-cm^{-1} band of nitrogen molecule). (d) Oscillogram taken with monochromator set at 387.5 nm (no Raman line is predicted from the atmosphere at this wavelength) (Melfi et al., 1969).

probing, and Fredriksson et al. (1981) have detailed a representative lidar real-time data-recording system.

The very brief spike at the far left of the photographs shown in Fig. 6.5 represents the zero-time marker mentioned earlier. The initial rise of the return signal results from the offset configuration (i.e., the field of view of the receiver optics increasingly overlaps the path of laser excitation with increasing range), and the fall of the signal is principally a manifestation of the inverse-square diminution of the signal with range. This $1/R^2$ dependence leads in many applications to a signal-amplitude dynamic range that extends over several decades. Wide-band logarithmic amplifiers and gain-switching techniques can be used to compress this range so that the signals are compatible with recording and display electronics (Uthe and Allen, 1975; Frush, 1975). Alternatively, Hirschfeld (1974) has suggested that a variable-focus system could be used to significantly reduce the range dependence of the signal, and we shall see later (in Chapter 8) that other optical techniques may be used to compress the dynamic range of the observed signal. When the return signal is very weak, as obtained from Raman experiments or studies of the upper atmosphere, integration of many pulses may be necessary, and under these circumstances photon counting (Poultney, 1972b) is usually employed. In this mode of operation the current pulses arising from the detection of single photons are recorded and counted. Such an approach lends itself naturally to digital data processing.

6.3. RECEIVER SYSTEMS

The basic characteristics instrumental in determining the choice of a photodetector include the spectral response, quantum efficiency, frequency response, current gain, and dark current. Sometimes, other considerations such as physical size, ruggedness, and cost may also be important. In most instances, the wavelength of the signal to be detected constitutes the primary factor in selecting the class of photodetector to be employed in any application. For wavelengths that lie between 200 nm and 1 μm (ultraviolet to near infrared), photomultipliers are generally preferred because of their high gain and low noise. Indeed, the single-photon detection capability of these devices has led to low-light-level detection schemes based on counting of individual photons (Poultney, 1972b).

In general, the performance of a photomultiplier is determined by (1) the spectral response of its photocathode, (2) the dark-current characteristics of its photocathode, (3) the gain of the dynode chain, (4) time dispersal effects of the electrons moving through the dynode chain, and (5) the transit time of the electrons between the last dynode and the anode. Table 6.2 provides the kind of data required for a selection of photomultipliers currently available. Since a photomultiplier represents a current source, the observed signal is limited at high frequencies to the voltage that can be generated across the cable impedance $Z_0 < 100\Omega$ (usually). Although cables with impedance $Z_0 > 100\Omega$ are

TABLE 6.2. REPRESENTATIVE CHARACTERISTICS FOR SOME PHOTOMULTIPLIER TUBES[a]

Anode pulse Risetime[b] (ns)	Spectral Response	Radiant Sensitivity: Anode (A W^{-1})	Radiant Sensitivity: Cathode (mA W^{-1})	Amplification[c] (10^6)	Anode Dark Current[c] (nA)	RCA Type No.
≦ 1.5	110 (S-20)	6,900	64	0.11	3	8644
	101 (S-1)	235	1.9	0.125	300	C31004A
	140	1,700	43	0.04	1	C31025K
	142	1,700	30	0.06	1	C31025N
≦ 2	101 (S-1)	310	2.8	0.11	800	C70102B
	110 (S-20)	6,900	64	0.11	3	8645
	125	3,000[d]	15[d]	0.2	0.5[e]	C70128
≦ 2.5	101 (S-1)	940	2.8	0.33	400	C70007A
	110 (S-20)	37,500	64	0.59	3	7326
	133	710,000	97	7.3	0.6	C31000M
	140	14,000	57	0.25	10	C31034B
	142	17,000	42	0.4	10	C31034D

[a] From RCA catalogue.
[b] At maximum supply voltage.
[c] Approximate.
[d] At 253.7 nm.
[e] At 3000 A W^{-1}

available, in general their high-frequency loss characteristics limit their usefulness. Consequently, current gain constitutes an important consideration when high-frequency response is required.

Photomultipliers with high gain reach almost ideal quantum-noise-limited sensitivity for the detection of weak light signals. The large range of photocathode materials currently available offer a wide choice of spectral response characteristics, as illustrated by the selection of curves presented in Fig. 6.6. Melchior (1972) and Ross (1966) have prepared comprehensive reviews on demodulation and photodetection techniques, and Poultney (1972a, b) has provided a detailed discussion of fast-response photomultipliers of interest in photon counting. Although fatigue in photomultipliers has long been recognized, the study of Lopez and Rebolledo (1981) has shown that in some

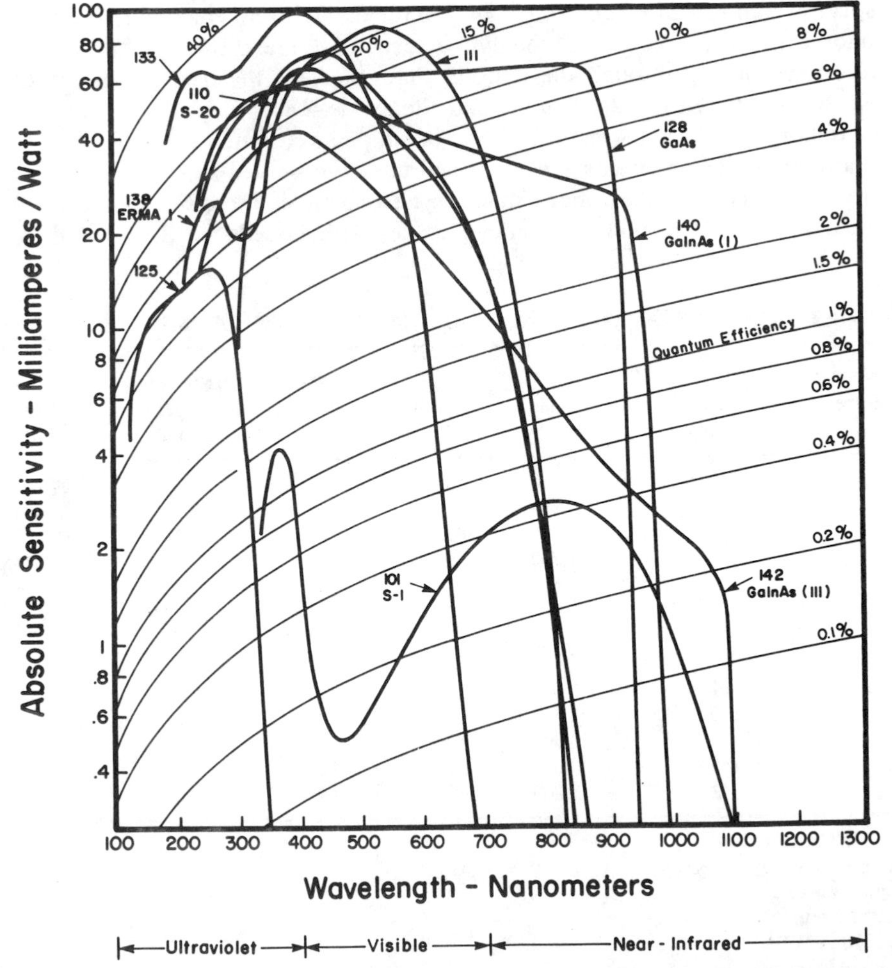

Fig. 6.6. Typical photocathode spectral response characteristics (RCA photomultiplier brochure).

photomultipliers the fatigue experienced in photon counting is comparable to that found with continuous light sources of the same mean irradiance.

Channel multipliers (Wolber, 1968; Leskovar, 1977) are tiny hollow glass tubes with a secondary-emitter coating deposited on the inner walls. When a voltage is applied to such a tube and an electron is injected (from a photocathode), current gains of 10^7 can be achieved. The small size, fast response, high gain, and low noise of these devices make them suitable for single-photon counting and offer exciting new possibilities with regard to photodetection arrays (Timothy and Bybee, 1975). Wafers comprising tens of thousands of such microchannels are commercially available. The recent development of curved microchannel plates promises improved performance, through suppression of ion feedback (Timothy, 1981).

Infrared detectors can be divided, broadly speaking, into two classes—photodetectors and thermal detectors. The most sensitive infrared detectors are semiconductors in which the incident radiation creates charge carriers via a quantum interaction. These photodetectors may be further divided into photovoltaic and photoconductive devices. Of these, photovoltaic devices (photodiodes) are the more popular for environmental sensing. Although some photodiodes can be used in the visible, they come into their own at longer wavelengths, where their high quantum efficiency (30–80%) becomes important. Unfortunately, the output of a photodiode must be externally amplified, so that its sensitivity is often limited by thermal noise. Altmann et al. (1980a) developed a *fast current amplifier* that enables a photovoltaic InSb detector, operated at zero bias voltage (for optimum detectivity), to achieve background-limited performance.

Some photodiodes, when operated at a high reverse bias, develop internal gain through a process of carrier multiplication. These avalanche photodiodes are similar to photomultipliers in the sense that their sensitivity is no longer determined by thermal noise of the detector and output circuit (Keyes and Kingston, 1972). One of the most widely used and most sensitive infrared detectors for the 1- to 5.5-μm range appears to be the liquid-nitrogen-cooled InSb photodiode (Wang, 1974; Shewchun et al., 1976). A useful overview of infrared detectors has been prepared by Emmons et al. (1975), and Lussier (1976a) has compiled several tables that summarize various infrared detector characteristics.

The spectrum analyzer is used to select the wavelength interval of observation and to provide adequate rejection of all off-frequency radiation, whether this be laser-scattered radiation, solar background radiation, or any other form of radiation having a wavelength different from that of the signal. In general, this is accomplished with the aid of one or more spectral components.

These components fall into one of three basic categories: absorption filters, interferometric elements, and dispersive systems. Absorption filters can take the form of colored glass, gelatine, or a liquid solution (Leonard, 1970), and are employed to attenuate the incident intensity, separate interfering spectral orders, or block all wavelengths that are shorter or longer than those of interest. Long-wavelength pass filters (sometimes called short-wavelength blocking filters) are available from 250 nm to 1 μm (Leonard, 1970; Parker, 1968; Jenaer Glasswerk Schott, Mainz, 1965). The problem of laser-induced fluorescence of short-wavelength cutoff filters has been addressed by Bristow (1979). Short-wavelength pass filters, however, are less plentiful.

The internal spectral transmittance characteristics of some 800 colored glasses are presented in the comprehensive review of Dobrowolski et al. (1977), and Res et al. (1977) have provided the transmittance curves for a number of new band-pass filters in the visible. A guide to infrared transmissive materials has been published by Lussier (1976b).

An important component of many environmental lidars is a vapor-deposited dielectric interference filter. These filters, made of alternating layers of high and low refractive index, are useful over most of the UV to intermediate IR region of the spectrum. Their transmission profiles are similar to those of a low-order Fabry–Perot interferometer, being dispersive in nature and having a bandwidth that can be close to 1 nm. In almost all instances interference filters are designed for use with collimated radiation incident normal to the surface of the filter. If the filter is tilted by 45°, the center passband can shift to shorter wavelengths by as much as 2–3%. These filters can also be temperature-sensitive. The side-band transparency of such a filter is close to 10^{-3} times the center-band transparency, giving an off-frequency stray-light rejection ratio of 10^{-3}. In many instances a stack of two or more such filters is used to improve this factor and narrow the spectral width of the passband. However, this is always achieved at the expense of the passband transmission coefficient and an enhanced sensitivity to tipping.

The best, but most expensive, narrowband filters are made of birefringent materials and are called Lyot filters (Walther and Hall, 1970). The wide field of view of this kind of narrowband filter is particularly useful in lidar work. These filters can also be tuned. Reviews of this capability of birefringent filters were prepared by Title and Rosenberg (1981) and Gunning (1981).

For those applications that require high spectral resolution, the choice often lies between a Fabry–Perot interferometer and a grating monochromator (Born and Wolf, 1964). Of these, the Fabry–Perot etalon is usually the cheaper, can provide the higher resolving power, and has the greater light throughput. Indeed, it would be ideal for many applications if it were not for its major drawback—many overlapping orders. This difficulty can be overcome by prefiltering with an interference filter, with a second, wider-passband Fabry–Perot interferometer, or with a dispersive element. Each approach has its own limitations. A detailed comparison of prism, grating, and Fabry–Perot etalon spectrometers has been given by Jacquinot (1954). The passband of a Fabry–Perot etalon can be scanned by varying the pressure of the gas between the interferometer plates (Girard and Jacquinot, 1967) or by displacing one plate relative to the other (Ramsay, 1962).

In those situations where measurement of a spectral profile is important or where many wavelengths are of interest, a grating monochromator offers some advantages. Of the wide array of monochromator systems available, the

Czerny–Turner arrangement appears to be one of the most popular, being typically capable of providing a stray-light rejection ratio of 10^{-6}. For applications where this feature is of critical importance (such as those involving Raman backscattered signals), double monochromators are often employed. Stray-light rejection ratios of 10^{-6} (single) and 10^{-12} (double) can be achieved for displacements of about 60 cm^{-1} from the exciting line (*The Spex Speaker*, XI, No. 4, 1966, Fig. 4). The recent development of holographic gratings has led to the manufacture of compact, simplified, and cheaper instruments possessing even higher stray light rejection capability (*Diffraction Gratings*, Jobin Yvon Optical Systems).

For applications where a fixed number of well-defined spectral features are to be monitored, the monochromator may be replaced with a polychromator that provides a series of preselected wavelength intervals. The poor optical coupling between the entrance slit of any dispersive system and the receiver optics constitutes one of the major weaknesses of this class of spectral analyzer. A significant improvement in the throughput has been achieved by employing shaped fiber-optic bundles (Hirschfeld and Klainer, 1970) or an image slicer (Klainer et al., 1970). Some degree of challenge to monochromators and spectrometers has come from wedge filters for which the passband wavelength varies linearly with position along the length, or around the circumference of a disc. The compactness, mechanical stability, in-line optics, and high throughput of such devices are particularly relevant features for certain classes of remote environmental lidars.

6.4. TYPES AND ATTRIBUTES OF LASERS RELEVANT TO REMOTE SENSING

Certain types of lasers are capable of emitting pulses of optical energy that possess very high peak power, narrow bandwidth, and short duration and that propagate with a low degree of divergence. Lasers of this nature are close to ideal for probing the environment, but must also be capable of operating at a high repetition rate for most airborne missions and for those atmospheric applications in which the return signal is very weak.

The range of lasers available for environmental sensing can be gauged from the selection of commercial pulsed lasers presented in Table 5.1. A much wider spectrum of possibilities exists if use is made of second-, third-, or even fourth-harmonic generation and parametric conversion. In the past, fixed-frequency solid-state lasers offered the highest values of peak power and could be thought of as the workhorses of atmospheric lidars. Such lasers were quite adequate for experiments involving Rayleigh, Mie, and Raman scattering (Hall, 1974; Collis and Uthe, 1972). However, the development of new lasers operating over a broader range of frequencies encouraged exploration of a variety of additional approaches, including resonance excitation, differential absorption, and fluorescence (see Table 6.1).

Of particular relevance to the airborne aspect of environmental sensing was the development of the high-power, high-repetition-rate nitrogen laser operating at 337.1 nm (Heard, 1963; Leonard, 1965). As seen in Chapter 5, this is really just a gas channel that is made to emit a pulse of amplified spontaneous emission by means of an ultrafast transverse discharge. The large divergence associated with the output of this (so-called) laser is a direct consequence of it being a single-pass system. Some improvement can be made at the expense of power by introducing a degree of mode control, (Leonard, 1974). The short wavelength of the nitrogen laser makes it suitable for exciting fluorescence in a variety of materials, and the high repetition rate makes airborne surveillance with good spatial resolution possible. Also, the divergence and wavelength are favorable from an eye-safety standpoint, while the short duration attained with Blumlein systems (Basting et al., 1972) make it suitable for the measurement of fluorescent decay times (Fantasia and Ingrao, 1974; Measures et al., 1974).

The development of tunable organic-dye lasers (Schäfer, 1973) provided the means to excite specific atomic and molecular electronic transitions and thereby exploit both resonance scattering and differential absorption for the purpose of remote sensing. As indicated in Table 5.1, tunable organic dyes are available commercially, covering the wavelength range from the near ultraviolet to the near infrared. A population inversion is created within the dye by optical pumping with either a flashlamp or another laser. For pulsed operation, a nitrogen, Nd–YAG, or rare-gas halide excimer laser is employed; cw operation is achieved by pumping with a tightly focused argon laser. Flashlamp-pumped dye lasers have, in general, provided the greater energy per pulse, but their duration is rather long (hundreds of nanoseconds) for measurements with reasonable spatial resolution. Nevertheless, such a system lends itself to the oscillator-amplifier mode of operation and is ideal for probing the outer regions of the atmosphere (Hake et al., 1972). Pumping with a nitrogen laser, on the other hand, produces a pulse of less energy but with a duration of only a few nanoseconds and a repetition rate of up to a thousand pulses per second. As indicated in Chapter 5, spectral narrowing and tuning across the broad emission band of a dye are achieved by means of a dispersive element such as a prism or a grating. Hänsch (1972) demonstrated that good spectral condensation, better than 10 pm (10^{-1} Å), could be achieved with a diffraction grating and an intracavity beam-expanding telescope. Further spectral narrowing can be achieved with the inclusion of a Fabry–Perot etalon in the dye-laser cavity.

A small degree of tunability can also be achieved with many high-pressure gas lasers and with some semiconductor lasers. A comprehensive review of high-pressure pulsed molecular lasers is provided by Wood (1974); Nill (1974) and Hinkley (1972) present useful reviews of tunable infrared lasers. For environmental sensing, tunable infrared lasers have the advantage that most materials possess vibrational–rotational transitions that can be selectively excited by infrared radiation (Murray et al., 1976). The impact of the new class of tunable near-infrared lasers based on alexandrite is yet to be determined.

At short wavelengths (below 300 nm) there is again a rich selection of transitions for a variety of materials; however, the lack of a convenient tunable laser limits the scope of remote-sensing applications. Tunable laser radiation down to 230 nm is available (Table 5.4) with a frequency-doubled dye laser, but such a system is somewhat elaborate and limited in its output power.

Rare-gas halide lasers, (Bhaumik et al., 1976; Ewing, 1978; Loree et al., 1979; Sze, 1979), are capable of supplying high power, with high efficiency, at wavelengths below 337 nm. Such excimer lasers are inherently tunable, albeit over a small spectral interval. For example, 100-MW pulses have been generated at 248.4 nm from a krypton fluoride laser, and it is believed that the output might be tunable across 4 nm (Bhaumik et al., 1976). Although such developments offer exciting new possibilities with regard to remote sensing, due to the absence of solar background at these wavelengths and the possibility of achieving resonance Raman scattering (Rosen et al., 1975), the extreme sensitivity of living material to this radiation (Koller, 1969) could prevent its realization, except in rather limited situations.

6.5. SOURCES OF NOISE

Of crucial importance to any discussion of remote sensing is the question of signal-to-noise ratio. Noise, in this context, may be thought of as false signals that can reduce the accuracy of a given measurement or even obscure the true signal completely. Noise, in general, can have either an optical or a thermal origin. In the context of laser environmental sensing, there are four important kinds of noise. These are listed in Table 6.3. The first three represent different forms of shot noise. Under daytime operation scattered solar radiation from either the sky or the ground can often dominate all other forms of noise. The solar spectral irradiance as observed from space and from the ground was shown in Fig. 1.1 and is available in tabulated form from Mecherikunnel and Duncan (1982). The spectral radiance of a clear sky is shown in Fig. 6.7. It is also important to realize that for both Raman and fluorescence measurements

TABLE 6.3. KINDS OF NOISE RELEVANT TO LASER ENVIRONMENTAL SENSING

Kind of Noise	Physical Mechanism
Noise in signal (quantum noise)	Statistical fluctuations of signal radiation
Background-radiation noise	Statistical fluctuations of background radiation
Dark-current noise	Thermal generation of current carriers in the absence of an optical signal
Thermal (Johnson, Nyquist) noise	Thermal agitation of current carriers

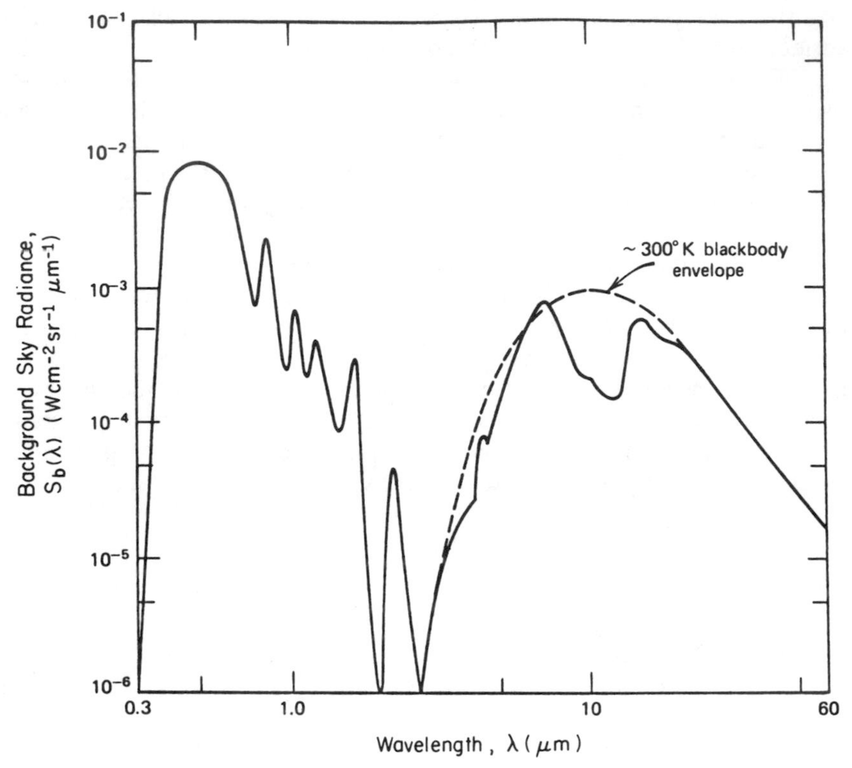

Fig. 6.7. Diffuse component of typical background radiance from sea level, zenith angle 45°, excellent visibility (Pratt, 1969).

the background radiation may include a laser-scattered component if adequate spectral rejection is not provided.

The increment of radiative energy (arising from natural sources) accepted by the receiving optics in the detection time τ_d can be expressed in the form (Pratt, 1969)

$$E_b^N(\lambda) = \int_{\Delta\lambda_0} S_b(\lambda')\xi(\lambda')\Omega_0 A_0 \tau_d \, d\lambda' \tag{6.1}$$

where $S_b(\lambda')$ represents the spectral radiance of the sky background (W cm^{-2} nm^{-1} sr^{-1}), and $\xi(\lambda')$ represents the receiver-system transmission efficiency at the wavelength λ' and includes the influence of any spectrally selecting components. $\Delta\lambda_0$ can be taken as the spectral window of the receiver system, A_0 its effective aperture and Ω_0 its acceptance solid angle. For a good optical system the *étendue matching condition* implies $\Omega_0 A_0 = \Omega_s A_s$, where Ω_s and A_s represent, respectively, the acceptance solid angle and the entrance aperture of the spectral element (such as a monochromator). For example, if a $f/7$ monochromator is used with an entrance slit of 0.1 cm^2 and $A_0 = 10^3$ cm^2, then for a well-matched system $\Omega_0 = 2 \times 10^{-6}$ sr. The spectral radiance of the clear daytime sky peaks in the visible (due to scattered solar radiation) and can attain a value of close to 10^{-5} W cm^{-2} nm^{-1} sr^{-1} (Ross, 1966; Pratt, 1969). Toward 300 nm, this background decreases rapidly because of attenuation within the ozone shield above the earth. The more gradual decline in the solar spectral radiance at the infrared end of the spectrum is punctuated by many absorption bands (see Fig. 6.7). The second hump, centered at about 10 μm, represents the thermal background radiation. More detailed distributions plotted on a linear scale were presented as Figs. 1.1 and 1.2.

In the case of downward-pointing airborne lidars, reflected and scattered solar radiation from the earth provides the background radiation required in calculating $E_b^N(\lambda)$. Plass et al. (1976) have calculated the upward radiance from the ocean surface taking account of sun glitter, reflected sky radiance, and the upwelling scattered radiation from various depths. They attempted to make their model fairly realistic by including the effect of surface waves and have also calculated the upward radiation as seen from the top of the atmosphere. This would be of relevance to lidars operating from space platforms.

In order to reduce the solar background, the receiver system bandwidth $\Delta\lambda_0$ is always adjusted to be as small as compatible with the spectral width of the signal of interest. In almost all instances (scattering or fluorescence) the spectral window of the receiver system is sufficiently narrow that we can neglect the variation of the solar spectral radiance over the range of wavelength integration. In that case we can write

$$E_b^N(\lambda) = S_b(\lambda)\Omega_0 A_0 \tau_d K_0(\lambda) \tag{6.2}$$

where λ is the center wavelength of the radiation of interest and

$$K_0(\lambda) \equiv \int_{\Delta\lambda_0} \xi(\lambda')\, d\lambda' \tag{6.3}$$

is termed the *filter function* of the receiver system. It can be thought of as the effective receiver bandwidth at unit transmission efficiency.

The noise associated with a photomultiplier is determined by several factors: the type of photomultiplier, the characteristics of the photocathode, the operating gain, and the usage history of the particular photomultiplier selected. The predominant form of photomultiplier noise is associated with the release of single electrons from the photocathode in the absence of any incident light. This so-called *dark current* arises from thermal and field-emission processes and can be anywhere from 10^{-13} to 10^{-17} A at room temperature. Cooling the device is a popular method of reducing this component of noise and has been reviewed by Foord et al. (1969). A good overall description of photomultiplier noise problems is provided by Poultney (1972).

6.6. SIGNAL-TO-NOISE RATIO

In a photomultiplier the arrival of the signal pulse of energy $E_s(\lambda)$ in the time interval t to $t + \tau_d$ gives rise to a momentary photocathode current

$$i_s(t) = \frac{\lambda\eta(\lambda)eE_s(\lambda)}{hc\tau_d} \tag{6.4}$$

where $\eta(\lambda)$ represents the quantum efficiency of the photocathode at wavelength λ, and e the electronic charge. It is worth noting that $\lambda E_s(\lambda)/hc\tau_d$ is the mean rate of arrival of signal photons (energy hc/λ) at the photocathode. The equivalent noise photocathode current during the same time interval is given by

$$i_n = \frac{e\sqrt{\delta n_e^2}}{\tau_d} \tag{6.5}$$

where $\sqrt{\delta n_e^2}$ is the root-mean-square fluctuation of the total number of photocathode electrons created in the detection interval τ_d.

If, as is most often the case, this burst of photocathode electrons can be described by Poisson statistics, then we can write (See section 8.3.1; Oliver, 1965)

$$\left[\delta n_e^2\right]^{1/2} = \left[\langle n_e\rangle\right]^{1/2} \tag{6.6}$$

where $\langle n_e\rangle$ is the mean number of photocathode-generated electrons produced in the interval τ_d, namely,

$$\langle n_e\rangle = \frac{1}{hc}\{\lambda\eta(\lambda)\left[E_s(\lambda) + E_b^N(\lambda)\right] + \lambda_L\eta(\lambda_L)E_b^L(\lambda_L)\} + \langle n_e^d\rangle \tag{6.7}$$

Here $\langle n_e^d\rangle$ is the mean number of photocathode electrons created in the interval τ_d in the absence of any light; $E_b^N(\lambda)$ and $E_b^L(\lambda_L)$ respectively represent the natural background and laser-scattered background pulses of radiant energy incident upon the photocathode in the period τ_d. If the wavelength of the photodetection system is set equal to λ_L, then of course the first and third terms within the braces of equation (6.7) coalesce.

The photocathode signal-to-noise ratio can thus be expressed in the form

$$(\mathrm{SNR})_c = \frac{e\lambda\eta(\lambda)E_s(\lambda)}{hc\tau_d\left\{\dfrac{e^2}{hc\tau_d^2}\left[\lambda\eta(\lambda)\{E_s(\lambda) + E_b^N(\lambda) + \nu_\eta E_b^L(\lambda_L)\}\right] + \dfrac{ei_d}{\tau_d}\right\}^{1/2}} \tag{6.8}$$

where i_d represents the mean photocathode dark current $(e\langle n_e^d\rangle/\tau_d)$ and $\nu_\eta \equiv \lambda_L\eta(\lambda_L)/\lambda\eta(\lambda)$. An alternative form of this expression is used by some authors (Inaba and Kobayasi, 1972; Poultney, 1972; Nakahara et al., 1972) who are concerned with electronically gating the photomultiplier in order to account for the natural-background and dark-current contributions to the noise of the system. This is of particular relevance in photon counting (Poultney, 1972a, b). Under these circumstances one assumes that the photomultiplier is gated on for a duration τ_g^s, during the period of signal return, and again for an additional period τ_g^b in the absence of any signal. In this case the gated signal-to-noise ratio is

$$(\text{SNR})_c^g = e\lambda\eta(\lambda)E_s(\lambda)\frac{\tau_g^s}{\tau_d}$$

$$\times \frac{1}{hc\tau_d}\left\{\frac{e^2}{hc\tau_d^3}\left[\lambda\eta(\lambda)\{E_s(\lambda)\tau_g^s + E_b^N(\lambda)\left(\tau_g^s + \tau_g^b\right) + \nu_\eta E_b^L(\lambda_L)\tau_g^s\}\right]\right.$$

$$\left. + \frac{ei_d}{\tau_d}\left(\frac{\tau_g^s + \tau_g^b}{\tau_d}\right)\right\}^{-1/2} \tag{6.9}$$

As indicated earlier, in equation (6.4), the equivalent signal current can be related to the signal energy received by the photodetector, and by the same reasoning the equivalent background current can be expressed in the form

$$i_b \equiv \frac{e\{\lambda\eta(\lambda)E_b^N(\lambda) + \lambda_L\eta(\lambda_L)E_b^L(\lambda_L)\}}{hc\tau_d} \tag{6.10}$$

If we introduce the photodetection bandwidth

$$B \equiv \frac{1}{2\tau_d} \tag{6.11}$$

then we can rewrite equation (6.8) in the more conventional form

$$(\text{SNR})_c = \frac{i_s}{\left[2eB(i_s + i_b + i_d)\right]^{1/2}} \tag{6.12}$$

where we see that the denominator represents the shot noise in the form first shown by Schottky (1918):

$$i_n = \left[2eB(i_s + i_b + i_d)\right]^{1/2} \tag{6.13}$$

If we introduce G as the *dynode chain gain*, then the anode signal current is

$$I_s = G\xi_e i_s \tag{6.14}$$

where ξ_e (< 1) is the collection efficiency of the electrostatic focusing, that is to say, ξ_e defines what fraction of the photoelectrons, created at the cathode, arrive at the first dynode. The corresponding noise current at the anode is given by

$$I_n = eG\delta \frac{[\xi_e \langle n_e \rangle]^{1/2}}{\tau_d} \tag{6.15}$$

where δ (≈ 1) accounts for the statistical fluctuations in the emission of secondary electrons from the dynodes (Topp et al., 1969) and can be thought of as a noise factor associated with the gain. Consequently, we may write the signal-to-noise ratio for the anode current as

$$(\mathrm{SNR})_a = \frac{I_s}{I_n} = \frac{G\xi_e i_s}{\left[2eBG^2F_G\xi_e(i_s + i_b + i_d)\right]^{1/2}} \tag{6.16}$$

where we have introduced F_G to represent δ^2.

It is clear that in the case of a photomultiplier the current multiplication factor G does not itself enter into the output SNR. However, a somewhat similar analysis in the case of an avalanche photodiode would produce an output SNR of the form (Melchior, 1972)

$$\mathrm{SNR} = \frac{G\xi_e i_s}{\left[(4kTB/R_{\mathrm{eq}}) + 2eBG^2F_G\xi_e(i_s + i_b + i_d)\right]^{1/2}} \tag{6.17}$$

where the first term within the square brackets represents the Johnson or thermal noise current associated with the equivalent load resistance of the output circuit (R_{eq}), and i_d in this case refers to the bulk leakage current of the avalanche photodiode; ξ_e can be taken to be very close to unity. The magnitude of G is so large (10^5–10^7) in the case of a photomultiplier that thermal noise never imposes a real limit. However, the value of F_G can range from 1 to about 2.5 for both high-gain photomultipliers and photodiodes (Melchior, 1972). In general, the signal-to-noise ratio improves by averaging over many pulses. This improvement is proportional to the square root of the number of pulses averaged.

The expression for the output current signal-to-noise ratio, equation (6.17), has found general application, although several effects are omitted. These include the contribution to the dark current from electrons that originate on the dynodes and the possibility of leakage current at the output socket of the device (particularly relevant to cooled photomultipliers). The influences of cosmic radiation and natural radioactivity have also been neglected—with justification, one might add, in most situations. These additional noise contributions can be taken into account by appropriately modifying the dark-current term (Topp et al., 1969).

In order to see where improvements in the SNR can be made, we shall express the output SNR (on the basis of a single pulse) in the general form

$$\mathrm{SNR} = \frac{E_s(\lambda)[\xi_e \lambda \eta(\lambda)]^{1/2}}{[F_G hc\{E_s(\lambda) + E_b^N(\lambda) + \nu_\eta E_b^L(\lambda_L) + (i_d + i_J)/2BS_d\}]^{1/2}} \tag{6.18}$$

where

$$S_d \equiv \frac{e\lambda\eta(\lambda)}{hc} \tag{6.19}$$

and

$$i_J \equiv \frac{2kT}{eG^2 \xi_e R_{eq} F_G} \tag{6.20}$$

represents the effective Johnson noise current. As we have seen, i_J is only important where there is negligible internal gain.

It is immediately apparent that a careful choice of the detector and its mode of operation can optimize the SNR, all other things being equal. In the case of a photomultiplier, the photocathode should be selected on the basis of maximum quantum efficiency at the wavelength of interest. The noise factor associated with the gain and the collection efficiency of the first dynode can both be optimized by application of a suitable voltage between the photocathode and the first dynode (Topp et al., 1969). Thermionic and field-emission release of electrons from the dynodes can represent an important component of the dark-current noise in photomultipliers. Consequently, at very low light levels where dark-current noise is important, photon counting can significantly improve the SNR because it discriminates against single electron pulses that have not acquired the full gain (Poultney, 1972; Topp et al., 1969). A further increase in the SNR under these conditions can be obtained by focusing the incident light so that only a small area of the photocathode is illuminated while ensuring that all electrons that originate from the unilluminated regions of the photocathode are defocused magnetically. Cooling also helps, since thermionic emission is strongly temperature-dependent. Topp et al. (1969) demonstrated that use of these techniques enabled signals with a power of less than 10^{-17} W (35 photons s^{-1}) to be detected with a time constant of 1 s at 650 nm.

The general output SNR (6.18) can be expressed in another form:

$$(\mathrm{SNR})^2 = \frac{E_s^2(\lambda)/E(\lambda)}{E_s(\lambda) + E_b^T(\lambda) + (i_d + i_J)E(\lambda)/2eB^*} \tag{6.21}$$

where

$$E_b^T(\lambda) \equiv E_b^N(\lambda) + \nu_\eta E_b^L(\lambda_L) \tag{6.22}$$

$$B^* \equiv \frac{BF_G}{\xi_e} \tag{6.23}$$

and

$$E(\lambda) \equiv \frac{hcF_G}{\lambda\eta(\lambda)\xi_e} \tag{6.24}$$

which corresponds to the signal energy that would just give a unit value to the SNR in the signal shot-noise limit to be defined shortly. Equation (6.21) represents a simple quadratic expression in terms of $E_s(\lambda)$ and has the solution

$$E_s(\lambda) = \tfrac{1}{2}(\mathrm{SNR})^2 E(\lambda) \times \left[1 + \left\{1 + \frac{4}{(\mathrm{SNR})^2}\left[\frac{E_b^T(\lambda)}{E(\lambda)} + \frac{i_d + i_J}{2eB^*}\right]\right\}^{1/2}\right] \tag{6.25}$$

In essence, there are four limiting situations that cover most laser-sensing situations. Three of them presuppose that the system's spectral discrimination is adequate to enable the laser scattered component of the background signal to be neglected, that is, $E_b^T(\lambda) \approx E_b^N(\lambda)$. We shall consider these first. It is also reasonable to assume that for virtually all lidar work detectors with internal gain will be used, and consequently i_J is usually neglected. In this case the highest sensitivity is achieved if the detectable signal is limited only by the quantum fluctuations of the signal itself, namely,

$$E_b^N(\lambda) + \frac{i_d E(\lambda)}{2eB^*} \ll E_s(\lambda)$$

Under these circumstances (*signal shot-noise limit*) the minimum detectable energy (MDE) is

$$E_s^{\min}(\lambda) \approx \left[(\mathrm{SNR})^{\min}\right]^2 E(\lambda) \tag{6.26}$$

If we assume that $F_G/\xi_e \approx 1$ and that an acceptable value for $(\mathrm{SNR})^{\min}$ is about 1.5, we find that for $\eta(\lambda) \approx 0.2$ the minimum detectable number of photons is close to 10 per pulse.

In daytime operation, the level of background radiation can be so high that

$$E_b^N(\lambda) \gg \frac{i_d E(\lambda)}{2eB^*} + E_s(\lambda) \tag{6.27}$$

and we speak of the *background-noise limit*. Under these circumstances the MDE is

$$E_s^{\min}(\lambda) = \left\{\frac{F_G hcK_0(\lambda)\Omega_0 A_0 \tau_d S_b(\lambda)}{\xi_e \lambda\eta(\lambda)}\right\}^{1/2} (\mathrm{SNR})^{\min} \tag{6.28}$$

and it is evident that the filter function $K_0(\lambda)$ plays an important role in permitting small values of $E_s(\lambda)$ to be detected.

In the *dark-current-limited situation* we assume that

$$i_d \gg \frac{2eB^* E_b^N(\lambda)}{E(\lambda)} + \frac{2eB^* E_s(\lambda)}{E(\lambda)}$$

This leads to a MDE given by

$$E_s^{\min}(\lambda) \approx \frac{hc}{\lambda\eta(\lambda)} \left\{ \frac{F_G i_d}{2eB\xi_e} \right\}^{1/2} (\text{SNR})^{\min} \tag{6.29}$$

The typical value for the photocathode dark current of a photomultiplier is around 10^{-15} A, which means that the inequality expressed above is not likely to be satisfied unless the photon arrival rate is less than about $10^3\ \text{s}^{-1}$. On the other hand, for practical bandwidths, most infrared detectors tend to be dark-current-limited. Thus, we shall introduce the term *detector detectivity D^** (Kruse et al., 1963) defined by

$$D^* \equiv \frac{S_d}{(2ei_d/A_d)^{1/2}} \tag{6.30}$$

where S_d, defined by equation (6.19), represents the detector sensitivity as defined by the relation

$$i_s \equiv \frac{S_d E_s(\lambda)}{\tau_d}$$

where A_d represents the detector area. In this case, using this equation with (6.19) and (6.30), we may express equation (6.29) in the form

$$E_s^{\min}(\lambda) = \frac{1}{D^*} \left\{ \frac{F_G A_d}{4B\xi_e} \right\}^{1/2} (\text{SNR})^{\min} \tag{6.31}$$

For most solid-state devices the typical detector areas are only a few square millimeters and the factor F_G/ξ_e can be replaced with unity. Representative detectivity (D^*) values for a number of such detectors have been given by Melchior (1972) and are reproduced here as Fig. 6.8.

Clearly equations (6.28), (6.29), and (6.31) represent limiting cases which primarily serve to indicate the relative importance of the various factors in determining the minimum detectable signal. In general, equation (6.25) should be employed, and under these circumstances the MDE for the background-dominated case characterized by

$$E_b^T(\lambda) \gg \frac{i_d E(\lambda)}{2eB^*} \tag{6.32}$$

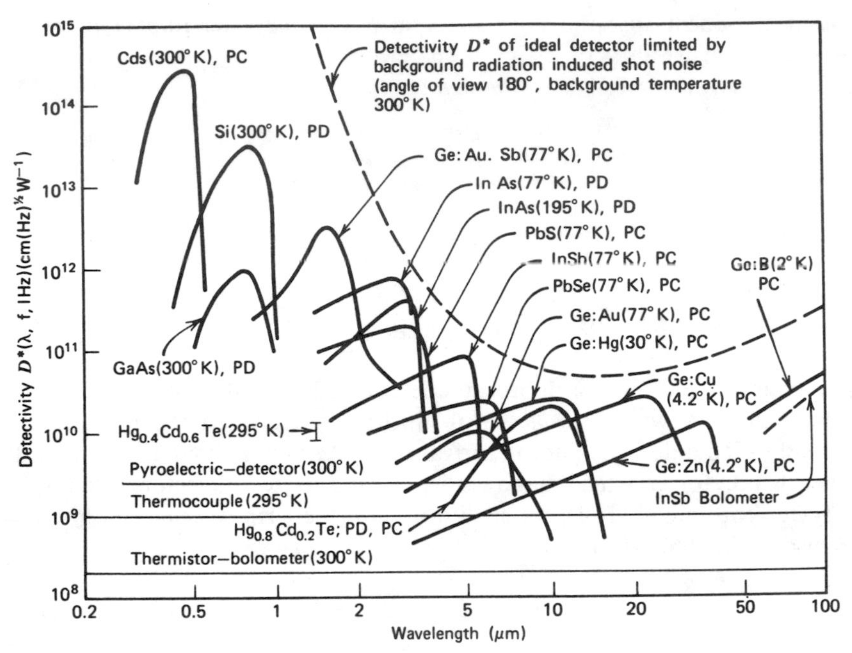

Fig. 6.8. Spectral dependence of the detectivity D^* for high-sensitivity photoconductors (PC) and photodiodes (PD). Representative values are given based on the literature and manufacturer's data (Melchior, 1972).

is given by the expression

$$E_s^{\min}(\lambda) \approx (\mathrm{SNR})^{\min}\left[\tfrac{1}{2}(\mathrm{SNR})^{\min} E(\lambda) + \{E(\lambda)E_b^T(\lambda)\}^{1/2}\right] \quad (6.33)$$

If laser backscattered radiation dominates the natural background component, then essentially $\nu_\eta E_b^L(\lambda_L)$ should replace $E_b^T(\lambda)$ in equation (6.33). Alternatively, if spectral rejection of the laser backscattered radiation is adequate, then

$$E_s^{\min}(\lambda) \approx (\mathrm{SNR})^{\min}\left[\tfrac{1}{2}(\mathrm{SNR})^{\min} E(\lambda) + \{E(\lambda)E_b^N(\lambda)\}^{1/2}\right] \quad (6.34)$$

A comparison of (6.28) with (6.34) reveals that in the more general situation the MDE includes an additional term which amounts to half of the signal-shot-noise MDE. A similar result is found for the more general dark-current dominated situation, where we can write

$$i_d \gg \frac{2eB^* E_b^T(\lambda)}{E(\lambda)} \quad (6.35)$$

and the MDE is

$$E_s^{\min}(\lambda) = (\mathrm{SNR})^{\min}\left[\tfrac{1}{2}(\mathrm{SNR})^{\min} E(\lambda) + \frac{1}{D^*}\left\{\frac{F_G A_d}{4B\xi_e}\right\}^{1/2}\right] \tag{6.36}$$

As we shall see in Chapter 7, $E_S(\lambda)$ can be related through the appropriate form of the lidar equation to the density of the target species and the output energy of the laser. Consequently, these various MDE expressions can be used to evaluate either the minimum laser pulse energy required for detection of a given density of species at a specific range, or the threshold density of a species that can be detected at a given range for a given transmitted laser energy—in various noise-dominated situations.

6.7. NOISE-REDUCTION TECHNIQUES

In many instances it is possible to improve the SNR by suppression of the noise. One of the first problems encountered in the early lidar work on the upper atmosphere involved near-field scattering of long-lived laser fluorescence being misinterpreted as return signals from high altitudes. Although sufficient separation of transmitter and receiver overcame this source of noise, fluorescence shutters tended to become the more usual method of avoiding this difficulty. A large separation (10 m) between the laser and the collection optics also prevents near-field scattering from overloading the photomultipliers. Alternatively, fast mechanical or electro-optical shutters can be used to avoid this problem (Poultney, 1972a). More recently, electronically gated photomultipliers have been used.

Spectral rejection of laser backscattered radiation where the signal wavelength differs from that of the laser represents an important example of such noise suppression; see Section 8.1. Where absorption is used to measure the density of a specific molecular constituent, operation at two wavelengths (λ_0 and λ_w), corresponding to adjacent regions of strong and weak absorption by the molecule of interest, enables an allowance to be made for attenuation by the background constituents. Measurements at λ_0 and λ_w should in principle be performed simultaneously to avoid errors associated with temporal changes in the absorption and scattering properties of the atmosphere.

As an alternative to employing two lasers (an expensive and cumbersome business), Brassington (1978) proposed to switch the wavelength of operation of a tunable dye laser between λ_0 and λ_w on consecutive output pulses. For the detection of SO_2 this involved rocking the birefringent filter (the main wavelength-selective element) and the frequency-doubling crystal in synchronism with the laser pulses in order to provide an output at 300.1 and 299.4 nm. Alternative techniques for obtaining a pair of laser pulses from a nitrogen laser pumped dye laser, with a wavelength separation that can range from 0 to 10 nm, were described by Kittrell and Bernheim (1976) and Inomata and Carswell (1977). In the infrared part of the spectrum Kanstad et al. (1977) have developed a CO_2 laser that can intermittently operate on any two lines from either the 9.4 μm or the 10.4-μm band, and Stewart and Bufton (1980) proposed to use two CO_2 lasers in their DIAL experiments.

The range and detection limit of a Raman lidar is usually limited by the presence of luminescence at the same wavelength as the Raman signal. This luminescence can have a natural origin such as solar scattered radiation (or moonlight), or it can arise as a result of the laser-induced fluorescence. Indeed, as pointed out by Hirschfeld (1977b), it is almost impossible to have zero fluorescence in association with Raman excitation. Much of this broadband radiation can be made negligible compared to the Raman signal through the use of narrowband filtering. Temporal discrimination is particularly effective against continuous radiation, and for local measurements some degree of suppression of laser-induced fluorescence can also be achieved in this manner (Burgess and Shepherd, 1977). Unfortunately, in laser remote sensing time-resolved measurements cannot, in general, be used to overcome laser-induced fluorescence; however, Morhange and Hirlimann (1976) have shown that considerable rejection of luminescence can be achieved through alternately switching between two laser lines while monitoring only at the Raman wavelength produced by one of them. If the wavelength separation of the two laser lines is small enough that the luminescence signal is nearly unchanged, then it can be subtracted from the other signal to yield the Raman-scattered component.

Long-path absorption of tunable infrared laser radiation has been found to be one of the most sensitive techniques for detecting low concentrations of molecular pollutants within the atmosphere (Hinkley, 1976; Hanst, 1976). Unfortunately, thermal fluctuations in the atmosphere were found to constitute the dominant source of noise in the received laser signal. In order to monitor trace concentrations of molecular pollutants over long atmospheric paths with high sensitivity, Ku et al. (1975) have developed a technique which they term *fast derivative spectroscopy*. This involves taking the derivative of the absorption spectra through modulating the output wavelength of the laser. In the case of a semiconductor diode laser this can be accomplished by modulating the diode current about its steady value.

Mathematically, the instantaneous received laser power $P_s(\nu, t)$ can be expressed in the form

$$P_s(\nu, t) = C_S P_L \mathrm{e}^{-2[N\sigma^A(\nu) + \bar{\kappa}_e(t)]R} \tag{6.37}$$

where P_L represents the laser power (assumed independent of frequency and time), N the average number density of the molecular constituent of interest along the path, and $\sigma^A(\nu)$ its absorption cross section at frequency ν. The total path over which the laser beam travels is taken as $2R$. C_S is a system parameter that takes account of time-independent losses, and $\bar{\kappa}_\varepsilon(t)$ is the atmospheric extinction coefficient exclusive of absorption within the species to be monitored. The time dependence of $\bar{\kappa}_\varepsilon(t)$ arises from atmospheric turbulence and changes in the aeroticulate concentration and distribution. Since turbulence and scattering are weak functions of frequency, $\bar{\kappa}_\varepsilon(t)$ is assumed to be essentially independent of ν over the frequency modulation range.

Under these circumstances, the derivative of the received laser signal with respect to the frequency is

$$\frac{dP_s(\nu, t)}{d\nu} = 2RNP_s(\nu, t)\frac{d\sigma^A(\nu)}{d\nu} \tag{6.38}$$

It is quite apparent from equation (6.38) that dividing this derivative by the direct signal eliminates the time-dependent turbulence and scattering parameter and produces a new signal

$$\mathscr{P}_s \equiv \frac{1}{P_s(\nu, t)}\frac{dP_s(\nu, t)}{d\nu} = 2RN\frac{d\sigma^A(\nu)}{d\nu} \tag{6.39}$$

that is directly proportional to the average density to be measured. Clearly for maximum sensitivity the ratioed signal $\mathscr{P}_s$ should be evaluated at the laser frequency where the slope of the absorption line is greatest. This conclusion would no longer be true in the event of a high burden of the pollutant, since then most of the laser power could be absorbed within the core of the line.

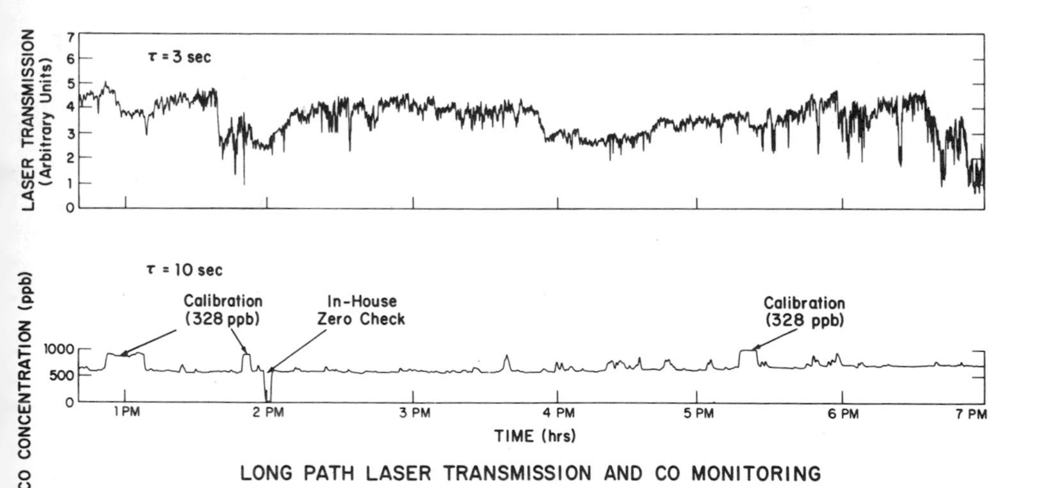

Fig. 6.9. Long-path (0.61-km) monitoring of CO over an extended period of time. The upper curve is the direct transmission signal using a 3-s time constant. The lower curve is the CO measurement using a 10-s time constant (Ku et al., 1975).

Under these circumstances a weaker line (possibly of an isotope) could be used, or the nominal laser frequency could be detuned from the line center.

At atmospheric pressure the lines are essentially Lorentzian, so that

$$\sigma^A(\nu) = \frac{\sigma^A(\nu_0)}{\left(\frac{\nu - \nu_0}{\Delta\nu_c}\right)^2 + 1} \tag{6.40}$$

where ν_0 is the line center frequency and $\Delta\nu_c$ is the collision-induced frequency half width (HWHM) for the line of interest. Using (6.40) in (6.39) yields

$$\mathscr{P}_s = \frac{4\sigma^A(\nu_0)(\nu - \nu_0)RN}{\Delta\nu_c^2\left[\left(\frac{\nu - \nu_0}{\Delta\nu_c}\right)^2 + 1\right]^2} \tag{6.41}$$

An example of the difference in the signals is presented in Fig. 6.9, where the upper curve shows the received laser signal with its attendant noise, while the lower curve displays the calibrated ratioed signal.

7

Laser-Remote-Sensor Equations

In this chapter we shall derive and study the basic equations of importance in the field of laser remote sensing. The form of equation to be used in any given situation depends upon the kind of interaction invoked by the laser radiation. This in turn is determined by the nature of the measurement to be undertaken. For those applications in which backscattering (elastic or inelastic) of the laser beam is utilized, the form of the lidar equation is fairly simple and is derived first. Most atmospheric probing, including those instances where differential absorption is employed, is covered by this equation.

For those situations involving laser-induced fluorescence, finite relaxation effects of the laser-excited species has to be taken into consideration. This leads to a more complex form of the lidar equation and to an optical-depth dependence of the target media, detector integration period, and laser-pulse shape and duration. In the limit of large optical depth this form of the lidar equation becomes identical to the laser fluorosensor equation which was specifically developed to cover airborne lidars that probe natural bodies of water.

In general, interpretation of the lidar signal is further complicated by geometrical considerations that include the degree of overlap between the laser beam and the field of view of the receiver optics as well as the details of the telescope. Nevertheless, it is possible to use one simplified form of lidar equation under a fairly wide range of conditions. This is discussed at the end of Section 7.3.

7.1. SCATTERING FORM OF THE LIDAR EQUATION

In the case of a pulsed, monostatic lidar, the increment of signal power $\Delta P(\lambda, R)$ received by the detector in the wavelength interval $(\lambda, \lambda + \Delta\lambda)$

from the element of range located in the interval $(R, R + \Delta R)$ is given by

$$\Delta P(\lambda, R) = \int J(\lambda, R, \mathbf{r})\, \Delta\lambda\, \Delta R\, p(\lambda, R, \mathbf{r})\, dA(R, \mathbf{r}) \qquad (7.1)$$

where

$J(\lambda, R, \mathbf{r})$ represents the laser-induced spectral radiance at wavelength λ, at position $\mathbf{r}$ in the target plane located at range R, per unit range interval;
$dA(R, \mathbf{r})$ represents the element of target area at position $\mathbf{r}$ and range R; and
$p(\lambda, R, \mathbf{r})$ represents the probability that radiation of wavelength λ emanating from position $\mathbf{r}$ at range R will strike the detector.

Many factors will affect this probability. These include geometrical considerations, atmospheric attenuation, the receiver optics, and spectral transmission characteristics. Fortunately, we can separate most of these influences and write

$$p(\lambda, R, \mathbf{r}) = \frac{A_0}{R^2} \times T(\lambda, R) \times \xi(\lambda) \times \xi(R, \mathbf{r}) \qquad (7.2)$$

where

A_0/R^2 represents the acceptance solid angle of the receiver optics (A_0 being the area of the objective lens or mirror);
$T(\lambda, R)$ represents the atmospheric transmission factor at wavelength λ over range R;
$\xi(\lambda)$ represents the receiver's spectral transmission factor and includes the influence of any spectrally selecting elements such as a monochromator; and
$\xi(R, \mathbf{r})$ represents the probability of radiation from position $\mathbf{r}$ in the target plane at range R reaching the detector, based on geometrical considerations.

For the moment we shall assume that $\xi(R, \mathbf{r})$ depends only upon the *overlap* of the area of laser irradiation with the field of view of the receiver optics. Consequently we shall refer to $\xi(R, \mathbf{r})$ as the *overlap factor*. Later, in Section 7.4, we shall return to this function and see that it is somewhat more involved and can depend quite critically upon the details of the receiver optics. For this reason it is often referred to as the *geometrical form factor* in the literature.

The *target spectral radiance* $J(\lambda, R, \mathbf{r})$ depends upon the nature of the interaction between the laser radiation and the target medium. In this section we shall consider a *scattering* (elastic or inelastic) medium. In this instance we may write

$$J(\lambda, R, \mathbf{r}) = \beta(\lambda_L, \lambda, R, \mathbf{r}) I(R, \mathbf{r}) \qquad (7.3)$$

where $I(R,\mathbf{r})$ is the laser irradiance at position $\mathbf{r}$ and range R, and

$$\beta(\lambda_L, \lambda, R, \mathbf{r}) = \sum_i N_i(R,\mathbf{r}) \left\{ \frac{d\sigma(\lambda_L)}{d\Omega} \right\}_i^s \mathscr{L}_i(\lambda) \tag{7.4}$$

is the *volume backscattering coefficient*, in which

$N_i(R,\mathbf{r})$ represents the number density of scatterer species i;
$\{d\sigma(\lambda_L)/d\Omega\}_i^s$ represents the differential scattering cross section under irradiation with laser radiation at wavelength λ_L; and
$\mathscr{L}_i(\lambda)\,\Delta\lambda$ represents the fraction of the scattered radiation that falls into the wavelength interval $(\lambda, \lambda + \Delta\lambda)$.

The total signal power received by the detector at the instant t $(= 2R/c)$, corresponding to the time taken for the leading edge of the laser pulse to propagate (at the velocity of light, c) to range R and the returned radiation to reach the lidar, can be expressed in the form

$$P(\lambda, t) = \int_0^{R=ct/2} dR \int_{\Delta\lambda_0} d\lambda \int J(\lambda, R, \mathbf{r}) p(\lambda, R, \mathbf{r})\, dA\,(R,\mathbf{r}) \tag{7.5}$$

The range integral is required to account for the fact that radiation reaching the detector at time t not only originates from the distance $ct/2$, but also from any position along the path of the laser pulse from which scattering arises. The range of wavelength integration extends over the lidar receiver's spectral window $\Delta\lambda_0$ centered about λ. If equations (7.2) and (7.3) are employed we can write

$$P(\lambda, t) = A_0 \int_0^{R=ct/2} \frac{dR}{R^2} \int_{\Delta\lambda_0} \xi(\lambda)\, d\lambda$$

$$\times \int \beta(\lambda_L, \lambda, R, \mathbf{r}) T(\lambda, R) \xi(R,\mathbf{r}) I(R,\mathbf{r})\, dA\,(R,\mathbf{r}) \tag{7.6}$$

In the case of a scattering medium the observed radiation is as narrowband as that of the laser radiation, and if we assume both to be much smaller than the receiver's spectral window $\Delta\lambda_0$, we can treat $\mathscr{L}_i(\lambda)$, and therefore β, as a delta function. If we also assume that the medium is homogeneous over the zone of overlap between the field of view and the laser beam, then we can write

$$P(\lambda, t) = A_0 \xi(\lambda) \int_0^{R=ct/2} \beta(\lambda_L, \lambda, R) T(\lambda, R) \frac{dR}{R^2} \int \xi(R,\mathbf{r}) I(R,\mathbf{r})\, dA\,(R,\mathbf{r}) \tag{7.7}$$

As mentioned above, we shall assume at this point that the probability $\xi(R, \mathbf{r})$ is essentially unity where the field of view of the receiver optics overlaps the laser beam and zero elsewhere. We shall also assume that the lateral distribution of the laser pulse is uniform over an area $A_L(R)$ at range R. In this case

$$\int \xi(R,\mathbf{r}) I(R,\mathbf{r})\, dA\,(R,\mathbf{r}) = \xi(R) I(R) A_L(R) \tag{7.8}$$

and

$$P(\lambda, t) = A_0 \xi(\lambda) \int_{R=0}^{R=ct/2} \beta(\lambda_L, \lambda, R) T(\lambda, R) \xi(R) I(R) A_L(R) \frac{dR}{R^2} \tag{7.9}$$

An additional simplification that is usually made in the literature is to approximate the temporal shape of the laser pulse by a rectangle of duration τ_L. Then the limits of the range integration in equation (7.9) are obviously $c(t - \tau_L)/2$ to $ct/2$. Furthermore, since the range of interest is generally much greater than the laser pulse length $c\tau_L$ (otherwise the resolution would be poor), we may treat the range-dependent parameters as constants over the small interval of range integration. Then the total scattered laser power received at a time $t = 2R/c$ can be expressed in the form

$$P(\lambda, t) = A_0 \xi(\lambda) \beta(\lambda_L, \lambda, R) T(\lambda, R) \xi(R) I(R) A_L(R) \frac{c\tau_L/2}{R^2} \tag{7.10}$$

More precisely the last factor should be $(c\tau_L/2)/[R(R - c\tau_L/2)]$, but as stated, $R \gg c\tau_L/2$.

For a rectangular-shaped laser pulse of duration τ_L,

$$I(R) = \frac{E_L T(\lambda_L R)}{\tau_L A_L(R)} \tag{7.11}$$

where E_L represents the output energy of the laser pulse, and $T(\lambda_L R)$ represents the atmospheric transmission factor at the laser wavelength to range R. It follows from the Beer–Lambert law, equation (4.80), that the transmission factors are

$$T(\lambda_L, R) \equiv \mathrm{e}^{-\int_0^R \kappa(\lambda_L, R)\, dR}$$

and

$$T(\lambda, R) \equiv \mathrm{e}^{-\int_0^R \kappa(\lambda, R)\, dR}$$

where $\kappa(\lambda_L, R)$ and $\kappa(\lambda, R)$ represent the atmospheric attenuation coefficients at the laser and detected wavelengths, respectively. It is evident that combining these leads to the total atmospheric transmission factor

$$T(R) \equiv T(\lambda_L, R)T(\lambda, R) = \mathrm{e}^{-\int_0^R \{\kappa(\lambda_L, R) + \kappa(\lambda, R)\} dR} \tag{7.12}$$

Although, the instantaneous power falling upon the detector is a useful quantity to evaluate, a more pertinent entity is the increment of radiative energy at wavelength λ received by the detector during the interval $(t, t + \tau_d)$, where τ_d is the integration period for the detector and $t = 2R/c$:

$$E(\lambda, R) = \int_{2R/c}^{2R/c+\tau_d} P(\lambda, t)\, dt \tag{7.13}$$

Combining equations (7.10), (7.11), and (7.12) with (7.13) yields the scattered laser energy received within the detector's response time τ_d:

$$E(\lambda, R) = E_L \xi(\lambda) T(R) \xi(R) \frac{A_0}{R^2} \beta(\lambda_L, \lambda, R) \frac{c\tau_d}{2} \tag{7.14}$$

This is often cited as the *basic scattering lidar equation.*

Implicit in the derivation of this equation is the assumption that $\tau_d \ll 2R/c$. The effective range resolution for such a system is limited to $c(\tau_d + \tau_L)/2$, as is clearly seen by reference to Fig. 7.1. If one species dominates the scattering and its scattering cross section is isotropic, the lidar equation can be expressed in the form

$$E(\lambda, R) = E_L \xi(\lambda) T(R) \xi(R) \frac{A_0}{R^2} N(R) \frac{\sigma^s(\lambda_L, \lambda)}{4\pi} \frac{c\tau_d}{2} \tag{7.15}$$

where $\sigma^s(\lambda_L, \lambda)$ represents the total cross section for scattering at λ for incident radiation of wavelength λ_L. Equation (7.15) represents the basic single-constituent-scattering form of the lidar equation.

In the case of a more general laser pulse shape we can write

$$I(R^*) = \frac{E_L T(\lambda_L, R^*)}{A_L(R^*)} \mathscr{I}(t^*) \tag{7.16}$$

where $\mathscr{I}(t^*)$ describes the temporal behavior of the laser pulse in a frame of reference fixed to the leading edge of the laser pulse. Clearly

$$t^* = \frac{2(R - R^*)}{c} \tag{7.17}$$

and represents the time taken for the leading edge of the laser pulse to

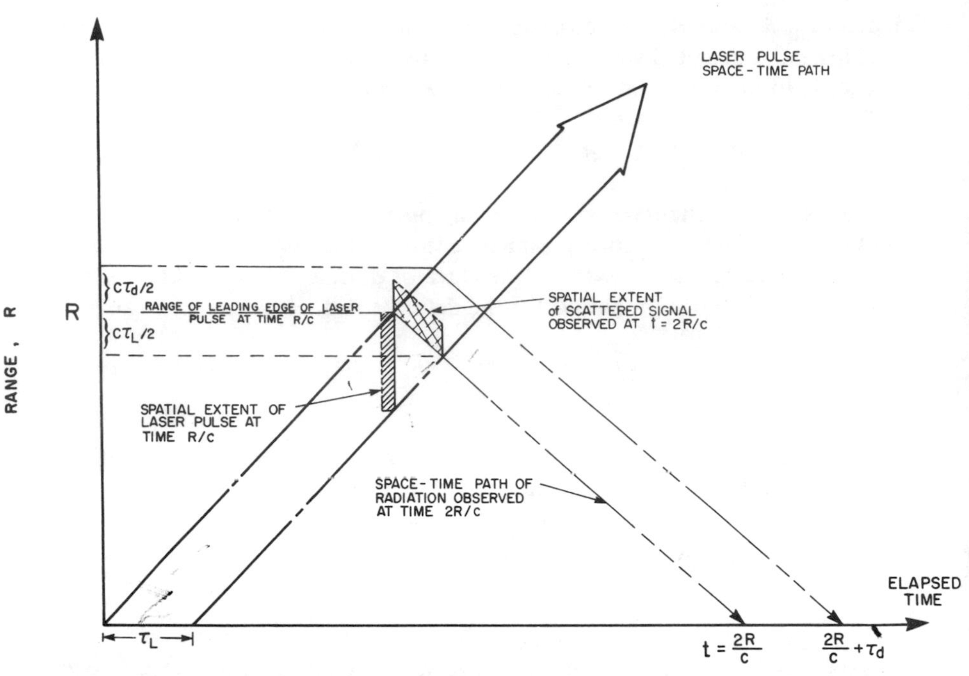

Fig. 7.1. Spatial resolution for scattering phenomena a. ~een from the space–time diagram of a propagating rectangular-shaped laser pulse (Measures, 1977).

propagate from R^* to R and the scattered radiation to return to R^*. In general,

$$\int_0^\infty \mathscr{I}(t^*)\, dt^* = 1 \tag{7.18}$$

Under these circumstances, the total signal power received after an elapsed time t corresponding to range R follows from equation (7.9):

$$P(\lambda, R) = E_L A_0 \xi(\lambda) \int_{R=0}^{R=ct/2} \beta(\lambda_L, \lambda, R^*) T(R^*) \xi(R^*) \mathscr{I}(R^*) \frac{dR^*}{R^{*2}} \tag{7.19}$$

where $\mathscr{I}(R^*)$ is used to represent $\mathscr{I}(t^*)$.

If again we assume that the effective laser pulse length, $c\tau_L$, is small compared to the range of interest, then we may treat the factor $\beta(\lambda_L, \lambda, R^*)T(R^*)\xi(R^*)/R^{*2}$ as constant over the small interval of range for which $\mathscr{I}(R^*)$ is finite:

$$P(\lambda, R) \cong E_L \frac{A_0}{R^2} \xi(\lambda) \beta(\lambda_L, \lambda, R) T(R) \xi(R) \int_{t=0}^{t=2R/c} \mathscr{I}(t^*) \frac{c\, dt^*}{2} \tag{7.20}$$

Since the upper limit of this integration corresponds to the time that scattered radiation from the leading edge of the laser pulse reaches the lidar system, it is evident from (7.18) that

$$\int_{t=0}^{t=2R/c} \mathscr{I}(t^*)\, dt^* \cong 1$$

and so we can write the total scattered laser power received at a time corresponding to the leading edge of the laser pulse propagating to a range R as

$$P(\lambda, R) = P_L \frac{A_0}{R^2} \xi(\lambda) \beta(\lambda_L, \lambda, R) \xi(R) \frac{c\tau_L}{2} \mathrm{e}^{-\int_0^R \kappa(R)\, dR} \tag{7.21}$$

where from equation (7.12) we have introduced

$$\kappa(R) = \kappa(\lambda_L, R) + \kappa(\lambda, R) \tag{7.22}$$

as the two-way attenuation coefficient and $P_L \equiv E_L/\tau_L$ as the average power in the laser pulse.

In the event that we are interested in *elastic* (Mie or Rayleigh) scattering, then the wavelength of observation is invariably the same as that of the laser and we may write

$$P(\lambda_L, R) = P_L \frac{A_0}{R^2} \xi(\lambda_L) \beta(\lambda_L, R) \xi(R) \frac{c\tau_L}{2} \mathrm{e}^{-2\int_0^R \kappa(\lambda_L, R)\, dR} \tag{7.23}$$

In summary, equation (7.21) is the scattering lidar equation most often quoted in the literature, and equation (7.23) represents a special case that is of growing interest due to the increasing popularity of differential absorption techniques. With regard to the radiative energy received within the detector integration period, equation (7.14) represents the more relevant form of the scattering lidar equation.

7.2. DIFFERENTIAL ABSORPTION LIDAR (DIAL) EQUATION

Differential absorption of laser radiation by a particular molecular species represents both a selective and a sensitive method of measuring specific atmospheric constituents. There are two ways in which such measurements can be undertaken. Both involve using two laser pulses of slightly different wavelength (one chosen to coincide with a strong absorption feature of the specific constituent of interest, the other detuned into the wing of this feature) and comparing the attenuation of the two pulses. The difference in the techniques stems from the mechanism chosen to return the laser radiation to the lidar receiver system. In one case elastic scattering from atmospheric aerosols and particulates is employed, and consequently we shall refer to this as the DAS (differential absorption and scattering) technique. The other approach relies on scattering of the laser radiation from some conveniently located topographical target. An extreme example of this uses a strategically positioned retroreflector.

In the DAS approach, two laser wavelengths, λ_0 and $\lambda_0 + \delta\lambda$, are selected such that λ_0 corresponds to the center wavelength of some prominent absorption line of the molecule of interest, while $\lambda_0 + \delta\lambda$ lies in the wing of this line. If we write λ_W for $\lambda_0 + \delta\lambda$, and use equation (7.23) (the elastic-scattering form of the lidar equation), then the ratio of the return power signals at the two wavelengths is

$$\frac{P(\lambda_0, R)}{P(\lambda_W, R)} = \frac{\xi(\lambda_0)\beta(\lambda_0, R)}{\xi(\lambda_W)\beta(\lambda_W, R)} e^{-2\int_0^R \{\kappa(\lambda_0, R) - \kappa(\lambda_W, R)\} dR} \tag{7.24}$$

where the output power of the laser is assumed to be the same at both wavelengths.

If we separate the absorption associated with the molecule of interest from the total attenuation coefficient, then we can write

$$\int_0^R N(R)\sigma_A(\lambda_0 : \lambda_W)\, dR = \tfrac{1}{2}\ln\left[\frac{P(\lambda_W, R)\xi(\lambda_0)\beta(\lambda_0, R)}{P(\lambda_0, R)\xi(\lambda_W)\beta(\lambda_W, R)}\right] - \int_0^R \{\bar{\kappa}(\lambda_0, R) - \bar{\kappa}(\lambda_W, R)\}\, dR \tag{7.25}$$

where we have introduced the *differential absorption cross section*,

$$\sigma_A(\lambda_0 : \lambda_W) \equiv \sigma^A(\lambda_0) - \sigma^A(\lambda_W) \tag{7.26}$$

and have assumed that in general the total attenuation coefficient is given by

$$\kappa(\lambda, R) = \bar{\kappa}(\lambda, R) + N(R)\sigma^A(\lambda) \tag{7.27}$$

$\bar{\kappa}(\lambda, R)$ is obviously the attenuation coefficient exclusive of the absorption contribution from the molecular species of interest, $N(R)$ represents the number density of these molecules at range R, and $\sigma^A(\lambda)$ their absorption cross section at wavelength λ.

In differential form equation (7.25) becomes

$$N(R) = \frac{1}{2\sigma_A(\lambda_0 : \lambda_W)}\left[\frac{d}{dR}\left\{\ln\left[\frac{P(\lambda_W, R)}{P(\lambda_0, R)}\right] - \ln\left[\frac{\beta(\lambda_W, R)}{\beta(\lambda_0, R)}\right]\right\} + \bar{\kappa}(\lambda_W, R) - \bar{\kappa}(\lambda_0, R)\right] \tag{7.28}$$

where we have assumed that the receiver's spectral transmission factor is effectively independent of wavelength over the small interval $\delta\lambda$:

$$\xi(\lambda_0) \approx \xi(\lambda_W)$$

Additional simplification can be attained if we also assume that the volume backscattering coefficient β and the residual attenuation coefficient $\bar{\kappa}$ are independent of wavelength over this small interval $\delta\lambda$. We shall consider this in more detail later in Chapter 8.

A considerable improvement in sensitivity can be achieved if this differential-absorption technique is used in conjunction with a *topographical* scatterer. However, this gain in sensitivity is achieved at the expense of range resolution, so that this technique is only applicable in situations where the integrated concentration of the trace constituents along the path of the laser beam is worth evaluating. Under these circumstances the signal power equation takes the form

$$P(\lambda_0, t) = \frac{E_L}{\tau_L}\frac{A_0}{R_T^2}\xi(\lambda_0)\xi(R_T)\frac{\rho^s}{\pi}\exp\left(-2\int_0^{R_T}\kappa(\lambda_0, R)\,dR\right) \quad (7.29)$$

where ρ^s represents the scattering efficiency of the topographical target and R_T is the range to the topographical target. Values of ρ^s can range from 0.1 in the visible to 1 in the infrared (Wolfe, 1966; Shumate et al., 1982; Grant, 1982).

The corresponding increment of radiative energy received within the detector's integration period τ_d is then

$$E(\lambda_0, R_T) = E_L\frac{A_0}{R_T^2}\xi(\lambda_0)\xi(R_T)\frac{\rho^s\tau_d}{\pi\tau_L}\exp\left(-2\int_0^{R_T}\kappa(\lambda_0, R)\,dR\right) \quad (7.30)$$

provided $\tau_d \leq \tau_L$. In the event that $\tau_d > \tau_L$, the factor τ_d/τ_L is replaced with unity. In order to have optimum temporal discrimination against any solar background illumination of the target, τ_d should be chosen to be as close to τ_L as possible. At locations of known pollution emission, a retroflector might be positioned so as to maximize the system sensitivity. Under these conditions the factor $\rho^s A_0/\pi R_T^2$ in equation (7.30) is replaced with ξ_0, the receiver collection efficiency. This can amount to an improvement of several orders of magnitude, depending primarily upon the range.

As with the DAS technique, two closely spaced laser wavelengths must be employed if contributions other than from the species of interest are to be eliminated. With the same reasoning as above for the case of a distributed scatterer, we can express the integrated concentration of the constituent along the path of the laser beam as

$$\int_0^{R_T} N(R)\,dR = \frac{1}{2\sigma_A(\lambda_0 : \lambda_W)}\left[\ln\left\{\frac{E(\lambda_W, R_T)}{E(\lambda_0, R_T)}\right\} + \bar{\kappa}(\lambda_W, R_T) - \bar{\kappa}(\lambda_0, R_T)\right] \quad (7.31)$$

In many cases of pollution monitoring the constituent of concern is normally present in the atmosphere, so that it is the increased loading of the atmosphere that is the entity to be measured. Under these circumstances an additional measurement has to be undertaken either prior to the release of pollution or at a different orientation so that the path of the laser beam misses the effluent plume. This second measurement provides the reference background level that has to be subtracted from the measurement across the plume.

In such cases Byer and Garbuny (1973) have indicated that the criterion for minimum transmitted energy does not correspond to the use of a laser wavelength that coincides with the peak absorption cross section. Their results show that some degree of detuning may be necessary for optimization. On the other hand, Measures and Pilon (1972) have drawn attention to the severe attenuation of the laser beam that can occur if the laser wavelength is chosen to maximize absorption.

7.3. LIDAR EQUATION IN THE CASE OF A FLUORESCENT TARGET

If the laser is capable of exciting fluorescence within the target, then finite relaxation effects have to be taken into consideration. Kildal and Byer (1971) were the first to recognize the significance of this effect and to illustrate its influence on the return signal. The following discussion stems from a more detailed analysis of this problem by Measures (1977).

In the case of fluorescence the signal power received by the lidar photodetector is of a similar form to that indicated in equation (7.5), except that the radiance of the target element arises from the emission of excited molecules:

$$J(\lambda, R, \mathbf{r}) = \sum_i \frac{N_i^*(R, \mathbf{r}) hc \mathscr{L}_i^F(\lambda)}{4\pi\lambda\tau_{\mathrm{rad}}^i} \tag{7.32}$$

where

$N_i^*(R, \mathbf{r})$ represents the number density of laser-excited molecules (or atoms) of species i, at position $\mathbf{r}$ of the target plane at range R, capable of undergoing fluorescence;

$\mathscr{L}_i^F(\lambda)\,\Delta\lambda$ represents the fraction of fluorescence emitted by species i into the wavelength interval $(\lambda, \lambda + \Delta\lambda)$;

τ_{rad}^i represents the radiative lifetime for the excited molecules (or atoms) of species i; and

h and c represent Planck's constant and the velocity of light in a vacuum, respectively.

We shall assume that only one molecular species is excited by the laser, and that these molecules can return to the ground level (or some other low-lying level) with the subsequent emission of radiation at some wavelength $\lambda(>\lambda_L)$, or they may be deexcited by some nonradiative process. If the laser power density is very high, then the excited molecule may be forced to return to its original level by the process of stimulated emission, and saturation effects can arise; see Section 4.3.

In the weak-beam limit, which is generally of interest for remote sensing, stimulated emission can be neglected, and from equation (4.21) the temporal variation of the excited-state number density $N^*(R, t)$ can be expressed in the form

$$\frac{dN^*(R,t)}{dt} = \frac{\lambda_L \sigma^A(\lambda_L)}{hc} N(R,t) I(R,t) - \frac{N^*(R,t)}{\tau} \tag{7.33}$$

where

$\sigma^A(\lambda_L)$ represents the absorption cross section per molecule for incident radiation at wavelength λ_L [see (3.174)];

$N(R, t)$ represents the ground-state number density in the target plane at range R and time t (we have dropped the **r**-dependence, as we shall assume that the target medium is homogeneous over the area of laser excitation);

τ represents the observed lifetime of the excited population and is given by $1/\tau = 1/\tau_{\text{rad}} + C_Q$, where τ_{rad} represents the radiative lifetime of the excited state, and C_Q represents the collision quenching rate for the excited state.

We shall restrict our attention to situations where the level of irradiation is sufficiently low that nonlinear effects can be neglected and that no appreciable depletion of the ground-state number density is produced. In addition we shall assume that the ground-state number density prior to irradiation is $N_0(R)$ and that $N^*(R,0) \equiv 0$. Then the solution of equation (7.33) yields the temporal variation of the excited-state population at range R and time t:

$$N^*(R,t) = \frac{\lambda_L N_0(R) \sigma^A(\lambda_L)}{hc} \mathrm{e}^{-t/\tau} \int_0^t I(R,x) \mathrm{e}^{x/\tau}\, dx \tag{7.34}$$

where $I(R, x)$ represents the laser irradiance at range R and at a *time* x after the leading edge of the laser pulse reaches this location (x is a dummy time variable).

At the instant that the leading edge of the laser pulse reaches range R, the target medium at range R' ($< R$) will have been exposed to laser radiation for a period $(R - R')/c$. Fluorescence induced by the leading edge of the laser pulse and propagating toward the detector will be reinforced by fluorescence emanating from the target medium at range R', having been exposed to laser radiation for a period $t' = 2(R - R')/c$. Consequently, the appropriate value of radiance to be used in the range integral of equation (7.5) is obtained by combining equations (7.32) and (7.34), namely,

$$J(\lambda, R') = \frac{N_0(R')\sigma^A(\lambda_L)\lambda_L \mathscr{L}^F(\lambda)}{4\pi\tau_{\text{rad}}\lambda} \mathbf{e}^{-t'/\tau} \int_0^{t'} I(R', x)\mathbf{e}^{x/\tau}\, dx \quad (7.35)$$

From Section 3.5.4, equation (3.186), it is apparent that we can introduce the *spectrally integrated fluorescence cross section*,

$$\sigma^F(\lambda_L) \equiv \sigma^A(\lambda_L)\frac{\tau}{\tau_{\text{rad}}} \quad (7.36)$$

If equations (7.35) and (7.36) are substituted into (7.5) and we make the same assumptions in regard to the probability factor $p(\lambda, R, \mathbf{r})$ as we did in the scattering section, we arrive at the fluorescence signal power received at the lidar detector after an clapsed time $t\ (= 2R/c)$:

$$P(\lambda, R) = \frac{A_0\sigma^F(\lambda_L)\lambda_L}{4\pi\tau\lambda} \int_0^R dR'\, \xi(R')A_L(R')\frac{N_0(R')}{R'^2}\mathbf{e}^{-t'/\tau}$$

$$\times \int_0^{t'} I(R', x)\mathbf{e}^{x/\tau}\, dx \int_{\Delta\lambda_0} \mathscr{L}^F(\lambda)\xi(\lambda)T(\lambda, R')\, d\lambda \quad (7.37)$$

The range integration in equation (7.37) is along the space–time path of the observed ray, path AB of Fig. 7.2.

In most cases of interest the variation of the atmospheric transmission factor $T(\lambda, R')$ over the spectral bandwidth $\Delta\lambda_0$ of the receiver system is small enough that it can be taken out of the wavelength integral in equation (7.37). Most practical situations can be approximated by one of two limiting cases. If the bandwidth of the receiver system is made adequate to accept the entire fluorescence profile, then we have a situation akin to that of scattering, where the wavelength integral can be replaced by the factor $T(\lambda, R')\xi(\lambda)$. Under these circumstances the wavelength λ corresponds to the center value of the fluorescence profile.

If, as is more often the situation, the spectral window of the receiver system is small compared to the spectral width of the observed fluorescence, the wavelength integral can be approximated in the following manner:

$$\int_{\Delta\lambda_0} \mathscr{L}^F(\lambda)\xi(\lambda)T(\lambda, R')\, d\lambda = T(\lambda, R')K_0(\lambda)\mathscr{L}^F(\lambda) \quad (7.38)$$

where, as indicated in equation (6.3),

$$K_0(\lambda) \equiv \int_{\Delta\lambda_0} \xi(\lambda')\, d\lambda'$$

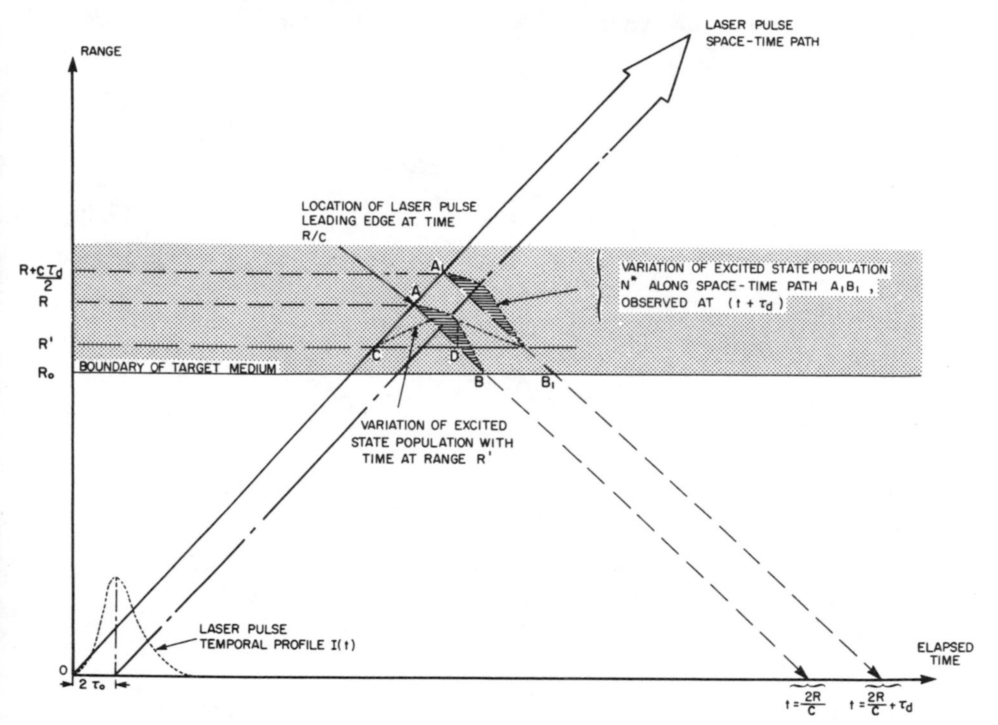

Fig. 7.2. Space–time view of laser pulse propagation and excitation of a fluorescent target medium (Measures, 1977).

and is termed the *filter function*. The wavelength λ indicated on the right-hand side (RHS) of equation (7.38) corresponds to the value at the center of the receiver bandwidth; $K_0(\lambda)$ represents the effective bandwidth which would transmit, with unit transmission efficiency, the same fraction of the fluorescence as achieved by the real system.

We shall again assume that the laser irradiance at range R' and time x is given by

$$I(R', x) = \frac{E_L T(\lambda_L, R')}{A_L(R')} \mathcal{I}(x) \qquad (\mathrm{W\ cm^{-2}}) \tag{7.39}$$

where, as before,

$$\int_0^\infty \mathcal{I}(x)\, dx = 1$$

Furthermore, if the total attenuation coefficient $\kappa_T \equiv \kappa(\lambda_L) + \kappa(\lambda)$ is regarded as a constant within the target medium, the boundary of which is located at range R_0, then the radiative energy received by the detector in the time interval $(t, t + \tau_d)$ is

$$E(\lambda, R) = E_L T(R_0) K_0(\lambda) \xi(R_0) \frac{A_0 N_0 \sigma^F(\lambda_L) \mathscr{L}^F(\lambda)}{4\pi R^2 \tau} \int_t^{t+\tau_d} H(R)\, dt' \tag{7.40}$$

where

$$H(R) = \int_{R_0}^{R} dR'\, \mathrm{e}^{-\kappa_T(R' - R_0)} \mathrm{e}^{-t'/\tau} \int_0^{t'} \mathscr{I}(x) \mathrm{e}^{x/\tau}\, dx \tag{7.41}$$

To obtain this relation we have expressed

$$T(\lambda, R') T(\lambda_L, R') = T(R_0) \mathrm{e}^{-\kappa_T(R' - R_0)} \tag{7.42}$$

where

$$T(R_0) = \exp\left(- \int_0^{R_0} \kappa(R)\, dR \right) \tag{7.43}$$

and have assumed that the overlap factor $\xi(R')$ is only weakly dependent upon the range, so that we can approximate its value by $\xi(R_0)$. We have also assumed that the physical extent of the fluorescent target is sufficiently small that the factor $1/R^2$ can be taken out of the range integration and that the factor λ_L/λ can be approximated by unity.

Clearly, equation (7.40) can be written in a form that bears a close resemblance to the scattering lidar equation as expressed by equation (7.15), namely,

$$E(\lambda, R) = E_L K_0(\lambda) T(R_0) \xi(R_0) \frac{A_0}{R^2} N_0 \frac{\sigma^F(\lambda_L, \lambda)}{4\pi} \frac{c\tau_d}{2} \gamma(R) \mathrm{e}^{-\kappa_T(R - R_0)} \tag{7.44}$$

where we have introduced the *fluorescence cross section*

$$\sigma^F(\lambda_L, \lambda) = \sigma^F(\lambda_L) \mathscr{L}^F(\lambda) \tag{7.45}$$

and the *fluorescence* (or *lifetime*) *correction factor*

$$\gamma(R) = \frac{2}{c\tau_d \tau} \mathrm{e}^{\kappa_T(R - R_0)} \int_t^{t+\tau_d} H(R)\, dt' \tag{7.46}$$

to account for the effects of the finite lifetime of the laser excited molecules, the optical depth of the target medium, the detector integration period, the laser pulse shape and duration.

It should also be evident that in the fluorescence case the filter function $K_0(\lambda)$ is used in place of $\xi(\lambda)$, and that since we have assumed the fluorescent medium only exists beyond range R_0, the atmospheric transmission factor $T(R)$ is replaced with its value at the boundary [i.e., $T(R_0)$] multiplied by the transmission factor within the target medium [i.e., $\mathrm{e}^{-\kappa_T(R-R_0)}$].

The fluorescence correction factor $\gamma(R)$ can be evaluated once the shape of the laser pulse has been defined. Measures (1977) has calculated the variation of this correction factor for rectangular-shaped laser pulses and more realistic pulses given by

$$\mathscr{I}(t) = \frac{(t/\tau_0)^n \mathrm{e}^{-t/\tau_0}}{\tau_0 \Gamma(n+1)} \tag{7.47}$$

where τ_0 is a characteristic time of the laser pulse (the peak amplitude occurs for $t = n\tau_0$, $\Gamma(n)$ represents the gamma function, and n is an integer. For the particular case of $n = 2$, the laser pulse duration τ_L^* is defined to be five times the characteristic time (i.e., $\tau_L^* = 5\tau_0$) and corresponds to approximately the time interval between the 20% points on the laser temporal profile. Under these circumstances the fluorescence or lifetime correction factor is

$$\gamma(z) = \frac{(1-q)\mathrm{e}^{\zeta z}}{\tau_d \Gamma(3) q^3} \int_t^{t+\tau_d} dt \int_0^z \mathrm{e}^{-(\zeta z' + (z-z')(1-q))}\, dz' \int_0^{(z-z')q} y^2 \mathrm{e}^{-y}\, dy \tag{7.48}$$

where $q = 1 - \tau_0/\tau$, $y = (x/\tau_0)q$, $\zeta \equiv \kappa_T c\tau_0/2$ (represents the nondimensional ratio of the laser characteristic length to the target attenuation length $1/\kappa_T$) and $z \equiv 2(R - R_0)/c\tau_0$ represents the penetration depth of the laser-pulse leading edge within the fluorescence medium in terms of the laser characteristic length $c\tau_0$.

Measures (1977) has shown that for an optically thin target the lifetime correction factor $\gamma(z)$ tends to approach unity for large penetration depths. This is illustrated in Fig. 7.3(a) and (b), where the variation in the lifetime correction factor with the normalized effective penetration depth $z^* \equiv (R - R_0)/L$ is shown. Here $L \equiv c\tau_L^*/2$ and represents the laser effective pulse length. $T^* \equiv \tau_L^*/\tau$ represents the ratio of the laser effective pulse duration to the fluorescence lifetime and can be seen to be an important parameter in determining the magnitude of $\gamma(z^*)$. It should be noted (Measures, 1977) that very similar results are obtained if we consider the simpler rectangular-shaped laser pulse of duration τ_L, and consequently we shall drop the asterisk from τ_L in the remaining discussion. For small values of T^* (i.e., $\tau_L \ll \tau$) it is apparent that $\gamma(z) \ll 1$ even for considerable penetration depths.

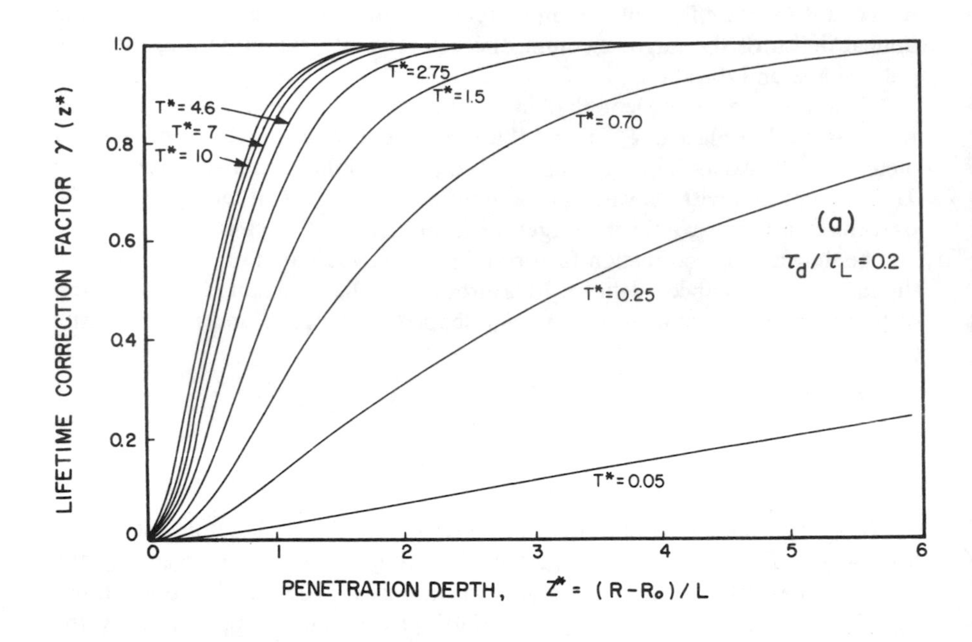

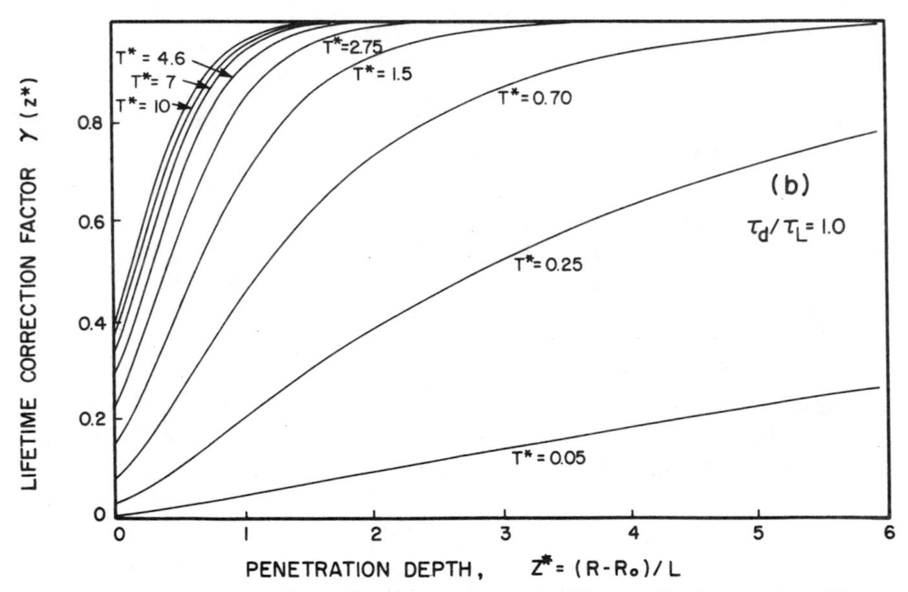

Fig. 7.3. Variation of optically thin correction factor $\gamma(z^*)$ with normalized penetration z^* into a fluorescent target for several values of T^* (ratio of laser pulse duration to fluorescence lifetime, τ_L/τ); R is the range of the laser pulse leading edge, R_0 is the range of the target boundary, and L is the laser pulse length ($c\tau_L/2$). Ratio of laser pulse length to attenuation length is $\kappa_T L = 0.005$ for each case (Measures, 1977).

If $\tau_L \gg \tau$ (i.e., T^* is large), the value of $\gamma(z)$ approaches unity for target penetration depths of about one laser pulse length. Nevertheless, the fact that the correction factor $\gamma(z)$ can be much less than unity even for large values of T^*—corresponding to the *scattered limit*—indicates that in effect the simple form of the scattering lidar equation [equation (7.15)] overestimates the expected signal in situations where the return arises from a real laser pulse close to a sharp boundary. The results presented in Fig. 7.3(a) assume $\tau_d/\tau_L \approx 0.2$. If, however, we set $\tau_d = \tau_L$, somewhat similar behavior is observed [Fig. 7.3(b)], except that $\gamma(z^*)$ has a finite value at $z = 0$. This can be understood in terms of the spatial resolution of the system.

In the *scattering limit* (i.e., $\tau \approx 0$), we saw that the spatial resolution is determined by $c(\tau_L + \tau_d)/2$. For those situations where the excited-state lifetime has to be taken into account, the range resolution is approximated by $c(\tau_d + \tau_L + \tau)/2$. This is illustrated for the case of a rectangular-shaped laser pulse in Fig. 7.4.

In the case of hydrographic work the optical depth of the target is usually large ($\kappa_T L \geq 1$). In this case, combining the exponential factor of equation (7.44) with the lifetime correction factor $\gamma(z^*)$, we derive a new correction factor termed the *optically thick correction factor*:

$$\bar{\gamma}(z^*) = \gamma(z^*)e^{-\kappa_T L z^*}$$

The variation of $\bar{\gamma}(z^*)$ with penetration depth z^* has been evaluated for a similar range of T^*-values, but with $\tau_d = 0.2\tau_L$ and optical-depth values $\kappa_T L$ of

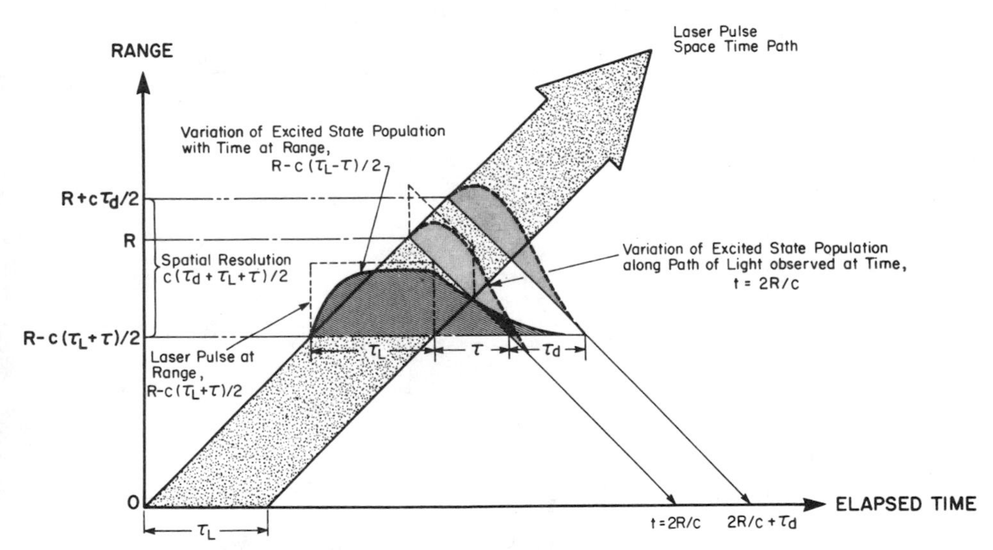

Fig. 7.4. Space–time view of the propagation and excitation of a fluorescent medium by a rectangular-shaped laser pulse. The spatial resolution attainable is clearly seen to be $c(\tau_d + \tau_L + \tau)/2$. The flattening of the excitation curve is intended to imply saturation.

2.5 and 25, respectively. The results are presented in Fig. 7.5(a) and (b). It is quite apparent that whereas $\gamma(z^*)$ tends to unity for large penetration depths in an optically thin medium, $\bar{\gamma}(z^*)$ reaches a maximum and then decays to zero for large penetrations. It is also evident from a study of Fig. 7.3(a) and (b) that the influence of long relaxation times is merely to prevent $\gamma(z^*)$ from reaching unity until considerable penetration of the target has been attained. By contrast, the maximum value of $\bar{\gamma}(z^*)$ is reduced considerably by long fluorescent lifetimes in an optically thick medium. For most atmospheric applications

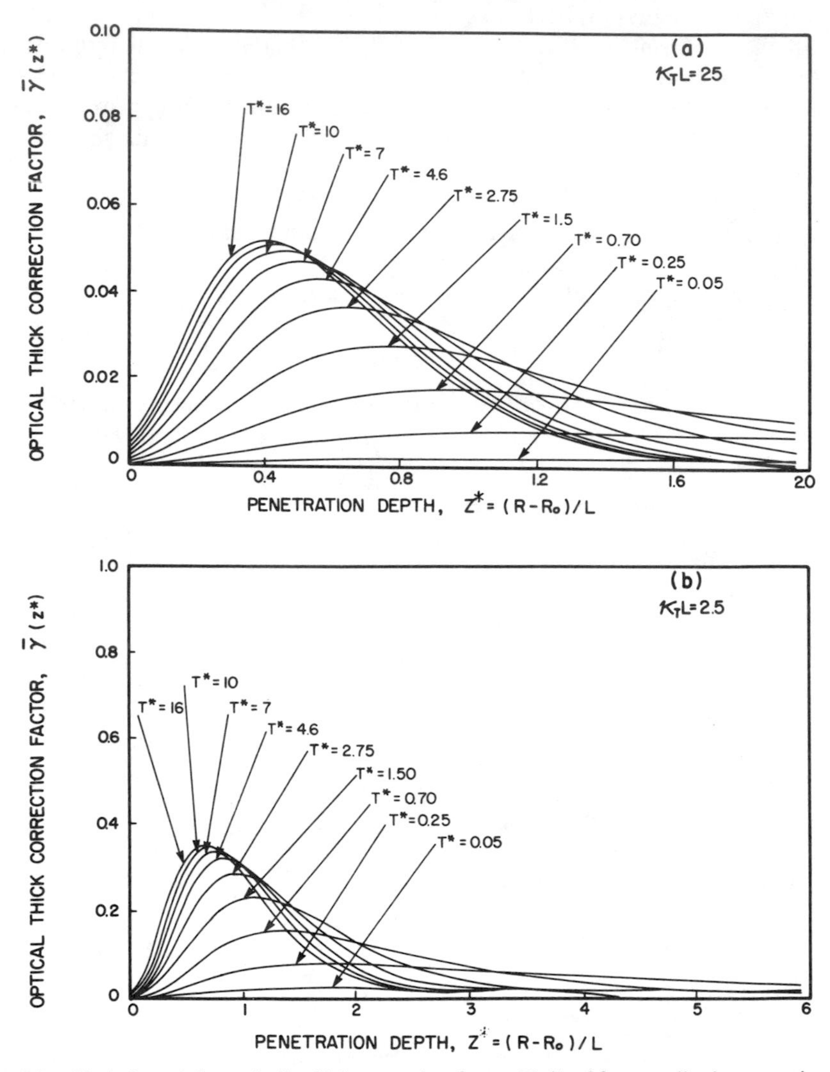

Fig. 7.5. Variation of the optically thick correction factor $\bar{\gamma}(z^*)$ with normalized penetration z^* into a fluorescent target, for several values of T^*; $\tau_d/\tau_L = 0.2$ for each case (Measures, 1977).

involving electronic transitions, $\gamma(z^*)$ will be close to unity, as typical values for τ are a few nanoseconds. This may not be true for infrared fluorescent studies, where the lifetime of vibrational–rotational transitions can be much longer.

In hydrographic work, the targets are nearly always optically thick, and Measures (1977) has shown that, if $\kappa_T L > 5$, $\tau_d > 0.2\tau_L$ and $\tau_L > \tau$ (short fluorescence lifetimes compared to laser pulse duration), the limiting value of the correction factor is given by

$$\bar{\gamma}(z^*) \cong \frac{2}{\kappa_T c \tau_d} \tag{7.49}$$

Under these circumstances the lidar equation, represented by equation (7.44), takes the form

$$E(\lambda, R) = E_L T(R_0) K_0(\lambda) \xi(R_0) \frac{A_0}{R^2} N_0 \frac{\sigma^F(\lambda_L)}{4\pi} \frac{\mathscr{L}^F(\lambda)}{\kappa(\lambda_L) + \kappa(\lambda)} \tag{7.50}$$

If we assume that $\kappa(\lambda_L) \gg \kappa(\lambda)$, as is often the case, and further that $\kappa(\lambda_L) \approx N_0 \sigma^A(\lambda_L)$, then we arrive at an equation of the form

$$E(\lambda, R) = E_L T(R_0) K_0(\lambda) \xi(R_0) \frac{A_0}{4\pi R^2} F(\lambda, \lambda_L) \tag{7.51}$$

where we have introduced the *target fluorescence efficiency*

$$F(\lambda, \lambda_L) \equiv Q^F \mathscr{L}^F(\lambda) \tag{7.52}$$

by using equations (3.182) and (3.186). Under such circumstances the lidar equation, as expressed by equation (7.51), becomes identical to the *laser fluorosensor equation* as formulated by Measures et al. (1975), and the return signal is no longer capable of providing any information regarding the concentration of the fluorescent species within the target zone. However, since the target fluorescence efficiency $F(\lambda, \lambda_L)$ is proportional to the emission profile $\mathscr{L}^F(\lambda)$, identification of the target is still possible from a spectral scan of the fluorescence return.

In the event that fluorescence decay-time measurements are required in order to better characterize the target medium (Measures et al., 1974), care must be used in selecting the values of τ_L and τ_d in relation to the expected values of τ, as the simple form of the lidar equation, expressed by equation (7.51), may not be adequate, and the more complete equation may have to be employed. In the case of oil slicks, where the need for characterization is greatest, the extremely small penetration depth of the laser radiation [tens of microns (Measures et al., 1973)] ensures that equation (7.51) is reasonable even when the detector integration time is made short enough for good decay-time resolution.

In summary, where laser-induced fluorescence is used to remotely probe the environment, we have shown that the appropriate form of the lidar equation (7.44) is more complex than for scattering through the introduction of the lifetime correction factor $\gamma(R)$. As a result considerable care is needed in the interpretation of the return radiation, particularly where the fluorescence lifetime greatly exceeds the laser pulse duration and the fluorescent medium possesses a fairly sharp boundary.

In most situations, the value of this correction factor tends to unity for target penetration greater than a few laser pulse lengths, provided the medium is optically thin and the laser pulse duration is greater than, or of the same order of magnitude as, the lifetime of the laser-excited molecule. For atmospheric work these conditions will often be fulfilled even if good spatial resolution is required.

Under these circumstances the fluorescence (energy) lidar equation, as given by equation (7.44), can be rewritten in a form that makes it remarkably similar to its scattering counterpart, equation (7.15), namely,

$$E(\lambda, R) = E_L K_0(\lambda) T(R) \xi(R) \frac{A_0}{R^2} N(R) \frac{\sigma^F(\lambda_L, \lambda)}{4\pi} \frac{c\tau_d}{2} \quad (7.53)$$

The essential differences are the use of the filter function $K_0(\lambda)$ in place of $\xi(\lambda)$ and of course the fluorescence cross section $\sigma^F(\lambda_L, \lambda)$ instead of the scattering cross section $\sigma^S(\lambda_L, \lambda)$. However, in the limit of great optical depth we have shown that all information on the concentration of the fluorescent species is lost and the lidar equation becomes identical to the laser fluorosensor equation (7.51), developed for hydrographic work.

7.4. GEOMETRY OF THE RECEIVER OPTICS

In the development of the lidar equations for both scattering and fluorescent targets we assumed that the *geometrical probability factor* $\xi(R, \mathbf{r})$ was unity where the field of view of the receiver optics overlapped the laser beam, and zero elsewhere. The distribution of laser irradiance across the target plane was also assumed to be uniform over the area of illumination $A_L(R)$. Although, these assumptions are reasonable for long-range measurements, their soundness is highly questionable for short-range work, and in this section we shall drop them and study the consequences for the lidar equation. Although we shall employ the *scattering* lidar equation for this purpose, the results obtained and the conclusions drawn would be equally applicable in the case of fluorescence work because, as shown earlier, the form of lidar equation is very similar.

We shall assume that the laser irradiance at position $\mathbf{r}$ in the target plane located at range R from the lidar is given by

$$I(R, r, \psi) = \frac{P_L T(\lambda_L, R)}{\pi W^2(R)} F(R, r, \psi) \quad (7.54)$$

where we assume azimuthal symmetry so that position **r** can be represented by (r, ψ). The origin of this coordinate system is taken to be the intersection of the target plane with the axis of the receiver optics (telescope). r is the radial displacement of the point of interest in the target plane from the telescope axis, and ψ is the corresponding azimuthal angle from a vertical plane passing through this axis. This is illustrated in Fig. 7.6. $W(R)$ represents the radius of the laser pulse in the target plane at the instant of interest, and P_L is the total output power of the laser at a time R/c earlier. $F(R, r, \psi)$ describes the distribution of this laser power over the target plane at the instant of interest. The two most common distributions used in this context are

$$F(R, r, \psi) = \mathscr{G}(R, r, \psi) = \mathrm{e}^{-[r^*/W(R)]^2} \tag{7.55}$$

and

$$F(R, r, \psi) = \mathscr{H}(R, r, \psi) = \begin{cases} 1 & \text{where the receiver-optics field of view} \\ & \text{and laser beam overlap} \\ 0 & \text{elsewhere} \end{cases} \tag{7.56}$$

The first corresponds to a Gaussian distribution about the laser beam axis, so that

$$r^* = \{r^2 + d^2 - 2rd\cos\psi\}^{1/2} \tag{7.57}$$

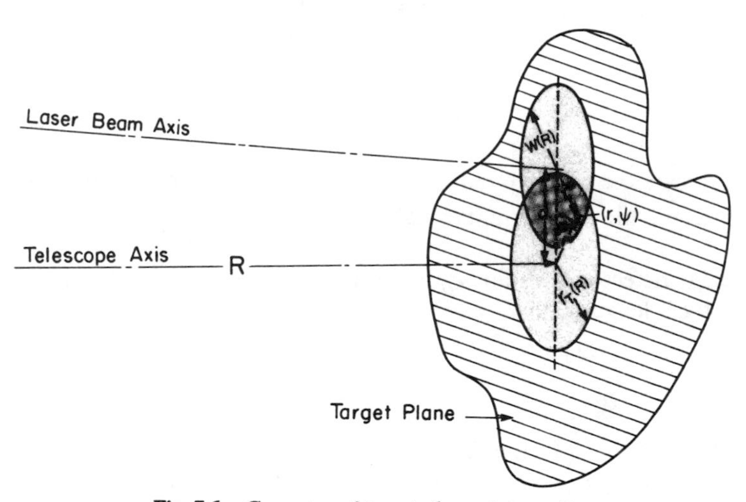

Fig. 7.6. Geometry of target plane at range R.

where d represents the separation of the laser and telescope axes in the target plane (see Fig. 7.6). This kind of laser pulse distribution is expected where the output of the laser is predominantly in the TEM_{00} mode. In this situation the laser beam radius

$$W(R) = \{W_0^2 + \theta^2 R^2\}^{1/2} \tag{7.58}$$

where

W_0 represents the laser output aperture radius; and
θ represents the laser's half divergence angle.

Clearly, for the Gaussian beam $W(R)$ corresponds to the exponential radius.

For a laser pulse that is dominated by high-order transverse modes, the *flat* distribution given by equation (7.56) is a reasonable approximation. In this case $W(R)$ represents the actual radius of the beam; nevertheless, we shall use equation (7.58) to define $W(R)$.

If we use equation (7.54) with equation (7.7) and assume a rectangular temporal profile of duration τ_L for the laser pulse, then the total scattered laser power received by the detector at the instant t $(= 2R/c)$ is

$$P(\lambda, t) = P_L \xi(\lambda) r_0^2 \int_{R=c(t-\tau_L)/2}^{ct/2} dR \, \frac{\beta(\lambda_L, \lambda, R) T(R)}{W^2(R) R^2} \times \int_{r=0}^{r_T} \int_{\psi=0}^{2\pi} \xi(R, r, \psi) F(R, r, \psi) r \, dr \, d\psi \tag{7.59}$$

In formulating this equation we have assumed that the field of view of the receiver optics in the target plane is a circle of radius

$$r_T(R) = r_0 + \phi R \tag{7.60}$$

where

r_0 represents the effective radius of the telescope lens (or mirror); and
ϕ represents the receiver-optics half opening angle.

If a circular detector of radius r_D is positioned on the axis of the telescope of effective focal length f, then

$$\phi = \frac{r_D}{f} \tag{7.61}$$

This will be discussed further in the next section.

We can simplify equation (7.59) in the same way that we had previously by taking account of the fact that the range of interest is always very much greater than the laser pulse length, that is, $ct/2 \gg c\tau_L/2$. Under these circumstances we can neglect dependence of all of the variables in comparison with the range integration, and the total scattered laser power received by the detector after an elapsed time t (corresponding to the leading edge of the laser pulse propagating to a range R) is given by

$$P(\lambda, t) = P_L \frac{c\tau_L}{2} \xi(\lambda) \frac{r_0^2}{R^2} \beta(\lambda_L, \lambda, R) \frac{T(R)}{W^2(R)} \times \int_{r=0}^{r_T} \int_{\psi=0}^{2\pi} \xi(R, r, \psi) F(R, r, \psi) r\, dr\, d\psi \tag{7.62}$$

Evaluation of the double integral in this equation requires detailed knowledge of the geometrical probability factor $\xi(R, r, \psi)$ as well as the laser irradiance distribution function $F(R, r, \psi)$. One of the approaches often adopted in the literature is to introduce the *effective telescope area*,

$$A(R) \equiv \frac{A_0}{\pi W^2(R)} \int_{r=0}^{r_T} \int_{\psi=0}^{2\pi} \xi(R, r, \psi) F(R, r, \psi) r\, dr\, d\psi \tag{7.63}$$

where A_0 $(= \pi r_0^2)$ represents the area of the telescope objective lens (or mirror). This enables the scattering (power) lidar equation to be written in the form

$$P(\lambda, R) = P_L \xi(\lambda) \beta(\lambda_L, \lambda, R) \frac{A(R)}{R^2} \frac{c\tau_L}{2} \mathrm{e}^{-\int_0^R \{\kappa(\lambda_L, R) + \kappa(\lambda, R)\}\, dR} \tag{7.64}$$

which can be seen to be identical to equation (7.21) except that $A(R)$ has replaced the product $A_0 \xi(R)$. Evidently, the geometrical form factor $\xi(R)$ introduced earlier can be defined according to the relation

$$\xi(R) \equiv \frac{1}{\pi W^2(R)} \int_{r=0}^{r_T} \int_{\psi=0}^{2\pi} \xi(R, r, \psi) F(R, r, \psi) r\, dr\, d\psi \tag{7.65}$$

In the case of a rectangular-shaped laser temporal profile P_L represents the average laser output power.

7.4.1. Simple Overlap Factor

The geometrical form factor for a coaxial lidar having no apertures (other than the objective lens or mirror of the telescope) or obstructions is unity, provided the divergence angle of the laser beam is less than the opening angle of the telescope. In reality the majority of lidar systems employ reflecting (Newtonian or Cassegrainian) telescopes. These invariably require some kind of mirror support structure which represents an obstruction to the return radiation. Furthermore, in the case of coaxial lidar systems a mirror is also required to merge the telescope and laser axes. Two possible arrangements are illustrated in Fig. 7.7.

The geometrical form factor can also be evaluated fairly easily for a biaxial lidar if the objective lens (or mirror) of the telescope represents the limiting aperture of the receiver optics, we neglect any obstruction, and we assume a flat laser distribution over the area of illumination. Under these circumstances the geometrical probability factor $\xi(R, r, \psi)$ is unity in the region of overlap between the laser beam and the field of view of the receiver optics, and zero elsewhere. The geometrical form factor, as defined by equation (7.65), becomes

$$\xi(R) = \frac{1}{\pi W^2(R)} \int_{r=0}^{r_T(R)} \int_{\psi=0}^{2\pi} \mathscr{H}(R, r, \psi) r\, dr\, d\psi \tag{7.66}$$

Under these conditions we can think of $\xi(R)$ as a simple overlap factor, and we can write

$$\xi(R) = \frac{\mathscr{A}\{r_T(R), W(R); d(R)\}}{\pi W^2(R)} \tag{7.67}$$

where $\mathscr{A}$ represents the *area overlap function*. $r_T(R)$, the radius of the receiver-optics field of view in the target plane, is given by equation (7.60), and we shall assume that the radius of the laser pulse in the target plane, $W(R)$, is described by equation (7.58). The separation of the telescope and laser axes in

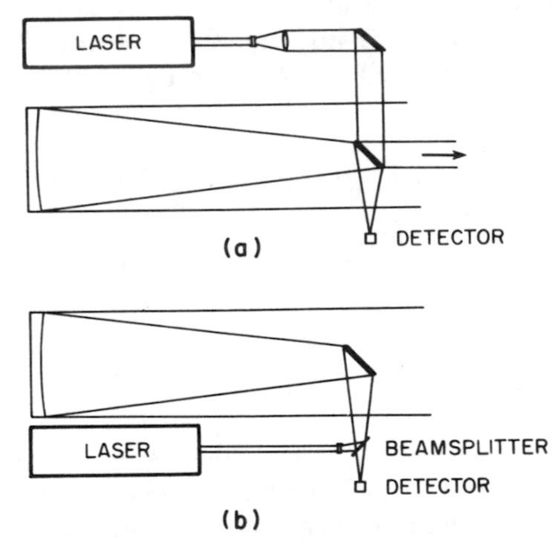

Fig. 7.7. Two possible coaxial lidar arrangements: (a) separate beam expansion; (b) beam expansion via receiving mirror (Harms et al., 1978).

the target plane is

$$d = d_0 - R\delta \tag{7.68}$$

where

d_0 represents the separation of the axes at the lidar, and
δ represents the inclination angle between the laser and telescope axes (see Fig. 7.8).

Three situations are possible:

1. The separation of the axes is too large for there to be any overlap between the receiver-optics field of view and the area of laser illumination [see Fig. 7.9(a)], that is, $\mathscr{A} = 0$ if $d > r_T + W$.
2. The separation of the axes is small enough that either the area of laser illumination lies totally within the receiver-optics field of view or vice versa. The former case is illustrated in Fig. 7.9(c). This amounts to saying that if $d < |r_T - W|$ then $\mathscr{A} = \pi \times$ (the smaller of r_T^2 or W^2).
3. The separation of the axes lies between these extremes:

$$|r_T - W| < d < r_T + W$$

This situation is illustrated in Fig. 7.9(b), and under these circumstances the area overlap function is

$$\mathscr{A}\{r_T, W; d\} = W^2\psi_W + r_T^2\psi_r - r_T d \sin\psi_r \tag{7.69}$$

where

$$\psi_W = \cos^{-1}\left[\frac{d^2 + W^2 - r_T^2}{2Wd}\right] \tag{7.70}$$

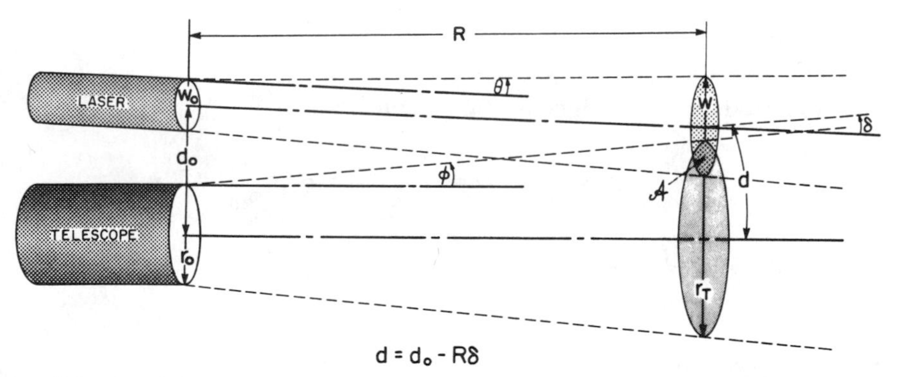

Fig. 7.8. Geometry of a biaxial lidar, where the separation of the laser and telescope axes is $d = d_0 - R\delta$ in the target plane. r_T is the radius of the circular field of view and W is the radius of the circular region of laser illumination.

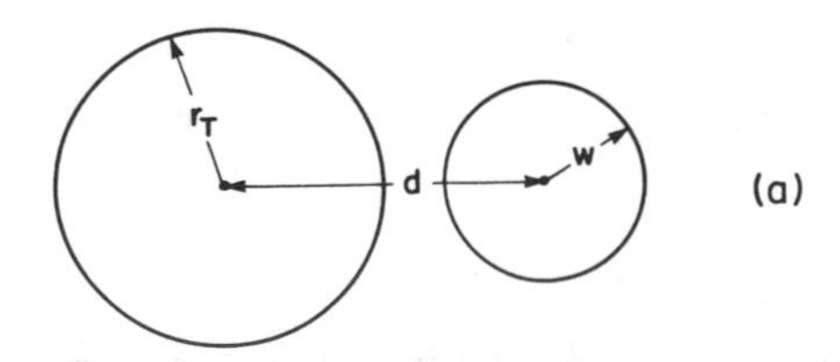

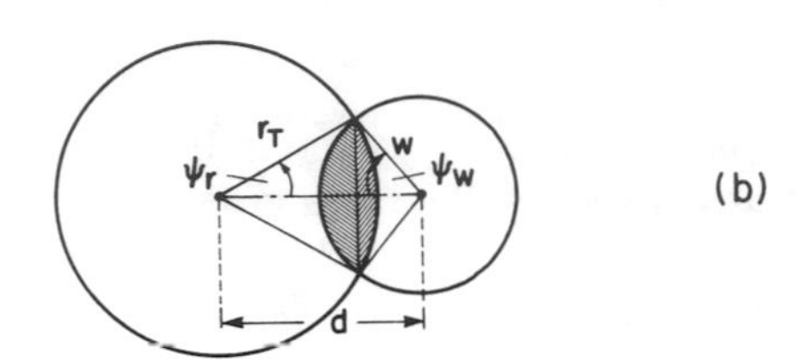

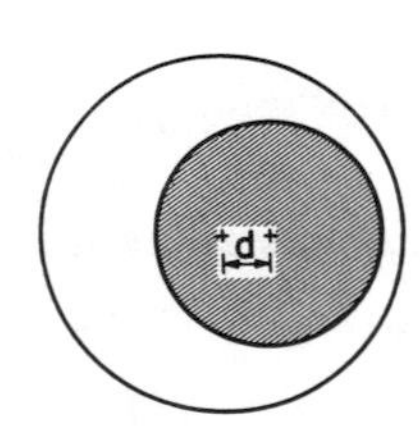

Fig. 7.9. Three overlap situations possible for a biaxial lidar. r_T is the radius of the circular field of view and W is the radius of the circular region of laser illumination.

and

$$\psi_r = \cos^{-1}\left[\frac{d^2 + r_T^2 - W^2}{2r_T d}\right] \tag{7.71}$$

If we introduce the nondimensional parameters

$$z \equiv \frac{R}{r_0}, \quad A = \frac{r_0}{W_0}, \quad D = \frac{d_0}{r_0}, \quad \rho(z, \phi) \equiv \frac{r_T}{r_0} = 1 + z\phi$$

$$s(z, \delta) \equiv \frac{d}{r_0} = D - z\delta, \qquad \omega(z, \theta) \equiv \frac{W}{W_0} = \{1 + z^2\theta^2 A^2\}^{1/2}$$

and

$$y(z, \theta, \phi) \equiv \frac{\omega^2(z, \theta)}{\rho^2(z, \phi) A^2}$$

then the *overlap factor* can be expressed in the form

$$\xi(z) = \frac{\psi_W(z)}{\pi} + \frac{1}{\pi y(z)}\left[\psi_r(z) - \frac{s(z)}{\rho(z)}\sin\psi_r(z)\right] \tag{7.72}$$

where

$$\psi_W(z) = \cos^{-1}\left[\frac{s^2(z) + y(z)\rho^2(z) - \rho^2(z)}{2s(z)\rho(z)\sqrt{y(z)}}\right] \tag{7.73}$$

and

$$\psi_r(z) = \cos^{-1}\left[\frac{s^2(z) + \rho^2(z) - y(z)\rho^2(z)}{2s(z)\rho(z)}\right] \tag{7.74}$$

Clearly, the overlap factor $\xi(z)$ will depend upon the normalized range z and will also depend upon the values of the angular parameters θ, ϕ, and δ and the scale parameters A and D. First we shall consider the case of a biaxial lidar system in which the laser and telescope axes are parallel (i.e., $\delta = 0$) and both the telescope opening angle and the laser divergence angle are 2 mrad (i.e., $\theta = \phi = 10^{-3}$). The variation of $\xi(z)$ with z for three values of D (1.0, 1.1, and 1.25) and two values of A (20 and 5) are presented in Fig. 7.10. Although $\xi(z)$

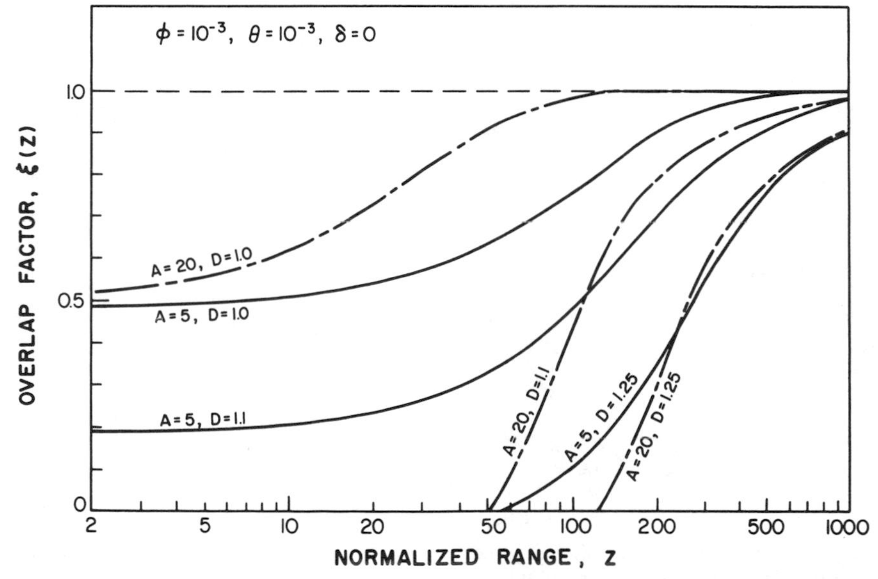

Fig. 7.10. Variation of the overlap factor $\xi(z)$ with the normalized range z for $A = 5$ and 20 (with $D = 1.0$, 1.1, and 1.25 for each value of A).

tends to approach unity for large values of z, it is quite apparent that a separation of 10% more than the telescope lens (or mirror) radius (i.e., $D = 1.1$) can prevent the overlap factor from attaining this value until large values of z. Indeed, for $A = 20$ and $D = 1.1$ the value of $\xi(z)$ can be very small ($< 1\%$) for $z < 50$. This feature of a biaxial arrangement can be useful in eliminating near-field scattering—an effect that can saturate or even damage the photodetector when the lidar system is designed for long-range measurements.

The sensitivity of $\xi(z)$ to a change in the opening angle of the telescope, ϕ [which can be thought of as a measure of the sensitivity of $\xi(z)$ to the detector radius] can be gauged by reference to Fig. 7.11, where ϕ is varied between 10^{-2} and 10^{-3}. The laser divergence half angle θ was assumed to be 10^{-3}, A was set equal to either 20 or 5, and D was assigned the value of 1.25.

The behavior of $\xi(z)$ in the case of a biaxial lidar with an inclination between the telescope and laser axes is radically different from that of the parallel-axis arrangement discussed above. In this instance $\xi(z)$ increases with z at first, attains the value of unity over some range interval, and then rapidly drops to zero for larger values of z. This is illustrated in Fig. 7.12, where $\xi(z)$ is plotted against z for three values of the inclination angle ($\delta = 0.05$, 0.01, and 0.001) and two values of A (20 and 5). In this set of examples, $D = 1.25$ and $\theta = \phi = 10^{-3}$.

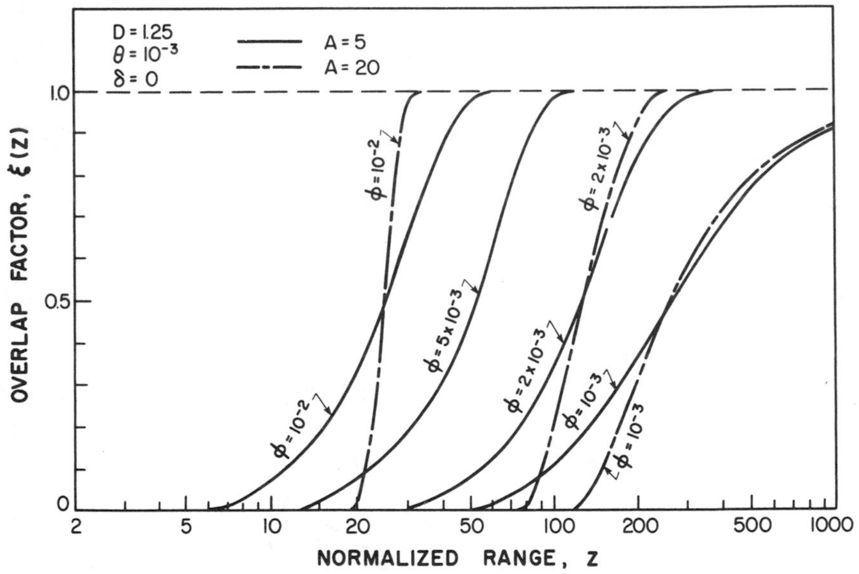

Fig. 7.11. Variation of the overlap factor $\xi(z)$ with z for $A = 5$ and 20 (with $\phi = 10^{-3}$, 2×10^{-3}, 5×10^{-3}, and 10^{-2} for each value of A).

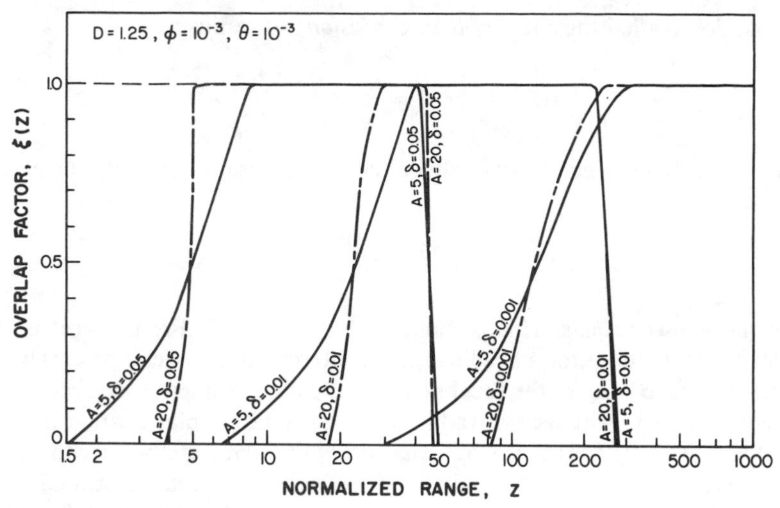

Fig. 7.12. Variation of the overlap factor $\xi(z)$ with z for $A = 5$ and 20 (with $\delta = 0.001$, 0.01, and 0.05 for each value of A).

These results clearly indicate that interpretation of short-range lidar measurements must take proper account of the geometrical factors involved. It is also quite apparent from Fig. 7.12 that misalignment of the optical system can lead to misinterpretation of even long-range measurements. For example, a misalignment of the telescope and laser axes of only 10 mrad would lead to a lack of signal return for $z > 300$ (i.e., $R > 300r_0$) irrespective of the size of the telescope or the power of the laser, supposing the separation of the two axes at the lidar was about 25% more than the telescope radius (i.e., $D = 1.25$) and both the half opening angle ϕ of the telescope and the half divergence angle θ of the laser were 1 mrad.

7.4.2. Geometrical Form Factor

As discussed earlier, the kinds of telescopes used for the receiver optics of most laser remote sensors have some form of central obstruction, and many also have limiting apertures that are smaller than the telescope objective lens (or mirror).

In order to be able to evaluate the geometrical form factor $\xi(R)$ for these more general situations it is necessary to consider the geometrical optics involved. If we represent the telescope by a single lens of radius r_0 and focal length f, then it is well known from geometrical optics that each radiating point (r, ψ) in the target plane at range R will give rise to a circle of illumination (called the *circle of least confusion*) of radius

$$r_c = \frac{r_0 f}{R} \tag{7.75}$$

in the focal plane. The center of this circle of confusion is radially displaced a distance

$$r_f = \frac{rf}{R} \tag{7.76}$$

from the telescope axis. This is illustrated in Fig. 7.13. From this it is fairly obvious that if a detector, of radius r_D, is centered on the telescope axis with its sensitive surface lying in the focal plane of the objective lens (or mirror) of the telescope, then it will receive radiation from a target-plane circle of radius $r_T(R) = r_0 + r_D R/f$, as given by equation (7.60). This enables us to see that the receiver's field of view is a circle of radius $r_0 + \phi R$, as illustrated in Fig. 7.14, where the half opening angle of the receiver optics is $\phi \equiv r_D/f$, as stated in equation (7.61).

The calculation of the geometrical form factor can be quite complicated for an actual lidar system. Nevertheless, a reasonable approximation can be obtained if a couple of simplifying assumptions are made: In the first place the limiting aperture is assumed to lie in the focal plane of the telescope, and second, the obstruction is imagined to reside in the plane of the objective lens (or mirror) of the telescope. Reference to the ray diagram presented as Fig. 7.15 suggests that under these circumstances the geometrical probability factor will be azimuthally symmetric and given by

$$\xi(R, r, r_a, r_b) = \frac{\mathscr{A}(r_a, r_c; r_f) - \mathscr{A}(r_a, r_b'; r_f)}{\pi r_c^2} \tag{7.77}$$

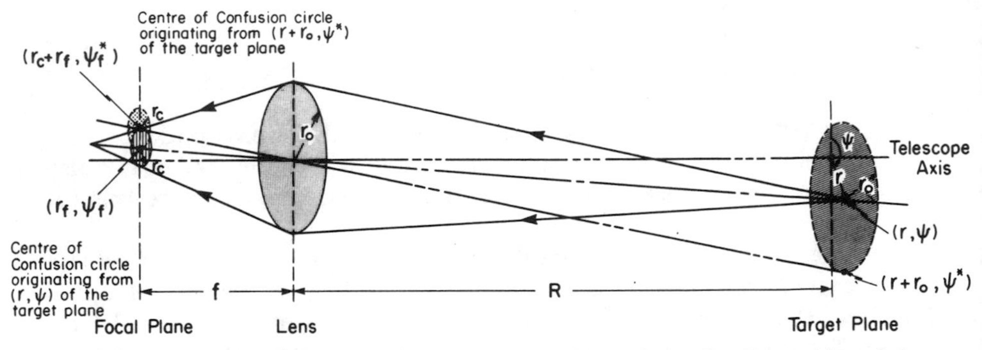

Fig. 7.13. Ray diagram illustrating that each point within a circle of radius r_0 (that of the telescope lens or mirror), centered about the point (r, ψ) in the target plane at range R, will contribute radiation to the point (r_f, ψ_f) in the focal plane, where $r_c = r_0 f/R$ and $r_f = rf/R$.

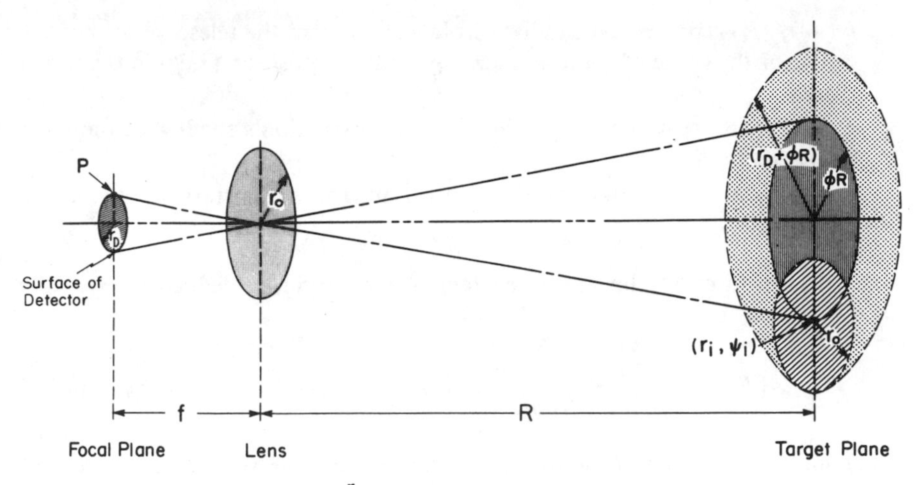

Fig. 7.14. Ray diagram illustrating that each point on the detector receives radiation from a circle of radius r_0 (that of the telescope lens or mirror) in the target plane. A point P on the edge of the detector receives radiation from a circle of radius r_0 centered at (r_i, ψ_i) in the target plane, where $r_i = r_D R/f = \phi R$. Thus the field of view of the receiver is a circle of radius $r_T = r_0 + \phi R$.

where

$\mathscr{A}(r_1, r_2; r_f)$ represents the *area overlap function*, described earlier in Section 7.4.1;

r_a represents the radius of the aperture—which in certain instances corresponds to the radius of the detector's sensitive surface;

r_c represents the *radius of confusion*;

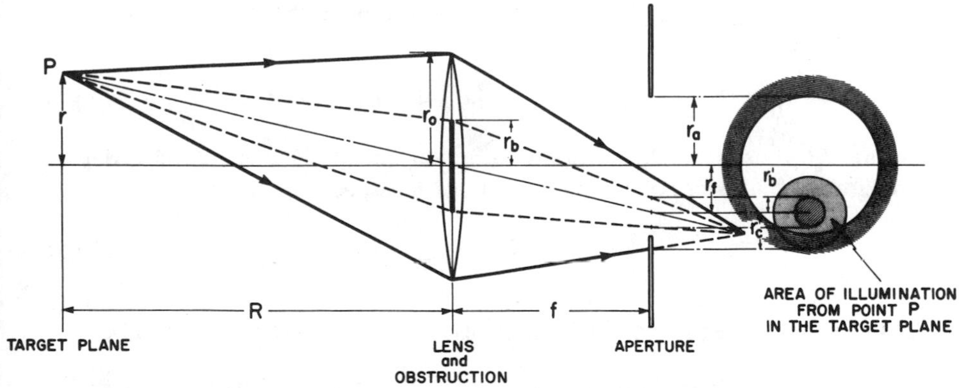

Fig. 7.15. Ray diagram for the calculation of the geometrical probability factor $\xi(R, r, \psi)$ where $r_f = rf/R$, $r_c = r_0 f/R$, $r_b' = r_b f/R$.

r_f $(= rf/R)$ represents the radial displacement from the telescope axis of the center of the circle of confusion arising from the position (r, ψ) in the target plane;

r_b' $(= r_b f/R)$ represents the radius of the obstruction's shadow in the focal plane; and

r_b represents the radius of the central mirror support structure of the telescope.

With this in mind, the telescope effective area $A(R)$, as defined by equation (7.63), becomes

$$A(R) = \frac{A_0}{\pi W^2(R)} \int_{r=0}^{r_T} \int_{\psi=0}^{2\pi} \xi(R, r, r_a, r_b) F(R, r, \psi) r \, dr \, d\psi \tag{7.78}$$

It should be noted that if the aperture radius is the same as that of the lens and there is no obstruction (i.e., $r_b = 0$), then $\xi(R, r, r_0) = \mathscr{A}(r_0, r_c; 0)/(\pi r_c^2) = 1$ and we have the situation described earlier in Section 7.4.1. If this is not the case, then we can write

$$A(R) = \frac{A_0}{\pi W^2(R)} \int_{r=0}^{r_T} \int_{\psi=0}^{2\pi} \xi(R, r, r_a, r_b) \mathscr{G}(R, r, \psi) r \, dr \, d\psi \tag{7.79}$$

for a Gaussian laser pulse distribution. $\mathscr{G}(R, r, \psi)$ was described by equation (7.55), and if this is incorporated into equation (7.79), then we have

$$A(R) = \frac{A_0}{\pi W^2(R)} \int_{r=0}^{r_T} \xi(R, r, r_a, r_b) \int_{\psi=0}^{2\pi} \mathrm{e}^{-\{r^2 + d^2 - 2rd\cos\psi\}/W^2(R)} \, d\psi \, r \, dr \tag{7.80}$$

A relation that is helpful to the evaluation of equation (7.80) can be found in a number of good mathematical texts (for example, Pipes, 1958, p. 361) and takes the form

$$\mathrm{e}^{x\cos\psi} = I_0(x) + 2I_1(x)\cos\psi + 2I_2(x)\cos\psi + \cdots \tag{7.81}$$

where $I_n(x)$ is a modified Bessel function of the first kind of order n. Integrating both sides of equation (7.81) with respect to ψ yields

$$\int_{\psi=0}^{2\pi} \mathrm{e}^{x\cos\psi} \, d\psi = 2\pi I_0(x) \tag{7.82}$$

If this is used in equation (7.80), the telescope effective area is seen to be

$$A(R) = \frac{r_0^2}{W^2} \int_{r=0}^{r_T} \xi(R, r, r_a, r_b) \mathrm{e}^{-(r^2+d^2)/W^2} 2\pi I_0 \left\{ \frac{2rd}{W^2} \right\} r \, dr \tag{7.83}$$

Halldórsson and Langerholc (1978) have evaluated equation (7.83) for a few representative situations and have also considered the case of uniform laser illumination. Their results indicate that for a coaxial lidar with a flat laser distribution, $A(R)$ becomes equal to the uncorrected receiver aperture,

$$A_{\text{tel}} \equiv \pi\left(r_0^2 - r_b^2\right) \tag{7.84}$$

for a range in excess of 300 m, provided the telescope opening angle is greater than the laser divergence angle. Smaller limiting values of $A(R)$ were found for a Gaussian laser distribution. If the opening angle is less than the divergence angle (e.g., $\phi = 10^{-4}$, while $\theta = 5 \times 10^{-4}$), then $A(R)$ appears to level off at a value of about $\frac{1}{30}$ that of A_{tel} for both kinds of laser distribution. On the other hand $A(R)$ is found always to decrease rapidly with decreasing R. This is attributed to the shadowing effect of the central obstruction they included in their calculations. Similar behavior was observed in the case of a biaxial lidar configuration with $D \approx 1$, except that the range for which $A(R)$ levels off is increased somewhat. This latter case is presented as Fig. 7.16, where $r_0 = 0.175$ m, $r_b = 0.04$ m, $d_0 = 0.2$ m, $\theta = 5 \times 10^{-4}$ rad, $W_0 = 0.01$ m, and $\delta = 0$.

Halldórsson and Langerholc (1978) also investigated the effects of coplanar misalignment for the biaxial lidar arrangement and found that the telescope effective area remained considerably below A_{tel}, even for long-range observations, once the inclination of the telescope and laser axes exceed the laser divergence half angle.

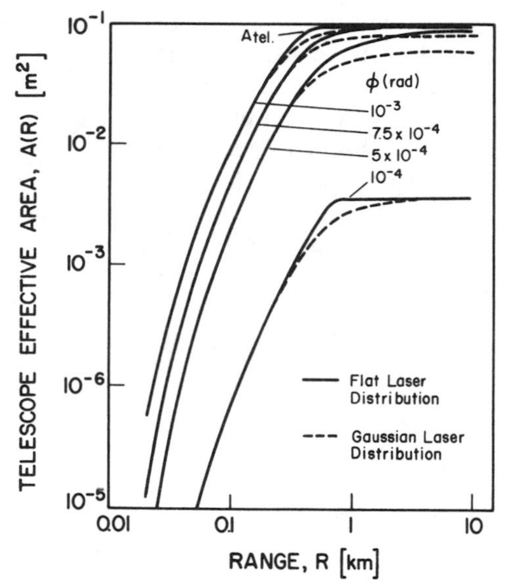

Fig. 7.16. Variation of the telescope effective area $A(R)$ with range R for several values of the opening angle ϕ, in the case of a biaxial configuration with $r_0 = 0.175$ m, $r_b = 0.04$ m, $\delta = 0$, $\theta = 5 \times 10^{-4}$ rad, $d_0 = 0.2$ m and $W_0 = 0.01$ m, Halldorsson and Langerholc (1978).

7.4.3. Small Detectors and Geometrical Compression of the Lidar Return Signal

There are in fact two kinds of radiation loss mechanism that affect the geometrical probability factor $\xi(R, r, \psi)$. The first includes the overlap of the laser pulse with the receiver optics field of view and the shadowing of the detector by a secondary mirror support structure of the telescope, and have been treated above. The second is concerned with the fact that radiation from a target plane located at short or intermediate ranges is not focused onto the focal plane. Instead it forms a diffuse region of illumination in the focal plane, so that a *small* detector positioned at that location will not receive all of the radiation expected on the basis of our above deliberations.

Both of these loss mechanisms can prevent the amount of scattered laser power incident upon the detector from decreasing in accordance with the $1/R^2$ prediction of the lidar equation (7.64). This turns out to be of considerable value where lidar measurements are required over a large range interval. For example, a laser remote sensor that is intended to be operated between 100 m and 10 km must be capable of handling a dynamic range of 5 orders of magnitude. Although there are electronic methods (see Section 6.2) of handling dynamic ranges of this magnitude, clearly there are some advantages in designing the receiver optics to perform the signal compression.

The large dynamic range predicted by the lidar equation presupposes that all of the backscattered radiation collected by the receiver optics is focused onto the detector. As we have indicated above, this may not always be the case, and we shall now determine the backscattered irradiance incident upon the detector for the case where the detector size is important. If the lidar is to be efficient for long-range observations, then the detector's sensitive surface should usually be located in the focal plane of the telescope.

As we have seen earlier (Fig. 7.13), each point (r, ψ) in the target plane at range R produces a circle of illumination of radius r_c $(= r_0 f/R)$ at the position (r_f, ψ_f) of the focal plane, where $\psi_f = \pi - \psi$ and r_f was given by equation (7.76). Consequently, the increment of irradiance at (r_f, ψ_f) arising from an element of area dA centered at (r, ψ) in the target plane is given by the relation

$$dI(r_f, \psi_f) = \frac{\xi(\lambda)T(\lambda, R)J(R, r, \psi)\Omega\, dA}{\pi r_c^2} \tag{7.85}$$

where

Ω $(= \pi r_0^2/R^2)$ represents the acceptance solid angle of the telescope;

$\xi(\lambda)$ represents the spectral transmission factor of the receiver system (ahead of the detector);

$J(R, r, \psi)$ represents the radiance of the target plane; and

$T(\lambda, R)$ represents the atmospheric attenuation factor for this radiation.

If the target radiance arises from volumetric scattering at range R of a pulse of laser radiation of duration τ_L and described by equation (7.54), then the total irradiance at (r_f, ψ_f) is

$$I(r_f, \psi_f) = P_L \frac{c\tau_L}{2} \xi(\lambda) T(R) \frac{\pi r_0^2}{R^2} \frac{\beta(\lambda_L, \lambda, R)}{\pi W^2(R)} \int_{A(r, \psi, r_0)} F(R, r^*, \psi^*) \frac{dA}{\pi r_c^2} \tag{7.86}$$

where, as before,

$$T(R) = \mathrm{e}^{-\int_0^R \kappa(R)\, dR}$$

with $\kappa(R) \equiv \kappa(\lambda_L, R) + \kappa(\lambda, R)$ from equation (7.12), and $\beta(\lambda_L, \lambda, R)$ is the appropriate volume backscattering coefficient. The area of integration $A(r, \psi, r_0)$ in equation (7.86) is taken as a circle of radius r_0 and centered at (r, ψ). The total laser scattered power received by a detector of area A_D and positioned in the focal plane of the telescope is

$$P_D(R) = \int_{A_D} I(r_f, \psi_f) r_f \, dr_f \, d\psi_f \tag{7.87}$$

Harms et al. (1978) have evaluated both the focal-plane irradiance and the total signal power received by several detectors of different size. They assumed a coaxial lidar configuration and a Gaussian laser beam, but neglected the shadowing effect of the secondary mirror structure needed for a coaxial arrangement. Their calculations of the focal-plane irradiance are illustrated in Fig. 7.17 and reveal that the irradiance decreases only slowly with increasing

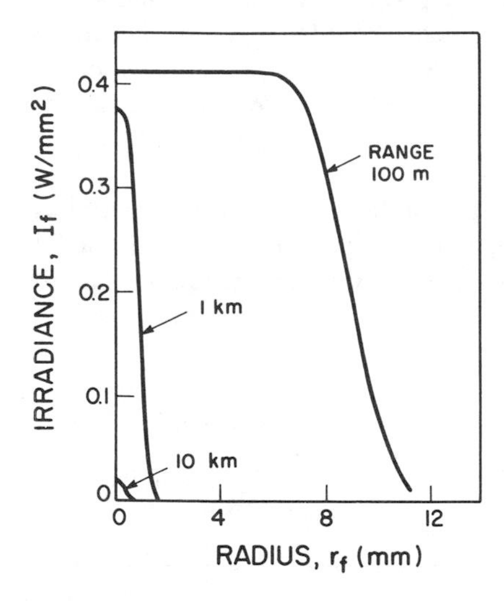

Fig. 7.17. Irradiance in the focal plane of a lidar telescope with $r_0 = 0.3$ m, $r_D = 2$ mm (corresponding to $\phi = 0.67$ mrad) and $W_0 = 0.075$ m (corresponding to $\theta = 0.125$ mrad) (Harms et al., 1978).

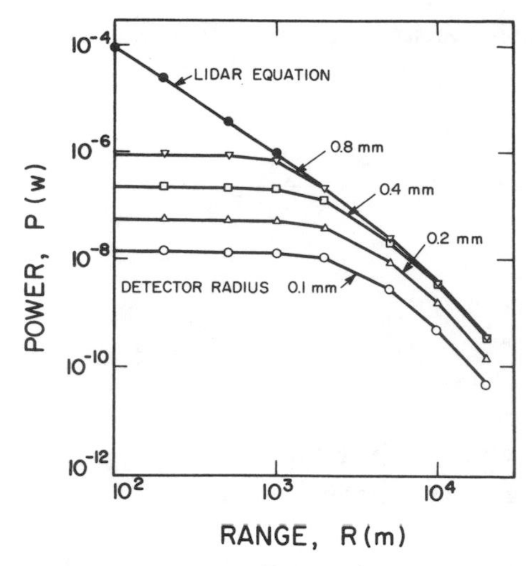

Fig. 7.18. Power incident on detectors of radii 0.1 to 0.8 mm for $r_0 = 0.30$ m, $W_0 = 0.075$ m (corresponding to $\theta = 0.125$ mrad) of a coaxial lidar arrangement (Harms et al., 1978). Optimum detector radius ≈ 0.4 mm.

range for short and intermediate values, then drops dramatically at greater values of the range. Clearly a small detector positioned on the axis will only intercept a tiny fraction of the incident radiation from short and intermediate ranges, while it receives most of the radiation backscattered from large distances. This geometrical signal compression is seen in Fig. 7.18, where it is evident that the optimum detector radius for the configuration considered by Harms et al. (1978) is probably around 0.4 mm. This size of detector would provide an almost flat response out to about 1 km, then a decreasing signal that would be in keeping with the predictions of the lidar equation beyond 4 km. Such a detector is providing both a *small dynamic range* and *maximum sensitivity* at large distances. The parameters used in these calculations are presented in Table 7.1. In addition $d_0 = \delta = 0$ for the coaxial configuration assumed.

Harms (1979) has extended this work to biaxial lidar configurations and has also included the effect of a central obstruction in both coaxial and biaxial arrangements. Under these circumstances the focal-plane irradiance about

TABLE 7.1. PARAMETERS ASSUMED BY HARMS ET AL. (1978) AND HARMS (1979)

$r_0 = 0.30$ m	$\theta = 1.25 \times 10^{-4}$ rad	$\kappa(R) = 5 \times 10^{-2}\ \mathrm{km}^{-1}$
$f = 3$ m	$\tau_L = 500$ ns	$\beta(R) = 10^{-3}\ \mathrm{km}^{-1}\ \mathrm{sr}^{-1}$
$W_0 = 0.075$ m	$P_L = 50$ kW	

(r_f, ψ_f) can be expressed in the form

$$I(r_f, \psi_f) = P_L \frac{c\tau_L}{2} \xi(\lambda) \frac{\beta(\lambda_L, \lambda, R)}{\pi W^2(R) f^2} e^{-\int_0^R \kappa(R)\, dR}$$

$$\times \int_{\Delta A(r, \psi, r_0 : r_s, \psi_s, r_b)} r^* \, dr^* \, d\psi^* \, e^{-(r^{*2} + d^2 - 2r^* d \cos \psi^*)/W^2} \tag{7.88}$$

where $\Delta A(r, \psi, r_0 : r_s, \psi_s, r_b)$ denotes the integration region, a circle of radius r_0 centered at (r, ψ) but excluding the circle with radius r_b (that of the obstruction) around the position (r_s, ψ_s). According to Harms (1979),

$$r_s = \frac{R - S_b}{R} r \quad \text{and} \quad \psi_s = \psi \tag{7.89}$$

S_b is taken as the axial distance of the central obstruction from the telescope objective lens (mirror in reality).

The parameters used by Harms (1979) are the same as given in Table 7.1 except the detector radius was assumed to be 0.4 mm. The resulting calculation of the focal-plane irradiance in the case of a biaxial configuration without an obstruction and for three target-plane ranges (100 m, 1 km, and 10 km) are presented as Fig. 7.19. A small detector centered on the axis will only receive

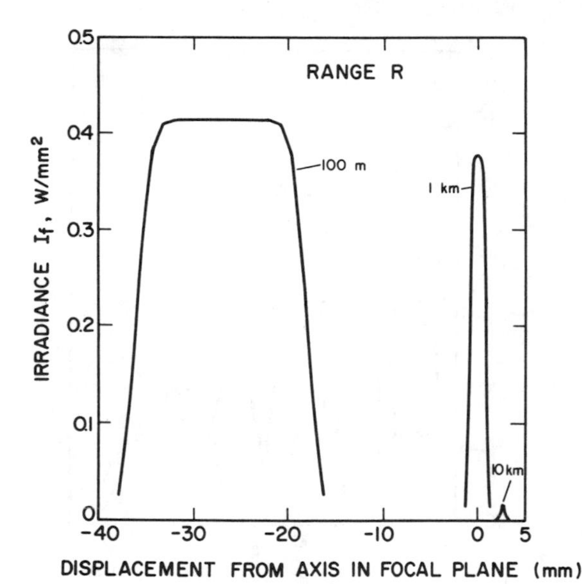

Fig. 7.19. Irradiance in the focal plane of a biaxial lidar system where the separation of the laser and telescope axes $d_0 = 1$ m and the inclination angle $\delta = 10^{-3}$ rad. The focal-plane displacement is taken to be in the plane of misalignment between the telescope and laser axes, and the other parameters assumed are provided in Table 7.1 (Harms, 1979).

backscattered radiation from a small range interval, in this instance around 1 km. This range interval will be displaced further and further from the lidar as smaller the inclination angle δ decreases. This is illustrated in Fig. 7.20, where the power received by a 0.4-mm detector is displayed as a function of range for several values of δ. A comparison with the corresponding unobstructed coaxial case of Harms et al. (1978) is included in this figure.

A reduction in the separation of the telescope and laser axes, d_0, results in a less steep rise of the curves. This can be understood in terms of the simple overlap behavior discussed earlier in Section 7.4.1. An example of this is presented in Fig. 7.21, where the separation is $d_0 = 0.375$ m and the inclination angle is $\delta = 10^{-3}$ rad. A comparison with the corresponding unobstructed coaxial case is again provided. Another important feature of Fig. 7.21 is the clear evidence of insensitivity of the detector's response curve, in the case of a biaxial configuration, to the presence of an obstruction of 100-mm radius. The explanation of this lies in the fact that in a biaxial arrangement the shadow cast by the obstruction appears off axis and therefore does not fall across the detector at short ranges, while radiation from large distances does not form a well-defined shadow.

In the case of a coaxial lidar configuration the effects of a central obstruction are significant, as can be seen by reference to Fig. 7.22, where the power received by a 0.4-mm-radius detector is plotted as a function of range F for

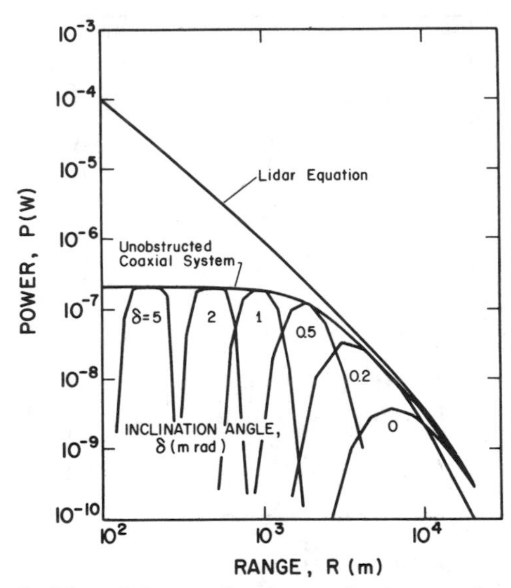

Fig. 7.20. Power received by a 0.4-mm-radius detector, in the case of an unobstructed biaxial configuration, as a function of range R, for several values of the inclination angle δ. A comparison with the corresponding unobstructed coaxial case is also provided. The parameters used are indicated in Table 7.1 (Harms, 1979).

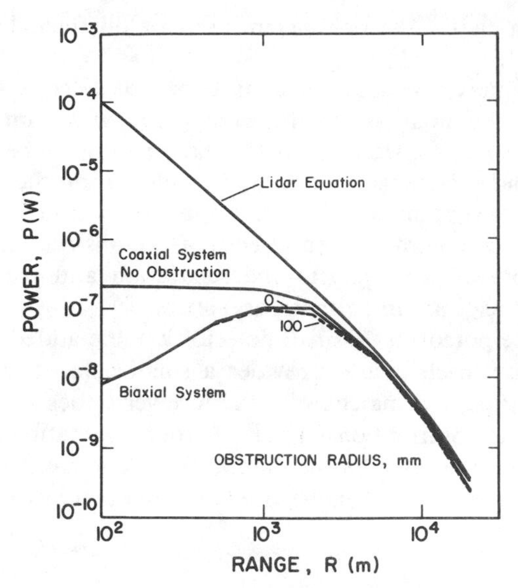

Fig. 7.21. Power received by a 0.4-mm-radius detector is plotted as a function of range R, for both an obstructed ($r_b = 100$ mm) and unobstructed ($r_b = 0$) biaxial configuration, where the separation between the telescope and laser axes is $d_0 = 0.375$ m and the inclination angle $\delta = 10^{-4}$ rad. The other parameters are provided in Table 7.1. For comparison the corresponding unobstructed coaxial response curve is also shown (Harms, 1979).

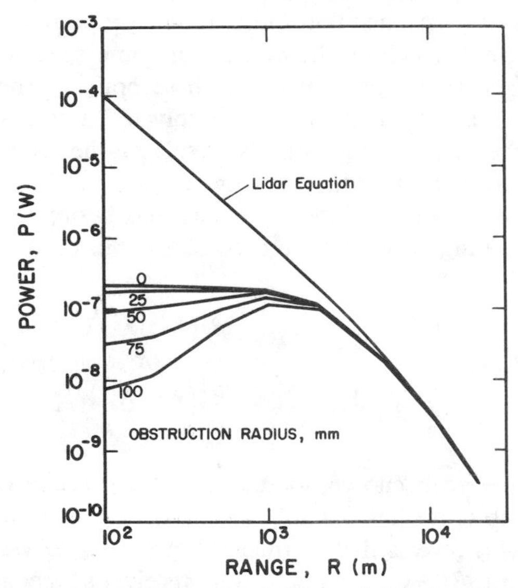

Fig. 7.22. Power received by a 0.4-mm-radius detector as a function of range R, in the case of a coaxial lidar, for several radii of the central obstruction. The parameters are provided in Table 7.1 (Harms, 1979).

several obstruction radii. The parameters used in this calculation are again provided in Table 7.1.

In summary, we may say that a judicious choice of detector size can lead to a compression of the lidar signal dynamic range with almost no loss of sensitivity for long-range measurements. This appears to be true for both coaxial or biaxial lidar systems. In a coaxial configuration the shadow cast by the secondary-mirror support structure results in a reduction of the near signal, while in a biaxial configuration the effects of an obstruction are not very important. In a biaxial arrangement the separation and inclination of the telescope and laser axes are important parameters.

It should also be noted that a small detector has the added advantage of a small inherent noise level. It also provides a small field of view that allows good spatial resolution and matching of the receiver optics to the laser beam, thereby reducing the contribution from background radiation arising beyond the lateral extent of the laser beam. Lastly, small detectors are available as linear photodiode arrays that can be used in simultaneous multiwavelength measurements.

7.5. SOLUTIONS OF THE LIDAR EQUATION

As we have seen, the appropriate lidar equation provides a means of relating the radiation returned from a probing laser beam to the relevant optical properties (such as the scattering or attenuation coefficients) of the target medium. It is implicitly assumed that these optical properties can be related to some physical property (such as the density of some specific constituent) of interest in the target. In order to evaluate these optical properties from the return signal, the lidar equation has to be solved. In this section we shall consider some of the most useful methods of solving the lidar equation.

In this regard we can introduce a form of the lidar equation that will reasonably apply to both long-range scattering and fluorescent measurements in the limit of good range resolution and optically thin targets, namely,

$$E(\lambda, R) = E_L \frac{c\tau_d}{2} \left\{ \begin{matrix} \xi(\lambda) \\ \text{or} \\ K_0(\lambda) \end{matrix} \right\} \frac{A(R)}{R^2} \left\{ \begin{matrix} \beta^s(\lambda_L, \lambda, R) \\ \text{or} \\ \beta^F(\lambda_L, \lambda, R) \end{matrix} \right\} \mathrm{e}^{-\int_0^R \kappa(R)\, dR} \qquad (7.90)$$

The conditions have been chosen so that the fluorescence correction factor $\gamma(R)$ is very close to unity (see Section 7.3). A comparison of equation (7.90) with (7.15) and (7.44) reveals that in the case of scattering we use the volume scattering coefficient $\beta^s(\lambda_L, \lambda, R)$ and the receiver's spectral transmission function $\xi(\lambda)$, while for fluorescence work we use the volume fluorescence coefficient $\beta^F(\lambda_L, \lambda, R)$ and the filter function $K_0(\lambda)$. $A(R)$ is the effective receiver area introduced earlier, in equation (7.63). In general we may write

$$\beta^{\alpha}(\lambda_L, \lambda, R) = \frac{N(R)\sigma^{\alpha}(\lambda_L, \lambda)}{4\pi} \tag{7.91}$$

where $N(R)$ is the number density of the species at range R responsible for returning radiation at wavelength λ when illuminated with laser radiation at λ_L, and $\sigma^{\alpha}(\lambda_L, \lambda)$ is the relevant cross section. In the subsequent analysis we shall drop the superscript and reference to wavelength for the sake of brevity. Consequently, we may express the lidar equation in the form

$$E(R) = E_D \frac{A(R)}{R^2} \beta(R) e^{-\int_0^R \kappa(R)\, dR} \tag{7.92}$$

where

$$E_D = E_L \frac{c\tau_d}{2} \left\{ \begin{matrix} \xi(\lambda) \\ \text{or} \\ K_0(\lambda) \end{matrix} \right\} \tag{7.93}$$

It is evident that equation (7.92) will have to be solved for one or both of the optical parameters, $\beta(R)$ and $\kappa(R)$.† The simplest approach is to assume that the two-way atmospheric transmission factor can be approximated by unity. This effectively reduces the lidar equation to only one unknown and has been useful for some aerosol work (see Chapter 8). Another approach involves making use of some kind of atmospheric model to specify $\kappa(R)$.

In general, however, both $\kappa(R)$ and $\beta(R)$ are unknown and it is necessary to assume some kind of relation between them. To eliminate system constants (such as the area or the spectral transmission factor of the receiver) and uncertainties associated with the laser pulse irreproducibility, we introduce a new, *range-normalized signal variable* defined by the relation‡

$$S(R) \equiv \ln\{E(R)R^2\} \tag{7.94}$$

The virtue of this new variable is seen by subtracting from it the S-value appropriate to some reference range $R_{\#}$ (this might, for example, correspond

†Recall that $\kappa(R)$ represents the two way attenuation coefficient, see equation (7.22).

‡In much of the literature (Davis, 1969; Johnson and Uthe, 1971), the function S is defined in terms of the range-corrected signal ratio in decibels:

$$S(R) = 10 \log\left\{ \frac{E(R)R^2}{E(R_{\#})R_{\#}^2} \right\}$$

This definition is closely related to the form of measurement using logarithmic amplifiers (Johnson and Uthe, 1971).

to the location of some monitor of β):

$$S(R) - S(R_{\#}) = \ln\left\{\frac{E(R)R^2}{E(R_{\#})R_{\#}^2}\right\}$$

$$= \ln\left\{\frac{\beta(R)A(R)}{\beta(R_{\#})A(R_{\#})}\right\} - \int_{R_{\#}}^{R} \kappa(R)\, dR \qquad (7.95)$$

In the case of long-range work with a lidar system that is well aligned ($\delta \approx 0$), we saw earlier (Section 7.4) that $A(R) \approx A_0$, so that

$$S(R) - S(R_{\#}) = \ln\left\{\frac{\beta(R)}{\beta(R_{\#})}\right\} - \int_{R_{\#}}^{R} \kappa(R)\, dR \qquad (7.96)$$

and this difference in the S-values only depends upon the atmospheric-dependent factors. The differential form of equation (7.96) is

$$\frac{dS(R)}{dR} = \frac{1}{\beta(R)}\frac{d\beta(R)}{dR} - \kappa(R) \qquad (7.97)$$

and we can see that for a homogeneous atmosphere, where $d\beta(R)/dR \approx 0$,

$$\kappa_{\text{hom}} \approx -\frac{dS(R)}{dR} \qquad (7.98)$$

This is the basis of the *slope method* of inversion, which is appropriate in good visibility conditions [$R_v \gtrsim 3$ km—see equation (4.84)], provided multiple scattering is unimportant (Collis and Russell, 1976). Viezee et al. (1973) attempted to employ this technique to assess slant visibility for aircraft landing operations through lidar observations at 694.3 nm, and Murray et al., (1978) made use of this approach in the measurement of atmospheric infrared extinction using a CO_2 laser. Unfortunately, the slope method becomes inappropriate for certain conditions that are sometimes important for lidar work, that is to say, remotely measuring atmospheric inhomogeneities like fogs, clouds, and smoke plumes (Klett, 1981).

There is a reasonable body of evidence which suggests that, in the case of elastic scattering (i.e., $\lambda = \lambda_L$), $\beta(R)$ can often be related to $\kappa(R)$ through a relation of the kind

$$\beta(\lambda_L, R) = \text{const}\, \kappa^{\mathfrak{g}}(\lambda_L, R) \qquad (7.99)$$

where $\mathfrak{g}$ depends on the lidar wavelength and the specific properties of the constituent of interest (Collis and Russell, 1976). Klett (1981) has indicated that the value of this exponent is generally in the range $0.67 \leq \mathfrak{g} \leq 1.0$, and as we have seen [equation (2.164)], for a single constituent in the absence of absorption $\mathfrak{g} = 1.0$ and const $= 1.0$. A relationship similar to (7.99), with $\mathfrak{g} = 1.0$, would also be expected in the case of fluorescence work; see equation (3.182).

Eliminating β in terms of κ, through equation (7.99), enables us to rewrite equation (7.97) in terms of only one unknown:

$$\frac{dS(R)}{dR} = \frac{\mathfrak{g}}{\kappa_L(R)} \frac{d\kappa_L(R)}{dR} - 2\kappa_L(R) \tag{7.100}$$

where we have introduced

$$\kappa_I(R) = \kappa(\lambda_L, R) \tag{7.101}$$

The factor of 2 in equation (7.100) is required by virtue of equation (7.22). Although this is a nonlinear ordinary differential equation, it has a well-known form, namely that of the Bernoulli or homogeneous Ricatti equation (Klett, 1981). The solution of equation (7.100) is made easier by introducing the reciprocal of the attenuation coefficient,

$$\mathfrak{u} \equiv \frac{1}{\kappa_L(R)} \tag{7.102}$$

On substitution of equation (7.102) into (7.100) we obtain

$$\frac{d\mathfrak{u}}{dR} = -\frac{1}{\mathfrak{g}}\left(2 + \mathfrak{u}\frac{dS}{dR}\right) \tag{7.103}$$

The solution of this equation is quite straightforward; however, Klett (1981) has shown that the limits of integration make a difference to the stability and accuracy of the solution. In particular, the *reference range* $R_{\#}$ should be taken as the upper limit of range integration, rather than the lower limit as assumed by earlier researchers. With this in mind the solution of (7.103) can be written in the form

$$\mathfrak{u}(R_{\#}) - \mathfrak{u}(R)\exp\left(-\frac{1}{\mathfrak{g}}\int_R^{R_{\#}} \frac{dS}{dR^*} dR^*\right) =$$

$$-\frac{2}{\mathfrak{g}}\int_R^{R_{\#}} dR^* \exp\left(-\frac{1}{\mathfrak{g}}\int_{R^*}^{R_{\#}} \frac{dS}{dR^{**}} dR^{**}\right) \tag{7.104}$$

where $\mathfrak{g}$ is assumed to be independent of range.

If we then draw on equation (7.102), we can arrive at an equation for the attenuation coefficient,

$$\kappa_L(R) = \frac{e^{-[S(R_\#) - S(R)]/g}}{\dfrac{1}{\kappa_L(R_\#)} + \dfrac{2}{g}\displaystyle\int_R^{R_\#} dR^* \, e^{-[S(R_\#) - S(R^*)]/g}} \tag{7.105}$$

Although, this solution appears to work reasonably well for conditions ranging from clear air to haze, in a turbid atmosphere (such as a fog or cloud) multiple scattering occurs and the above solution is not valid. In these instances a more sophisticated formulation of the lidar equation is necessary (Collis and Russell, 1976).

Lastly, a word of caution should be added in connection with the assumption that the range dependence of the effective telescope area $A(R)$ can be neglected. As we have seen in Section 7.4, this is highly questionable for short-range measurements or if the telescope and laser axes are inclined to each other.

8

Analysis and Interpretation of Lidar Return Signals

In order to aid in the interpretation of lidar return signals and to draw attention to some of their less obvious features, it has sometimes been found useful to calculate the expected signal for a given situation. Simulations of this kind have also served as the basis of comparisons between the different lidar techniques used in the detection of trace atmospheric molecular constituents. One of the first such comparative studies was undertaken by Measures and Pilon (1972). Their analysis clearly indicated the potential superiority of differential absorption and scattering (DAS) over laser-induced fluorescence (F) or Raman backscattering (R) for the purpose of remotely detecting small concentrations of molecules in the lower regions of the atmosphere.

An example of their results is presented as Fig. 8.1, where the number of photoelectrons created at the photocathode of a lidar photomultiplier is computed as a function of range for the three techniques indicated above. The molecular gas considered was SO_2, and the calculations were undertaken for three homogeneous concentrations (10, 1, and 0.1 ppm). The atmospheric attenuation coefficient assumed ($\kappa = 0.45\ \mathrm{km}^{-1}$) corresponded to a hazy day with a visibility of about 10 km.

In the troposphere, quenching and redistribution of the emitted energy into a wide spectral interval can reduce the effective fluorescence cross section by a factor of 10^{-5} to 10^{-6} (see Fig. 6.1). Furthermore, the resulting broad wavelength band of detection makes a fluorescence-based lidar more susceptible to both solar background radiation and laser backscattered radiation. The latter can be understood in terms of the difficulty of combining broadband detection with efficient spectral rejection (see next section) and the high value of the atmospheric backscattering coefficient observed at low altitudes (Fig. 2.25). On the other hand, the DAS approach benefits from a moderately large value of β.

In the stratosphere the atmospheric density is much lower and there is a relative absence of dust and aerosols. Consequently, the detrimental effects alluded to above, in connection with laser-induced fluorescence, are practically nonexistent, and as a result fluorescence lidar measurements of trace atmospheric constituents from the Space Shuttle look quite attractive (McGee and McIlrath, 1979; Heaps, 1980; McIlrath, 1980; see Chapter 9.)

It is apparent from Fig. 8.1 that high background concentrations of SO_2 (10 ppm) can lead to a premature cutoff in both the DAS and fluorescence signal returns (indicated by the crossover of the curves) due to strong attenuation of the laser beam. Since detuning of the laser radiation can reduce this excessive attenuation, a careful choice of the laser wavelength will be required in order to optimize both the range and the sensitivity of the system. Attenuation can also lead to complications when the species of interest is localized, as in the plume from a chimney or the exhaust cloud over a heavily used highway.

An example of this was also provided by Measures and Pilon (1972) and is illustrated in Fig. 8.2. In this instance the fluorescence return signal from a Lorentzian spatial distribution of NO_2 (with a peak concentration of 100 ppm and a plume of HWHM of 20 m) was calculated and plotted as a function of range. Attenuation within the plume is seen to lead to a severe distortion of the return signal. This distortion results in both a considerable reduction in

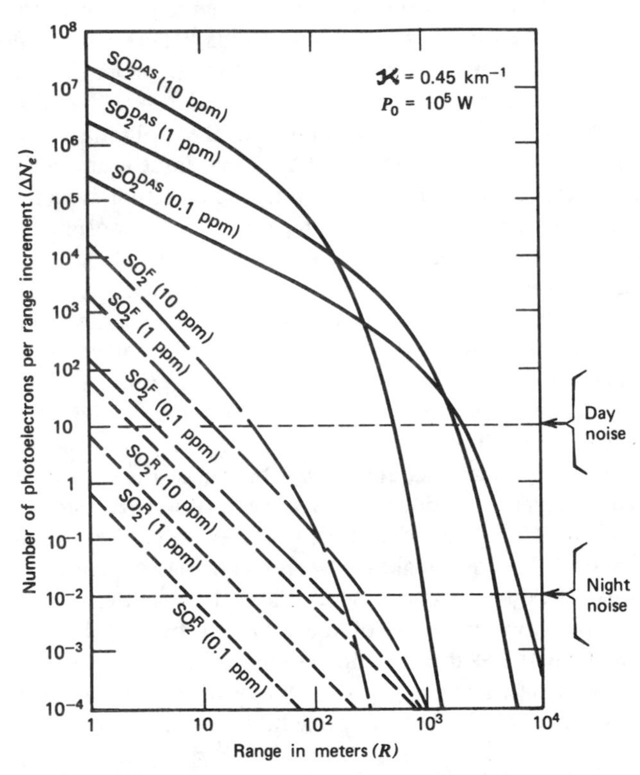

Fig. 8.1. Fluorescence, Raman, and DAS lidar signals versus range for sulfur dioxide (Measures and Pilon, 1972).

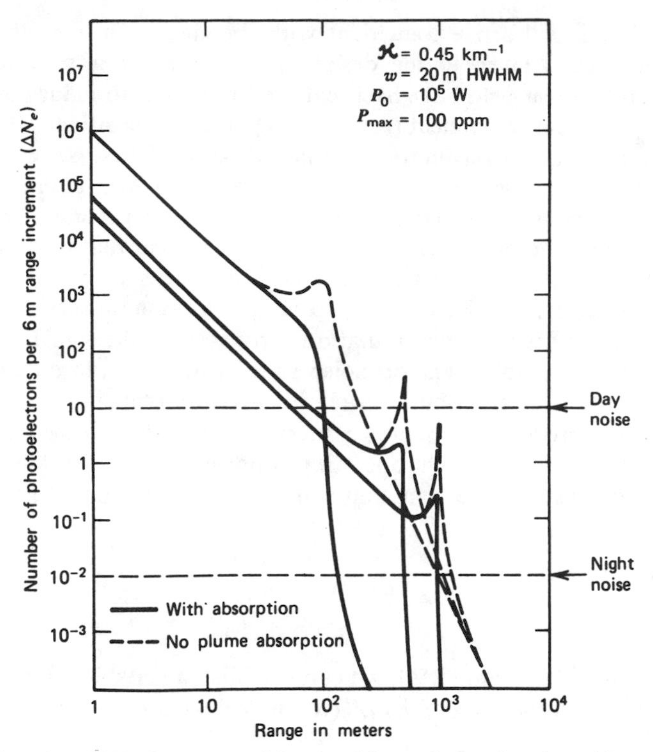

Fig. 8.2. Distortion of NO_2 fluorescence lidar signal from a Lorentzian plume (due to absorption within the plume) centered at one of three locations: $R_0 = 10^2$, 5×10^2, and 10^3 m (Measures and Pilon, 1972).

the peak amplitude and an apparent shift of the peak concentration towards the observer. The sharp drop in the background signal subsequent to the plume might, in principle, alert the observer to the problem and avoid misinterpretation of the return signal. Unfortunately, this sharp drop in the signal may well be masked by either background or instrument noise.

It is evident that the interpretation of lidar return signals will require careful analysis and sufficient information if misleading conclusions are to be avoided. In this chapter we shall attempt to lay the foundation of this analysis and determine those factors that limit both the sensitivity and accuracy of lidar measurements.

8.1. SPECTRAL REJECTION OF LASER BACKSCATTERED RADIATION

In certain instances, namely when fluorescence or Raman scattering is being employed, laser backscattered radiation can limit the sensitivity of the lidar system in two ways. If insufficient spectral rejection is used, some fraction of this laser return will arrive coincident with the signal. On the other hand, if inadequate care is given to the design of the lidar, it is possible at short wavelengths for near-field laser backscattered radiation to induce fluorescence within some optical component (such as a lens) that is positioned ahead of the spectral analyzer, or to saturate the photodetector. These situations can be avoided by the inclusion of a narrowband laser-blocking filter ahead of all vulnerable components. A biaxial lidar configuration is generally less susceptible to the latter kind of problem. Nevertheless it is clear that considerable care must be given to it in the design of a laser remote sensor.

An estimate of the degree of spectral rejection necessary to avoid the problem of laser backscattered radiation arriving with the signal, and thereby reducing the achievable signal-to-noise ratio, can be made on the basis of equation (6.18). The criterion relevant to any given situation will of course depend on the predominant source of noise competing with the backscattered laser radiation. Under daylight operating conditions, where background radiation tends to constitute the dominant source of noise, the necessary criterion is simply

$$E_b^L(\lambda_L) \ll \frac{E_b^N(\lambda)}{\nu_\eta} \tag{8.1}$$

Using equations (6.2) and (7.14), we can establish a suitable criterion in terms of the *spectral rejection ratio* $\xi(\lambda_L)/\xi(\lambda)$, namely,

$$\frac{\xi(\lambda_L)}{\xi(\lambda)} \ll \frac{2S_b(\lambda)\Omega_0 R^2\,\Delta\lambda_0}{E_L\beta(\lambda_L, R)\nu_\eta\xi(R)c}\,e^{2\int_0^R \kappa(\lambda_L, R)\,dR} \tag{8.2}$$

where we have assumed that we can express the filter function $K_0(\lambda)$ in terms of the spectral transmission factor at the wavelength of the signal and an appropriate linewidth $\Delta\lambda_0$, viz.,

$$K_0(\lambda) = \xi(\lambda)\,\Delta\lambda_0 \tag{8.3}$$

The range dependence of the criterion given by the inequality (8.2) arises as a result of using the backscattered laser-radiation return that originates from the same location as the signal. In principle, any earlier component can be discriminated against by temporal means.

Alternatively, the spectral rejection ratio could be calculated on the basis that the laser backscattered radiation has to be much less than the signal arriving at the same time. From a Raman lidar this leads to the criterion

$$\frac{\xi(\lambda_L)}{\xi(\lambda)} \ll \frac{N(R)\sigma^R(\lambda_L, \lambda)}{4\pi\beta(\lambda_L, R)\nu_\eta}\,e^{\int_0^R \{\kappa(\lambda_L, R) - \kappa(\lambda, R)\}\,dR} \tag{8.4}$$

In the case of fluorescence, a similar criterion can be obtained, except that in this instance the filter function has to be used again, due to the relatively broadband nature of the interaction, and an allowance made for the lifetime correction factor $\gamma(R)$; see equation (7.44). Under these circumstances the spectral rejection criterion becomes

$$\frac{\xi(\lambda_L)}{\xi(\lambda)} \ll \frac{N(R)\sigma^F(\lambda_L, \lambda)\gamma(R)\,\Delta\lambda_0}{4\pi\beta(\lambda_L, R)\nu_\eta} e^{\int_0^R \{\kappa(\lambda_L, R) - \kappa(\lambda, R)\}\,dR} \tag{8.5}$$

It might be worth reiterating at this point that the Raman cross section $\sigma^R(\lambda_L, \lambda)$ used in the inequality (8.4) is in essence spectrally integrated, whereas the fluorescence cross section $\sigma^F(\lambda_L, \lambda)$, used in the inequality (8.5), refers to a unit wavelength interval [see equation (3.182)]. This is the reason we have to include in (8.5) the effective spectral window $\Delta\lambda_0$ of the lidar receiver system.

The value of the spectral rejection ratio required for Raman scattering is in general very small. We can demonstrate this by considering the representative case of a Raman lidar operating at 337 nm and attempting to monitor the concentration of SO_2 emitted from a chimney at a range of 200 m. From Table 3.4, we find that the appropriate differential Raman cross section is $\sigma^R(337, 350.8)/4\pi \approx 1.7 \times 10^{-29}$ cm^2 sr^{-1}. We also see that the observer wavelength of 350.8 nm is displaced by 13.8 nm from the laser wavelength. If we assume that the lidar is expected to operate under light-haze conditions, then from Fig. 2.26 we might expect an atmospheric volume backscattering coefficient $\beta(\lambda_L)$ of about 10^{-7} cm^{-1} sr^{-1}. Unfortunately, the SO_2 in the plume is likely to be associated with a higher value of β, so we shall assume that β(337 nm, 200 m) = 10^{-6} cm^{-1} sr^{-1}. Thus we require

$$\frac{\xi(337)}{\xi(350.8)} \ll \frac{N_{SO_2} \times 1.7 \times 10^{-29}}{10^{-6}}$$

provided the difference in the wavelength is small enough that we can neglect the exponential factor in (8.4). We have also assumed that $\nu_\eta \approx 1$. If the threshold density of SO_2 to be detected is 2.5×10^{14} cm^{-3}, corresponding roughly to 10 ppm at sea level, then we see that the minimum value of the spectral rejection ratio $\xi(337)/\xi(350.8)$ should be about $\frac{1}{10}$ of $2.5 \times 10^{14} \times 1.7 \times 10^{-29}/10^{-6}$, or 4.42×10^{-10}. This extremely small value of the spectral rejection ratio can be achieved with a double monochromator, and therefore Raman lidars usually have such instruments incorporated into their receiver system.

8.2. SIGNAL-TO-NOISE LIMITS OF DETECTION FOR MOLECULAR CONSTITUENTS

As we saw earlier [equation (6.21)], it is possible to express the lidar detector (output) signal-to-noise ratio in terms of the increment of radiative signal energy $E_s(\lambda, R)$ and the total radiative background energy $E_b^T(\lambda)$ received by

TABLE 8.1. LIDAR MINIMUM DETECTABLE ENERGY FOR THREE LIMITING SNR CASES[a]

Limit	Criterion	MDE ($E_s^{min} \approx$)
Signal shot-noise	$E_s \gg E_b^N + E_d$	$[(\mathrm{SNR})^{min}]^2 E(\lambda)$
Background-noise	$E_b^N \gg E_s + E_d$	$[E(\lambda)E_b^N(\lambda)]^{1/2}(\mathrm{SNR})^{min}$
Dark-current-noise	$E_d \gg E_s + E_b^N$	$[\frac{1}{2}\tau_d A_d]^{1/2}(\mathrm{SNR})^{min}/D^*$

[a] Neglecting laser backscattered radiation.

the detector within its response time:

$$(\mathrm{SNR})^2 = \frac{E_s^2(\lambda, R)/E(\lambda)}{E_s(\lambda, R) + E_b^T(\lambda) + (i_d + i_J)E(\lambda)/2eB^*} \tag{8.6}$$

The last term in the denominator represents the sum of the dark-current and Johnson-current noise contributions. For most laser remote sensors the detector will have some degree of internal gain, and consequently the Johnson-noise term can be neglected. As indicated in Chapter 6, there are three limiting SNR cases for lidar systems monitoring a return signal wavelength sufficiently displaced from the laser wavelength for successful spectral discrimination against laser backscattered radiation. These are summarized in Table 8.1†.

8.2.1. Signal Shot-Noise Limit

The greatest sensitivity is achieved when the minimum detectable energy (MDE) is limited only by the quantum fluctuations of the signal itself. This *signal shot-noise limit* arises when

$$E_s(\lambda, R) \gg E_b^T(\lambda) + E_d \tag{8.7}$$

where

$$E_d = \frac{i_d E(\lambda)}{2eB^*} \tag{8.8}$$

and $E(\lambda)$ was defined by equation (6.24). Under these conditions the MDE is

$$E_s^{min}(\lambda, R) \approx \left[(\mathrm{SNR})^{min}\right]^2 E(\lambda) \tag{8.9}$$

where $(\mathrm{SNR})^{min}$ represents the smallest credible value for the SNR. By using the appropriate form of the lidar equation, equation (8.9) can be translated into a relation for the threshold number density of a molecular species that can be detected with a given lidar configuration.

†It should be noted that although the expressions provided in Table 8.1 are approximate (see the end of section 6.6) nevertheless for most situations they are adequate.

By way of illustration let us consider the case of Raman scattering from a rectangular-shaped (both spatially and temporally) laser pulse of energy E_L and duration τ_L. In this instance equation (7.15) represents the relevant form of lidar equation and we can write

$$E_s(\lambda, R) = E_L\xi(\lambda)\frac{A_0\xi(R)}{R^2}N(R)\frac{\sigma^R(\lambda_L, \lambda)}{4\pi}\frac{c\tau_d}{2}e^{-\int_0^R \kappa(R)\,dR} \quad (8.10)$$

where we have also used equations (7.12) and (7.22), and $\sigma^R(\lambda_L, \lambda)$ is taken as the appropriate Raman scattering cross section for the molecule of interest. Substitution of equation (8.10) into (8.9) yields the threshold number density $N^{\min}$ that can be detected at range R, averaged over the range interval $c(\tau_L + \tau_d)/2$, under conditions of signal shot-noise limit, namely,

$$[N(R)]^{\min} \cong \frac{R^2\left[(\mathrm{SNR})^{\min}\right]^2 e^{\int_0^R \kappa(R)\,dR}}{E_L\xi(R)\left[\sigma^R(\lambda_L, \lambda)/4\pi\right]U(\lambda)} \quad (8.11)$$

where we have introduced the (wavelength-sensitive) *system parameter*

$$U(\lambda) \equiv \frac{A_0\xi(\lambda)\tau_d\lambda\eta(\lambda)\xi_e}{2hF_G} \quad (8.12)$$

Clearly, this parameter should be made as large as possible if a high sensitivity is to be achieved.

It is evident from Table 3.4 that the differential Raman cross section $\sigma^R(\lambda_L, \lambda)/4\pi$ for a wide range of constituents has a value that lies between 10^{-30} and 10^{-29} $\mathrm{cm^2\ sr^{-1}}$ for excitation at 337 nm. In order to indicate the kind of performance that might be expected from a representative lidar system that is operating in the signal shot-noise limit we shall evaluate the threshold number density of carbon monoxide molecules that may be detected with a lidar system having the following characteristics: $A_0 = 5000$ $\mathrm{cm^2}$, $\tau_d = 20$ ns, $\xi(\lambda) = 0.5$, $\eta(\lambda) = 0.2$, $\xi_e = 1.0$, and $F_G = 1.0$. For $\lambda_L = 308$ nm (XeCl laser), we have $\lambda = 330$ nm and $\sigma^R(\lambda_L, \lambda) = 5.8 \times 10^{-30}$ $\mathrm{cm^2\ sr^{-1}}$ for CO, using Table 3.4 and taking account of the λ^{-4} dependence of σ^R. If we also assume that the receiver's optical efficiency corresponding to range R [i.e., $\xi(R) = 1$; see Section 7.4 for the validity of this assumption] and that $(\mathrm{SNR})^{\min} = 1.5$, then $U(\lambda) = 2.49 \times 10^{23}$ $\mathrm{cm^3\ J^{-1}}$ and

$$[N(R)]_{\mathrm{CO}}^{\min} \approx \frac{1.70 \times 10^{10} R^2}{E_L} e^{\int_0^R \kappa(R)\,dR} \quad (8.13)$$

where R is in m, E_L in J, and $\kappa(R)$ in $\mathrm{m^{-1}}$.

If we assume reasonably clear weather conditions (i.e., visibility of about 10 km), then $\kappa(R) \cong 10^{-3}$ $\mathrm{m^{-1}}$ at 308 nm (see Fig. 4.9). In Fig. 8.3 we have plotted the threshold concentration (in ppm) of CO that could be detected as a function of range R for a XeCl laser having an output energy E_L of 0.30, 0.75, and 1.90 J (full curves). We have taken the sea-level density of air to be 2.55×10^{19} cm^{-3} in this set of calculations. It is apparent from Fig. 8.3 that Raman scattering can be viewed as potentially useful for remote monitoring of pollution sources where the concentrations are likely to be fairly high and ranges of 100 to 300 m would be acceptable. Alternatively, a Raman lidar could be employed to measure the density of major atmospheric constituents, such as N_2, O_2, and H_2O, for altitudes of a few kilometers.

8.2.2. Background-Noise Limit

A laser remote sensor operating under daylight conditions is likely to be subject to levels of background illumination so high that

$$E_b^N(\lambda) \gg E_s(\lambda, R) + E_d \tag{8.14}$$

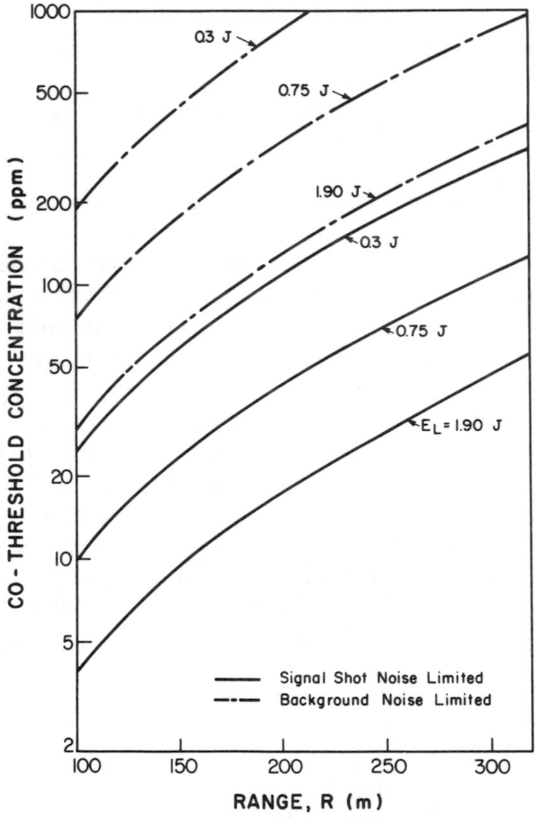

Fig. 8.3. Lidar threshold concentration of carbon monoxide as a function of range, based on Raman scattering, for laser energies E_L of 0.3, 0.75, and 1.90 J. Full curves: signal shot-noise limit; broken curves: background-noise limit.

Nevertheless, signal extraction is still possible provided the fluctuation in this background radiation is sufficiently small and that the time-averaged level of the radiation is not so great as to saturate the detector. The MDE under these circumstances was given by equation (6.28) and can be expressed in the form

$$E_s^{\min}(\lambda, R) \approx \left[E(\lambda)E_b^N(\lambda)\right]^{1/2}(\mathrm{SNR})^{\min} \tag{8.15}$$

This can be translated into a threshold density for a particular atmospheric molecular constituent through the use of the appropriate lidar equation. For example, if we again imagine using the Raman lidar discussed earlier, then we use equation (8.10) together with (8.15) to yield the threshold number density,

$$[N(R)]^{\min} \approx \frac{2R^2 \mathrm{e}^{\int_0^R \kappa(R)\,dR}}{E_L \xi(R)\left[\sigma^R(\lambda_L, \lambda)/4\pi\right]} \left\{\frac{S_b(\lambda)\Omega_0 \Delta\lambda_0}{2U(\lambda)c}\right\}^{1/2} (\mathrm{SNR})^{\min} \tag{8.16}$$

where we have again used the system parameter $U(\lambda)$ defined by equation (8.12), and equation (8.3) for the filter function $K_0(\lambda)$.

It is evident that improvement in sensitivity for the background-noise-limited situation necessitates using as small as possible values of the receiver's acceptance solid angle Ω_0 and the effective spectral window $\Delta\lambda_0$. The system parameter should also be made as large as practical.

If the same Raman lidar is considered, then the system parameter $U(\lambda)$ is again 2.49×10^{23} cm^3 J^{-1}. However, in the background-noise-limited situation we also have to know the receiver-optics acceptance solid angle and system's effective spectral window. We shall assume $\Omega_0 = 2 \times 10^{-6}$ sr, $\Delta\lambda_0 = 1$ nm, and again $(\mathrm{SNR})^{\min} \cong 1.5$. If we also take a fairly conservative value of 4×10^{-6} W cm^{-2} nm^{-1} sr^{-1} for the sky spectral radiance $S_b(\lambda)$ at 330 nm, then we obtain

$$[N(R)]_{\mathrm{CO}}^{\min} \approx \frac{1.3 \times 10^{11} R^2}{E_L} \mathrm{e}^{\int_0^R \kappa(R)\,dR} \tag{8.17}$$

where, as before, R is in m, E_L is in J, and $\kappa(R)$ is in m^{-1}. If we assume the same weather conditions so that $\kappa(R) \approx 10^{-3}$ m^{-1} at 308 nm, then we arrive at the broken curves in Fig. 8.3 for the same set of laser energies. A comparison of the carbon monoxide threshold-detection concentration curves for the same lidar system operating under signal shot-noise- and background-noise-limited situations clearly illustrates the superiority of the former. Unfortunately, the radiance of sunlit clouds can, for longer wavelengths, be close to one order of magnitude larger than the above figure (Pratt, 1969). On the other hand, the radiance of a clear, moonless night sky is typically 10^{-7} smaller.

8.2.3. Dark-Current Limit

In the dark-current-limited situation,

$$i_d \gg \frac{2eB^*}{E(\lambda)}\left[E_s(\lambda, R) + E_b^T(\lambda)\right] \tag{8.18}$$

The dark current associated with photomultipliers is rarely large enough to satisfy (8.18), and even in those situations where the dark current is excessive, cooling the photomultiplier will more often than not alleviate the problem. The MDE of solid-state detectors, on the other hand, is often determined by this criterion:

$$E_s^{\min}(\lambda, R) \approx \frac{1}{D^*}\left\{\frac{F_G A_d}{4B\xi_e}\right\}^{1/2} (\text{SNR})^{\min} \tag{8.19}$$

Representative values of the detectivity D^*, defined by equation (6.30), of a number of solid-state detectors range from 10^{13} cm Hz$^{1/2}$ W^{-1} in the visible to 10^{10} cm Hz$^{1/2}$ W^{-1} in the infrared part of the spectrum (see Fig. 6.8). In most instances these devices are employed only for lidar systems that operate in the infrared, and consequently it is unlikely that they would be used in laser remote sensors based on Raman scattering or fluorescence.

8.2.4. Differential-Absorption Detection Limit

Probably the most important role played by solid-state detectors will be in the expanding area of infrared differential absorption. This technique has considerable potential for laser remote sensors due to the wide range of molecular species that are detectable in this part of the spectrum. In the case of the DIAL technique it is essentially the difference between two backscattered laser signals at slightly different wavelength that is related to the density of a specific molecular constituent. In fact, the incremental decrease in the observed signal at λ_0, associated with the range increment ΔR and arising from the attenuation of the specific molecular constituent for which the laser is tuned, is given by

$$\Delta E_0^* = \Delta E_0 - \Delta E_w \tag{8.20}$$

where

$$\Delta E_0 = E(\lambda_0, R) - E(\lambda_0, R + \Delta R) \tag{8.21a}$$

represents the total decrease in the signal at λ_0, and

$$\Delta E_w = E(\lambda_w, R) - E(\lambda_w, R + \Delta R) \tag{8.21b}$$

effectively represents the incremental decrease in the signal due to nonspecific attenuation. λ_0 represents the laser wavelength chosen to be close to the peak of the absorption line of the molecule of interest, and λ_w the wavelength of the laser pulse detuned to lie in the wing of the absorption line; see Section 7.2 and Fig. 8.4.

As a first-order approximation we shall assume that the change in the "off" energy signal, ΔE_w, is small compared to the change in the "on" energy signal over the small range interval ΔR. Under these circumstances detection of the incremental decrease of the backscattered energy signal at λ_0 requires that

$$E(\lambda_0, R) - E(\lambda_0, R + \Delta R) > \frac{E(\lambda_0, R + \Delta R)}{\text{SNR}} \tag{8.22}$$

which states, in effect, that this incremental energy signal should be greater than the noise in the λ_0-signal from range $R + \Delta R$. If we use the backscattering lidar equation with (7.27)

$$E(\lambda_0, R) = E_L \xi(\lambda_0) \frac{A_0 \xi(R)}{R^2} \beta(\lambda_0, R) \frac{c\tau_d}{2} e^{-2\int_0^R \{\bar{\kappa}(R) + N(R)\sigma^A(\lambda_0)\}\, dR} \tag{8.23}$$

in the inequality (8.22), then we can arrive at a criterion for the threshold number density of the molecular constituent that can be detected by means of the differential absorption and scattering:

$$[N(R)]^{\min} = \frac{1}{2\sigma^A(\lambda_0)\,\Delta R} \ln\left[\left\{1 + \frac{1}{\text{SNR}}\right\}\left(\frac{R}{R + \Delta R}\right)^2\right] \tag{8.24}$$

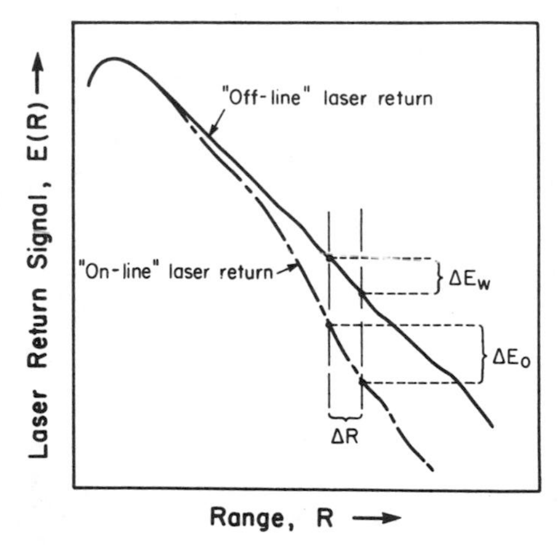

Fig. 8.4. Schematic of differential-absorption and scattering laser return signals.

In arriving at this result we have assumed that ΔR is sufficiently small that the factor $\xi(R)\beta(\lambda_0, R)$ can be regarded as a constant over this range increment and that the term $\bar{\kappa}(\lambda_0, R)/\sigma^A(\lambda_0)$ is very much less than the right-hand side of equation (8.24), where $\bar{\kappa}(\lambda_0, R)$ represents the residue attenuation coefficient after allowing for absorption by the molecular constituent of interest. This assumption is tantamount to stating that $N(R)\sigma^A(\lambda_0) \gg \bar{\kappa}(\lambda_0, R)$ at the point of interest.

Clearly, the smaller the signal-to-noise ratio, the larger the threshold density given by equation (8.24). A conservative form of equation (8.24), which also allows for good spatial resolution, can be written

$$[N(R)]^{\min} = \frac{1}{2\sigma^A(\lambda_0)\,\Delta R} \ln\left\{1 + \frac{1}{(\mathrm{SNR})^{\min}}\right\} \tag{8.25}$$

If $(\mathrm{SNR})^{\min} \cong 1.5$, then we have

$$[N(R)]^{\min} \approx \frac{1}{4\sigma^A(\lambda_0)\,\Delta R} \tag{8.26}$$

It may at first seem rather surprising that the DIAL threshold density appears to be independent of virtually everything other than the absorption cross section and the range increment. If one reflects for a moment on the means by which the presence of the molecular constituent of interest is detected, one will understand this state of affairs. In essence, detection of the species is assured if the incremental change in the signal is discernible against the noise in the signal. An implicit assumption in all of this is that the laser backscattered energy is always larger than the MDE for the detector, so that from equation (8.19)

$$E(\lambda_0, R + \Delta R) \gtrsim \frac{1}{D^*}\left\{\frac{F_G A_d}{4B\xi_e}\right\}^{1/2} (\mathrm{SNR})^{\min} \tag{8.27}$$

This, in effect, provides us with a criterion for the minimum laser output energy required for a given lidar system to comply with this assumption. The magnitude of this threshold laser energy is determined by using equation (8.23):

$$E_L^{\min} \approx \frac{2R^2(\mathrm{SNR})^{\min}}{\beta(\lambda_0, R)\xi(R)U^*(\lambda_0)} \mathrm{e}^{2\int_0^R \kappa(\lambda_0, R)\,dR} \tag{8.28}$$

where we have introduced the modified system parameter,

$$U^*(\lambda_0) = A_0\xi(\lambda_0)c\tau_d D^*\left\{\frac{4B\xi_e}{F_G A_d}\right\}^{1/2} \tag{8.29}$$

This can be related to the system parameter previously introduced, in equation (8.12), through the substitution of equations (6.19) and (6.30). That is to say,

$$U^*(\lambda_0) = 2U(\lambda_0)\left\{\frac{2eBF_G}{i_d\xi_e}\right\}^{1/2} \tag{8.30}$$

In the case of a solid-state detector we can generally set $F_G/\xi_e \approx 1$, and if we draw upon equation (6.11) we can write

$$U^*(\lambda_0) = \frac{A_0\xi(\lambda_0)cD^*}{\{A_dB\}^{1/2}} \tag{8.31}$$

In order to gain some experience with the magnitude of these quantities, we shall consider the representative DIAL system of Murray (1977). The characteristics of this system are listed in Table 8.2. In general, the threshold concentration (in ppm) of a molecular constituent of interest can be written

$$C_i^{\min} = \frac{N_i^{\min} \times 10^6}{N_{\text{atm}}} \tag{8.32}$$

where $N_i^{\min}$ is the threshold number density of species i, and N_{atm} is the total number density of molecules in the atmosphere under conditions of interest. Nominally, at sea level, $N_{\text{atm}} = 2.55 \times 10^{19}$ cm^{-3} (where the Loschmidt number is 2.69×10^{19} cm^{-3}). The attenuation coefficient for the molecule of interest is

$$\kappa_A^i(\lambda_0) = \sigma_i^A(\lambda_0)N_{\text{atm}} \qquad (\text{cm}^{-1}\ \text{atm}^{-1}) \tag{8.33}$$

TABLE 8.2. CHARACTERISTICS OF A REPRESENTATIVE INFRARED DIAL SYSTEM[a]

DF Laser	
Energy, E_L	150 mJ
Pulse width, τ_L	1.0 μs
Beam divergence, 2θ	1.0 mrad
Wavelength, λ_L	3.6 to 3.9 μm
Receiver	
Telescope area, A_0	792 cm^2
Field of view, 2ϕ	3.0 mrad
HgCdTe detector: size	1 × 1 mm
$D^*(3.7\ \mu\text{m})$	10^{10} cm Hz$^{-1/2}$ W^{-1}
τ_d	75 ns

[a] Murray (1977).

so that if we use equations (8.26), (8.32), and (8.33) we can express the threshold concentration for HCl in the form

$$C_{\text{HCl}}^{\min} = \frac{10^6}{4\kappa_A^{\text{HCl}}(\lambda_0)c(\tau_L + \tau_d)/2} \tag{8.34}$$

From Table 8.2, $\tau_L = 10^{-6}$ s and $\tau_d = 7.5 \times 10^{-8}$ s. Murray (1977) also gives κ_A^{HCl} (3.64 μm) = 5.64 cm^{-1} atm^{-1}. Then

$$C_{\text{HCl}}^{\min} = 2.75 \text{ ppm} \tag{8.35}$$

As stated earlier, this presupposes that the laser energy is adequate to provide a backscattered laser signal, from the range of interest, that is greater than the MDE for the DIAL system. This threshold laser energy was given by equation (8.28). To determine its magnitude we must calculate the value of the modified system parameter, as given by equation (8.31), namely,

$$U^*(3.64\ \mu\text{m}) = \frac{792 \times 3 \times 10^{10} \times 10^{10}}{\left[0.01/(2 \times 7.5 \times 10^{-8})\right]^{1/2}}$$

$$= 9.2 \times 10^{20} \text{ cm}^3 \text{ J}^{-1}$$

where we have assumed $\xi(3.64\ \mu\text{m}) \approx 1$. If the DIAL system is designed to operate under low-altitude haze conditions, then from Figs. 2.26 and 2.27 we can assume that the atmospheric volume backscattering coefficient $\beta(3.64\ \mu\text{m}) = 8 \times 10^{-9}$ cm^{-1} sr^{-1} and the atmospheric attenuation coefficient $\kappa(R) = 5 \times 10^{-7}$ cm^{-1}. Consequently

$$E_L^{\min} = 4.08 \times 10^{-13} R^2 \times e^{10^{-6}R} \quad \text{J} \tag{8.36}$$

assuming $\xi(R) = 1$ and $(\text{SNR})^{\min} = 1.5$. For operation up to a 10-km range,

$$E_L^{\min} \approx 1.1 \quad \text{J}$$

while operation up to maximum range of only 3 km lowers this energy to a value

$$E_L^{\min} \approx 50 \quad \text{mJ}$$

8.3. LIDAR ERROR ANALYSIS

Any quantitative measurement has errors and uncertainties associated with it. The purpose of an error analysis is therefore twofold. First, we require to estimate the possible error incurred in a given measurement in order that the accuracy of that measurement can be ascertained, and second the dependence of errors and uncertainties upon the measurement system parameters needs to be identified so that the technique can be optimized.

The terms *accuracy* and *precision* as related to experimental observations are sometimes confused. Bevington (1969) has distinguished between them as follows: *accuracy* is the measure of how close an experimental result comes to the true value, while *precision* is a measure of how exactly a result is determined, without reference to any true value. At this point it may also be well to recall the difference between *systematic* errors and *random* errors. The former arise when faulty equipment or techniques are used, while the latter can be thought of as a measure of the irreproducibility of a given observation.

8.3.1. Basic Statistical Definitions

Let us assume that we undertake a series of N measurements of a variable u, where u_i represents the ith measurement. The (sample) *mean* value of u as ascertained from this series of measurements:

$$\langle u \rangle \equiv \frac{1}{N} \sum_{i=1}^{N} u_i \tag{8.37}$$

where the *deviation* of the ith measurement from the mean is

$$d_i \equiv u_i - \langle u \rangle \tag{8.38}$$

The (sample) *variance* is

$$\delta u^2 \equiv \left\langle |u_i - \langle u \rangle|^2 \right\rangle = \frac{1}{N} \sum_{i=1}^{N} |u_i^2 - 2u_i \langle u \rangle + \langle u \rangle^2| \tag{8.39}$$

that is,

$$\delta u^2 = \frac{1}{N} \sum_{i=1}^{N} u_i^2 - \langle u \rangle^2 \tag{8.40}$$

The uncertainty in a measurement u is represented by the *standard deviation*,

$$\delta u \equiv \{\delta u^2\}^{1/2} \tag{8.41}$$

In general for a function $f(u, v, \dots)$ derived from measured variables $u, v, \dots$, the uncertainty in f can be obtained from the relation

$$\delta f^2 = \delta u^2 \left(\frac{\partial f}{\partial u}\right)^2 + \delta v^2 \left(\frac{\partial f}{\partial v}\right)^2 + 2C_{uv}^2 \left(\frac{\partial f}{\partial u}\right)\left(\frac{\partial f}{\partial v}\right) + \cdots \tag{8.42}$$

where δu and δv are the respective uncertainties in the variables u and v, and

$$C_{uv}^2 = \lim_{N \to \infty} \left[\frac{1}{N} \sum_{i=1}^{N} [u_i - \langle u \rangle][v_i - \langle v \rangle] \right] \tag{8.43}$$

represents the *covariance* between the measured variables u and v. C_{uv}^2 vanishes when the measurement errors $u_i - \langle u \rangle$ and $v_i - \langle v \rangle$ are uncorrelated. Equation (8.42) expresses the propagation of errors associated with a measurement that depends upon several parameters.

In laser remote sensing we are primarily concerned with the detection of small amounts of radiant energy. From the quantum standpoint this burst of energy can be viewed as a bunch of photons. These photons generate a pulse of electrons within the detector. If a series of measurements were undertaken under essentially identical circumstances the inherent quantum nature of the radiation would mean that the number of photon-generated electrons would fluctuate in such a way that the probability of n_e electrons arising in any particular measurement was

$$P(n_e, \langle n_e \rangle) = \frac{\langle n_e \rangle^{n_e}}{n_e!} \mathrm{e}^{-\langle n_e \rangle} \tag{8.44}$$

where $\langle n_e \rangle$ represents the mean number of photon-created electrons in this series of measurements. Equation (8.44) represents a *Poisson distribution*, and it is easily shown (Bevington, 1969, p. 78) that for such a statistical distribution, the uncertainty δn_e in the mean number of electrons in a series of m measurements is approximately given by the relation

$$\delta n_e^2 \approx \frac{\langle n_e \rangle}{m} \tag{8.45}$$

This implies that the estimated error in a series of lidar measurements decreases as the inverse square root of the number of observations.

8.3.2. Error Analysis for Elastic-Backscattering Lidar

The role of aerosols and particulates in our climate is known to be important. It follows that accurate measurements of the distribution of these aeroticulates in the atmosphere could lead to a better understanding of climatic processes. Elastic-backscattering lidar represents one of the most promising tools for attaining this information remotely. In this section we shall examine the uncertainties associated with such measurements and try to determine which of the lidar parameters need to be optimized in order to minimize the measurement errors.

The number $n_e(R)$ of signal-generated electrons created within the detector integration period and arising from elastic backscattering of laser radiation from range R can be obtained from equations (6.4) and (7.14):

$$n_e(R) = \frac{E_L D_s(\lambda_L)\beta(\lambda_L, R)}{R^2} e^{-2\int_0^R \kappa(\lambda_L, R)\,dR} \tag{8.46}$$

where we have introduced the *lidar system parameter*

$$D_s(\lambda_L) = \frac{A_0 \xi(\lambda_L)\xi_D(\lambda_L)\tau_d \lambda_L}{2h} \tag{8.47}$$

and the *detector efficiency factor* $\xi_D(\lambda_L)$. In the case of a photomultiplier

$$\xi_D(\lambda_L) = \eta(\lambda_L)\xi_e G \tag{8.48}$$

where

$\eta(\lambda_L)$ represents the quantum efficiency of the photocathode at λ_L;
ξ_e represents the probability of an electron liberated from the photocathode reaching the first dynode; and
G the dynode gain of the photomultiplier—see Section 6.6.

In arriving at equation (8.46) we have also assumed that the design of the lidar system, its alignment, and the range of interest are such that we can set $\xi(R) = 1$ (see Chapter 7). If this were not the case, our analysis would be much more complicated.

In reality both the volume elastic-backscattering coefficient $\beta(\lambda_L, R)$ and the volume attenuation coefficient $\kappa(\lambda_L, R)$ can be divided into molecular (m) and aeroticulate (a) contributions:

$$\beta(\lambda_L, R) = \beta_m(\lambda_L, R) + \beta_a(\lambda_L, R) \tag{8.49}$$

and

$$\kappa(\lambda_L, R) = \kappa_m(\lambda_L, R) + \kappa_a(\lambda_L, R) \tag{8.50}$$

It is often found convenient (Russell et al., 1979) to introduce the *total-to-molecular backscattering ratio*

$$B(R) \equiv \frac{\beta(\lambda_L, R)}{\beta_m(\lambda_L, R)} \tag{8.51}$$

which on invoking equation (8.46) becomes

$$B(R) = \frac{n_e(R)R^2}{E_L D_s(\lambda_L)\beta_m(\lambda_L, R)T_L(R)} \tag{8.52}$$

where we have introduced

$$T_L(R) = e^{-2\int_0^R \kappa(\lambda_L, R)\, dR} \tag{8.53}$$

In solving equation (8.52), $n_e(R)$ is provided by the lidar measurements, $T_L(R)$ is often evaluated from an atmospheric model (which is often updated with the lidar data), and $\beta_m(\lambda_L, R)$ is determined from radiosonde (or satellite) density data or from a model. The system parameter $D_s(\lambda_L)$ could be calibrated, but is often eliminated through the introduction of the minimum value of the backscattering ratio, $B_{\min}(R_\#)$. Clearly, $R_\#$ represents some reference range at which the minimum in $B(R)$ is achieved. Equation (8.52) can consequently be rewritten in the form

$$B(R) = \frac{n_e(R) R^2 \beta_m(\lambda_L, R_\#) T_L(R_\#)}{n_e(R_\#) R_\#^2 \beta_m(\lambda_L, R) T_L(R)} B_{\min}(R_\#) \tag{8.54}$$

and from equations (8.49) and (8.51), the aeroticulate backscattering coefficient is

$$\beta_a(\lambda_L, R) = \{B(R) - 1\}\beta_m(\lambda_L, R) \tag{8.55}$$

For the purpose of the error analysis and to simplify the computation, we shall introduce the nondimensional parameters

$$n \equiv \frac{n_e(R)}{n_e(R_\#)}, \qquad x \equiv \frac{R^2}{R_\#^2}$$
$$q \equiv \frac{T_L(R_\#)}{T_L(R)}, \qquad B_\# \equiv B_{\min}(R_\#) \tag{8.56}$$

and the abbreviated notation

$$\beta_m = \beta_m(\lambda_L, R), \qquad \beta_m^\# = \beta_m(\lambda_L, R_\#), \quad \text{and} \quad \beta_a = \beta_a(\lambda_L, R) \tag{8.57}$$

Using these, we can write the aeroticulate contribution to the volume backscattering coefficient,

$$\beta_a = nxq\beta_m^\# B_\# - \beta_m \tag{8.58}$$

The uncertainty in the measurement of the aeroticulate backscattering coefficient, $\delta\beta_a$, can be expressed in terms of the uncertainty of the variables n,

x, q, $\beta_m^{\#}$, $B_{\#}$, and β_m in accordance with equation (8.42). We may thus write

$$\delta\beta_a^2 = \delta n^2\left(\frac{\partial\beta_a}{\partial n}\right)^2 + \delta x^2\left(\frac{\partial\beta_a}{\partial x}\right)^2 + \delta q^2\left(\frac{\partial\beta_a}{\partial q}\right)^2 + \delta\beta_m^{\#2}\left(\frac{\partial\beta_a}{\partial\beta_m^{\#}}\right)^2$$

$$+\delta B_{\#}^2\left(\frac{\partial\beta_a}{\partial B_{\#}}\right)^2 + \delta\beta_m^2\left(\frac{\partial\beta_a}{\partial\beta_m}\right)^2 + 2C_{\beta_m\beta_m^{\#}}^2\left(\frac{\partial\beta_a}{\partial\beta_m}\right)\left(\frac{\partial\beta_a}{\partial\beta_m^{\#}}\right) \tag{8.59}$$

The form of equation (8.59) presupposes that no correlation exists between the variables n, x, q, and $B_{\#}$, due to their different methods of evaluation. On the other hand, some correlation might exist between β_m and $\beta_m^{\#}$ which would lead to a finite value for the corresponding covariance factor $C_{\beta_m\beta_m^{\#}}$. Using equation (8.58) to determine the appropriate partial derivatives for equation (8.59), we obtain

$$\delta\beta_a^2 = \delta n^2\{xq\beta_m^{\#}B_{\#}\}^2 + \delta x^2\{nq\beta_m^{\#}B_{\#}\}^2 + \cdots$$

which on substituting equation (8.54), namely

$$B(R) = \frac{nxq\beta_m^{\#}B_{\#}}{\beta_m}$$

yields

$$\left(\frac{\delta\beta_a}{\beta_a}\right)^2 = \left\{\frac{\beta_m B(R)}{\beta_a}\right\}^2\left[\left(\frac{\delta n}{n}\right)^2 + \left(\frac{\delta x}{x}\right)^2 + \left(\frac{\delta q}{q}\right)^2 + \left(\frac{\delta\beta_m^{\#}}{\beta_m^{\#}}\right)^2 + \left(\frac{\delta B_{\#}}{B_{\#}}\right)^2\right.$$

$$\left. + \frac{1}{B^2(R)}\left(\frac{\delta\beta_m}{\beta_m}\right)^2 - \frac{2C_{\beta_m\beta_m^{\#}}^2}{\beta_m\beta_m^{\#}B(R)}\right] \tag{8.60}$$

The relative error in the aeroticulate backscattering coefficient is seen to have contributions from the signal measurement error, the range error, the two-way transmission error, the molecular-density error, and the error in the assumed value of $B_{\#}$.

Russell et al. (1979) discussed each of these sources of error and evaluated the range dependence of both the backscattering ratio and the aeroticulate backscattering coefficient expected for two lidar systems based on several atmospheric models. They also calculated the range dependence of the error associated with each of these parameters. The two lidars were assumed to be very similar, except that one employed a ruby laser operating at 694.3 nm, and the other a neodymium glass laser operating at 1.06 μm. Also in their analysis they assumed that there was no uncertainty in the range.

An example of one of their atmospheric models is presented as Fig. 8.5. The molecular density and aeroticulate extinction profiles were derived from a dustsonde measurement made at Pt. Barrow, Alaska in November 1973, during nonvolcanic conditions. Their simulation of the lidar backscattering-ratio profiles for both the ruby- and Nd-laser lidars is presented as Fig. 8.6. Included with these profiles are the expected error bars evaluated in accordance with the above analysis. These calculations assume that both lidar systems are zenith-viewing and flying at an altitude of 4 km with a vertical range increment of 0.25 km.

Figure 8.7 shows a comparison between the simulated measurement (dots) of the aeroticulate backscattering coefficient and the model profile (full curves) used in these calculations. These simulated measurements included the above-mentioned random errors, excluding the range uncertainty, which was set to zero. The scatter of the simulated data points is seen to lie approximately within the error bars. It can be seen from Fig. 8.7 that the relative errors in the Nd backscattering-coefficient profile are about half as large as the relative

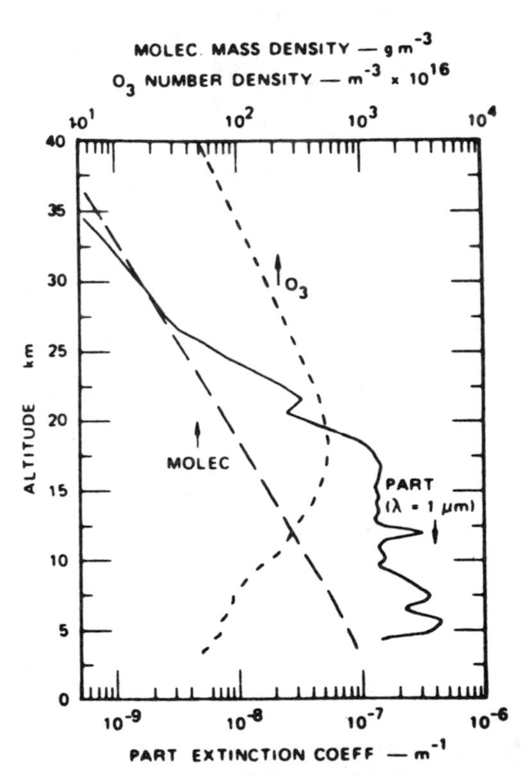

Fig. 8.5. Molecular density, particulate extinction, and ozone concentration for the Arctic nonvolcanic model atmosphere used by Russell et al. (1979) in their lidar simulations.

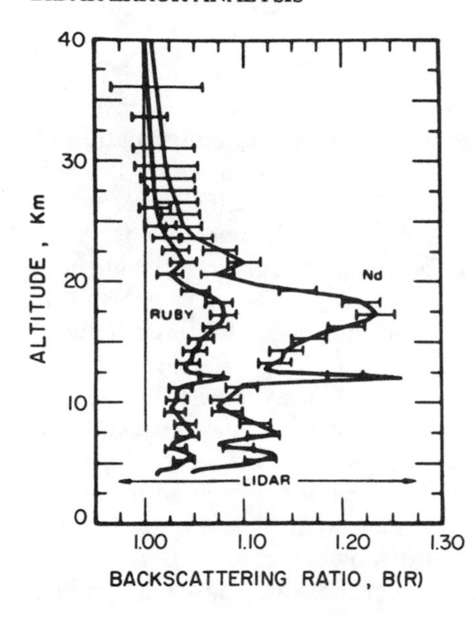

Fig. 8.6. Variation of the simulated laser backscattering measurements (including expected errors) with altitude for both ruby and neodymium lidar systems for the Arctic nonvolcanic model atmosphere indicated in Fig. 8.5 (Russell et al., 1979).

errors for the ruby wavelength profile. This can be understood in terms of the greater contribution of molecular scattering to the total backscattering coefficient at the shorter wavelength and the fact that the molecular-density profile is not known very accurately. This suggests that a Nd–YAG-based lidar would be more accurate for measurements of the aeroticulate distribution than one based on a ruby laser.

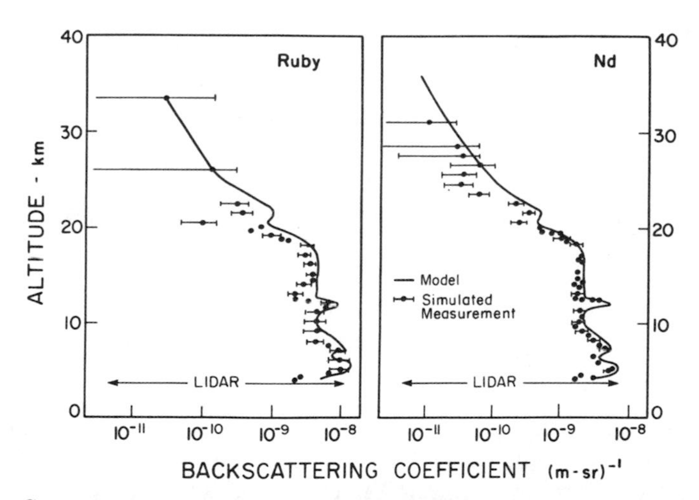

Fig. 8.7. Comparison of simulated elastic laser backscattering from Arctic nonvolcanic model atmosphere (indicated in Fig. 8.5) for ruby and neodymium lasers. The magnitude of the expected random errors is indicated by the length of the bars (Russell et al., 1979).

8.3.3. Error Analysis for Fluorescence Lidar

The application of laser-induced fluorescence to the field of remote sensing is, as discussed earlier, limited by the effects of quenching and redistribution of the emitted radiation. Nevertheless, there are two arenas in which this approach can be used to study or probe the environment. In the upper atmosphere the molecular density is low enough that these detrimental effects are reduced and the efficiency of laser-induced fluorescence approaches its intrinsic high value. In addition, there are a number of situations in which the concentration of the fluorescing agent is sufficiently high that laser-induced fluorescence is still attractive.

In the case of fluorescence the wide spectral interval required to capture an appreciable fraction of the emitted radiation tends to make background radiation the dominant source of noise. In order to overcome this problem the lidar usually operates in two modes: First, the detector is permitted to sample the background radiation for some period τ_d^*; then, after the laser fires ($t = 0$), the detector again accepts radiation during some time interval $(t, t + \tau_d)$ chosen to coincide with the return of fluorescence emanating from a target range R $(= ct/2)$.

The number of photon-generated electrons created within the detector during the first time interval, n_b, can be ascertained from equations (6.2) and (6.4):

$$n_b = \frac{D_K(\lambda)\tau_d^* \nu_R S_b(\lambda)}{4\pi R^2 \tau_d} \tag{8.61}$$

where we have introduced

$$D_K(\lambda) \equiv \frac{A_0 K_0(\lambda) \xi_D(\lambda) \tau_d \lambda}{2h} \tag{8.62}$$

and

$$\nu_R = \frac{8\pi\Omega_0 R^2}{c} \tag{8.63}$$

$D_K(\lambda)$ can be seen to represent a modified form of the lidar system parameter previously defined by equation (8.47). In particular the filter function $K_0(\lambda)$ is used in place of the receiver's spectral transmission factor $\xi(\lambda)$.

The number of detector electrons produced by the combination of fluorescence return radiation and background illumination during the second period is

$$n_f = \frac{D_K(\lambda)}{4\pi R^2}\left[E_L T(R)\xi(R)N(R)\sigma^F(\lambda_L, \lambda) + \nu_R S_b(\lambda)\right] \qquad (8.64)$$

In obtaining this result we have used equation (7.53), which represents the fluorescence form of the lidar equation that assumes the lifetime correction factor is close to unity. We have also assumed that there is no fluorescence generated within any interfering species, that the spectral rejection of the receiver system is sufficient for us to neglect backscattered laser radiation, and that the number of dark-current electrons created within the times of interest (τ_d and τ_d^*) is negligible.

Combining equations (8.61) and (8.64) enables us to write

$$N = \zeta(n_f - n_b \iota) \qquad (8.65)$$

where

$$\zeta = \frac{4\pi R^2}{E_L D_K(\lambda)T(R)\xi(R)\sigma^F(\lambda_L, \lambda)} \qquad (8.66)$$

$$\iota = \frac{\tau_d}{\tau_d^*} \qquad (8.67)$$

and we have used the abbreviated notation N for $N(R)$. The uncertainty in the measured density of the fluorescent species δN can be evaluated from a propagation-of-errors analysis:

$$\delta N^2 = \delta n_f^2\left(\frac{\partial N}{\partial n_f}\right)^2 + \delta n_b^2\left(\frac{\partial N}{\partial n_b}\right)^2 + \delta\zeta^2\left(\frac{\partial N}{\partial \zeta}\right)^2 + \delta\iota^2\left(\frac{\partial N}{\partial \iota}\right)^2 \qquad (8.68)$$

We have assumed that none of these four variables are correlated and so we can omit the covariance terms. Furthermore, we shall assume that the variance $\delta\iota^2$ is negligible. Then using equation (8.65) to obtain the partial derivatives for equation (8.68), we can write

$$\frac{\delta N^2}{N^2} = \frac{1}{(n_f - n_b\iota)^2}\left[\delta n_f^2 + \iota^2 \delta n_b^2\right] + \frac{\delta\zeta^2}{\zeta^2} \qquad (8.69)$$

If we further assume that the electron bursts are described by Poisson statistics, so that

$$\delta n_f^2 \approx n_f \quad \text{and} \quad \delta n_b^2 \approx n_b \qquad (8.70)$$

then we can write the *relative error* in the density measurement as

$$\varepsilon_N \equiv \frac{\delta N}{N} = \left\{\frac{n_f + \iota^2 n_b}{(n_f - \iota n_b)^2} + \varepsilon_\zeta^2\right\}^{1/2} \qquad (8.71)$$

where ε_ζ ($\equiv \delta\zeta/\zeta$) represents the relative error in the variable ζ.

From equations (8.61) and (8.64) it is apparent that we can write

$$\frac{n_f}{n_b} = (f_b + 1)\iota \tag{8.72}$$

where f_b represents the ratio of the fluorescence to background contributions to the total signal as given by equation (8.64). Substitution of equation (8.72) in equation (8.71) leads to

$$\varepsilon_N = \left\{ \frac{(f_b + 1)(f_b + 1 + \iota)}{n_f f_b^2} + \varepsilon_\xi^2 \right\}^{1/2} \tag{8.73}$$

In reality the relative error ε_ξ will have contributions from the uncertainty in each of the factors that constitute the lidar system parameter $D_K(\lambda)$ as well as from the other factors in equation (8.66). Nevertheless, careful design and calibration can keep most of these to a minimum. Since it is instructive to consider the limiting case of $\varepsilon_\xi = 0$, we shall assume that ε_ξ is negligible compared to the first term within the square root of equation (8.73). Under these circumstances we can write

$$\varepsilon_N \sqrt{n_f} = \frac{[(f_b + 1)(f_b + 1 + \iota)]^{1/2}}{f_b} \tag{8.74}$$

In Fig. 8.8 we have plotted $\varepsilon_N \sqrt{n_f}$ against f_b for the two extreme cases of $\iota = 0$ and 1. The former corresponds to having a very long interval with which to determine the background level of illumination, while in the latter case this interval is the same as the period chosen to sample the fluorescence return. Clearly, when the signal associated with the fluorescence radiation is much greater than that of the background (i.e., $f_b \gtrsim 10$), then we have the limiting situation given by

$$\varepsilon_N \approx \frac{1}{\sqrt{n_f}} \tag{8.75}$$

For most practical situations a more rigorous analysis would have to be undertaken. This should account for the possibility of errors associated with the lidar system parameters in addition to the uncertainties of the two-way transmission factor $T(R)$, the overlap factor $\xi(R)$, and the possibility that the laser beam might excite fluorescence within a number of species other than the one of interest. Heaps (1981) has made an initial attempt at considering this problem. Unfortunately, his analysis is somewhat limited in that he assumes no change of the fluorescence efficiency from the interfering species when the laser wavelength is considered to be shifted sufficiently for there to be negligible fluorescence from the species of interest.

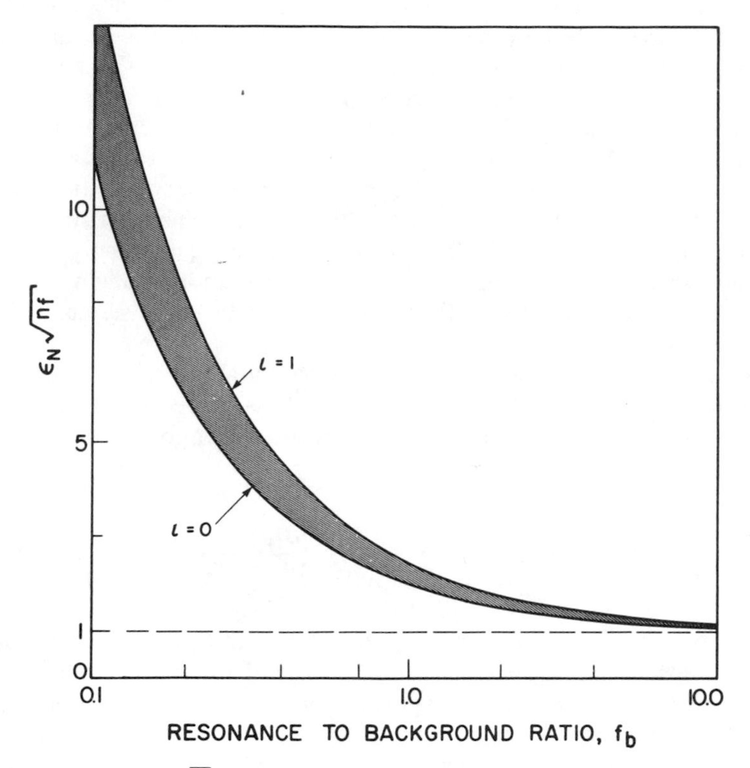

Fig. 8.8. Variation of $\epsilon_N\sqrt{n_f}$ with the resonance-to-background ratio f_b for two limiting cases of the PMT gate time ratio ι (0 and 1).

8.3.4. Error Analysis for Differential-Absorption Lidar

As we have seen earlier, laser sensing based on differential absorption and scattering appears to have very little competition when it comes to remote three-dimensional mapping of atmospheric trace constituents in the troposphere. However, its domain of application is not restricted to the lower regions of the atmosphere. Indeed, Remsberg and Gordley (1978), Uchino et al. (1979), and Megie and Menzies (1980) have indicated that it could be used well into the stratosphere. Thus it is clearly an important technique, and we need to consider carefully those factors which determine its accuracy and limit its sensitivity.

For any particular molecule differential absorption can be exploited at any wavelength where there exists a suitable (sharp) absorption feature. Although virtually all molecules possess strong, electronic absorption lines, very few of them have such features at wavelengths long enough (Section 4.5) for them to be accessible to laser probing within the atmosphere. On the other hand, the infrared region of the spectrum is rich with vibrational–rotational lines of molecules. Unfortunately, in the lower atmosphere collision broadening tends to make it difficult to find sharp absorption lines. Nevertheless, a study of the possibility of probing the atmosphere with a spaceborne laser by Megie and Menzies (1980) indicated that UV and IR–DIAL measurements could be complementary. Their analysis suggested that when factors such as the efficiencies of the laser systems and the scattering processes, the extinction due to atmospheric gases and aeroticulates, the spectral characteristics of the transitions involved, and the optimum values of the signal-to-noise ratio are taken into consideration, it appears that UV laser soundings might work best at stratospheric altitudes, while IR probing would be more suitable for the troposphere.

In Section 7.2 we introduced the basic differential-absorption lidar equation. For the purpose of streamlining the error analysis to be undertaken in this section we shall simplify the notation. We rewrite equation (7.25) using this new notation:

$$\int_0^{R_1} N(R)\sigma_{12}^A \, dR = \tfrac{1}{2}\ln\left[\frac{P_{21}\xi(\lambda_1)\beta_{11}}{P_{11}\xi(\lambda_2)\beta_{21}}\right] - \int_0^{R_1} \bar{\kappa}_{12} \, dR \tag{8.76}$$

where the two wavelengths are now denoted by λ_1 and λ_2 (corresponding to λ_0 and λ_W, respectively), $\sigma_{12}^A \equiv \sigma^A(\lambda_1) - \sigma^A(\lambda_2)$, and $\bar{\kappa}_{12} \equiv \bar{\kappa}(\lambda_1) - \bar{\kappa}(\lambda_2)$. We define P_{jk} as the backscattered laser power at wavelength λ_j ($j = 1$ or 2) from range R_k ($k = 1$ or 2), and β_{jk} as the volume backscattering coefficient corresponding to wavelength λ_j and range R_k.

If we define the molecular number density, averaged over the range increment $\Delta R \equiv R_2 - R_1$, by the relation

$$N \equiv \frac{1}{\Delta R}\int_{R_1}^{R_2} N(R) \, dR \tag{8.77}$$

then from equation (8.76) it follows that

$$N = \frac{1}{2\sigma_{12}^A \, \Delta R}\left[\ln\left\{\frac{P_{21}P_{12}}{P_{22}P_{11}}\right\} + \ln\left\{\frac{\beta_{11}\beta_{22}}{\beta_{12}\beta_{21}}\right\} + \tau_\varepsilon\right] \tag{8.78}$$

where

$$\tau_\varepsilon \equiv \int_{R_1}^{R_2} \bar{\kappa}_{12} \, dR \tag{8.79}$$

represents the *differential-extinction optical depth* for the atmosphere exclusive of absorption within the molecular species of interest.

The backscattered laser power P_{jk} can be related to the total number of current carriers, n_{jk}, created within the detector during the period concomitant with the range increment ΔR:

$$P_{jk} = \rho_\lambda n_{jk} - P_n \tag{8.80}$$

where

$$\rho_\lambda \equiv \frac{hc}{\lambda\eta(\lambda)(2\,\Delta R/c)} \tag{8.81}$$

and P_n represents the noise power and is given by the sum of the background radiation power P_b and the instrumental noise power—which in photomultipliers, or devices with internal gain, consists of the equivalent dark-current power P_d.

If we use equation (8.80) in (8.78), we can write

$$N = \frac{1}{2\sigma_{12}^A\,\Delta R}\left[\ln\left\{\frac{(\rho_\lambda n_{21} - P_n)(\rho_\lambda n_{12} - P_n)}{(\rho_\lambda n_{22} - P_n)(\rho_\lambda n_{11} - P_n)}\right\} + \ln\left\{\frac{\beta_{11}\beta_{22}}{\beta_{12}\beta_{21}}\right\} + \tau_\varepsilon\right] \tag{8.82}$$

and the uncertainty (or error) in the measurement of the number density δN can be determined from the propagation-of-errors equation (see section 8.3.1):

$$\delta N^2 = \sum_{j=1}^{2}\sum_{k=1}^{2}\left\{\delta n_{jk}^2\left(\frac{\partial N}{\partial n_{jk}}\right)^2 + \delta\beta_{jk}^2\left(\frac{\partial N}{\partial \beta_{jk}}\right)^2\right\} + \delta\tau_\varepsilon^2\left(\frac{\partial N}{\partial \tau_\varepsilon}\right)^2 \tag{8.83}$$

where

δn_{jk} represents the uncertainty in the number of detector electrons associated with signal jk;

$\delta\beta_{jk}$ represents the uncertainty in the corresponding value of the backscattering coefficient; and

$\delta\tau_\varepsilon$ represents the uncertainty in the differential-extinction optical depth.

It is evident from equation (8.83) that we are assuming a negligible error in the differential absorption cross section, in the coefficient ρ_λ, and in the noise power P_n.

The partial derivatives in equation (8.83) can be evaluated by reference to equation (8.82):

$$\delta N^2 = \frac{1}{(2\sigma_{12}^A\,\Delta R)^2}\left[\sum_{j=1}^{2}\sum_{k=1}^{2}\left\{\frac{\rho_\lambda^2\,\delta n_{jk}^2}{(\rho_\lambda n_{jk} - P_n)^2} + \left(\frac{\delta\beta_{jk}}{\beta_{jk}}\right)^2\right\} + \delta\tau_\varepsilon^2\right] \tag{8.84}$$

If we assume that the number of electrons created within the time $2\,\Delta R/c$ is described by Poisson statistics, then equation (8.45) enables us to write

$$\delta n_{jk}^2 = \frac{n_{jk}}{m_{jk}} \tag{8.85}$$

where m_{jk} represents the *sample number* of laser shots on which this measurement is based. Substitution of equations (8.80) and (8.85) into (8.84) leads to the relation

$$\delta N^2 = \frac{1}{\left(2\sigma_{12}^A\,\Delta R\right)^2}\left[\sum_{j=1}^{2}\sum_{k=1}^{2}\left\{\frac{\rho_\lambda\left(P_{jk}+P_b+P_d\right)}{m_{jk}P_{jk}^2}+\left(\frac{\delta\beta_{jk}}{\beta_{jk}}\right)^2\right\}+\delta\tau_\varepsilon^2\right] \tag{8.86}$$

If we introduce the *local differential-absorption optical depth*,

$$\Delta\tau_A \equiv N\sigma_{12}^A\,\Delta R \tag{8.87}$$

we can write the relative error in the molecular number density

$$\varepsilon_N \equiv \frac{\delta N}{N} = \frac{1}{2\,\Delta\tau_A}\left[\sum_{j=1}^{2}\sum_{k=1}^{2}\left\{\frac{\rho_\lambda\left(P_{jk}+P_b+P_d\right)}{m_{jk}P_{jk}^2}+\left(\frac{\delta\beta_{jk}}{\beta_{jk}}\right)^2\right\}+\tau_\varepsilon^2\left(\frac{\delta\tau_\varepsilon}{\tau_\varepsilon}\right)^2\right]^{1/2} \tag{8.88}$$

There are several limiting situations that one can envisage. One of the most favorable is the signal shot-noise limit, wherein $P_{jk} \gg P_n$. If we further assume that

$$m_{jk}P_{jk} \gg \rho_\lambda\left[\left(\frac{\delta\beta_{jk}}{\beta_{jk}}\right)^2+\tau_\varepsilon^2\left(\frac{\delta\tau_\varepsilon}{\tau_\varepsilon}\right)^2\right]^{-1} \tag{8.89}$$

and that all of the scattering coefficient relative error terms are comparable in magnitude, then

$$\varepsilon_N \approx \frac{1}{2\,\Delta\tau_A}\left[4\left(\frac{\delta\beta}{\beta}\right)^2+\tau_\varepsilon^2\left(\frac{\delta\tau_\varepsilon}{\tau_\varepsilon}\right)^2\right]^{1/2} \tag{8.90}$$

Under these circumstances the minimum detectable density arises for $\delta N/N \approx 1$, namely,

$$N_{\min} \approx \frac{1}{2\sigma_{12}^A\,\Delta R}\left[4\left(\frac{\delta\beta}{\beta}\right)^2+\tau_\varepsilon^2\left(\frac{\delta\tau_\varepsilon}{\tau_\varepsilon}\right)^2\right]^{1/2} \tag{8.91}$$

Uchino et al. (1979) have examined the feasibility of measuring the ozone distribution up to altitudes of around 30 km using a single-wavelength XeCl laser operating at 308 nm. They estimated that for $\Delta\tau_\varepsilon/\tau_\varepsilon \approx \Delta\beta/\beta \approx 10^{-2}$ (determined by radiosonde) the ultimate detectable column density would amount to 10^{17} cm^2 based on equation (8.91) and $\sigma_A(308 \text{ nm}) \approx 1.3 \times 10^{-19}$ cm^2. That is to say, for a 0.5-km range increment, $N_{\text{min}} \approx 2 \times 10^{12}$ cm^{-3}.

Schotland (1974), in the first detailed error analysis of differential-absorption lidar, revealed that accurate spatial-distribution measurements of a gas density required knowledge regarding the uncertainties of many parameters, including the absorption coefficient, the power measurement, the exact laser wavelength, and the atmospheric parameters involved. In the case of water-vapor measurements, Schotland indicated that for observations below 2 km the greatest source of error was instability in the laser wavelength, while above that altitude it was the uncertainty in the power measurements.

In a DIAL system it is very important that the time between consecutive laser pulses at λ_1 and λ_2 be kept to a minimum in order to avoid the influence of atmospheric turbulence upon both the backscattering coefficient and the extinction coefficient. In the early development of such differential-absorption lidar systems a single laser was used for measurements at one wavelength; it was then returned to the second wavelength and other series of measurements undertaken. The averages of the two series of measurements were then compared. Unfortunately, this procedure led to substantial errors due to variations in the atmospheric properties.

Killinger and Menyuk (1981) discussed this problem and undertook a comparison of single-laser and dual-laser DIAL systems. Their results indicated that the observed increase in the measurement accuracy was in agreement with that predicted from a theory which considered the statistical and temporal character of DIAL returns.

8.3.5. Accuracy Optimization for DIAL Systems

The relative error in molecular-number-density measurement by a differential-absorption lidar system was given in equation (8.88). In this section we shall consider the possibility of optimizing the accuracy of such measurements. We shall follow closely the analysis of Remsberg and Gordley (1978) and Megie and Menzies (1980).

Reference to the elastic (power) backscattering lidar equation (7.23) enables us to write the following relationships:

$$P_{11} = P_{22}\mathbf{e}^{-2\tau_A} \tag{8.92}$$

$$P_{12} = P_{22}\mathbf{e}^{-2(\tau_A + \Delta\tau_A)} \tag{8.93}$$

where

$$\tau_A = \int_0^R N(R)\sigma_{12}^A \, dR \tag{8.94}$$

represents the integrated differential absorption optical depth and $\Delta\tau_A$ the local value. In arriving at equations (8.92) and (8.93) we have also assumed that

$$P_{21} = P_{22} \tag{8.95}$$

which implies that

$$1 \approx \exp\left(2\int_{R_1}^{R_2} \bar{\kappa}(R)\, dR + 2\int_{R_1}^{R_2} N(R)\sigma^A(\lambda_2)\, dR\right)$$

and that the quantity $\xi(\lambda)\beta(\lambda, R)\xi(R)/R^2$ is fairly insensitive to small changes in λ (i.e., from λ_1 to λ_2) and R (i.e., from R_1 to R_2). If we also introduce

$$x_p \equiv \frac{P_n}{P_{22}} \tag{8.96}$$

$$K_\tau \equiv \frac{\tau_A}{\Delta\tau_A} \tag{8.97}$$

and assume that we can neglect the uncertainties in β and τ_ε, then the relative error in the molecular density is

$$\varepsilon_N \equiv \frac{\delta N}{N} = \frac{K_\tau}{2\tau_A}\left[\frac{\rho_\lambda}{m_1 P_{22}}\left\{e^{2\tau_A}\left(1 + e^{2\tau_A/K_\tau}\right) + x_p e^{4\tau_A}\left(1 + e^{4\tau_A/K_\tau}\right)\right\} + \frac{2\rho_\lambda(1 + x_p)}{m_2 P_{22}}\right]^{1/2} \tag{8.98}$$

where we have also assumed that, $m_{11} = m_{12} = m_1$ and $m_{21} = m_{22} = m_2$.

The signal-to-noise ratio for the laser return from range R_2 at the off (or wing) wavelength λ_2 is

$$(\mathrm{SNR})_{22} \equiv \frac{n_{22}}{\delta n_{22}} = \left[\frac{P_{22}}{\rho_\lambda(1 + x_p)}\right]^{1/2} \tag{8.99}$$

Substitution of this relation into equation (8.98) yields

$$\varepsilon_N = \frac{K_\tau}{2\tau_A(\mathrm{SNR})_{22}(1 + x_p)^{1/2}}\left[\frac{1}{m_1}\left\{e^{2\tau_A}\left(1 + e^{2\tau_A/K_\tau}\right) + x_p e^{4\tau_A}\left(1 + e^{4\tau_A/K_\tau}\right)\right\} + \frac{2(1 + x_p)}{m_2}\right]^{1/2} \tag{8.100}$$

If we assume that we keep the total number of laser shots constant, (i.e., $m_1 + m_2 = m_0$), we can minimize ε_N by taking the derivative of equation (8.100) with respect to m_1. Furthermore, if we assume $K_\tau \gg 3$ (not unreasonable if results are to be range-resolved), then

$$\mathbf{e}^{2\tau_A/K_\tau} \approx \mathbf{e}^{4\tau_A/K_\tau} \cong 1$$

and we can write

$$\varepsilon_N^2 = (\varepsilon_N^*)^2 \left\{ \frac{a}{m_1} + \frac{b}{m_0 - m_1} \right\} \tag{8.101}$$

where

$$\varepsilon_N^* \equiv \frac{K_\tau}{2\tau_A (\mathrm{SNR})_{22} (1 + x_p)^{1/2}} \tag{8.102}$$

$$a = 2\mathbf{e}^{2\tau_A}\left(1 + x_p \mathbf{e}^{2\tau_A}\right) \tag{8.103}$$

$$b = 2(1 + x_p) \tag{8.104}$$

Differentiation of equation (8.101) with respect to m_1 leads to

$$2\varepsilon_N \frac{d\varepsilon_N}{dm_1} = (\varepsilon_N^*)^2 \left\{ -\frac{a}{m_1^2} + \frac{b}{(m_0 - m_1)^2} \right\} \tag{8.105}$$

Clearly, ε_N is a minimum for

$$\frac{m_1}{m_2} = \sqrt{\frac{a}{b}} = \mathbf{e}^{\tau_A} \left[\frac{1 + x_p \mathbf{e}^{2\tau_A}}{1 + x_p} \right]^{1/2} \tag{8.106}$$

which leads to

$$\left(\frac{m_1}{m_2} \right)_{\mathrm{opt}} = \mathbf{e}^{\tau_A} \tag{8.107}$$

in thc shot-noise limit, when $x_p = 0$, and to

$$\left(\frac{m_1}{m_2} \right)_{\mathrm{opt}} = \mathbf{e}^{2\tau_A} \tag{8.108}$$

in the background-noise limit, when $x_p \to \infty$. These values for the optimum ratio of laser shots on and off the line can be quite different from the usual experimental choice of unity.

In order to determine the optimum value for the differential-absorption optical depth, we differentiate equation (8.101) with respect to τ_A, assuming an appropriate value for m_1/m_2. If we then equate $d\varepsilon_N/d\tau_A$ to zero, we arrive at a transcendental equation of the form

$$\tau_A = \frac{e^{2\tau_A}\left(1 + x_p e^{2\tau_A}\right) + \left(1 + x_p\right)m_1/m_2}{e^{2\tau_A}\left(1 + 2x_p e^{2\tau_A}\right)} \tag{8.109}$$

Megie and Menzies (1980) have evaluated this relation to obtain the optimum value of the differential-absorption optical depth as a function of the parameter x_p ($\equiv P_n/P_{22}$) for two values of the *on*- to *off*-line laser-shot ratio m_1/m_2. Their results are presented as Fig. 8.9. It is evident that for $m_1 = m_2$, the signal shot-noise ($x_p = 0$) limiting value of τ_A^{opt} corresponds to 1.1. Under these circumstances,

$$\frac{\delta N}{N} \approx \frac{2.87K_\tau}{(\text{SNR})_{22}} \frac{1}{\sqrt{m_0}} \tag{8.110}$$

where m_0 represents the total number of laser shots (on-line + off-line).

On the other hand, if we assume $m_1 = m_2 e^{\tau_A}$, then in the signal shot-noise limit, $\tau_A^{\text{opt}} = 1.28$ and

$$\frac{\delta N}{N} \approx \frac{2.54K_\tau}{(\text{SNR})_{22}} \frac{1}{\sqrt{m_0}} \tag{8.111}$$

which corresponds to a 10% increase in accuracy over that given by equation (8.110). However, it is questionable under most conditions of interest whether this small improvement in accuracy is worth the added technical difficulties.

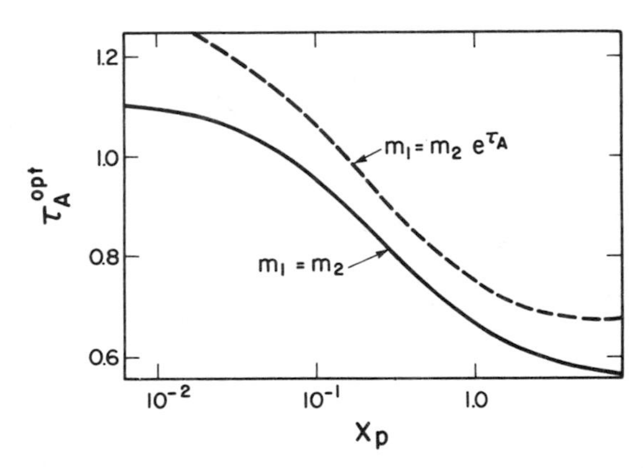

Fig. 8.9. Optimum absorption optical depth τ_A^{opt} as a function of the ratio of the noise power to the off-line signal power, x_p ($\equiv P_n/P_{22}$), for two values of the *on*- to *off*-line laser-shot ratio, m_1/m_2 (Megie and Menzies, 1980).

Indeed, there are many other considerations that could be of greater importance.

The wavelength dependence of K_τ (the ratio of the integrated to the local differential-absorption optical depth), determined by the range resolution of the measurement, is an important consideration in the accuracy optimization. In the ultraviolet region of the spectrum, Doppler broadening of the lines (arising from vibronic transitions) tends to make the absorption almost continuous. By comparison, pressure broadening tends to dominate Doppler broadening in the infrared part of the spectrum—at least in the troposphere—and this leads to a well-defined spectral structure for the vibrational–rotational transitions of molecules.

Megie and Menzies (1980) illustrate this point for ozone by evaluating K_τ for different altitudes as a function of frequency offset from the center of the $P(12)$ CO_2 laser line in the 9.6-μm band. Their results are presented as Fig. 8.10. The corresponding K_τ values for UV excitation at around 300 nm are also provided in Fig. 8.10.

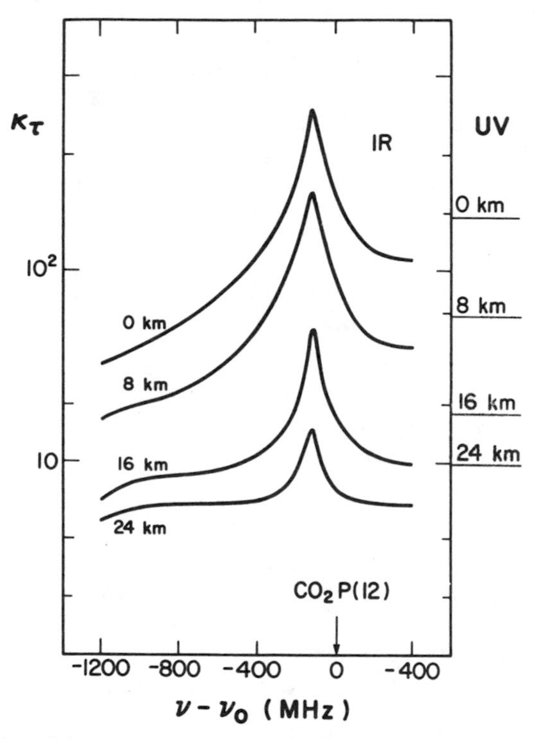

Fig. 8.10. Variation of the ratio of integrated to local absorption optical depth, κ_τ ($\equiv \tau_A/\Delta\tau_A$), for O_3 as a function of the frequency offset from the center of the $P(12)$ CO_2 laser line in the 9.6-μm band for several altitudes. The corresponding frequency-independent κ_τ-values at around 300 nm are also presented on the same scale (Megie and Menzies, 1980).

8.4. LASER-LINEWIDTH LIMITATIONS OF DIAL

For differential-absorption lidar measurements undertaken at ultraviolet wavelength it is not too difficult to arrange for the laser spectral width to be narrow in comparison with that of the molecular constituents under investigation. As we have indicated, DIAL measurements in the infrared are viewed as an important area of growth, due to the large selection of molecules having suitable vibrational–rotational absorption bands in that part of the spectrum. Unfortunately, the spectral width of such lines tend to be about two orders of magnitude narrower than those associated with UV wavelength (electronic) transitions, and this means that the laser linewidth is often comparable to that of the absorption feature.

Consequently, the laser radiation used for remote sensing at infrared wavelengths can no longer be regarded as monochromatic, and the possibility of systematic errors has to be considered. Equation (8.78) serves as the starting point of this analysis. If we assume that the changes in wavelength are sufficiently small that we may neglect variations in both the extinction and backscattering coefficients, then from equation (8.87) we may write

$$\Delta\tau_A = \tfrac{1}{2}\ln\left\{\frac{P_{21}P_{12}}{P_{22}P_{11}}\right\} \tag{8.112}$$

Monochromaticity of the laser emission was implicitly assumed in deriving this equation. If we relax this assumption, the elastic lidar equation, from which equation (8.112) was derived, has to be reconsidered. For a broadband laser with its emission centered at λ_1, equation (7.23) has to be rewritten into the form

$$P(\lambda_1, R) = \frac{c\tau_L}{2}\frac{A_0\xi(R)}{R^2}\int_{\lambda_1-\Delta\lambda_0/2}^{\lambda_1+\Delta\lambda_0/2} d\lambda\, \xi(\lambda)\beta(\lambda, R)P_L(\lambda)\mathrm{e}^{-2\int_0^R \kappa(\lambda, R)\,dR} \tag{8.113}$$

where $P_L(\lambda)$ represents the *power spectral distribution* emitted from the laser and $\Delta\lambda_0$ the receiver system's transmission linewidth. If P_0 represents the spectrally integrated laser power, it follows that

$$P_0 = \int P_L(\lambda)\,d\lambda \tag{8.114}$$

If we assume that we can divide the total attenuation coefficient into two parts, as indicated in equation (7.27), and that the range of wavelength integration in equation (8.113) is sufficiently narrow that we may treat both the backscattering coefficient and the extinction coefficient (exclusive of the species absorption component) as constants, then we can write

$$P(\lambda_1, R) = \frac{c\tau_L}{2}\frac{A_0\xi(R)\beta(\lambda_1, R)}{R^2}\mathrm{e}^{-2\int_0^R \bar{\kappa}(\lambda_1, R)\,dR} \times\left[\int_{\lambda_1-\Delta\lambda_0/2}^{\lambda_1+\Delta\lambda_0/2} d\lambda\, \xi(\lambda)P_L(\lambda)\mathrm{e}^{-2\int_0^R \kappa_A(\lambda, R)\,dR}\right] \tag{8.115}$$

where

$$\kappa_A(\lambda, R) \equiv N(R)\sigma^A(\lambda) \tag{8.116}$$

represents the absorption coefficient due to the species of interest.

If we use this form of the lidar equation in equation (8.112) and employ the same notation that we introduced earlier, then it is easily seen that

$$\frac{P_{21}P_{12}}{P_{22}P_{11}} = \frac{\beta_{21}\beta_{12}}{\beta_{22}\beta_{11}} e^{-2\tau_e} \frac{\int^{(\lambda_2, \Delta\lambda_0)} d\lambda\, \xi(\lambda) P_L(\lambda) \exp\left(-2\int_0^{R_1} \kappa_A(\lambda, R)\, dR\right)}{\int^{(\lambda_2, \Delta\lambda_0)} d\lambda\, \xi(\lambda) P_L(\lambda) \exp\left(-2\int_0^{R_2} \kappa_A(\lambda, R)\, dR\right)}$$
$$\times \frac{\int^{(\lambda_1, \Delta\lambda_0)} d\lambda\, \xi(\lambda) P_L(\lambda) \exp\left(-2\int_0^{R_2} \kappa_A(\lambda, R)\, dR\right)}{\int^{(\lambda_1, \Delta\lambda_0)} d\lambda\, \xi(\lambda) P_L(\lambda) \exp\left(-2\int_0^{R_1} \kappa_A(\lambda, R)\, dR\right)} \tag{8.117}$$

where we have used the notation $(\lambda_j, \Delta\lambda_0)$ to indicate integration over the wavelength interval, $\lambda_j + \Delta\lambda_0/2$ to $\lambda_j - \Delta\lambda_0/2$.

We shall restrict our attention to a range increment $\Delta R \equiv R_2 - R_1$ and a wavelength interval, $\lambda_2 + \Delta\lambda_0/2$ to $\lambda_1 - \Delta\lambda_0/2$, that are sufficiently small that

$$\frac{\beta_{21}\beta_{12}}{\beta_{22}\beta_{11}} e^{-2\tau_e} \approx 1$$

In order to ascertain the influence of both the laser linewidth and the receiver spectral window upon a differential-absorption lidar measurement, we assume that all three spectral functions $\sigma^A(\lambda)$, $P_L(\lambda)$, and $\xi(\lambda)$ are Gaussian, namely,

$$\sigma^A(\lambda) = \sigma_A G_A^A(\lambda) \tag{8.118}$$

$$P_L(\lambda) = \begin{cases} P_L G_1^L(\lambda) & \text{for} \quad \lambda_1 - \Delta\lambda_0/2 < \lambda < \lambda_1 + \Delta\lambda_0/2 \\ P_L G_2^L(\lambda) & \text{for} \quad \lambda_2 - \Delta\lambda_0/2 < \lambda < \lambda_2 + \Delta\lambda_0/2 \end{cases} \tag{8.119}$$

and

$$\xi(\lambda) = \begin{cases} \xi_0 G_1^0(\lambda) & \text{for} \quad \lambda_1 - \Delta\lambda_0/2 < \lambda < \lambda_1 + \Delta\lambda_0/2 \\ \xi_0 G_2^0(\lambda) & \text{for} \quad \lambda_2 - \Delta\lambda_0/2 < \lambda < \lambda_2 + \Delta\lambda_0/2 \end{cases} \tag{8.120}$$

where

$$G_\beta^\alpha(\lambda) \equiv \frac{1}{\Delta_\alpha \sqrt{\pi}} e^{-[(\lambda_\beta - \lambda)/\Delta_\alpha]^2} \tag{8.121}$$

$$\Delta_\alpha \equiv \frac{\Delta\lambda_\alpha}{(\ln 2)^{1/2}} \tag{8.122}$$

and $\Delta\lambda_\alpha$ represents the appropriate HWHM linewidth. In (8.121) and (8.122) α and β represent either A, L, 2, 1 or 0.

Under these conditions equation (8.117) becomes

$$\frac{P_{21}P_{12}}{P_{22}P_{11}} = \frac{\int^{(\lambda_2,\,\Delta\lambda_0)} d\lambda \exp\left\{-\left(\frac{\lambda_2-\lambda}{\Delta}\right)^2 - 2\tau_1^A \mathbf{e}^{-[(\lambda_A-\lambda)/\Delta_A]^2}\right\}}{\int^{(\lambda_2,\,\Delta\lambda_0)} d\lambda \exp\left\{-\left(\frac{\lambda_2-\lambda}{\Delta}\right)^2 - 2\tau_2^A \mathbf{e}^{-[(\lambda_A-\lambda)/\Delta_A]^2}\right\}}$$

$$\times \frac{\int^{(\lambda_1,\,\Delta\lambda_0)} d\lambda \exp\left\{-\left(\frac{\lambda_1-\lambda}{\Delta}\right)^2 - 2\tau_2^A \mathbf{e}^{-[(\lambda_A-\lambda)/\Delta_A]^2}\right\}}{\int^{(\lambda_1,\,\Delta\lambda_0)} d\lambda \exp\left\{-\left(\frac{\lambda_1-\lambda}{\Delta}\right)^2 - 2\tau_1^A \mathbf{e}^{-[(\lambda_A-\lambda)/\Delta_A]^2}\right\}} \tag{8.123}$$

where we have introduced the effective *lidar spectral width* Δ, defined by the relation

$$\frac{1}{\Delta^2} = \frac{1}{\Delta_L^2} + \frac{1}{\Delta_0^2} \tag{8.124}$$

and the appropriate absorption optical depth

$$\tau_j^A = \int_0^{R_j} N(R)\sigma_A \, dR \tag{8.125}$$

Here, σ_A is the spectrally integrated absorption cross section defined in equation (8.118); thus it should be noted that τ_1^A is not the same as τ_A defined by equation (8.94).

In order to reduce equation (8.123) to a more manageable form we shall use the following expansion:

$$\exp\left[-2\tau_j^A \mathbf{e}^{-[(\lambda_A-\lambda)/\Delta_A]^2}\right] = \sum_{m=0}^{\infty} \frac{\left(-2\tau_j^A\right)^m}{m!} \mathbf{e}^{-m[(\lambda_A-\lambda)/\Delta_A]^2} \tag{8.126}$$

and also assume that λ_1 is selected to coincide with the center wavelength for absorption:

$$\lambda_1 = \lambda_A \tag{8.127}$$

At the same time we shall assume that λ_2 is sufficiently detuned from the absorption line center that

$$\mathbf{e}^{-[(\lambda_A-\lambda)/\Delta_A]^2} \approx 0 \tag{8.128}$$

over the wavelength interval, $\lambda_2 - \Delta\lambda_0/2 < \lambda < \lambda_2 + \Delta\lambda_0/2$. Under these circumstances both $(\lambda_2, \Delta\lambda_0)$ integrals in equation (8.123) cancel and we can write

$$\frac{P_{21}P_{12}}{P_{22}P_{11}} \approx \frac{\displaystyle\sum_{m=0}^{\infty} \frac{\left(-2\tau_2^A\right)^m}{m!} \int^{(\lambda_A, \Delta\lambda_0)} d\lambda\, \mathbf{e}^{-[(\lambda_A-\lambda)/\Delta_m]^2}}{\displaystyle\sum_{m=0}^{\infty} \frac{\left(-2\tau_1^A\right)^m}{m!} \int^{(\lambda_A, \Delta\lambda_0)} d\lambda\, \mathbf{e}^{-[(\lambda_A-\lambda)/\Delta_m]^2}} \tag{8.129}$$

where

$$\frac{1}{\Delta_m^2} \equiv \frac{1}{\Delta^2} + \frac{m}{\Delta_A^2} \tag{8.130}$$

If we introduce

$$y_m \equiv \frac{\lambda - \lambda_A}{\Delta_m} \tag{8.131}$$

then

$$\int^{(\lambda_A, \Delta\lambda_0)} d\lambda\, \mathbf{e}^{-[(\lambda_A-\lambda)/\Delta_m]^2} = \Delta_m \int_{-\Delta\lambda_0/2\Delta_m}^{\Delta\lambda_0/2\Delta_m} dy_m\, \mathbf{e}^{-y_m^2} = \sqrt{\pi}\,\Delta_m \operatorname{erf}\left\{\frac{\Delta\lambda_0}{2\Delta_m}\right\} \tag{8.132}$$

and so

$$\frac{P_{21}P_{12}}{P_{22}P_{11}} = \frac{\displaystyle\sum_{m=0}^{\infty} \frac{\left(-2\tau_2^A\right)^m}{m!} \Delta_m \operatorname{erf}\left\{\frac{\Delta\lambda_0}{2\Delta_m}\right\}}{\displaystyle\sum_{m=0}^{\infty} \frac{\left(-2\tau_1^A\right)^m}{m!} \Delta_m \operatorname{erf}\left\{\frac{\Delta\lambda_0}{2\Delta_m}\right\}} \tag{8.133}$$

The *relative error* ε_L in a DIAL measurement arising from the nonmonochromaticity of the laser emission can be defined by the relation

$$\varepsilon_L = \frac{1}{\Delta\tau_A^*}\left[\Delta\tau_A^* - \tfrac{1}{2}\ln\left\{\frac{P_{21}P_{12}}{P_{22}P_{11}}\right\}\right] \tag{8.134}$$

where $\Delta\tau_A^*$ is the adopted increment of differential-absorption optical depth, the value of which will depend upon the model assumed in calculating the expected DIAL signal return.

Cahen and Megie (1981) have evaluated this DIAL relative error, in the limit of very wide receiver linewidth (i.e., $\Delta_0 \gg \Delta_L$ or Δ_A), for three models of the system. In the limit of infinite receiver spectral window equation (8.133) reduces to the form

$$\frac{P_{21}P_{12}}{P_{22}P_{11}} = \frac{\displaystyle\sum_{m=0}^{\infty} \frac{(-2)^m\left(\tau_1^A + \Delta\tau^A\right)^m}{m!\{1 + mx_\Delta^2\}^{1/2}}}{\displaystyle\sum_{m=0}^{\infty} \frac{(-2)^m\left(\tau_1^A\right)^m}{m!\{1 + mx_\Delta^2\}^{1/2}}} \tag{8.135}$$

where we introduce the local value of the spectrally integrated absorption optical depth,

$$\Delta\tau^A \equiv \int_{R_1}^{R_2} N(R)\sigma_A \, dR \tag{8.136}$$

and the ratio of the laser to the absorption linewidth,

$$x_\Delta \equiv \frac{\Delta_L}{\Delta_A} \tag{8.137}$$

The three adopted values of the increment of differential-absorption optical depth used by Cahen and Megie (1981) are: (1) $\Delta\tau_A^a = \Delta\tau_A(\lambda_A)$ when the laser emission was assumed to be monochromatic at the center wavelength for absorption; (2) $\Delta\tau_A^b = \langle\Delta\tau_A(\lambda)\rangle$ when the spectral line shape of the laser emission was taken into account in deriving the mean value of the local absorption optical depth over this spectral profile; and (3) $\Delta\tau_A^c = -\frac{1}{2}\ln\langle\mathbf{e}^{-2\,\Delta\tau_A(\lambda)}\rangle$ when the mean value of the local absorption $\mathbf{e}^{-2\,\Delta\tau_A(\lambda)}$ was averaged over the laser spectral profile. These three values translate into the relations

$$\Delta\tau_A^a = \Delta\tau_A(\lambda_A) \tag{8.138}$$

$$\Delta\tau_A^b = \frac{\Delta\tau_A(\lambda_A)}{\{1 + x_\Delta^2\}^{1/2}} \tag{8.139}$$

$$\Delta\tau_A^c = \sum_{m=0}^{\infty} \frac{(-2)^m}{m!} \frac{\Delta\tau_A(\lambda_A)}{\{1 + m x_\Delta^2\}^{1/2}} \tag{8.140}$$

where

$$\Delta\tau_A(\lambda_A) = \frac{1}{\Delta_A}\left\{\frac{\ln 2}{\pi}\right\}^{1/2} \Delta\tau_A \tag{8.141}$$

assuming that $\sigma^A(\lambda_2) \ll \sigma^A(\lambda_A)$.

The variation of the DIAL relative error ε_L with the linewidth-ratio parameter x_Δ for several values of the line-center absorption optical depth, defined by the relation

$$\tau_1^A(\lambda_A) = \frac{1}{\Delta_A}\left\{\frac{\ln 2}{\pi}\right\}^{1/2} \Delta\tau_1^A \tag{8.142}$$

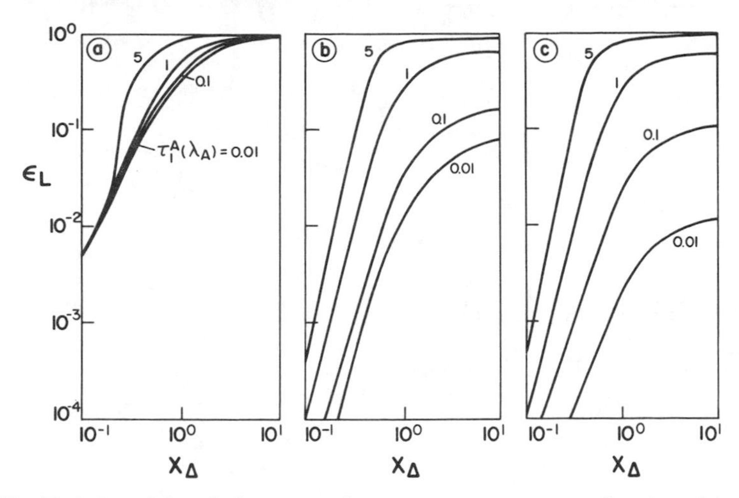

Fig. 8.11. Variation of the relative error ε_L in a DIAL measurement as a function of the ratio of the laser linewidth to the absorption linewidth x_Δ, for several values of the line-center absorption optical depth $\tau_1^A(\lambda_A)$. The three cases (a), (b), and (c) refer to the adopted local differential-absorption optical depth discussed in the text (Cahen and Megie, 1981).

has been evaluated by Cahen and Megie (1981) for the three models proposed above. Their results are presented as Fig. 8.11. As expected, ε_L increases with x_Δ—that is to say, the poorer the assumption of monochromaticity, the greater the error in the measurement of the local optical depth. It is also quite evident that for both cases (b) and (c), ε_L approaches unity for $x_\Delta \gtrsim 1.0$ provided the absorption optical depth $\tau_1^A(\lambda_A) \gtrsim 1$. As we saw earlier (Section 8.3.5), the optimum value for the differential-absorption optical depth tends to lie close to unity for large values of the signal-to-noise ratio (i.e., small x_p, figure 8.9).

This implies that the systematic error associated with a DIAL measurement, and arising from the finite linewidth of the laser emission, cannot be neglected in the infrared region of the spectrum. Cahen and Megie (1981) suggest that accurate tropospheric DIAL measurements at about 10-μm wavelength will require laser bandwidths of about 100 MHz, while in the stratosphere laser bandwidths as low as 10 MHz may be required. Bandwidths of this magnitude can be achieved through the use of intracavity spectral narrowing or injection frequency locking.

9

Atmospheric Lidar Applications

The development of laser-remote sensing techniques holds the promise of substantially improving man's understanding of the environment in which he is immersed. Although the application of lasers to studying the atmosphere is still in its infancy, it is readily apparent that the increased scope and quantity of measurements made possible by this means represents an effective quantum jump in our knowledge of the nature and behavior of the atmosphere.

No attempt has been made to summarize the vast literature of atmospheric lidar measurements within this chapter. Indeed, there would be little point, since the field is advancing to such an extent that anything included would soon be out of date. Consequently, I have chosen to provide a broad overview of the subject that will indicate the range of atmospheric applications that are amenable to remote laser sensing. In this undertaking I am acutely aware of some of the omissions. One of the most significant is atmospheric laser Doppler velocimetry, which includes measurements of wind shear, clear-air turbulence, aircraft wake vortices, tornadoes, severe storms, and global wind patterns. For this subject I can recommend the excellent text of Durst, Melling, and Whitelaw (1976) and the review by Bilbro (1980).

9.1. ATMOSPHERIC STUDIES

The oxygen–nitrogen balance of the atmosphere represents an important measurement in view of the potential reduction in the sources of free oxygen (namely, decline in marine life and vegetation) and the increase in the rate of oxygen consumption (population growth and corresponding increased fossil fuel combustion). Schwiesow and Derr (1970) have indicated that a precise measurement of the O_2–N_2 balance in the atmosphere should be possible using laser Raman scattering. They point out that with this technique a two-order-of-magnitude improvement in precision over other techniques should be possible (leading to an accuracy of 0.3 ppm for the O_2/N_2 ratio and 0.006 ppm for the CO_2/N_2 ratio). This would be adequate to determine the magnitude of any long-term drifts in the oxygen–carbon dioxide–nitrogen balance of the atmosphere.

9.1.1 Density Measurements

Leonard (1967) was the first to use the nitrogen laser to observe Raman backscattering from nitrogen (at 365.9 nm) and oxygen (at 355.7 nm) at a range of around 1 km. Of particular significance at the time was the fact that the peak output of the laser was only 100 kW. Leonard pointed out that in order to achieve similar results with a ruby laser, the peak power would have to be in excess of 21 MW due to the strong wavelength dependence of both the Raman cross section [equation (3.165)] and the quantum efficiency of the detector photocathode, $\eta(\lambda)$. However, care must be used in making such a comparison in view of the increased attenuation (due primarily to elastic scattering) at the shorter wavelength. Although Boudreau (1970) has studied this problem, his conclusions may be limited in their usefulness because he neglects the Mie-scattering contribution to the attenuation coefficient and his calculations do not take account of the improved red-response photocathodes currently available (see Fig. 6.6). Nevertheless, the nitrogen laser's high repetition rate represents another important advantage—commercial N_2 lasers are available that operate at 1000 pps.

The first Raman measurement of the gas density profile was undertaken by Cooney (1968) using a 25-MW, Q-switched ruby laser. The Raman vibrational–rotational return from nitrogen was observed, at night, up to an altitude of 3 km. Cooney used a combination of a 694.3-nm blocking filter and a 15-nm-passband interference filter to provide a net spectral rejection of 10^7. This was more than adequate to overcome the intensity factor of 500 between the elastically backscattered return at 694.3 nm and the nitrogen Raman return at 828.5 nm. Recently, Garvey and Kent (1974) have extended the range of Raman investigations of atmospheric nitrogen well into the stratosphere (to a height of at least 40 km) and have obtained good agreement between their observations, balloon-mounted radiosonde measurements, and the U.S. Standard Atmosphere (U.S. Government Printing Office, Washington, D.C., 1962).

As we have seen in equation (4.84), the relationship between the atmospheric attenuation coefficient and the visual range is at best approximate, and in strong-scattering situations is fraught with ambiguities. Several researchers have endeavored to circumvent this difficulty by attempting to separate the laser return due to the gaseous constituents from that of the aeroticulates. Cooney et al. (1969) and Cooney (1975) attempted to determine the Mie backscattering coefficient using the following relation:

$$\beta^{M}(\lambda_L, R) = \frac{E(\lambda_L, R) - E^{\text{Ray}}(\lambda_L, R)}{E^{\text{Ray}}(\lambda_L, R)} \beta^{\text{Ray}}(\lambda_L, R) \tag{9.1}$$

where

$$E(\lambda_L, R) = E_L \frac{c\tau_d}{2} \frac{A_0}{R^2} \frac{\xi(\lambda_L)}{4\pi} [\beta^{M}(\lambda_L, R) + \beta^{\text{Ray}}(\lambda_L, R)] \mathrm{e}^{-2\int_0^R \kappa(\lambda_L, R)\, dR} \tag{9.2}$$

from equation (7.14), represents the total elastic return from range R at the laser wavelength. $E^{\text{Ray}}(\lambda_L, R)$ is the Rayleigh component of $E(\lambda_L, R)$, while $\beta^{M}(\lambda_L, R)$ and $\beta^{\text{Ray}}(\lambda_L, R)$ are the respective Mie and Rayleigh backscattering coefficients. $\xi(R)$ was set equal to unity in this instance. In general,

$$\beta^{\text{Ray}}(\lambda_L, R) = \sum_i N_i(R) \sigma_i^{\text{Ray}}(\pi, \lambda_L) \tag{9.3}$$

Here, $N_i(R)$ represents the appropriate number density, and $\sigma_i^{\text{Ray}}(\pi, \lambda_L)$ the Rayleigh backscattering cross section for the ith constituent of the atmosphere. In reality there is no direct way of ascertaining the Rayleigh contribution to $E(\lambda_L, R)$. However, Cooney et al. (1969) and Cooney (1975) attempted to evaluate the Rayleigh component from the nitrogen Raman return as follows:

$$E^{\text{Ram}}(\lambda, R) = E_L \frac{c\tau_d}{2} \frac{A_0}{R^2} \frac{\xi(\lambda)}{4\pi} N_\alpha(R) \sigma_\alpha^{\text{Ram}}(\pi, \lambda) \mathrm{e}^{-\int_0^R \{\kappa(\lambda_L, R) + \kappa(\lambda, R)\}\, dR} \tag{9.4}$$

where $N_\alpha(R)$ represents the nitrogen number density at range R and $\sigma_\alpha^{\text{Ram}}(\pi, \lambda)$ the corresponding Raman backscattering cross section. In this case we can write

$$\beta^{M}(\lambda_L, R) = \frac{E(\lambda_L, R) - Y(\lambda_L/\lambda) E^{\text{Ram}}(\lambda, R)}{Y(\lambda_L/\lambda) E^{\text{Ram}}(\lambda, R)} \beta^{\text{Ray}}(\lambda_L, R) \tag{9.5}$$

where

$$Y\left(\frac{\lambda_L}{\lambda}\right) = \sum_i \left[\frac{N_i(R) \sigma_i^{\text{Ray}}(\pi, \lambda_L)}{N_\alpha(R) \sigma_\alpha^{\text{Ram}}(\pi, \lambda)}\right] \frac{\xi(\lambda_L)}{\xi(\lambda)} \mathrm{e}^{-\int_0^R \{\kappa(\lambda_L, R) - \kappa(\lambda, R)\}\, dR} \tag{9.6}$$

and can be assumed to be independent of R if

$$\kappa(\lambda_L, R) \approx \kappa(\lambda, R) \tag{9.7}$$

and the mixing ratio of the gaseous constituents is a constant, independent of altitude.

Whereas Cooney et al. (1969) normalized the elastic return with the nitrogen vibrational–rotational Raman return, Cooney (1975) employed the anti-Stokes wing of the nitrogen pure rotational Raman-backscattered return at 691.2 nm (excitation 694.3 nm) for this purpose. A spectral rejection of 10^5 was found to be adequate for this work. The latter approach has two obvious advantages: (1) the rotational Raman shift is only $\approx 50\ cm^{-1}$, as opposed to $2330\ cm^{-1}$ for the vibrational case, so that the assumption represented by equation (9.7) is more likely to be justified; (2) there should be a marked improvement in the magnitude of the inelastic signal due to both a larger cross section and a better photocathode response. However, it should be noted that care must be used in choosing the filter characteristics in order to avoid excessive temperature sensitivity. It is clear that the beauty of this scheme lies in the way the Raman measurement eliminates the need for an absolute calibration. Indeed, Leonard and Caputo (1974) have shown that it is possible to build a reliable monostatic atmospheric transmissometer based upon the measurement of the nitrogen Raman-backscattered signal.

9.1.2. Water-Content Measurements

The first remote laser measurement of the vertical water-vapor profiles in the atmosphere was based on differential absorption and scattering and was made by Schotland (1966) using a thermally tuned ruby laser. The earliest laser Raman measurements to yield the spatial distribution of water vapor in the atmosphere were performed by Melfi et al. (1969) and Cooney (1970). Each used a frequency-doubled, Q-switched ruby laser and normalized their water-vapor return with the nitrogen vibrational Raman return. Melfi et al. (1969) and Cooney (1971) were also able to demonstrate good agreement between their lidar-evaluated profiles and measurements undertaken by radiosondes. A representative set of backscattered profiles for nitrogen, water vapor, and aerosols is presented in Fig. 9.1, while an example of the close agreement obtained between the lidar water-vapor ratio and the balloon sonde is shown in Fig. 9.2, taken from Melfi (1972). Similar results have also been obtained by Pourny et al. (1979).

Although the general agreement obtained in the above work was excellent, the comparisons were made rather far apart in space and time. Strauch et al. (1972) avoided this problem by making a direct comparison of the Raman lidar measurements with that of a standard humidity meter mounted on a tower some 30 m above the ground. Their results show an excellent correlation between the two measurements and indicate that their system should be capable of determining the water-vapor profile to a range of about 4 km. An interesting observation made by these authors is that 365.4-nm radiation from mercury street lights falls very close to the N_2 Raman line at 365.8 nm (when excited with the nitrogen laser at 337.1 nm). They also point out that although

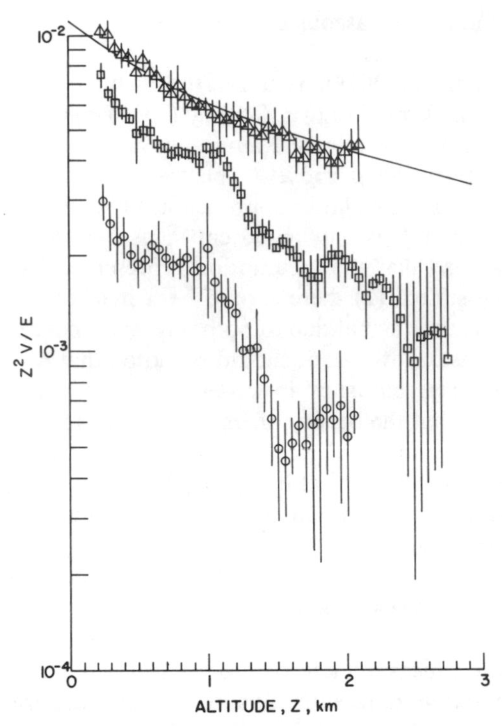

Fig. 9.1. Typical backscattering profiles normalized for range Z and laser energy E (Z^2V/E) from the atmosphere: △, nitrogen, average of eight oscillograms; ○, water vapor, average of six oscillograms; □, aerosol, average of six oscillograms, with two No. 1 n.d. filters; (solid curve), calculated return for nitrogen (Z^2V). V represents the voltage signal from the photomultiplier (Melfi, 1972).

allowance can be made for leakage of the N_2 Raman line into the H_2O observation, broadband emission from the laser discharge and fluorescence induced in the optics, the interference filters, and dust on the mirror surfaces by near-field laser scattered radiation can present serious sources of error and must be carefully avoided. Laser-induced fluorescence of near-field aerosols (Gelbwachs and Birnbaum, 1973) can also provide false signals unless adequate spectral discrimination is provided.

Most Raman-based lidar measurements are restricted to nighttime operation because of the strong daytime sky background radiance. One way of avoiding this form of noise is to operate between 230 and 300 nm. Stratospheric ozone absorbs the incoming solar radiation within this spectral interval, and consequently this is termed the *solar-blind* region of the spectrum. Unfortunately, operation at these wavelengths is something of a double-edged sword—for the absorption by ozone that is responsible for the solar-blind region also causes attenuation of both the outgoing laser pulse and the

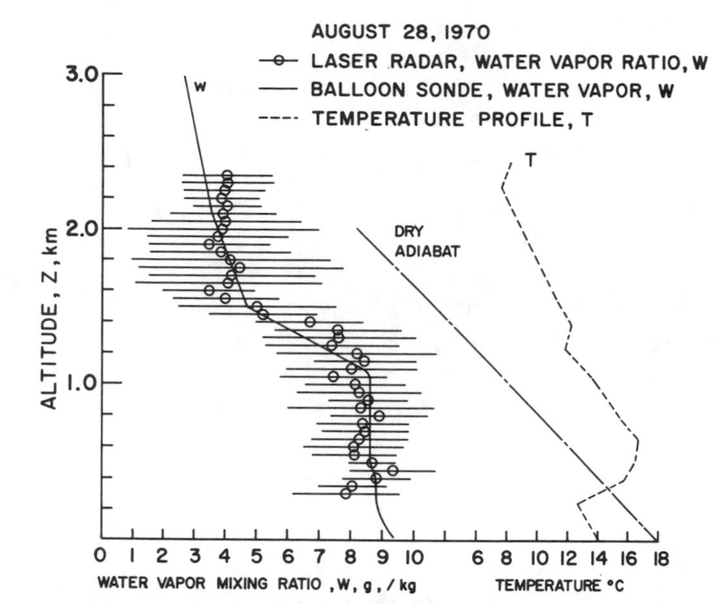

Fig. 9.2. Lidar measurement of water-vapor mixing ratio compared with standard balloon-sonde data (Melfi, 1972).

Raman-backscattered return. This problem is further aggravated by the strong wavelength dependence (Hartley bands) of this ozone absorption. Renaut et al. (1980) and Petri et al. (1982) have both attempted to use multiwavelength, solar-blind Raman lidars to remotely measure atmospheric water-vapor content and temperature. Renaut et al. (1980) employed a quadrupled Nd–YAG laser while Petri et al. (1982) compared the results from such a laser with that of a doubled dye laser and with that of two excimer lasers. They concluded that the convenience and tunability afforded by the dye laser permitted optimization of the SNR. Unfortunately the solar-blind region possesses the greatest eye hazard (see Section 5.9), and this will probably limit its use in laser remote sensing.

To date all measurements of the atmospheric water-vapor content based on Raman scattering have been limited to an operational range of less than 2 km. Although this could be extended with higher-power lasers and more sophisticated detection schemes (particularly better spectral rejection of background radiation), eye-safety considerations (especially in the solar-blind region) will probably limit the improvement attained in practice. On the other hand, elastic (Rayleigh and Mie) backscattering is much more intense (typically 10^5 to 10^7 times greater) than the H_2O Raman backscattering (see Fig. 9.3) and lends itself to a much better method of long-range concentration mapping through differential absorption and scattering (Schotland, 1966).

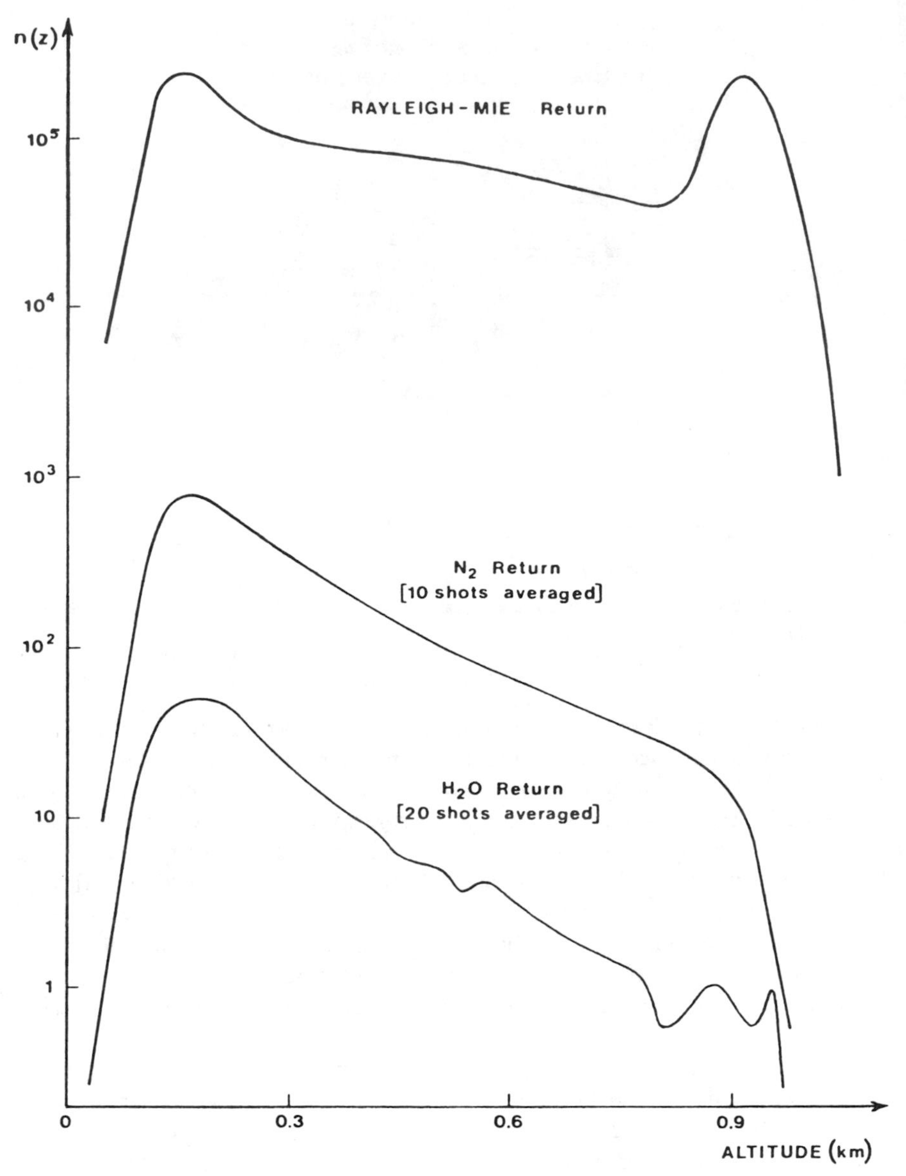

Fig. 9.3. Elastic backscattering and Raman lidar signals (photoelectron count per 30-m range increment) obtained in the presence of a stable status layer (Pourny et al., 1979).

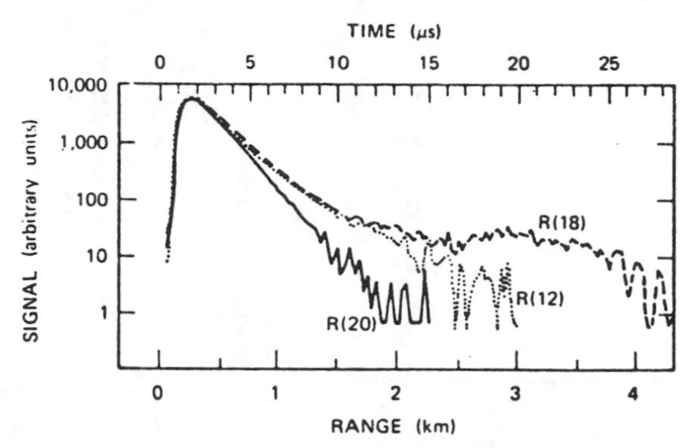

Fig. 9.4. Lidar backscattering signal for $R(12)$, $R(18)$, and $R(20)$ lines in the 10-μm band of a CO_2 laser. The $R(20)$ is more strongly absorbed by ambient H_2O vapor than is $R(12)$ or $R(18)$ (Murray et al., 1976).

Murray et al. (1976) used a 1-J/pulse CO_2 TEA laser to measure the water-vapor content of the atmosphere over a horizontal range of about 1 km using the DAS technique. The differential absorption between the R(20), R(12), and R(18) rotational–vibrational transitions of the CO_2 laser is clearly seen in Fig. 9.4. The corresponding water-vapor concentration profile, deduced from 100 paired pulses on the R(18) and R(20) lines, is shown as Fig. 9.5. It is apparent from these figures that, although the two return signals appear free of noise to about 1.5 km, the derived concentration becomes noisy around the 1-km range.

A more sensitive differential-absorption lidar (DIAL) operating in the near infrared—on the 724.37-nm H_2O absorption line—was developed by Browell

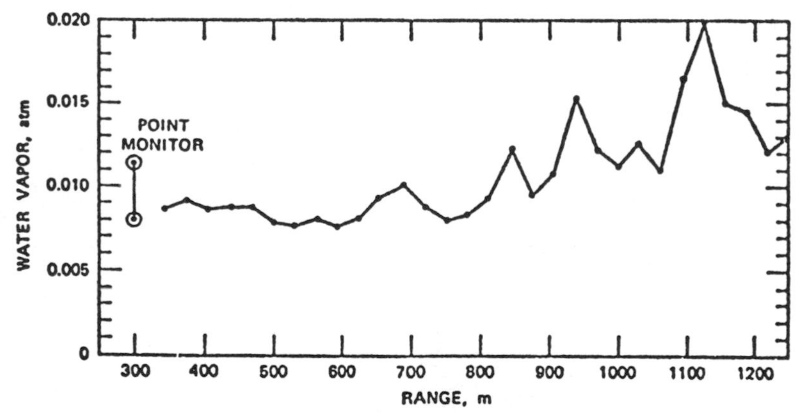

Fig. 9.5. Measured value of water-vapor concentration as a function of range. A 60-m-range cell was used in the data reduction. The point monitor was a calibrated recording hygrothermograph located near the laser line of sight (Murray et al., 1976).

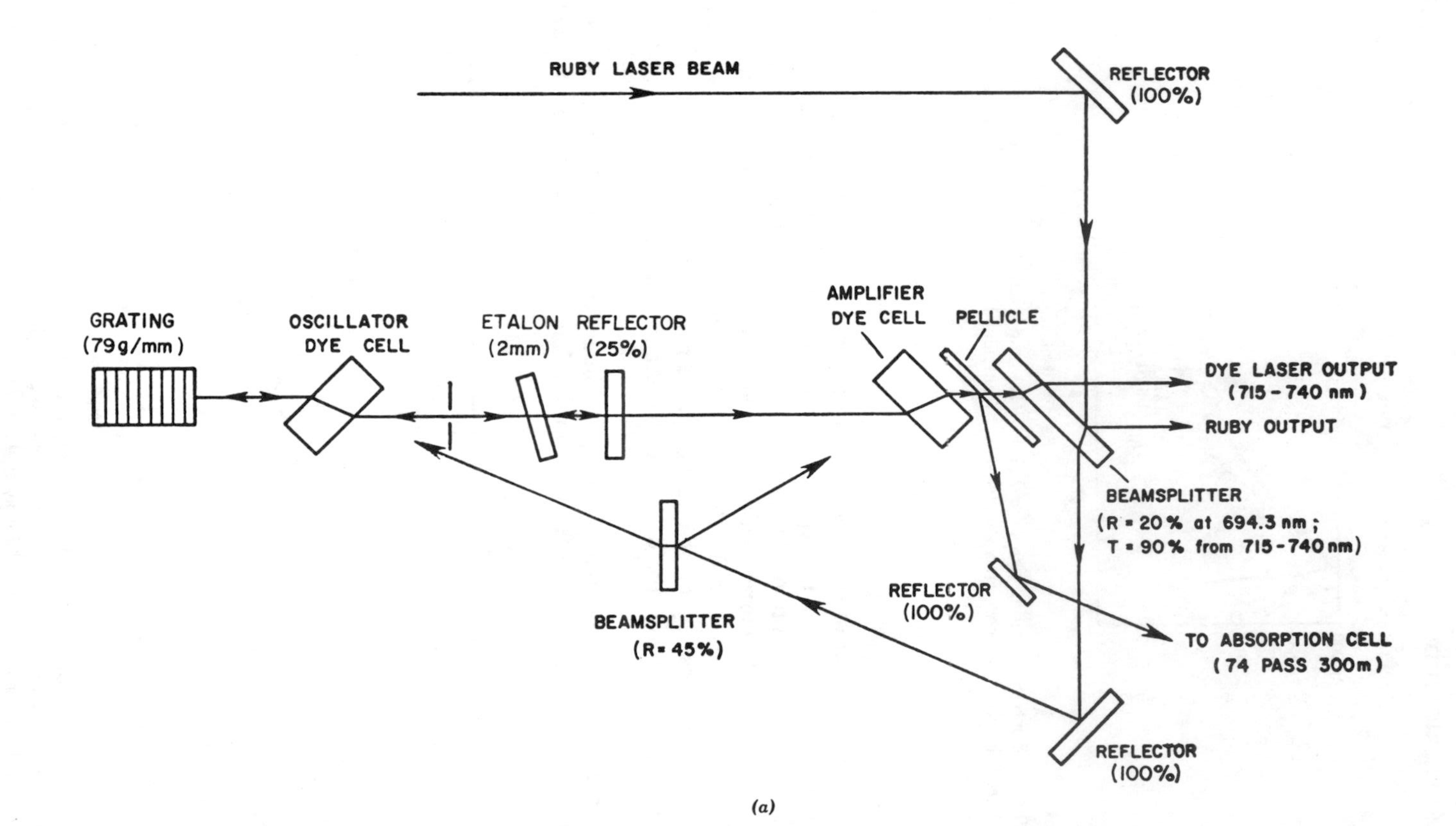

(a)

Fig. 9.6. (a) Schematic diagram of the ruby-laser-pumped dye-laser transmitter of the NASA water vapor DIAL system; (b) photograph of the transmitter (Browell et al., 1979).

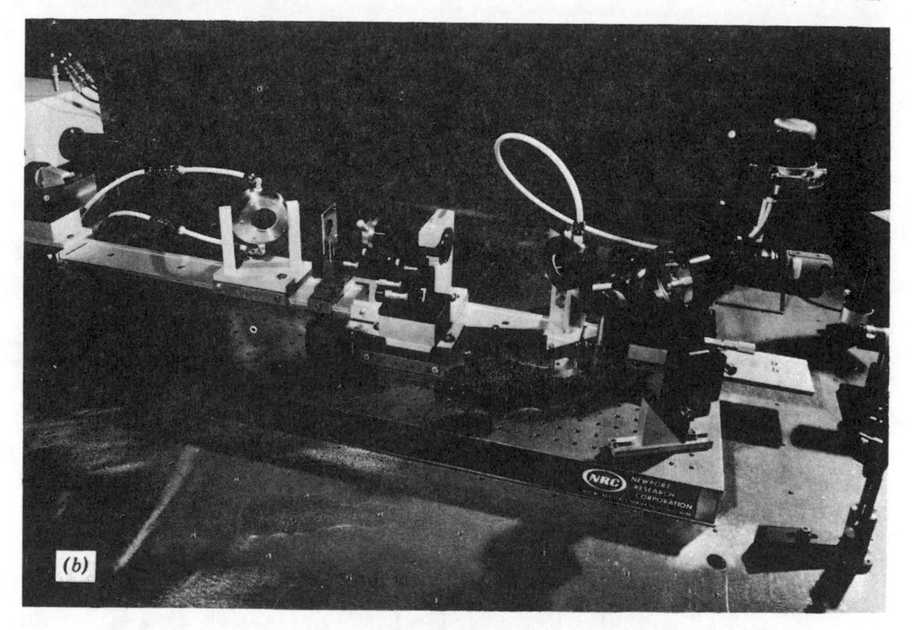

Fig. 9.6. (*Continued*)

et al. (1979). Their system employed a 1.5-J *Q*-switched (Holobeam) ruby laser to provide simultaneously both the off-line laser pulse (at 694.3 nm) and the on-line laser pulse (at 724.37 nm). This was accomplished through the use of a beamsplitter which allowed 0.25 J to be transmitted and the remaining 1.25 J to pump a dye oscillator–amplifier arrangement. A 1.2×10^{-4} *M* solution of DTDC in dimethyl sulfoxide was used to provide tuning between 715 and 740 nm, and a grating-plus-etalon cavity configuration led to an output of about 0.165 J (with a divergence of less than 3.5 mrad) in a 30-ns pulse of approximately 0.008-nm (8-pm) linewidth.

Schematics and photographs of their laser arrangement and receiver system are presented as Figs. 9.6 and 9.7 respectively. RCA-7265 photomultipliers were employed with two Biomation model 8100 transient digitizers. An important feature of this DIAL system was the incorporation of a multipass H_2O absorption cell for exact calibration of the dye laser's spectral output on *each shot*. This determination of the H_2O absorption for each dye-laser pulse eliminates a major source of uncertainty in such DIAL measurements. A scan of the appropriate region of the H_2O absorption spectrum obtained with this system was previously shown as Fig. 3.27. A representative vertical atmospheric H_2O profile (with comparison radiosonde data) is presented as Fig. 9.8. This successful demonstration of a ground-based system represented the initial step in the development of an airborne and ultimately Shuttle-borne H_2O DIAL.

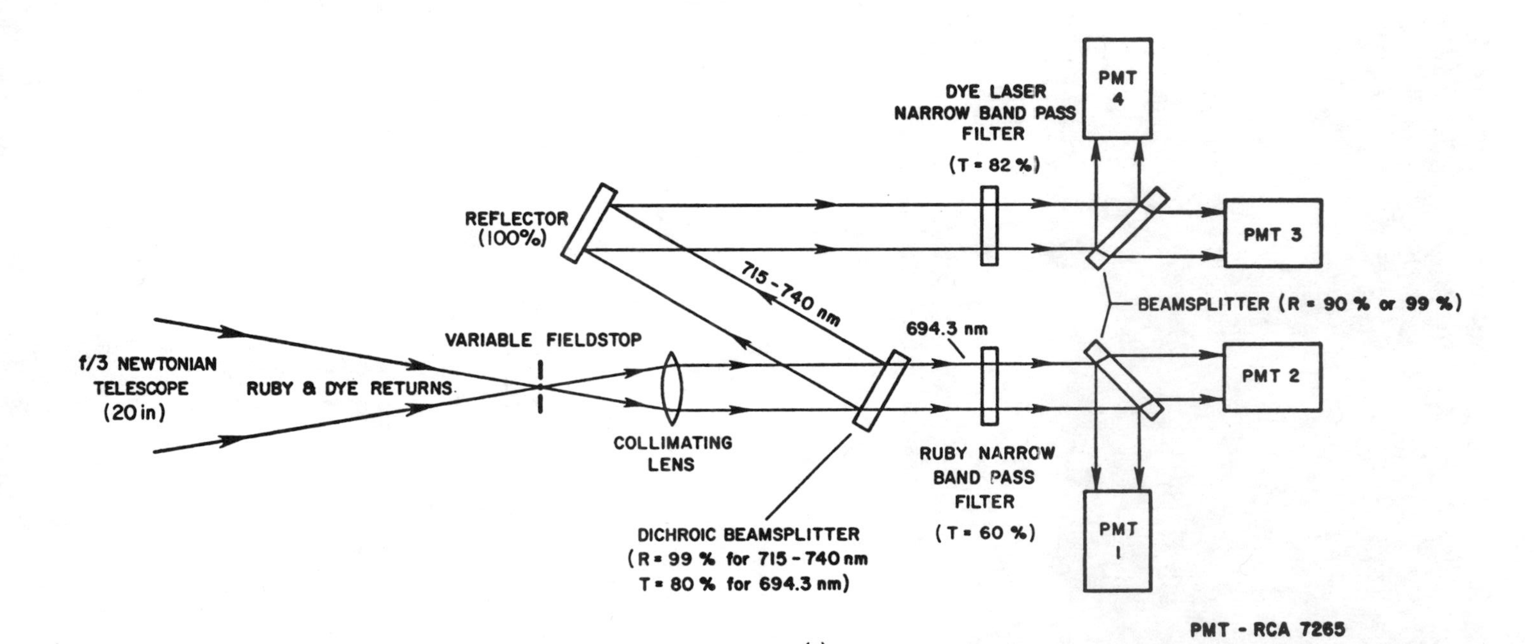

(a)

Fig. 9.7. (a) Schematic diagram and (b) photograph of the NASA water-vapor DIAL detection system (Browell et al., 1979).

Fig. 9.7. (*Continued*)

An aircraft-mounted H_2O DIAL would allow important studies of macro- and micrometeorology, air-mass modification over bodies of water, aerosol growth, and tropospheric–stratospheric exchange mechanism; Shuttle-derived H_2O profiles would provide inputs for meteorological forecasting models and improve our understanding of atmospheric radiative processes. Browell (1982b) has reported on the successful preliminary trials of an airborne DIAL system that has measured vertical water-vapor profiles with a flight range resolution of about 4 km.

9.1.3. Temperature Measurements

Temperature profiles are obviously important in both climate modeling and weather forecasting. For example, the severity of storms is often related to the temperature lapse rate, and the cloud ceiling and visibility are influenced by a combination of humidity and temperature profiles. It has also been shown that the formation of a temperature inversion over an urban area is often responsible for the most severe pollution episodes.

The idea of using a light beam to determine the atmospheric molecular-density profile and thereby derive the temperature profile was first proposed by Elterman (1953, 1954). With a searchlight he attempted to evaluate the temperature between 10 and 67 km by assuming that beyond 10-km altitude

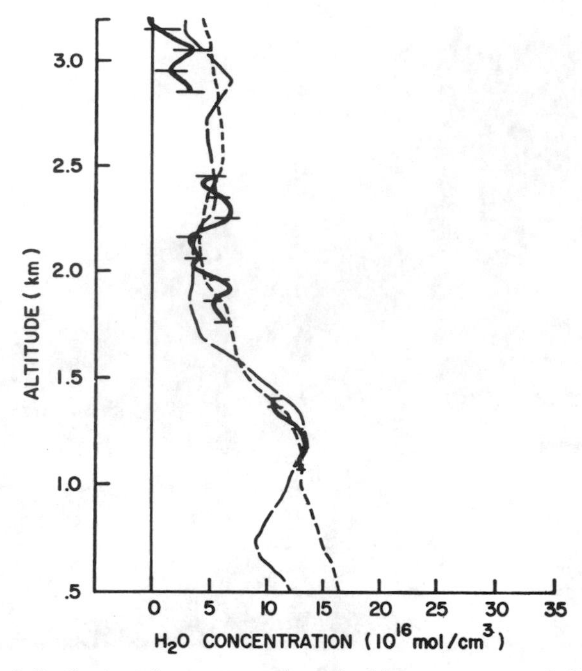

Fig. 9.8. Vertical distribution of water vapor determined by a DIAL system (solid line), involving 100 laser firings, a 100-m-range cell, and a H_2O absorption cross section of 5.2×10^{-23} m^2. Data from rawinsondes launched before (short dashes) and after (long dashes) DIAL measurements are also displayed (Browell et al., 1979).

the return was determined principally by Rayleigh scattering and that the ideal gas law could be combined with a hydrostatic relation to give the temperature in terms of the density change. Sandford (1967) also used elastic scattering, but from a powerful laser, to evaluate the density change and thereby the temperature according to the relation

$$T(z) = \frac{N(z_1)kT(z_1) + \int_z^{z_1} W_m(z)\,dz}{N(z)k} \tag{9.8}$$

where $T(z)$ is the temperature and $N(z)$ is the density at altitude z. $W_m(z)$ is the weight of air per cm^3 (the molecular mass, 4.8×10^{-22} g, times the gravitational acceleration times the density) at z. Note that z_1 represents the maximum altitude. $T(z_1)$ is often estimated and $T(z)$ calculated step by step downward from z_1. The value of T_1 rapidly becomes a little consequence. Figure 9.9 presents the upper-atmospheric temperature profile deduced by Sandford (1967) and found to be in reasonable agreement with the U.S. Standard Atmosphere.

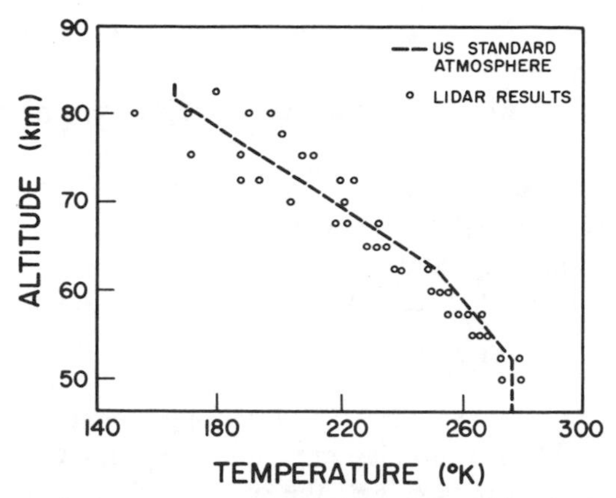

Fig. 9.9. Upper-atmospheric temperature profile deduced from lidar elastic-backscattering measurements at Wakefield, England, by Sandford (1967).

Strauch et al. (1971) have suggested that Raman backscattering from N_2 could be used to determine the molecular-density profile and thereby the atmospheric temperature, given the ground-level pressure. Although we have seen that Raman scattering can be used to evaluate the density profile, this is achieved by normalizing with respect to the nitrogen Raman return so that the effects of aeroticulate attenuation, $1/R^2$ corrections, $\xi(R)$ uncertainties, and E_L fluctuations are all canceled. An absolute measurement of the molecular-density profile would be vulnerable to all these difficulties. Strauch et al. (1971) were able to demonstrate a good correlation between the variations in the amplitude of the Raman return from nitrogen and the temperature fluctuations measured by a thermistor located in a tower.

Cooney (1972) proposed that rotational Raman scattering from N_2 might serve as a more convenient vehicle for ascertaining the temperature profile of the atmosphere. He based his argument on the stronger scattering intensity and the temperature sensitivity of the rotational Raman spectrum.

Cooney (1972) also suggested that a differential technique might be the best way of ensuring adequate temperature sensitivity and also canceling out most of the extraneous factors, such as atmospheric transmission and the spectral response of the photodetector. To see this we write the rotational Raman-backscattered power received by the detection system of a lidar having a detection linewidth $\Delta\lambda_1$ centered about the wavelength λ_1. In recognition of the finite detection linewidth $\Delta\lambda_1$, we replace $\xi(\lambda)$ in equation (7.21) with the filter function $K_1(\lambda)$:

$$P(\lambda_1) = P_L \xi(R) \frac{c\tau_L A_0}{2R^2} T(\lambda_L, R) \sum_i \sum_J \frac{N_J^i \sigma_i^{\mathrm{RR}}(\lambda_J) K_1(\lambda_J) T(\lambda_J, R)}{4\pi} \tag{9.9}$$

where

> P_L is the output power of the laser (assuming a very narrow laser bandwidth);
>
> N_J^i represents the number density of the ith species of molecule in the rotational state described by the quantum number J;
>
> $\sigma_i^{RR}(\lambda_J)$ represents the ith-species total rotational Raman cross section for the transition $J \to J \pm 2$, at wavelength λ_J and excited by radiation at wavelength λ_L.

As before, $c\tau_L/2$ represents the effective laser pulse length, and $T(\lambda_L, R)$ and $T(\lambda_J, R)$ the atmospheric transmission factors at λ_L and λ_J, respectively. A_0 is the area of the receiver optics, $\xi(R)$ the overlap factor, and of course R the range corresponding to detection at the moment of interest.

If we assume that the rotational states are in equilibrium at temperature T, then from (3.86)

$$N_J^i = N_i(2J + 1)\frac{\theta^i}{T}e^{-\theta^i J(J+1)/T} \tag{9.10}$$

where N_i represents the total density of molecules of the ith species in their vibrational and electronic ground state. $\theta^i (= hcB_e^i/k)$ represents the *characteristic rotational temperature* for the ith species and is about 2.88 K for N_2. The representative spectrum for N_2 (with excitation at 488 nm) has been given by Salzman (1974) and was reproduced as Fig. 3.24.

The ratio of the received rotational Raman-backscattered power at wavelength λ_1 and λ_2 is

$$\frac{P(\lambda_1)}{P(\lambda_2)} = \frac{\sum_i \sum_J K_1(\lambda_J)(2J + 1)\sigma_i^{RR}(\lambda_J)e^{-\theta^i J(J+1)/T}}{\sum_i \sum_J K_2(\lambda_J)(2J + 1)\sigma_i^{RR}(\lambda_J)e^{-\theta^i J(J+1)/T}} \tag{9.11}$$

provided that the atmospheric attenuation factor $T(\lambda_J, R)$ is reasonably constant over the small range of wavelengths involved. Reference to equation (9.11) reveals that this signal ratio is dependent primarily upon the temperature, the choice of filter-function wavelengths λ_1 and λ_2, and of course the molecular constants $\sigma_i^{RR}(\lambda_J)$ and θ^i. Cohen et al. (1976) have studied this approach and have shown that in the case of N_2 and laser excitation at 694.3 nm the optimum wavelengths are $\lambda_1 = 692.7$ nm and $\lambda_2 = 689.9$ nm, for a 1.0-nm-FWHM single-Gaussian filter function.

Salzman (1974) demonstrated the feasibility of this approach by making temperature measurements from -20 to 30°C at a range of 100 m (indoors) with a resolution of 5 m and an accuracy of ± 3°C. He used two interference filters with circular apertures. These sampled the rotational anti-Stokes spectrum out in the wing and close to the exciting line. The ratio of the resulting intensities is, in principle, solely dependent upon the gas temperature.

It should be noted, however, that if lines of equal J on the Stokes and anti-Stokes wings are compared, their ratio is rather insensitive to the gas temperature. Although the high spectral rejection ratio (10^6) required to eliminate interference from the elastically scattered (on-line) return is possible through the use of a multipass interferometer (Sandercock, 1970) or an iodine gaseous filter (Kobayasi et al., 1974), the transmission of such filters is very low. This leads to small signals and poor accuracy. Cohen et al. (1976) have suggested that direct measurement of the on-line intrusion would allow the elastic component to be subtracted from the two Raman returns and thereby remove this constraint on the filter rejection. This in turn should lead to larger signals and an improvement in accuracy. Alternatively, Armstrong (1975) has undertaken an analysis which indicates that the interferometric technique of Barrett and Myers (1971) could considerably improve the range and sensitivity of this approach by utilizing all of the rotational Raman lines simultaneously.

The temperature of the atmosphere can also be ascertained from the molecular-level populations by means of differential-absorption measurements. Mason (1975) first proposed that the DAS technique could be applied to such temperature evaluations, and Schwemmer and Wilkerson (1979) undertook an initial error analysis. This approach can be understood if we imagine the wavelength of the laser pulse to be selected so that, over a range R, absorption is primarily due to one constituent. The absorption optical depth—or *absorbance* as it is sometimes called—is

$$\tau(\lambda_\alpha, T) \equiv 2N_\alpha(T)\sigma^A(\lambda_\alpha)R = \kappa(\lambda_\alpha, T)R \tag{9.12}$$

where $N_\alpha(T)$ is the spatially averaged molecular density responsible for absorption, $\sigma^A(\lambda_\alpha)$ is the relevant cross section, and $\kappa(\lambda_\alpha, T)$ is the appropriate volume extinction coefficient. This peak cross section can be expressed in terms of the spectrally integrated cross section σ_α^A and the appropriate linewidth $\Delta\lambda_\alpha$:

$$\sigma^A(\lambda_\alpha) = \frac{\sigma_\alpha^A}{\pi\Delta\lambda_\alpha} \tag{9.13}$$

The temperature dependence of the linewidth can be described by a relation of the form

$$\Delta\lambda_\alpha(T) = \Delta\lambda_\alpha(T_0)\left\{\frac{T_0}{T}\right\}^n \tag{9.14}$$

where T_0 is some reference temperature and $n = \frac{1}{2}$ for Doppler broadening. In general, however, n depends upon the transition. Under equilibrium conditions the molecular density at temperature T can be related to that at the reference temperature through the Boltzmann relation

$$N_\alpha(T) = N_\alpha(T_0)\frac{Z(T)}{Z(T_0)}\exp\left\{\frac{\mathscr{E}_\alpha}{k}\left(\frac{1}{T_0} - \frac{1}{T}\right)\right\} \tag{9.15}$$

where $Z(T)$ represents the partition function at temperature T, and $\mathscr{E}_\alpha$ represents the energy of the lower level of the absorbing transition. The product $N_\alpha(T)\sigma_\alpha^A$ is often termed the *strength* of the absorption line.

In the event that $N_\alpha(T_0)$ is likely to vary with range, then it can be evaluated as a function of the temperature by monitoring the absorption on two lines at wavelengths λ_1 and λ_2. These lines are so chosen that λ_1 is close to λ_2 but the population densities of the two lower levels of the transitions are quite different due to the large separation of these levels compared to the mean thermal energy of the medium. Under these circumstances the ratio of the absorbances is

$$\frac{\tau(\lambda_1, T)}{\tau(\lambda_2, T)} = \frac{\kappa(\lambda_1, T_0)}{\kappa(\lambda_2, T_0)}\left\{\frac{T}{T_0}\right\}^{\Delta n}\exp\left\{\frac{\Delta\mathscr{E}}{k}\left(\frac{1}{T_0} - \frac{1}{T}\right)\right\} \tag{9.16}$$

where $\Delta n = n_1 - n_2$ is the difference of the exponents describing the temperature dependence of the linewidths and $\Delta\mathscr{E} = \mathscr{E}_1 - \mathscr{E}_2$ is the difference in the lower-level energies. Equation (9.16) can be converted into a relation for the temperature T in terms of the reference temperature T_0:

$$T = T_0\left\{1 - \frac{kT_0/\Delta\mathscr{E}}{1 + kT_0\Delta n/\Delta\mathscr{E}}\left[\ln\left(\frac{\tau(\lambda_1, T)}{\tau(\lambda_2, T)}\right) + \ln\left(\frac{\kappa(\lambda_2, T_0)}{\kappa(\lambda_1, T_0)}\right)\right]\right\}^{-1} \tag{9.17}$$

provided a linear approximation can be made for the logarithm of T/T_0 (i.e., $T - T_0 \ll T_0$).

If we can further assume that the second term within the square brackets is much less than unity, then we can write

$$T \approx C\ln\left\{\frac{\tau(\lambda_1, T)}{\tau(\lambda_2, T)}\right\} + D \tag{9.18}$$

where

$$C = \frac{kT_0^2}{\Delta\mathscr{E}}\left\{1 + \frac{kT_0\Delta n}{\Delta\mathscr{E}}\right\}^{-1} \quad \text{and} \quad D = C\ln\left\{\frac{\kappa(\lambda_2, T_0)}{\kappa(\lambda_1, T_0)}\right\} + T_0 \tag{9.19}$$

The constant D can be found by using spectroscopic data, or it can be determined empirically with calibrated temperature measurements. Each absorbance can be determined by a differential absorption lidar (DIAL), if a third laser pulse at a wavelength λ_3 (chosen to be close to λ_1 and λ_2 but to avoid any absorption line of the constituent of interest) is employed. Such measurements can be undertaken either using topographic targets as the noncooperative backscatterers or Rayleigh–Mie scatterers in the form of aeroticulates. The latter approach allows local evaluation of the absorbance, while the former only provides a long-path average measurement. From equation (7.25) we can write the absorption optical depth (or absorbance)

$$\tau(\lambda_\alpha, T) = \ln\left\{\frac{P(\lambda_3, R)}{P(\lambda_\alpha, R)}\right\} \tag{9.20}$$

where $P(\lambda_3, R)$ and $P(\lambda_\alpha, R)$ represent the backscattered laser powers incident on the receiver system from range R and at wavelengths λ_3 and λ_α (where $\alpha = 1, 2$) respectively. In converting equation (7.25) to (9.20) we have assumed that the wavelength difference $\lambda_3 - \lambda_\alpha$ is sufficiently small that

$$\bar{\kappa}(\lambda_3, R) \approx \bar{\kappa}(\lambda_\alpha, R)$$

$$\beta(\lambda_3, R) \approx \beta(\lambda_\alpha, R)$$

$$\xi(\lambda_3) \approx \xi(\lambda_\alpha)$$

Once the temperature is estimated, the density of the absorbing species can be evaluated from the measured absorbance if the absolute magnitude of the absorption cross section is known. This means that both the temperature and the concentration of the absorbing molecules can be determined from the same lidar data.

Endemann and Byer (1980, 1981) were able to show (through simulation) that the temperature should indeed be proportional to the logarithm of the ratio of the absorbances, as predicted from (9.18), over the range -10 to 30°C. This would be adequate for work in the lower troposphere. Their simulation involved the wavelengths $\lambda_1 = 1.7695$ μm, $\lambda_2 = 1.7698$ μm, and $\lambda_3 = 1.7696$ μm, and was based on absorption within the wing of the 1.9-μm band of water vapor. These wavelengths were selected after careful examination of the Air Force Cambridge Research Laboratory tapes of McClatchey et al. (1973). In a preliminary set of experiments, Endemann and Byer (1981), used a Nd–YAG laser to pump an optical parametric oscillator. This provided a tuning range from 1.4 to 4.0 μm and an output energy of 5 mJ in a 10-ns pulse. Their measurements relied on noncooperative scattering of these laser pulses from a building located 775 m from the telescope receiver and therefore only provided the average values of the temperature and humidity over that range. The relative temperature uncertainty was 1.5°C, while the absolute accuracy was

2.3°C. The relative error in the humidity data was 1.5%. However, the absolute error was estimated to be 20% and was due to uncertainties in the available cross-section data for the selected H_2O lines. Endemann and Byer (1981) suggest that an appreciable improvement could be expected if all three laser pulses were fired within the atmospheric correlation time of a few milliseconds (Menyuk and Killinger, 1981). It is obvious that greater laser pulse energy and better detector sensitivity would also contribute to more accurate observations from an extended range.

Kalshoven et al. (1981) have considered the case of an atmospheric molecular constituent for which the density can be assumed to be independent of location. Under these circumstances the temperature can be evaluated from an absorption measurement on a single temperature-sensitive line—this is taken to mean that the lower level of the transition is elevated above the ground level by about the mean thermal energy of the atmospheric constituents. Oxygen was chosen by Kalshoven et al. (1981) because of its virtually uniform mixing ratio throughout the atmosphere and the availability of convenient high-resolution, high-energy lasers that are capable of being tuned to the temperature-sensitive oxygen A-band at 770 nm.

The DIAL system used by Kalshoven et al. (1981) comprised two dye lasers that were pumped by a cw krypton laser. One dye laser was tuned to the 768.38-nm line of O_2 and had a linewidth of better than 5.9×10^{-5} nm; the other dye laser provided the reference laser radiation and was displaced by about 5.9×10^{-2} nm from the absorption-line center. Its linewidth was about 5.9×10^{-3} nm. Their results indicate that the average temperature of a 1-km path can be determined to better than 1.0°C with a noise level of 0.3°C.

A temperature measurement technique that is particularly applicable to the stratosphere has been proposed by McGee and McIlrath (1979). Their approach involves comparing the ratio of laser-induced fluorescence on two lines of the OH radical and is based on the fact that the chemical lifetime of this trace constituent of the stratosphere is long enough for its ground vibronic–rotational-state populations to be in thermodynamic equilibrium. The OH radical was chosen because it is readily excited by current high-power tunable lasers and its emission is strong and well defined.

As we saw earlier [equation (4.28)], the volume rate of excitation of an essentially two-level system ($|n\rangle$ and $|m\rangle$) subject to a pulse of intense, appropriately tuned radiation is

$$\frac{dN_n(t)}{dt} = N_0 R_{mn}(t) - \frac{N_n(t)}{\tau(t)} \tag{9.21}$$

where $N_n(t)$ is the population density of the upper level of the excited transition, N_0 the initial population density of the combined levels, and

$$\frac{1}{\tau(t)} = (1 + g) R_{nm}(t) + \frac{1}{\tau_n} \tag{9.22}$$

Here, g is the upper-to-lower degeneracy ratio, and τ_n the lifetime of the excited state prior to irradiation. The rate of excitation $R_{mn}(t)$ and the rate of stimulated emission were given by (4.24) and (4.25) respectively.

The general solution of (9.21) subject to excitation which commences at $t = 0$ takes the form

$$N_n(t) = e^{-\int_0^t dt^*/\tau(t^*)}\left[N_n(0) + N_0\int_0^t R_{mn}(t^*)e^{\int_0^{t^*} dt^{**}/\tau(t^{**})}\,dt^*\right] \quad (9.23)$$

If the radiation field is weak (i.e., far from saturation) so that $(1+g)R_{nm} \ll 1/\tau_n$ and there is virtually no excitation prior to irradiation, then

$$N_n(t) \approx N_m(0)e^{-t/\tau_n}\int_0^t \frac{B_{mn}}{4\pi}\int_{-\infty}^{\infty} I^L(\nu, t^*)\mathscr{L}(\nu)\,d\nu\, e^{t^*/\tau_n}\,dt^* \quad (9.24)$$

If we further assume that the period of excitation is short compared to the excited-state relaxation time τ_n, then for times of interest the exponential factors can be approximated by unity and the order of integration can be reversed. In this case we can express the total excited-state density in terms of the temporally integrated spectral photon density,

$$\phi(\nu) \equiv \int_0^{\tau_L} \frac{I^L(\nu, t)\,dt}{h\nu}$$

and the spectrally integrated absorption cross section,

$$\sigma_{mn}^A \equiv \frac{h\nu B_{mn}}{4\pi}$$

as

$$N_n(\tau_L) = N_m(0)\sigma_{mn}^A\int_{-\infty}^{\infty} \phi(\nu)\mathscr{L}(\nu)\,d\nu \quad (9.25)$$

We see that the excited-state density attained is proportional to the population density in the lower level prior to irradiation. Consequently, if two transitions (originating on two well-separated rotational levels of the ground electronic–vibrational state) are excited, then the ratio of the resulting fluorescences into a common vibrational band should be directly proportional to the ratio of the initial lower-level populations and thereby, from (3.83), should provide a measure of the temperature.

McGee and McIlrath (1979) have evaluated the expected temperature dependence of the broadband fluorescence observed when OH is excited by laser pulses at 282.06 or at 282.67 nm. Since the difference in excitation wavelength is only 0.6 nm and the same band of fluorescence wavelengths is observed, many of the wavelength-dependent factors such as the atmospheric and receiver-optics transmission coefficients should cancel. McGee and McIlrath suggest that a 10% measurement in the ratio of the fluorescence signals would correspond to about a 10-K measurement accuracy in temperature. Furthermore, they also show how comparing the fluorescence of the (1, 1) and (0, 0) vibrational bands could lead to an indication of the atmospheric pressure through the process of collisional transfer of excitation between vibrational states of the OH radical.

9.1.4. Minor-Constituent Detection

The atmosphere contains a number of naturally occurring minor (trace) constituents that play an important role in the scheme of things. Some of the best-known examples are CO_2, O_3, and OH. The first has a strong influence on the thermal balance of the atmosphere, the second forms an extremely effective shield against the life-damaging short-wavelength ($\lambda < 300$ nm) radiation from the sun, and the last appears to play a crucial role in many of the atmosphere's photochemical reactions. Since man's activities could disturb the natural balance of these important constituents it is evident that we must have adequate means of monitoring their concentrations in order to detect such perturbations before they can become significant. The review of Schofield (1977) provides an extensive evaluation of atomic and molecular fluorescence potential for *in situ* monitoring of the concentration of a large number of minor species within the stratosphere.

Although laser-induced fluorescence can, in principle, be used for the troposphere, the large quenching rates associated with the high density in that region make it rather impractical for remote-sensing applications. In the stratosphere the severity of the quenching rates is greatly reduced and fluorescence lidar looks attractive for the detection of a number of species. McIlrath (1980) has provided a list of potential species, their respective wavelengths for excitation and detection, and their appropriate cross sections. This information is reproduced here as Table 9.1. He also provides the transition probabilities for a selection of OH transitions.

The hydroxyl free radical OH is considered to be one of the most chemically active trace constituents of the atmosphere and is believed to control the worldwide conversion of CO to CO_2 as well as being an important intermediary in the photochemical formation of smog. In the stratosphere OH takes part in catalytic ozone destruction processes which regulate the concentration of O_3 in both the lower mesosphere and the upper stratosphere. Laser-induced fluorescence has been shown to be capable of detecting OH at the low concentrations of importance ($\approx 10^6$ cm^{-3}) (Baardsen and Terhune, 1972); Davis et al., 1979). Although the latter work involved an aircraft-mounted system, both groups were involved with *in situ* techniques. McGee and McIlrath (1979) and McIlrath (1980) have suggested that laser-induced fluorescence could be used to remotely monitor the stratospheric concentration of OH from a balloon. Recently, Heaps et al. (1982) have successfully undertaken

TABLE 9.1. SPECIES OF POTENTIAL INTEREST FOR FLUORESCENCE LIDAR WITH ASSOCIATE WAVELENGTHS AND CROSS SECTIONS[a]

Species	Transition	Absorption λ (nm)	Fluorescence λ (nm)	Cross Section (cm²)
Li	$2s$–$2p$	670.78	670.78	1.3×10^{-11}
Na	$3s$–$3p$	588.995	588.995	1.8×10^{-11}
K	$4s$–$4p$	766.491	766.491	3.1×10^{-11}
Ca II	$4s$–$4p$	393.366	393.366	1.6×10^{-11}
Fe	$4s^2$–$4s4p$	371.994	371.994	1.1×10^{-12}
Cl	$3p^5$–$3p^4 4s$	134.724	134.724	8.6×10^{-13}
O	$2p^4$–$2p^3 3s$	130.217	130.217	1.5×10^{-13}
OH	$^2\Pi$–$A^2\Sigma^{+}$ [b]	282	308–314	6.2×10^{-15}
NO	$^2\Pi$–$A^2\Sigma^{+}$ [b]	215	215–260	3×10^{-15}
BaO	$^1\Sigma$–$A^1\Sigma$ [c]	535	535	4.8×10^{-14}
NO_2	X–A	532	633–646	1.6×10^{-21}

[a] McIlrath (1980).
[b] $(v', v'') = (1, 0)$.
[c] $(v', v'') = (4, 0)$.

such a mission, and their results reveal a temporal variation of the hydroxyl-radical concentration over the 34–37-km altitude range, from 40 ppt shortly after noon to around 5 ppt 2 hours after sunset.

The balloon-borne lidar facility of Heaps et al. (1982) is illustrated in Fig. 9.10(a), while Fig. 9.10(b) reveals the layout of the laser system and transmission optics employed. A 400-mJ Nd–YAG laser constitutes the primary source, and a 30-cm-diameter Cassegrain telescope serves as the collector for the receiver subsystem. An onboard LSI-11 computer controlled the operations during the flight and for low-background low-signal regimes a photon-counting system was utilized. In high-background situations fast transient digitizers were used to convert the analog signals from the photomultiplier tubes to a form suitable for data reduction by the computer.

In addition to the OH measurements undertaken with this facility, Heaps et al. (1982) also used it to evaluate the distribution of ozone in the stratosphere. For this work the system was operated in a differential-absorption mode. However, the uniform density of backscatterers in the stratosphere enabled observations to be made using only one emission wavelength (i.e., 282 nm). Figure 9.11 shows the ozone profile determined during the ascent of the balloon. The lidar results are plotted as crosses with the horizontal crossbar extending twice the standard deviation on either side of the measured value. The averaging period for each measurement was 6–7 min, during which the balloon ascended between 500 and 700 m. The solid line is a plot of the value obtained by a Dasibi *in situ* ozone analyzer located aboard the gondola. Range-resolved ozone profiles taken on the ascent at an altitude of about 23

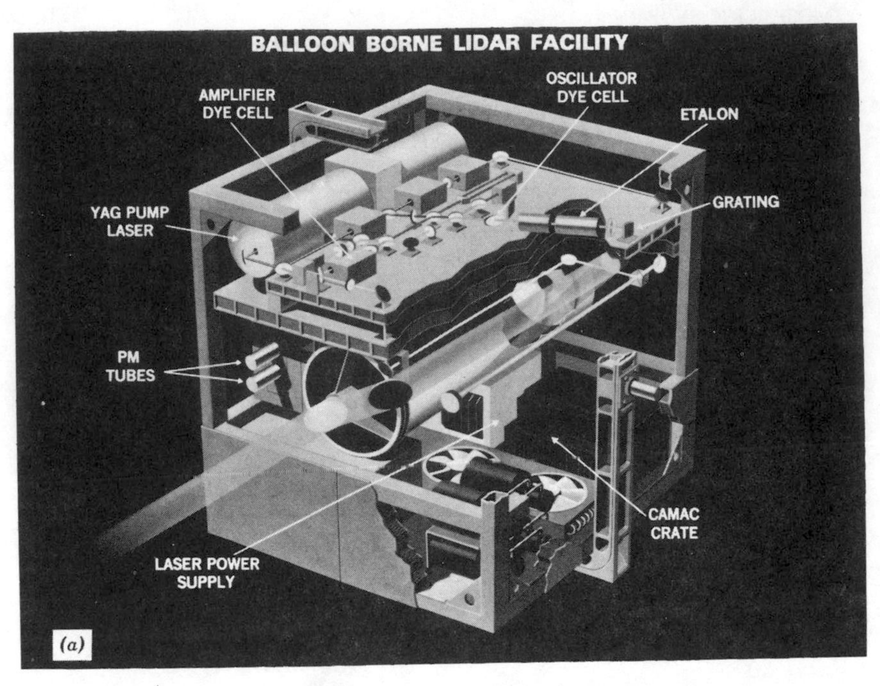

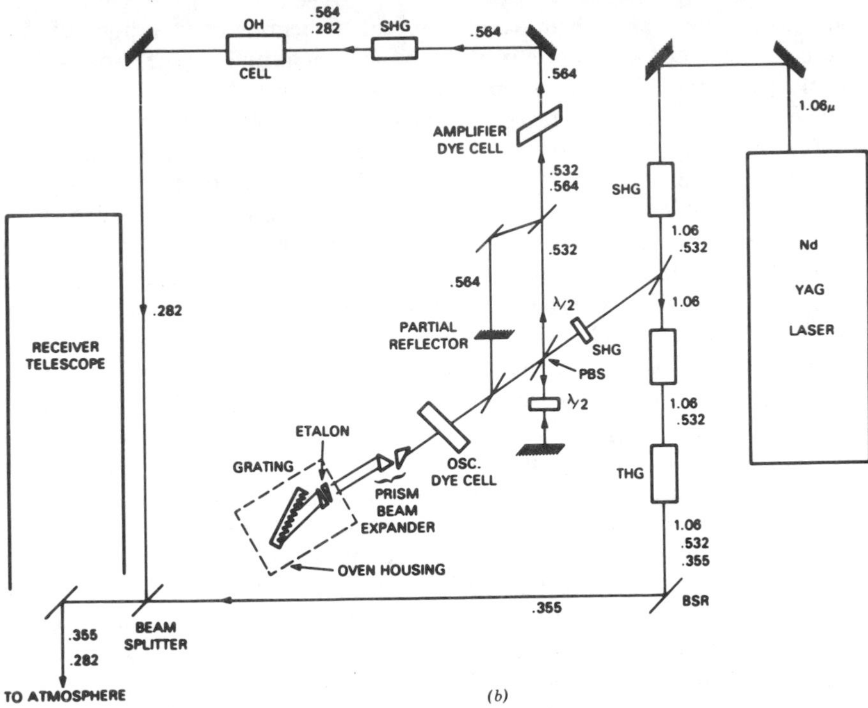

Fig. 9.10. Balloon-borne lidar facility: (a) artist's representation of the layout; (b) schematic diagram of transmitter system. All wavelengths are in micrometers (Heaps et al., 1982).

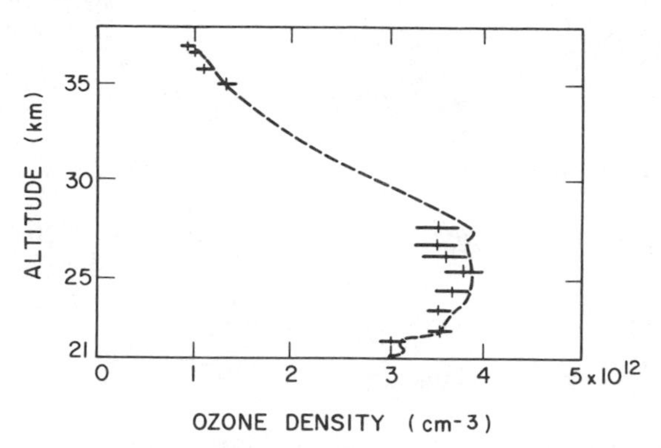

Fig. 9.11. Ozone density profile measured by balloon-mounted lidar (two standard deviation error bars) and Dasibi *in situ* ozone analyzer (dashed curve) during the ascent (Heaps et al., 1982).

km revealed an unexpected horizontal structure. This structure had more or less disappeared at an altitude of 37 km.

The coincidence of the 308-nm emission of a XeCl laser with an absorption band within O_3 enabled Uchino et al. (1978) to undertake the first experimental measurements of the ozone distribution in the 15- to 30-km section of the stratosphere with a ground-based lidar system. The average ozone concentration $[O_3(R)]$ within the range (altitude) interval $(R, R + \Delta R)$ can be seen from equation (7.23) to be given by

$$[O_3(R)] = \frac{1}{2\sigma_{O_3}^A \Delta R} \times \left[\ln\left\{\frac{P(R)}{P(R+\Delta R)}\right\} - \ln\left\{\frac{\beta(R)/R^2}{\beta(R+\Delta R)/(R+\Delta R)^2}\right\} - 2\bar{\kappa}(R)\Delta R \right] \tag{9.26}$$

where $\sigma_{O_3}^A$ represents the ozone absorption cross section at the laser wavelength, $P(R)$ is the total backscattered laser radiation received from the range (altitude) R, $\beta(R)$ is the appropriate atmospheric backscattering coefficient from this range, and $\bar{\kappa}(R)$ is the average extinction coefficient over the range interval $(R, R + \Delta R)$ due to aeroticulate scattering and molecular absorption (exclusive of ozone). In general we can write

$$\beta(R) = \beta^{\mathrm{Ray}}(R) + \beta^M(R)$$

and

$$\bar{\kappa}(R) = \bar{\kappa}^{\text{Ray}}(R) + \bar{\kappa}^{M}(R)$$

The Rayleigh components are

$$\beta^{\text{Ray}}(R) = N^{\text{Ray}}(R)\sigma_{\pi}^{\text{Ray}} = 5.95 \times 10^{-27} N^{\text{Ray}}(R) \quad \text{cm}^{-1}\ \text{sr}^{-1}$$

and

$$\bar{\kappa}^{\text{Ray}}(R) = N^{\text{Ray}}(R)\sigma_{\text{Ray}} = 5.07 \times 10^{-26} N^{\text{Ray}}(R) \quad \text{cm}^{-1}$$

where $N^{\text{Ray}}(R)$ represents the molecular number density at altitude R and we have used (2.145), which can be written as

$$\sigma_{\text{Ray}} = \frac{8\pi}{3} \times \sigma_{\pi}^{\text{Ray}} \left\{ \frac{6 + 3\delta_p}{6 - 7\delta_p} \right\} \tag{9.27}$$

In this relation the Rayleigh backscattering cross section was given by (2.134):

$$\sigma_{\pi}^{\text{Ray}} = 5.45 \left\{ \frac{550}{\lambda(\text{nm})} \right\}^4 \times 10^{-28} \quad \text{cm}^2\ \text{sr}^{-1}$$

and the depolarization ratio δ_p for air was taken (from Table 2.4) to be 0.042. The atmospheric molecular density is calculated from radiosonde pressure and temperature data.

The aeroticulate components at 308 nm were estimated from the *scattering ratio*

$$K_S(R) \equiv \frac{\beta(R)}{\beta^{\text{Ray}}(R)} \tag{9.28}$$

and observed on the same day by means of a frequency-doubled Nd–YAG laser at $\lambda = 532$ nm, using

$$\beta^{M}(R) = A^{M} N^{\text{Ray}}(R)\{K_S(R) - 1\}$$

and

$$\bar{\kappa}^{M}(R) = B^{M} \overline{N}^{\text{Ray}} \Delta R \{K_S(R) - 1\}$$

where $A^M = 1.13 \times 10^{-27}$ cm^2 sr^{-1} and $B^M = 6.50 \times 10^{-26}$ cm^2, which are determined by a typical stratospheric aeroticulate model using Mie scattering theory (Uchino et al., 1980).

The XeCl laser used by Uchino et al. (1980) transmitted an energy pulse of about 50 mJ, spread over the three vibrational bands at 307.6, 307.9, and 308.2 nm, with a duration of about 16 ns and a beam divergence of 1 mrad.

Backscattered laser radiation was collected by a 50-cm-diameter Newtonian telescope. A comparison between the lidar observations taken at Fukuoka and ozonesondes launched the same day from adjacent locations of Tateno and Kagoshina is presented as Fig. 9.12. The agreement is seen to be quite good over the 15- to 25-km range of altitudes. Also a high correlation was obtained between the mean ozone density at 17.25 km and the total vertical column density of atmospheric ozone observed by a Dobson spectrophotometer at Tateno. The influence of the stratospheric aeroticulate layer on the lidar measurement accuracy was generally < 10% and was about 20% when the scattering ratio was 1.2.

An airborne DIAL system developed at the NASA Langley Research Center has the capability to investigate the spatial distribution of many tropospheric gases and aeroticulates (Browell, 1982a). This system has the flexibility to operate in the UV for temperature and pressure measurements of O_3 or SO_2, in the visible for NO_2, and in the near IR for H_2O. Also, aeroticulate backscattering investigations in the visible and the near IR can be conducted simultaneously with the DIAL measurements. This lidar system employs two frequency-doubled Nd–YAG lasers to pump two dye lasers. The "on" and "off" (or wing) wavelength laser pulses at 286 and 300 nm respectively are

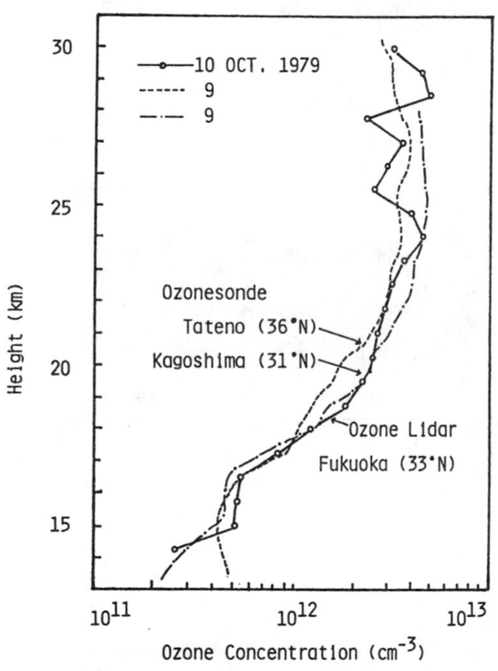

Fig. 9.12. Ozone-density comparison between a XeCl-laser-based lidar and an ozonesonde (Uchino et al., 1980).

Fig. 9.13. (a) The NASA DIAL system aboard the Wallops Flight Centre Electra aircraft; (b) schematic of the system (Browell, 1982b).

sequentially produced and separated by less than 100 μs. Dielectric-coated steering optics are used to direct the dye-laser outputs through a 40-cm-diameter quartz window (used for high UV transmittance) in the bottom or top of the aircraft. The receiver system is composed of a 36-cm-diameter Cassegrain telescope and gatable photomultiplier tubes. A photograph and schematic view of this airborne DIAL system are presented in Fig. 9.13, while its characteristics are indicated in Table 9.2.

The first remote measurement of tropospheric ozone profiles from an aircraft was obtained with this system in May 1980 and the results compared with *in situ* measurements obtained from an instrumented Cessna 402 aircraft. A major field experiment with the U.S. Environmental Protection Agency was directed at studying the large-scale pollution events in the northeast sector of the U.S. The objectives of this program included the characterization of persistent elevated pollution episodes (PEPE) and the evaluation of a four-layer regional oxidant model. A comparison of O_3 measurements made with the airborne DIAL system at an altitude of 3200 m and the *in situ* instruments on the Cessna is presented as Fig. 9.14. The lidar-observed variation in O_3 concentration, from 42 ppb above the mixed layer to 100 ppb within the mixed layer, is in fairly good agreement with the *in situ* Cessna measurements.

An even more interesting result is displayed in Fig. 9.15, where the NASA UV airborne DIAL system reproduces fairly faithfully the narrow double layer of enhanced ozone concentration detected with an ozonesonde launched from

TABLE 9.2. AIRBORNE DIAL SYSTEM CHARACTERISTICS[a]

	UV (≈ 300 nm)	Near IR (≈ 730 nm)
Transmitter:		
Two pump lasers—Quantel Model 482		
Pulse separation: 100 μs		
Pulse energy: 350 mJ at 532 nm		
Repetition rate: 10 Hz		
Pulse length: 15 ns		
Two dye lasers—Jobin Yvon Model HP-HR		
Dye output energy (mJ/pulse)	157 (near 600 nm)	63
Doubled dye output energy (mJ/pulse)	47	–
Transmitted laser energy (mJ/pulse)	40(near 300 nm) 80(near 600 nm)	50
Laser linewidth (pm)	< 4	< 2
Receiver		
Area of receiver (m^2)	0.086	0.086
Receiver efficiency to PMT (%)	28	29
PMT quantum efficiency (%)	29	4.8
Total receiver efficiency (%)	8.1	1.4
Receiver field of view (mrad)	2	2

[a] Browell (1982a).

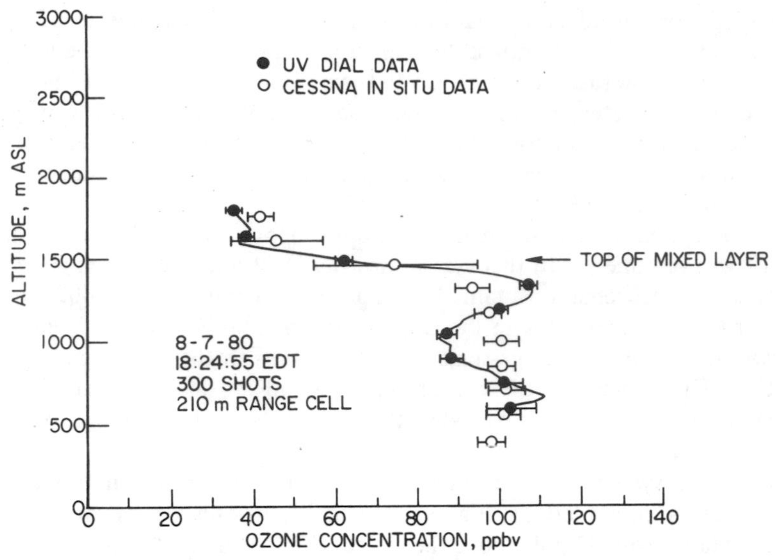

Fig. 9.14. Comparison of airborne DIAL and *in situ* O_3 data obtained in the vicinity of Salisbury, Maryland, during EPA field experiments (Browell, 1982a).

Caribou at 2209 EDT (Browell, 1982b). The altitude agreement for these layers is within 150 m, which is very good, considering that the DIAL data represent the average O_3 profile obtained from 300 lidar measurements along a 6-km flight path.

Besides the UV absorption bands of O_3, strong absorption bands also exist in the infrared, centered about 9.6 μm. This makes ozone amenable to detection by a CO_2-laser-based DIAL system. Unfortunately, this region of the spectrum is rich in absorption bands of other molecules, so that great care has to be taken in selecting the appropriate CO_2 laser lines in order to avoid interference from other atmospheric constituents, principally H_2O and CO_2. Asai et al. (1979) employed the $P(14)$ line in the (00°1–02°0) band of CO_2 as the on-resonance line and the $P(24)$ line in the same band as the off-resonance line in their DIAL measurements.

At the wavelengths of these laser lines, the differential-absorption coefficient of CO_2 is 0.011 km^{-1} for 330 ppm at STP (Menzies and Shumate, 1976, 1978). This value corresponds to an ozone concentration of 9 ppb. On the other hand, the differential-absorption coefficients for CO_2 and H_2O gives rise to a zero-offset contribution to the inferred O_3 concentration of about 12–13 ppb. In the experiments of Asai et al. (1979) the results were averaged over 50 shots on both the $P(14)$ and $P(24)$ laser lines. A typical result for the signal return against range is presented as Fig. 9.16. The corresponding range-resolved distribution of O_3 concentration is shown in Fig. 9.17. Close agreement was obtained with an *in situ* monitor located 1.5 km from the lidar system, and a

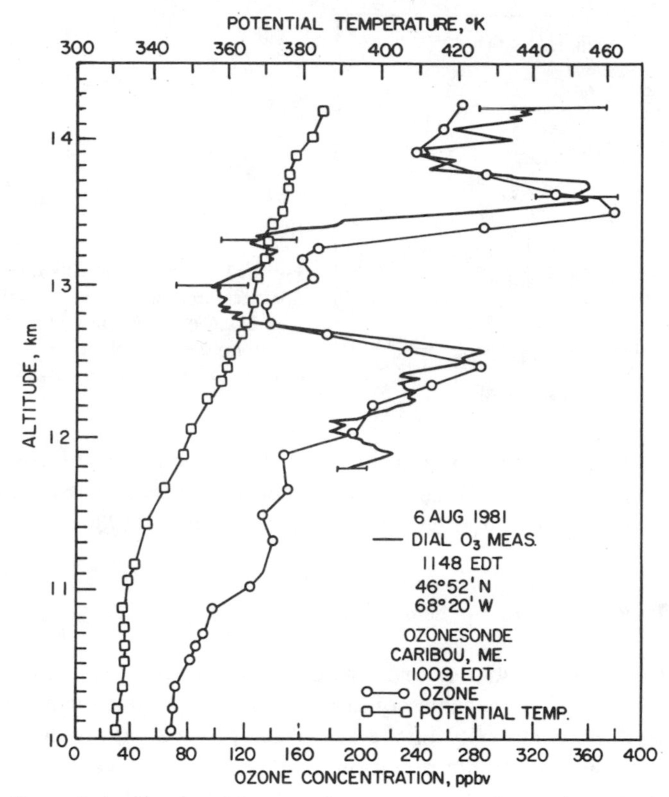

Fig. 9.15. Comparison of DIAL and ozonesonde measurements of ozone layers in the vicinity of the tropopause (Browell, 1982b).

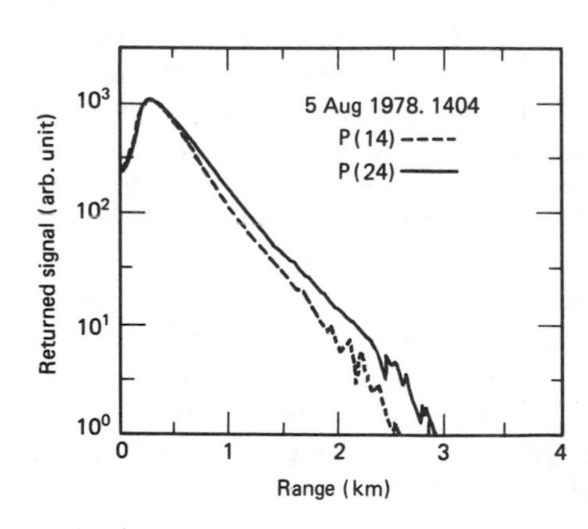

Fig. 9.16. DIAL return signals from the atmospheric aerosols at the $P(14)$ O_3 on-line (dashed curve) and $P(24)$ off-line (solid curve) wavelengths of the CO_2 laser (Asai et al., 1979).

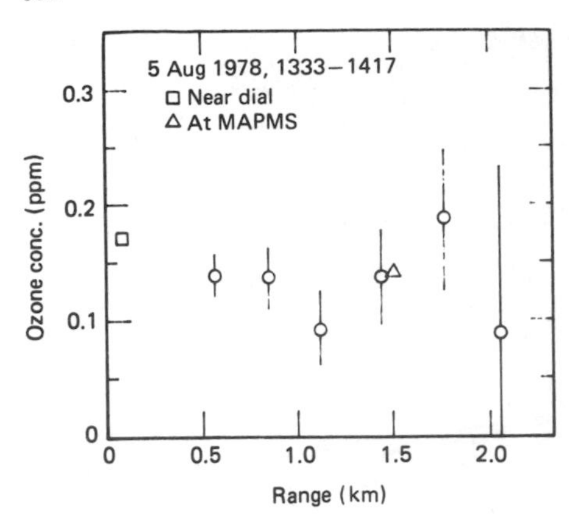

Fig. 9.17. Range-resolved ozone concentration measurements obtained with the CO_2-laser-based DIAL system (⏀) and two *in situ* monitors, one near the DIAL system (□) the other (△) situated 1.5 km away (Asai et al., 1979).

measurement uncertainty of around ± 40 ppb was predicted for the DIAL results at that range. More recently, an airborne DIAL system, based on a pair of CO_2 lasers, has been described by Stewart and Bufton (1980).

9.1.5. Cloud Studies

The amplitude of the lidar returns from clouds can be large, and so for the most part relatively high values of the signal-to-noise ratio can be attained. Nevertheless, the complexity of the scattering process within clouds and the consequent difficulty of interpretation tended for many years to limit lidar experiments to the study of structural features such as ceiling and base heights and the location of discontinuities. A good example of one of the earliest measurements of cloud height and structure was provided by Northend et al. (1966) and is reproduced here as Fig. 9.18. More recently depolarization studies of the elastically backscattered laser radiation have been able to provide more insight into the structure and composition of clouds, haze, and fog.

Pal and Carswell (1973) were among the first to experimentally study the variation of the depolarization ratio δ_p, defined by equation (2.144), through the atmosphere. For single scattering of plane-polarized radiation by either Rayleigh particles (radius smaller than the laser wavelength) or Mie particles (dielectric spheres with radii comparable to or greater than the laser wavelength) the polarization of the backscattered radiation is essentially the same as that of the incident radiation, and so $\delta_p \approx 0$ (Table 2.4 provides the observed values of δ_p for a number of molecules found in the atmosphere). Depolarization values in excess of about 0.02 can arise from spherical particles (such as water drops) in the event of multiple scattering or if the scattering particle is nonspherical (such as ice crystals). Depolarization studies can thus provide some insight into the distribution of ice and water within clouds and thereby lead to investigations of cloud formation and dynamics.

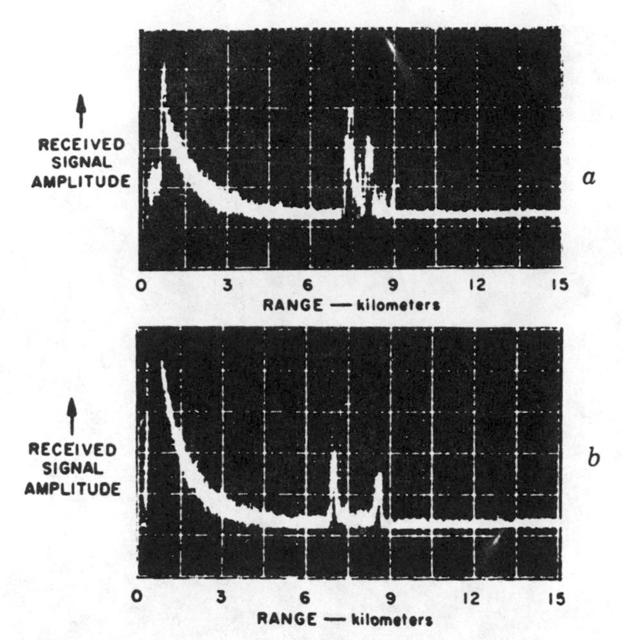

Fig. 9.18. Backscattered lidar return signals: (a) from an overhead tenuous cloud at an altitude of about 6.9 km and (b) from two layers of overhead cirrus cloud at 6.6 and 8.1 km (Northend et al., 1966).

In general the depolarization ratio is

$$\delta_p = \frac{(P_\perp^S)_A + (P_\perp^S)_W + (P_\perp^S)_I + (P_\perp^S)_M}{(P_\parallel^S)_A + (P_\parallel^S)_W + (P_\parallel^S)_I + (P_\parallel^S)_M}$$

where $(P_\perp^S)$ and $(P_\parallel^S)$ refer to the components of backscattered laser power that are respectively polarized perpendicular and parallel to the plane of polarization of the incident laser, and the subscripts indicate the atmospheric constituent responsible for the scattering: A, aeroticulates; W, water drops; I, ice crystals; and M, molecules. Lidar depolarization studies of clouds have revealed values of δ_p that range from near zero to 0.5, the larger values tending to be associated with dust and ice particles. There have also been some measurements that suggest values of the depolarization ratio that are in excess of unity (McNeil and Carswell, 1975; Smiley and Morley, 1981). McNeil and Carswell (1975) suggested that such polarization characteristics might be expected when an anisotropic layer of preferentially oriented particles resulting from wind shear or other alignment processes is responsible for the scattering.

The lidar system used by McNeil and Carswell (1975) and Pal and Carswell (1973, 1976, 1978) involved a Q-switched ruby laser with outputs at both the

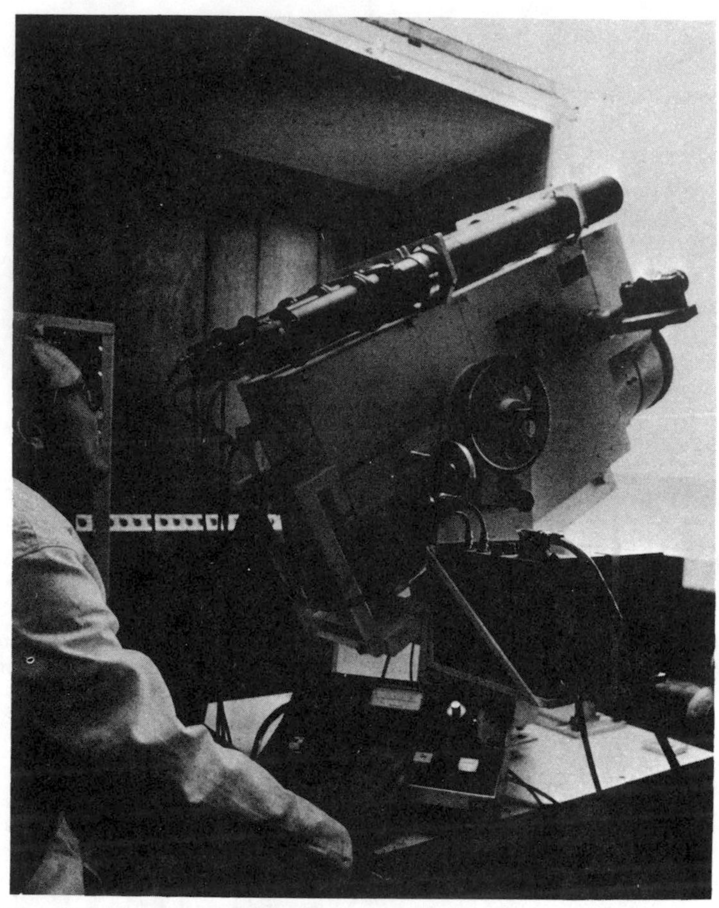

Fig. 9.19. York University elastic backscattering lidar system (courtesy of A. I. Carswell).

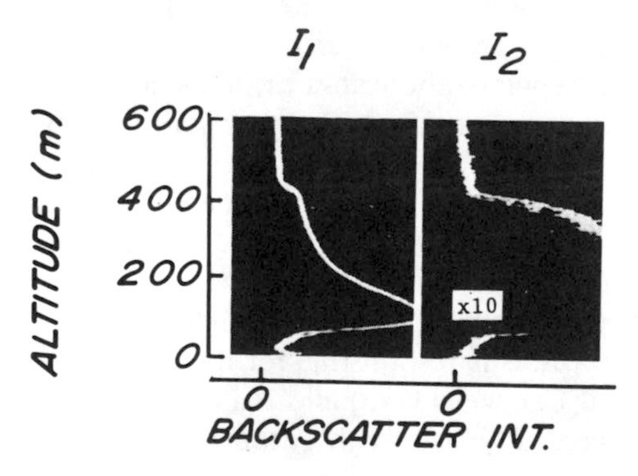

Fig. 9.20. Examples of 694-nm lidar returns at both polarizations, showing discontinuities at the boundary of an urban mixing layer (McNeil and Carswell, 1975).

fundamental and the second-harmonic frequency. Peak powers of up to 150 MW at 694.3 nm and 15 MW at 347.2 nm were available at repetition rates of 10 ppm and a pulse duration of about 15 ns. Four receiver channels were employed, so that the depolarization ratio at both wavelengths could be measured for each laser pulse. This mobile system was housed in a trailer and is shown in Fig. 9.19.

Lidar signatures are particularly adept at detecting sudden changes in the atmosphere, even in relatively clear air. An example of mixing-layer boundary detected under such conditions is presented as Fig. 9.20. The two oscilloscope traces correspond to the 694.3-nm lidar returns that are polarized parallel (left side) and perpendicular (right side) to the outgoing laser pulse. Cloud returns revealed the largest variability in depolarization of any meteorological situa-

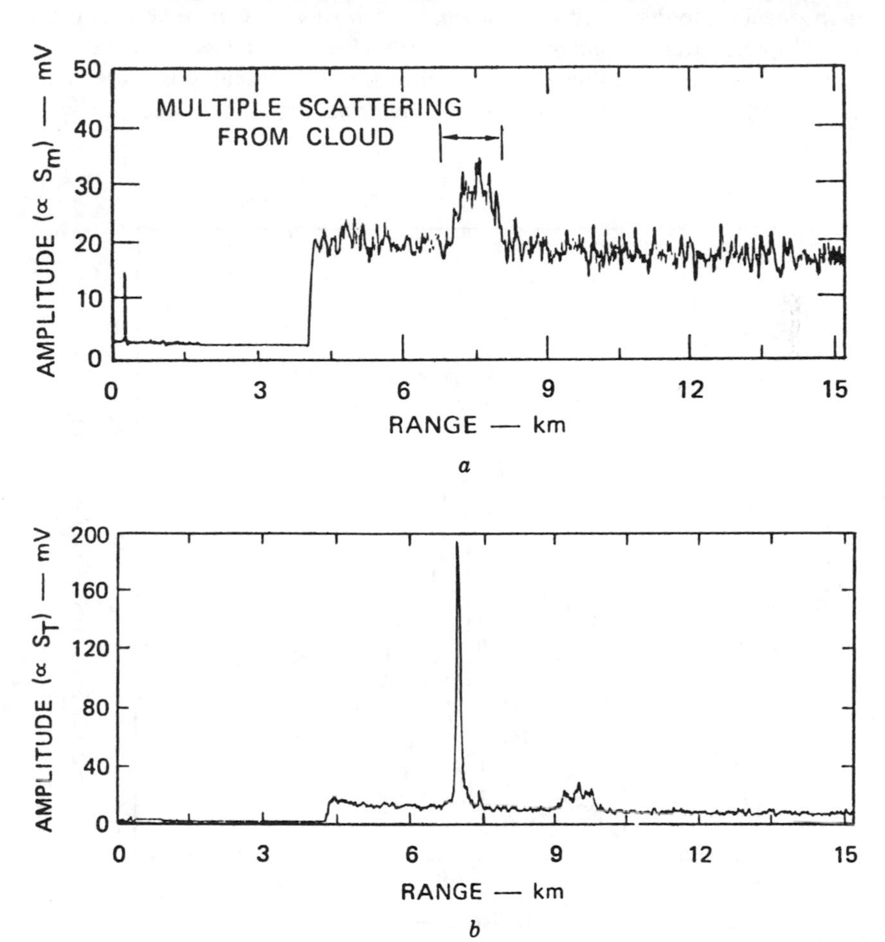

Fig. 9.21. (a) Multiple-scattered return from a cirrus cloud, (a) with and (b) without the use of a 2-mrad blocked-aperture field stop (2-MHz electronic filter and 10-mrad clear-aperture field stop) (Allen and Platt, 1977).

tion. McNeil and Carswell (1975) demonstrated that the depolarization ratio usually increases with penetration depth into low-altitude cumulous clouds. Values of 0.5 and above are often observed and suggest the presence of ice crystals towards the top of the clouds. By contrast the lowest values of δ_p ever recorded arose from the region near the base of such clouds and are in keeping with scattering from a dilute assembly of small spherical drops of water. In the case of cirrus clouds depolarization values of 0.3 and above are recorded, independent of penetration depth. These observations are consistent with the existence of substantial quantities of ice crystals throughout these clouds.

Allen and Platt (1977) have also undertaken lidar depolarization experiments on clouds using a *Q*-switched ruby laser. However, they employed special center-blocked field stops to restrict the receiver field of view to the region outside the diverging laser beam. In this way they were able to detect the multiple-scattered components of the backscattered laser radiation. An example of a multiple-scattered lidar return from a cirrus cloud is presented as Fig. 9.21(a), while the total scattered signal from the same cloud (a slightly

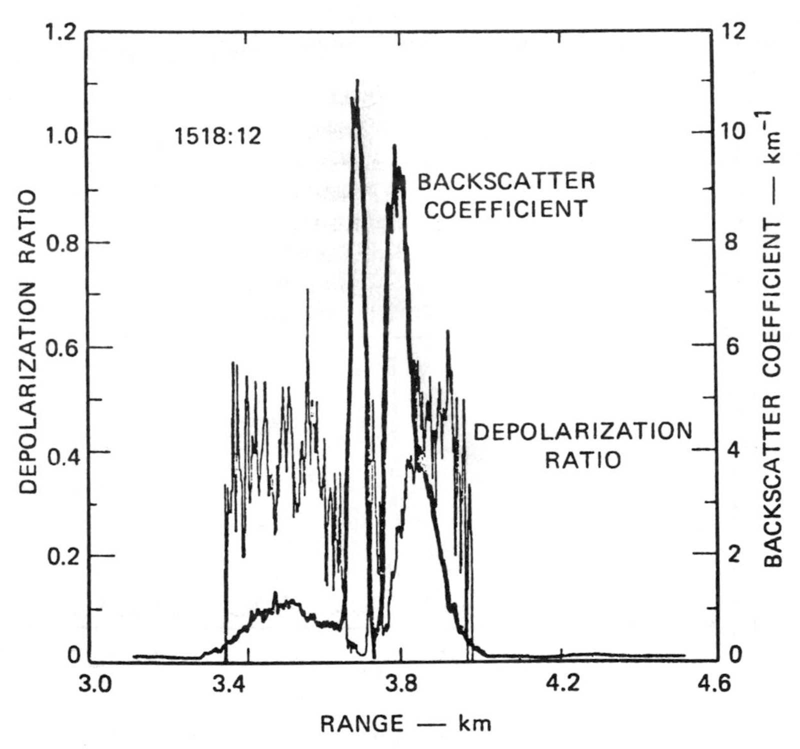

Fig. 9.22. Backscatter coefficient from mixed-phase (ice and water) altostratus clouds together with the depolarization ratio (Allen and Platt, 1977).

different region, due to the time delay between the laser pulses) is presented as Fig. 9.21(b). It is quite evident that within the cloud there are two dense scattering layers. The strongest return is seen to emanate from a region located about 7 km above the ground. The higher layer was apparently too weak to show in the multiple-scattered return.

The range variation of both the parallel backscattering coefficient $\beta_{\parallel}^{s}$ and the depolarization ratio δ_p for an altostratus cloud is presented as Fig. 9.22. The complex patterns of δ_p and $\beta_{\parallel}^{s}$ illustrate rather well the characteristic

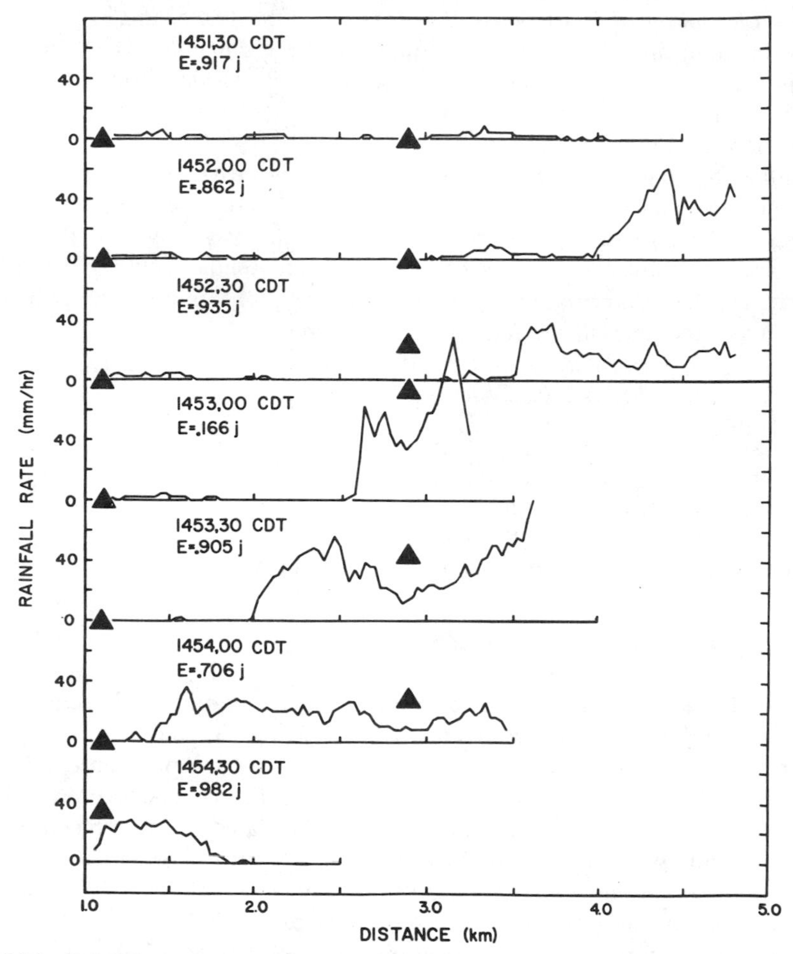

Fig. 9.23. Rainfall-rate–range profiles from the leading edge of a thunderstorm as a function of time: κ_e profiles were found by successive approximations and were transformed into rainfall rate. Output laser energy is given in joules. Rain-gage measurements are given to the nearest bucket tip interval at the appropriate ranges (large triangles) (Shipley et al., 1974).

anticorrelation between these two parameters. The lowest cloud layer, where $\delta_p \approx 0.35$ to 0.4, is typical of an ice cloud. The next, highly reflecting layer has $\delta_p \approx 0.03$, which is characteristic of small water drops. In the top layer, δ_p shows a steady increase with altitude. Allen and Platt (1977) speculate that this is due to a highly attenuating water layer at this altitude, since considerable multiple-scattered radiation was collected by their wide-field-of-view receiver from this region.

Lidar depolarization studies of the atmosphere at the South Pole were undertaken by Smiley and Morley (1981). The main purpose of their work was to determine where ice crystals were forming and growing, to discriminate between ice and water particles in the Antarctic atmosphere, and in general to assess the applicability of lidar techniques to the polar regions. Their measurements indicate that volume backscattering coefficients ranged from 2.4×10^{-8} cm^{-1} (for an optically thin ice-crystal layer) to about 2.4×10^{-6} cm^{-1} (for a representative water drop layer).

Although radar has been used for a number of years to estimate rainfall rates, the accuracy of the correlation between the microwave extinction coefficient and the rainfall rate has been limited by the wavelength of the radar transmission. Shipley et al. (1974) established that lidar could be used to determine a fairly precise relation between the optical extinction coefficient κ_ε (km^{-1}) and the rainfall rate R_R (mm hr^{-1}), namely,

$$\kappa_\varepsilon = 0.16 R_R^{0.74}$$

Using this relation, they were able to map the movement of the leading edge of a thunderstorm cell and compare their lidar observations with tipping-bucket rain gages deployed under the path of the laser beam. An example of a sequence of range–rainfall-rate profiles is presented as Fig. 9.23.

9.1.6. Stratospheric-Dust Observations

The effectiveness of elastic backscattering of laser radiation at detecting dust and aerosol layers in the atmosphere is indicated by the fact that such observations were involved in some of the earliest lidar experiments, (Fiocco and Smullin, 1963; Fiocco and Colombo, 1964; Fiocco and Grams, 1964; McCormick et al., 1966). The influence of aeroticulates on the atmosphere is not fully understood at present. Nevertheless, it is well known that they can affect regional and global climates, both through their role in forming clouds and stimulating precipitation and through their direct influence on the atmosphere's interaction with solar radiation. The aeroticulate characteristics of prime interest include the total mass density, composition, size distribution, and shape. Kerker and Cooke (1976) have shown how information on particle size could be ascertained from scattering experiments involving radiation that is varied over a broad range of wavelengths.

Although the presence of aeroticulates near or below the mesopause had been established by numerous visual observations of *noctilucent* clouds, it was the early lidar experiments that provided clear evidence of the extraterrestrial origin (micrometeoric fragmentation) of this material. The layer of aeroticulates that exists between 15- and 25-km altitude has also been observed primarily by lidar techniques. The existence of a semipermanent layer of aeroticulates was first established by Junge and Manson (1961). The first lidar experiments on this *Junge layer* were conducted by Fiocco and Grams (1964).

In much of the lidar work involving elastic backscattering from aeroticulates it is necessary to find a way of separating the Mie-scattering component from the total return signal. This is often accomplished by introducing the scattering ratio, defined earlier in (9.28):

$$K_S(\lambda_L, R) \equiv 1 + \frac{\beta^M(\lambda_L, R)}{\beta^{\text{Ray}}(\lambda_L, R)}$$

where $\beta^M(\lambda_L, R)$ and $\beta^{\text{Ray}}(\lambda_L, R)$ represent the Mie and Rayleigh volume backscattering coefficients at wavelength λ_L and range R. Since atmospheric attenuation in the stratosphere (at least for wavelengths in the red or near-infrared part of the spectrum) tends to be dominated by Rayleigh (molecular) scattering (Elterman, 1968), it follows from the lidar equation (9.2) that

$$K_S(\lambda_L, R) \approx \frac{E(\lambda_L, R)}{E^{\text{Ray}}(\lambda_L, R)} \tag{9.29}$$

where $E^{\text{Ray}}(\lambda_L, R)$ represents the lidar return that would be obtained in the absence of any Mie (aeroticulate) scattering. In practice $E^{\text{Ray}}(\lambda_L, R)$ could be calculated from the U.S. Standard Atmosphere, but Northam et al. (1974) suggest better accuracy is achieved if pressure and temperature profiles measured by balloon-borne *in situ* sensors are utilized.

An example of lidar scattering data obtained by Northam et al. (1974) is presented as Fig. 9.24(a). The lidar return has been *range-corrected* by plotting $R^2V(R)/E_L$ against R, where $V(R)$ is the photomultiplier voltage corresponding to backscattered laser radiation originating from range R. The dashed line in Fig. 9.24(a) represents the expected molecular return based on the temperature and pressure profiles measured from a balloonsonde. Figure 9.24(b) presents a plot of the scattering ratio corresponding to the data of Fig. 9.24(a).

From equation (9.28) it is apparent that the aeroticulate volume backscattering coefficient can be determined from the experimentally measured scattering ratio and a knowledge of the Rayleigh (molecular) volume backscattering coefficient:

$$\beta^M(\lambda_L, R) = \{K_S(\lambda_L, R) - 1\}\beta^{\text{Ray}}(\lambda_L, R) \tag{9.30}$$

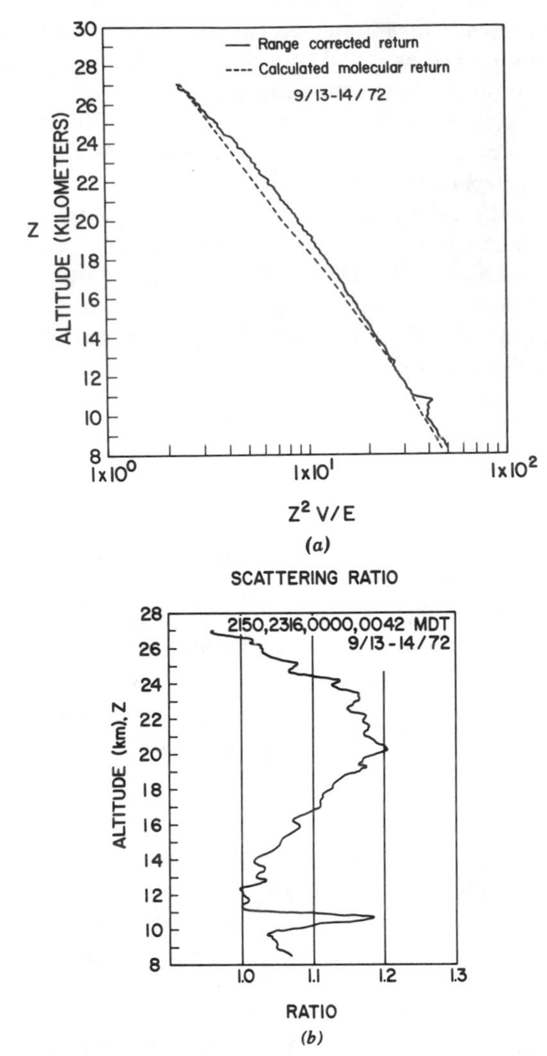

Fig. 9.24. (a) Range-corrected lidar backscattered return and (b) the corresponding scattering ratio as a function of altitude (Northam et al., 1974).

A comparison of the aeroticulate volume backscattering profile derived from the lidar experiments with that of the aeroticulate concentration measured by a balloon-borne dustsonde is presented as Fig. 9.25. The best agreement for these two profiles were obtained by equating an aeroticulate density (measured by a balloon-borne counter) of 1 particle cm^{-3} to an aeroticulate volume backscattering coefficient (derived from the lidar experiment) of 8×10^{-9} m^{-1} sr^{-1}. Both measurements indicate the presence of a subvisible cirrus cloud at an altitude of 9 to 11 km, a clean (relatively free of aeroticulate) region at about 12 km, and a fairly extensive dust layer extending over an altitude range of 15 to 24 km.

The ability of an elastic backscattering lidar to monitor the atmosphere continuously and map, with high sensitivity, the altitude distribution of low concentrations (subvisible) of aeroticulates enabled McCormick and Fuller (1975), Fegley and Ellis (1975), and Russell and Hake (1977) to detect a rapid infusion of dust in the stratosphere following the violent eruption of Volcán de Fuego in Guatamala during October of 1974. McCormick et al. (1978), using a 1- to 2-J *Q*-switched ruby laser and a 1.22-m Cassegrain ($f/10$) telescope (shown in Fig. 9.26 and schematically illustrated in Fig. 9.27), were able to follow the evolution and decay of this stratospheric dust layer over a 22-month period. A selection of the scattering-ratio and temperature profiles provided by McCormick et al. (1978) is presented as Fig. 9.28. They also introduced the *integrated aeroticulate backscattering* between altitudes R_1 and R_2, defined as

$$\int_{R_1}^{R_2} \{K_S(\lambda_L, R) - 1\} \beta^{\mathrm{Ray}}(\lambda_L, R)\, dR$$

and plotted the time history of this parameter (see Fig. 9.29) for several layers

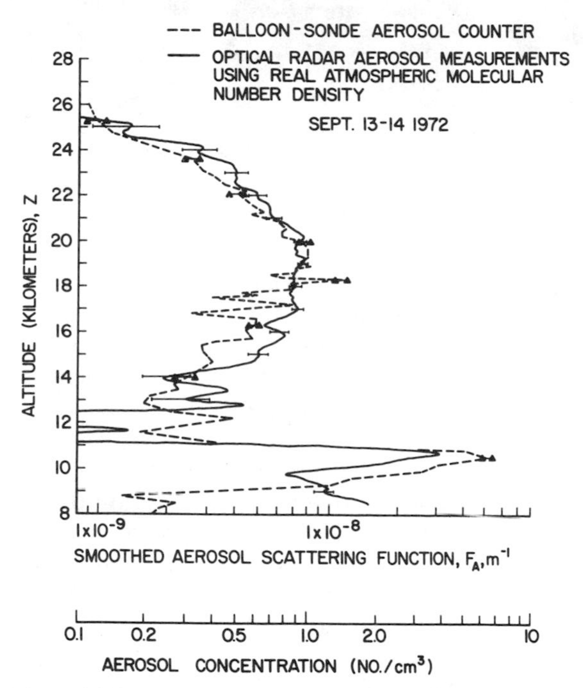

Fig. 9.25. Comparison of dustsonde particle concentration and lidar scattering function (Northam et al., 1974).

Fig. 9.26. NASA Langley atmospheric lidar system with a receiver aperture of 1.22 m (Melfi et al., 1973).

of the atmosphere. The Rayleigh (molecular) backscattering coefficient $\beta^{\mathrm{Ray}}(\lambda_L, R)$ used in this evaluation was calculated from near real-time rawinsonde data. For the 19-month period from January 1975 to July 1976, McCormick et al. (1978) estimated that the effective exponential $(1/e)$ decay time for the integrated aeroticulate backscattering was close to 12 months.

Recently the U.S. National Oceanic and Atmospheric Administration (NOAA) initiated a long-term lidar measurement program at the Table Mountain field site north of Boulder, Colorado. The objective of this program is to determine the average backscattering profile of the atmosphere and study its statistical variability. A 10.6-μm CO_2 laser with an output of from 60 to 120 mJ in a 3- to 6-μs pulse is the transmitter for this lidar. Evidence of volcanic dust, possibly from the Kurile Island (U.S.S.R.) eruption of 28 April 1982, has been detected with this system (Post et al., 1982), and is presented in Fig. 9.30. Volcanic dust from the May 1980 eruption of Mount St. Helens (U.S.) has also been detected by a lidar system in France (Lefrere et al., 1981), and recently

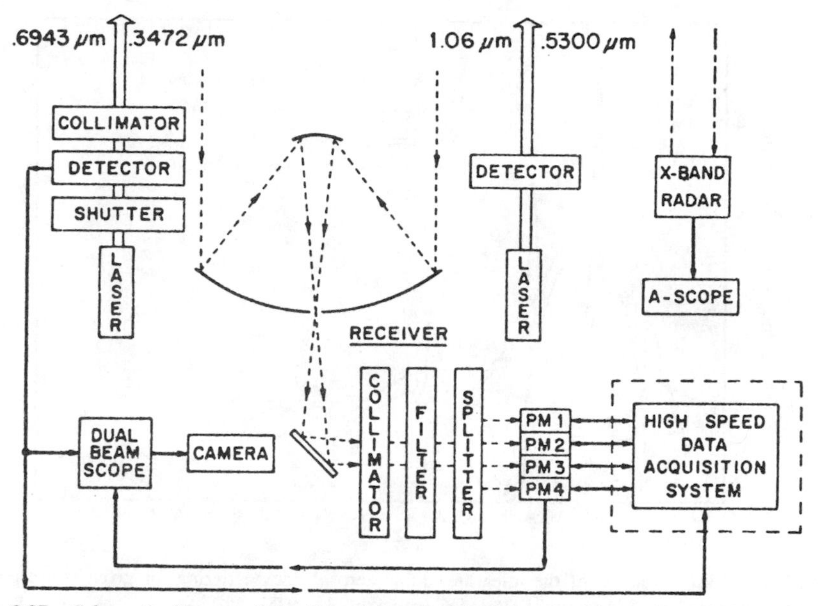

Fig. 9.27. Schematic diagram of the 48-in. (1.22-m) NASA Langley lidar system (McCormick and Fuller, 1975).

McCormick (1982) has published results of both ground-based and aircraft-mounted lidar studies of the worldwide movement of stratospheric aeroticulates released by this eruption. These observations suggest that dust layers below 20 km moved in an easterly direction while those above 20 km moved in a westerly direction. This can be seen by reference to Fig. 9.31, where the early movements of this material is indicated with the height of the intense layers.

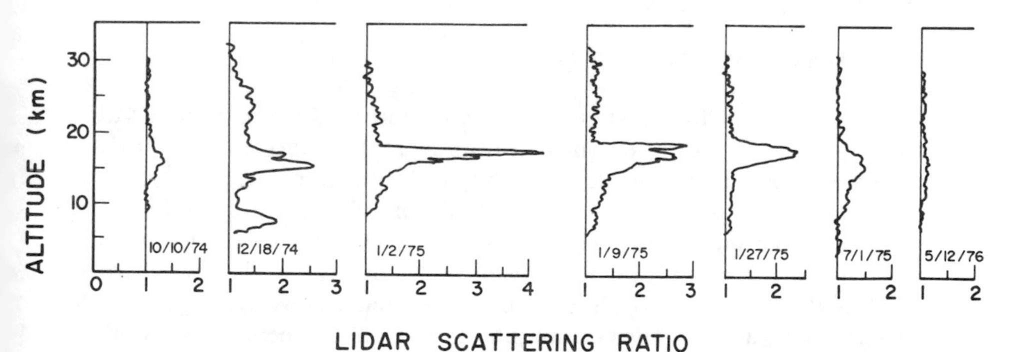

Fig. 9.28. Selected vertical profiles of lidar scattering ratios for the period October 1974 to December 1976, illustrating the injection of dust into the stratosphere following the eruption of the Volcán de Fuego, as measured by a *Q*-switched ruby laser (McCormick et al., 1978).

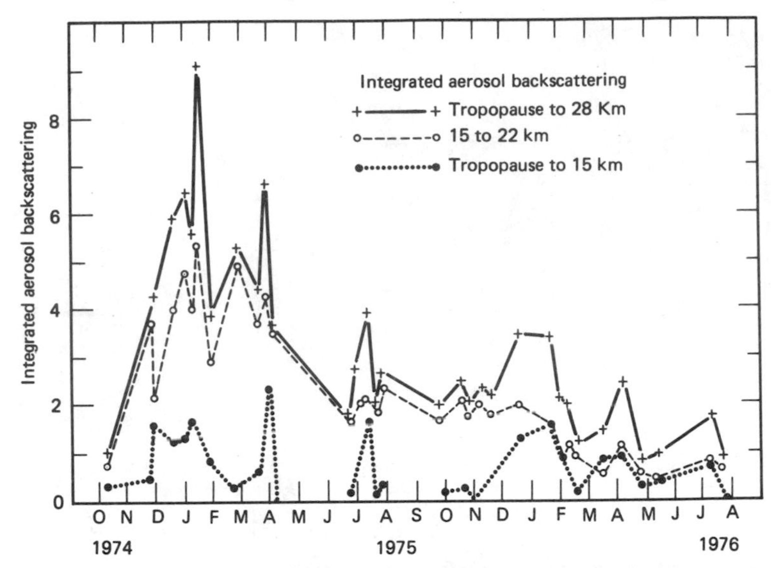

Fig. 9.29. Time variation of the integrated lidar aerosol backscattering for given stratospheric layers from October 1974 to July 1976 (McCormick et al., 1978).

Three days after the Mount St. Helens eruption McCormick (1982) reported that peak scattering ratios as high as 100 were recorded at a wavelength of 1.06 μm for a layer that was about 1 km thick at an altitude of 13.6 km and at a location south of Lake Erie.

These combined ground-based and airborne lidar measurements indicated that Mount St. Helens ejected approximately 0.5×10^6 metric tons of material into the stratosphere. Even though this represents a 200% increase in the aeroticulate loading of the Northern Hemisphere, no significant long-term climatic effects are expected.

9.1.7. Upper-Atmospheric Probing

Under normal conditions elastically backscattered radiation from altitudes between 30 and 90 km corresponds very closely to that expected from Rayleigh scattering by atmospheric molecules. This is illustrated in Fig. 9.32, where we present a typical atmospheric backscattering profile obtained by Kent and Wright (1970) using a *Q*-switched ruby laser. Noteworthy is the high precision and close fit to a standard-atmosphere scattering curve. The existence of noctilucent clouds at high altitudes in the summer, however, suggested the presence of dust in the upper atmosphere, and the pioneering lidar work of Fiocco and Smullin (1963) provided some evidence of this dust.

There are three probable sources of mesospheric dust: passage of the earth through the plane of a comet's orbit, micrometeor fragmentation, and vertical

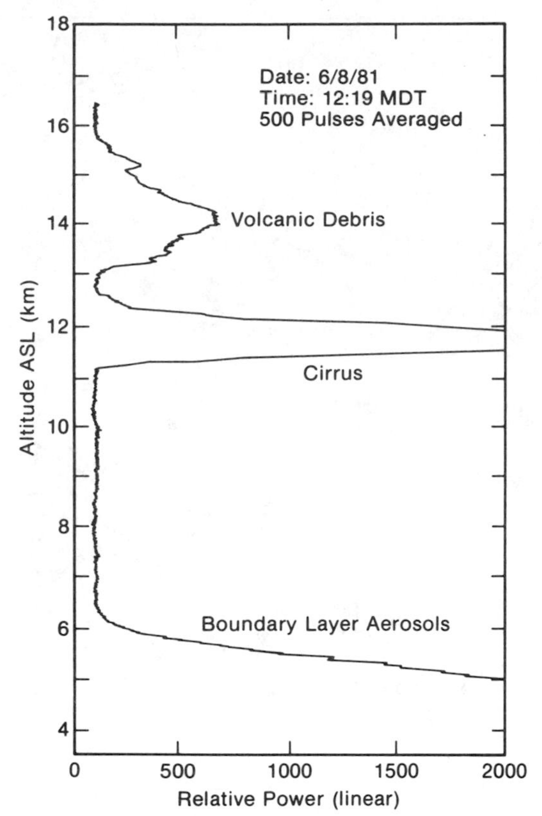

Fig. 9.30. Raw data (signal power versus altitude) for the strongest observed volcanic-debris lidar return. Aerosols are responsible for the return from lower altitude, and the debris is seen through a tenuous cirrus layer (Post et al., 1982).

transport of volcanic material. Poultney (1972a) indicates that the most probable cause of the occasional enhanced lidar returns from high altitudes is dust having a cometary origin. The lidar measurements of Kent et al. (1971), made during the passage of the earth through the plane of the orbit of the comet Bennett, show a sudden increase in the scattering signal from between 40 and 90 km. These observations revealed that dust entering the Earth's atmosphere from the comet descended more rapidly than if it were falling in a static atmosphere, reaching an altitude of 40 km within a day or so. This suggests strong vertical mixing in the upper atmosphere.

The development of high-energy, flashlamp-pumped, tunable dye lasers made it possible to map certain trace constituents in the upper atmosphere through the process of laser-induced (resonance) fluorescence. Bowman et al. (1969) undertook the first ground-based measurement of the sodium-atom

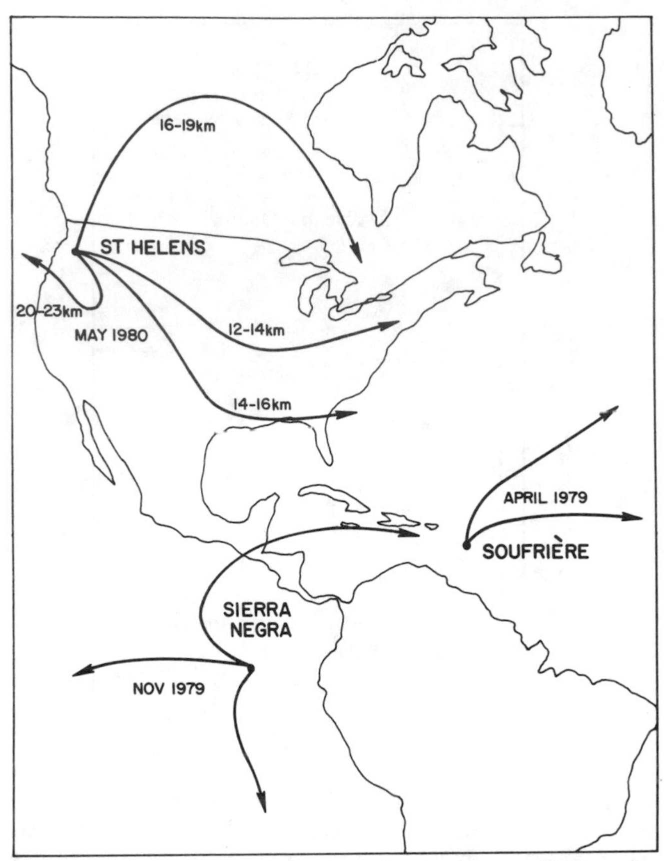

Fig. 9.31. Movement of material injected into the stratosphere at different altitudes by the eruptions of Soufrière, Sierra Negra and St. Helens (McCormick et al., 1982).

concentration in the tenuous outer regions of the atmosphere. More detailed measurements, including the seasonal variations of this sodium concentration, were obtained by Gibson and Sandford (1971).

Hake et al. (1972) observed a fourfold increase in the sodium-layer content during the maximum of the Geminids meteor shower on the night of 13–14 December 1971. This observation, presented as Fig. 9.33, lends support to the idea that meteor ablation represents an important source of this material. The laser used by Hake et al. is representative of those used in these kinds of experiments and comprised a two-stage (oscillator-amplifier) Rhodamine 6G dye laser with an output of 0.5 J in 300 ns at 589.0 nm. The spectral linewidth was less than 0.005 nm, and the final beam was collimated to better than 0.5 mrad. Additional confirmation of the meteor production theory was obtained

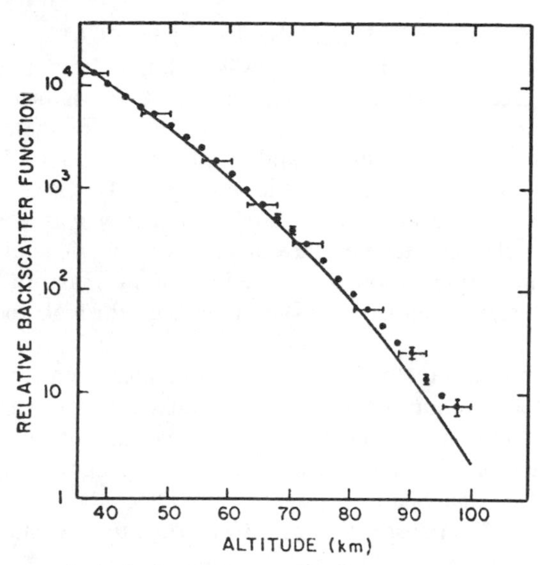

Fig. 9.32. Lidar atmospheric backscattering profile, Kingston, Jamaica, 25 March 1969 (300 shots). Points (with error bars): experimental results; curve: fitted standard atmosphere (Poultney, 1972a).

by Aruga et al. (1974). Although most observations of this sodium layer were made at night, Gibson and Sandford (1972) were able to modify their system sufficiently to map the spatial distribution of the sodium layer during the day and thereby dispel the notion of daytime enhancement.

Nighttime studies by Blamont et al. (1972) revealed a stratification of the sodium layer. They also employed an (oscillator–amplifier) flashlamp-pumped

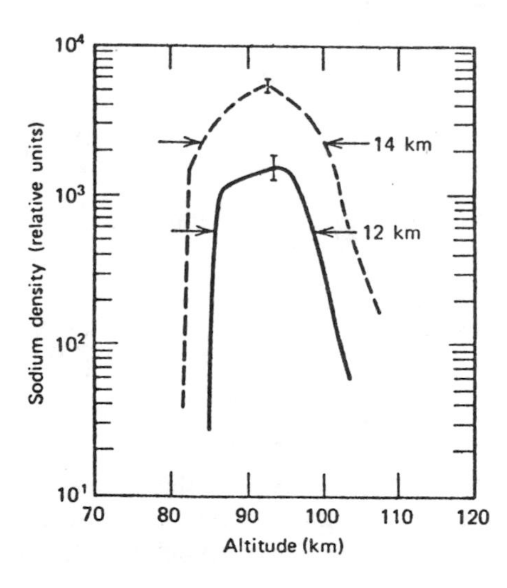

Fig. 9.33. Altitude profiles of the density of free atomic sodium, obtained at 2315–2400 PST (solid line) and 0255–0310 PST (dashed line), before and slightly after transit of the radiant of the Germinids meteor shower on 13–14 December 1971 (Hake et al., 1972).

dye laser, the output of which was about 0.3 J in a 0.013-nm linewidth. The pulse duration was 1 μs, and the beam was collimated by an afocal system to 0.3 mrad. The receiver system comprised an 80-cm-diameter telescope, an afocal optical arrangement for reducing the beam diameter to that of a 2-nm-linewidth interference filter, and a photomultiplier. A sodium-vapor scattering cell was used to check the tuning of the laser and provide a measurement of its energy. This sodium-cell facility was also employed to measure the translational temperature of the mesospheric sodium atoms. This was accomplished by measuring the reduction in magnitude of the lidar resonance fluorescence return signal after passage through the sodium-vapor cell.

To understand this method of temperature measurement we invoke the radiative-transfer equation for a two-level atom, equation (4.31). For the long times associated with the resonance-fluorescence return signal we can assume a steady-state situation. We can also assume that the magnitude of the radiation in this instance is inadequate to substantially effect the atomic level populations within the sodium absorption cell. Consequently, we can write

$$\frac{dI(\nu, z)}{dz} = -k(\nu_0, z)I(\nu, z)\frac{\mathscr{L}^A(\nu)}{\mathscr{L}^A(\nu_0)} \tag{9.31}$$

where $I(\nu, z)$ represents the spectral irradiance of the resonance-fluorescence return signal, $\mathscr{L}^A(\nu)$ is the absorption-line profile function, and $k(\nu_0, I)$ is the volume absorption coefficient at the line center frequency ν_0 of the sodium vapor at a depth z within the absorption cell. The solution of (9.31) yields the transmitted spectral irradiance:

$$I(\nu, z_A) = I(\nu, 0)\mathrm{e}^{-\tau_A \mathscr{L}^A(\nu)/\mathscr{L}^A(\nu_0)} \tag{9.32}$$

where

$$\tau_A = \int_{z=0}^{z=z_A} k(\nu_0, z)\, dz \tag{9.33}$$

represents the optical depth of the sodium-vapor cell at the line center frequency. We introduce the *reduction factor* for the cell,

$$R_A^M \equiv \frac{\int_{-\infty}^{\infty} I(\nu, z_A)\, d\nu}{\int_{-\infty}^{\infty} I(\nu, 0)\, d\nu} \tag{9.34}$$

and assume that the return resonance fluorescence is Doppler-broadened according to the translational temperature T_M of the sodium atoms in the mesosphere and that the absorption-line profile is also Doppler-broadened according to the sodium-vapor temperature T_A in the cell. Then we can write

$$I(\nu, 0) = \frac{I_0}{\Delta\nu_M \pi^{1/2}} \exp\left\{ - \left(\frac{\nu - \nu_0}{\Delta\nu_M} \right)^2 \right\} \tag{9.35}$$

for the resonance fluorescence incident on the cell, where

$$\Delta\nu_M = \left\{ \frac{2kT_M \nu_0^2}{mc^2} \right\}^{1/2} \tag{9.36}$$

represents the Doppler width [see equation (3.149)] for this radiation. We can also write

$$\mathscr{L}^A(\nu) = \frac{1}{\Delta\nu_A \pi^{1/2}} \exp\left\{ - \left(\frac{\nu - \nu_0}{\Delta\nu_A} \right)^2 \right\} \tag{9.37}$$

for the absorption-line profile, where in this instance the Doppler width is

$$\Delta\nu_A = \left\{ \frac{2kT_A \nu_0^2}{mc^2} \right\}^{1/2} \tag{9.38}$$

Under these circumstances, the reduction factor for the cell can be expressed in the form

$$R_A^M = \frac{\int_{-\infty}^{\infty} \exp\left\{ -x^2 - \tau_A e^{-x^2 T_M / T_A} \right\} dx}{\int_{-\infty}^{\infty} \exp\{-x^2\} \, dx} \tag{9.39}$$

where

$$x = \frac{\nu - \nu_0}{\Delta\nu_M}. \tag{9.40}$$

Evidently, R_A^M is only a function of the experimental quantities τ_A and T_M/T_A. The temperature T_A of the absorption cell is known, and its optical depth τ_A can be evaluated using a sodium resonance lamp of known temperature. Consequently, T_M can be determined; the values so obtained by Blamont et al. (1972) are presented as Fig. 9.34.

More recently Megie and Blamont (1977) have reported on an extensive program of nighttime atmospheric-sodium measurements. They have developed a dynamic photochemical model of the upper atmosphere and used it to simulate the behavior of the atmospheric sodium. The experimental results were then compared with this simulation and seem to confirm that photochemical equilibrium between

$$Na + O_3 \rightarrow NaO + O_2$$

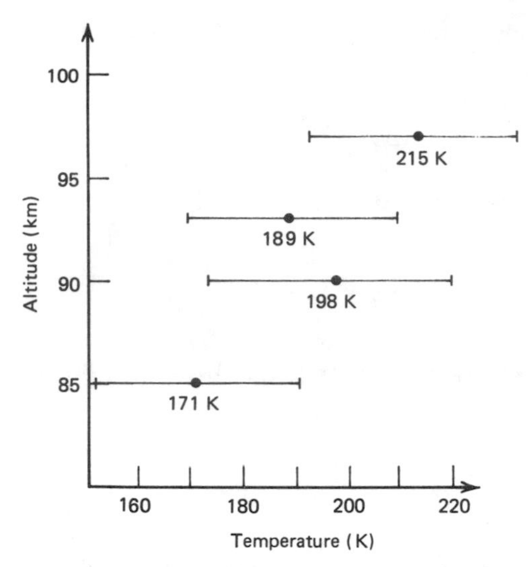

Fig. 9.34. Lidar-measured temperature of the upper atmosphere (Blamont et al., 1972).

and

$$NaO + O \rightarrow Na + O_2$$

exists during the night. The behavior of this sodium layer is then surmised to be determined by dynamical processes related essentially to eddy diffusion mixing and to the strength of a permanent source of sodium. The stratifications in the density profiles first observed by Blamont et al. (1972) were also confirmed and correlated with the propagation of gravity waves at mesospheric heights.

Although sporadic enhancement of the sodium content of the atmosphere seems to have been related to meteoritic showers, Megie and Blamont (1977) stated that the influence of strong eddy mixing makes it impossible to draw a conclusion concerning the meteoric origin of *all* the atmospheric sodium. They contended that a simultaneous measurement of the sodium and potassium concentrations in the upper atmosphere would permit the influence of vertical transport to be separated from that of meteoric deposition, and later reported on such observations (Megie et al., 1978).

The characteristics of the lidar system employed by Megie et al. (1978) are summarized in Table 9.3. It is apparent that the characteristics of the two lasers are very similar except for their pulse durations, their repetition rates, and the low divergence of the laser tuned to the potassium resonance line. These differences arise from the respective mode of pumping. A flashlamp is used to pump the dye laser tuned to the sodium resonance line, while a Q-switched ruby laser is used to pump the one tuned to the potassium line. The

TABLE 9.3 CHARACTERISTICS OF TWO LIDAR SYSTEMS[a]

Emitter:	Sodium (λ = 589 nm)	Potassium (λ = 769.9 nm)
Output energy	1J	1J
$\Delta\lambda$ output	8.5 pm	8 pm
Pulse duration	3 μs	30 ns
Repetition rate	0.5 Hz	0.1 Hz
Divergence	5×10^{-3} rad	5×10^{-4} rad
Divergence[b]	5×10^{-4} rad	
Receiver:		
Telescope diameter	0.818 m	
Telescope area	0.515 m^2	
Field of view	3×10^{-3} rad	
Bandwidth	0.5 nm	

[a]Megie et al. (1978).
[b]After collimation.

difference in the pulse repetition rate combined with the lower abundance of potassium necessitated a much longer integration time for the potassium measurement. The typical sampling period for sodium was about 30 min, while for potassium 2 to 3 hr were required.

In order to derive the absolute values of the sodium and potassium abundances together with their ratio, a detailed analysis of the accuracy of the experiment and calibration was performed. Of particular concern was the possibility that the much higher-power DOTC dye laser might lead to saturation effects in the potassium layer. To follow the reasoning of Megie et al. (1978), we draw upon equation (4.55). This can be taken to express the ratio of the resonance fluorescence received by the lidar to that which would be received in the absence of saturation and can be compared directly with their equation (9) if allowance is made for their nomenclature. In terms of their parameters, the laser irradiance is

$$I^L = \frac{N_L h\nu T(R)}{\tau_L R^2 \Omega} \tag{9.41}$$

where N_L is the number of photons in the laser pulse, $h\nu$ is the energy of each photon, $T(R)$ is the atmospheric transmission factor, R is the range from which the resonance fluorescence originates, and Ω is the solid angle of laser emission:

$$\Omega = \frac{\pi\theta_L^2}{4} \tag{9.42}$$

in which θ_L represents the full divergence angle of the laser beam.

The saturated irradiance [see equation (4.40)], using their parameters, becomes

$$I^S = \frac{h\nu}{2\sigma_E \tau_{21}} \tag{9.43}$$

where σ_E represents the spectrally integrated emission cross section and τ_{21} the radiative lifetime of the resonance level. The factor of 2 arises from assuming equal degeneracies for the ground and resonance levels. Under these conditions the *saturation parameter*,

$$S_I = \frac{2N_L \sigma_E \tau_{21} T(R)}{\tau_L R^2 \Omega} \tag{9.44}$$

is obviously inversely proportional to θ_L^2, and the *saturation time* [equation (4.51)] is given by

$$\tau_S = \frac{R^2 \Omega \tau_L}{2\sigma_E N_L T(R)} \tag{9.45}$$

Consequently, the *saturation correction factor* [equation (4.55)] can be written

$$\mathscr{J}(\tau_L, \theta_L) = \frac{\tau_{21}}{\tau_L(1 + S_I)} \left[\frac{\tau_L}{\tau_{21}} + \frac{S_I}{1 + S_I} \{1 - e^{-(\tau_L/\tau_{21})(1+S_I)}\} \right] \tag{9.46}$$

Megie et al. (1978) have chosen to express this as the ratio of fluorescence photons with and without saturation and have plotted this ratio as a function of the divergence angle θ_L for two values of the pulse duration, $\tau_L = 25$ ns and $\tau_L = 10$ ns. They have also considered two other laser pulse shapes and have found that $\mathscr{J}(\tau_L, \theta_L)$ is not very sensitive to this change. Their results are reproduced here as Fig. 9.35. The values of θ_L, τ_L, and N_L, as measured for their DOTC laser in a routine experiment, are

$$\theta_L > 7 \times 10^{-4} \text{ rad}, \qquad \tau_L = 30 \text{ ns}, \quad \text{and} \quad N_L \lesssim 3 \times 10^{18} \text{ photons.}$$

This leads to a saturation time $\tau_S \lesssim 150$ ns and a value for the ratio $\mathscr{J}(\tau_L, \theta_L) \geq 0.92$. Thus if saturation were neglected, the measured potassium density would not be expected to be out by more than 10%.

Laser soundings of the mesosphere for sodium and potassium have been performed for 55 and 15 nights, respectively, over the period from July 1975 to September 1976 at the Haute Provence Observatory by Megie et al. (1978). Their results are presented as Fig. 9.36 and indicate a peak *column density* of $(8 \pm 2) \times 10^9$ atoms cm^{-2} for sodium during the November–December period.

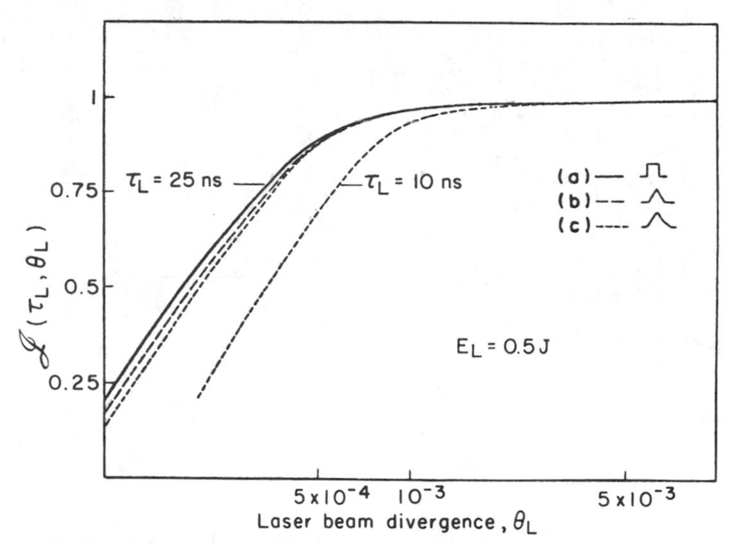

Fig. 9.35. Variation of the saturation correction factor $\mathcal{J}(\tau_L, \theta_L)$ with laser beam divergence θ_L (for a given laser energy), illustrating the saturation effects that can arise in the lidar return signal for short excitation time (Megie et al., 1978).

Representative examples of the sodium and potassium distributions in the mesosphere as determined by Megie et al. (1978) are shown as Fig. 9.37. It is apparent that the potassium layer is about 20% thinner than the sodium layer and its peak density is nearly an order of magnitude smaller.

The seasonal variations in the sodium and potassium abundances show that their ratio increases by a factor of 4–5 during the winter, so that the ratio varies between a minimum value of about 10 (compatible with a meteoritic origin) and a maximum of 40 to 50 (compatible with a terrestrial origin). These observations led Megie et al. (1978) to suggest that there are, in fact, two sources for the alkali content of the upper atmosphere:

A meteoritic source that is constant over the year and responsible for the low value of the abundance ratio in the summer. The correlation between sporadic meteoritic showers and increases in the alkali content bears witness to this source.

A terrestrial source due to the vertical transport of salt particles at high latitudes, which works only in winter, when the circulation pattern of the polar stratosphere breaks down.

More recently, Granier and Megie (1982) reported on daytime measurements of the mesospheric sodium layer with an improved lidar system. The characteristics of this system are presented in Table 9.4. A signal-to-noise ratio of better than 50 was obtained with this system, primarily as a result of

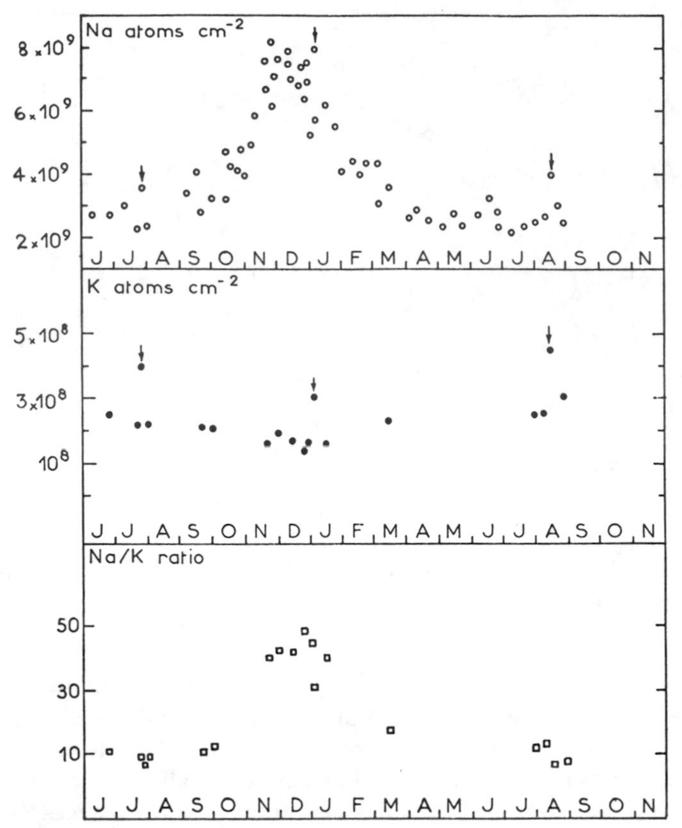

Fig. 9.36. Seasonal variations of sodium and potassium total contents and of their abundance ratio, as measured by a ground-based lidar (Megie et al., 1978).

reducing the receiver bandwidth to 20 pm (2×10^{-2} nm) and the field of view to 0.3 mrad. This improved performance enabled Granier and Megie to follow the time evolution of the sodium layer throughout a full diurnal cycle, as the SNR of 50 was achieved for resonance-fluorescence returns from 90 km with an integration period of less than one hour. A representative example of the average altitude distribution of the sodium layer for daytime and nighttime conditions integrated over a consecutive 3-day period is presented as Fig. 9.38. These measurements confirm the lack of large variation of the sodium abundance between day and night—a result earlier suggested by Gibson and Sandford (1972). This lack of regular variation of the sodium-layer characteristics—such as peak concentration and topside and bottomside scale heights—during the diurnal cycle or during day–night transitions suggests that the photochemical processes do not play a dominant role in the behavior of the mesospheric sodium atoms.

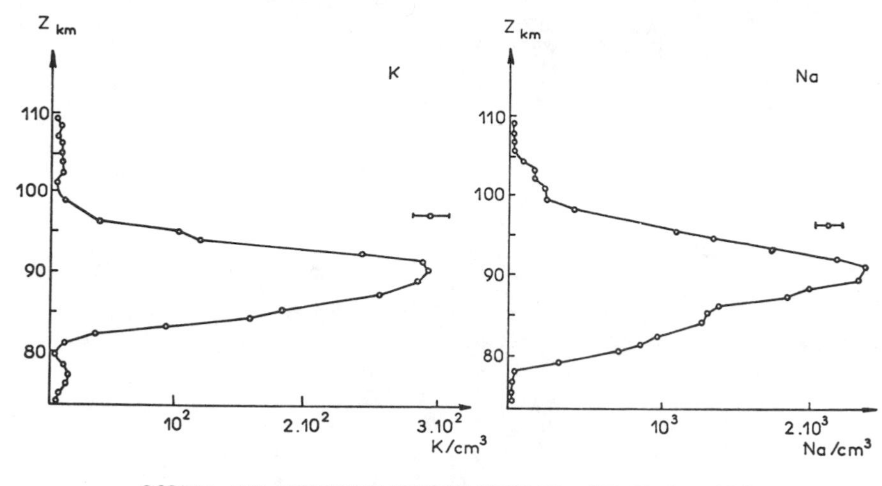

Fig. 9.37. Nominal profile of the sodium and potassium layers as determined by a ground-based lidar (Megie et al., 1978).

This conclusion is reinforced by studies of the evolution of the layer during twilight. This evolution is observed to be highly variable from one day to another, as seen by reference to Figs. 9.39 and 9.40. Figure 9.39 presents the lidar measurements of the sodium layer observed during the evening twilight on 14 June 1979, while Fig. 9.40 presents the same transition on 3 December 1979. Granier and Megie (1982) surmise that the three sodium profiles seen in Fig. 9.39 are due to the propagation of an internal gravity wave in the 80- to 100-km range rather than to any photochemical effects. By comparison, a relatively stable sodium layer is observed in Fig. 9.40. Granier and Megie

TABLE 9.4. LIDAR SYSTEM CHARACTERISTICS[a]

Emitter:	
Output energy	0.5 J
Wavelength	589 nm
Linewidth	6 pm
Repetition rate	1 Hz
Beam divergence[b]	2×10^{-4} rad
Receiver:	
Diameter	0.81 m
Field of view	3×10^{-4} rad
Bandwidth	20 pm
Separation of emitter and receiver	0.75 m

[a]Granier and Megie (1982).
[b]After collimation.

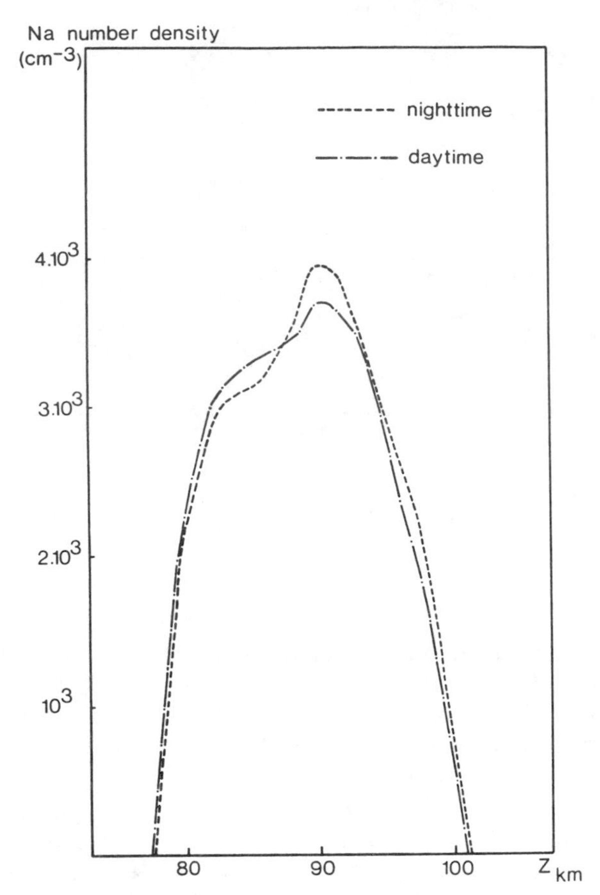

Fig. 9.38. Average altitude distribution of the sodium layer for nighttime and daytime conditions integrated over a 3-day period in December 1980 as determined by a ground-based lidar system (Granier and Megie, 1982).

conclude that the dynamical effects such as horizontal transport and organized vertical motions dominate the regular variations induced by the solar diurnal cycle as predicted by the photochemical models of the sodium layer (Megie and Blamont, 1977).

9.2. SPACEBORNE LIDAR OPERATION

A lidar system in orbit around the earth would be capable of gobal surveillance of the atmosphere. It would also possess a number of advantages over ground-based or even airborne lidars. These include an ability to probe the atmosphere with wavelengths that could not propagate through the lower

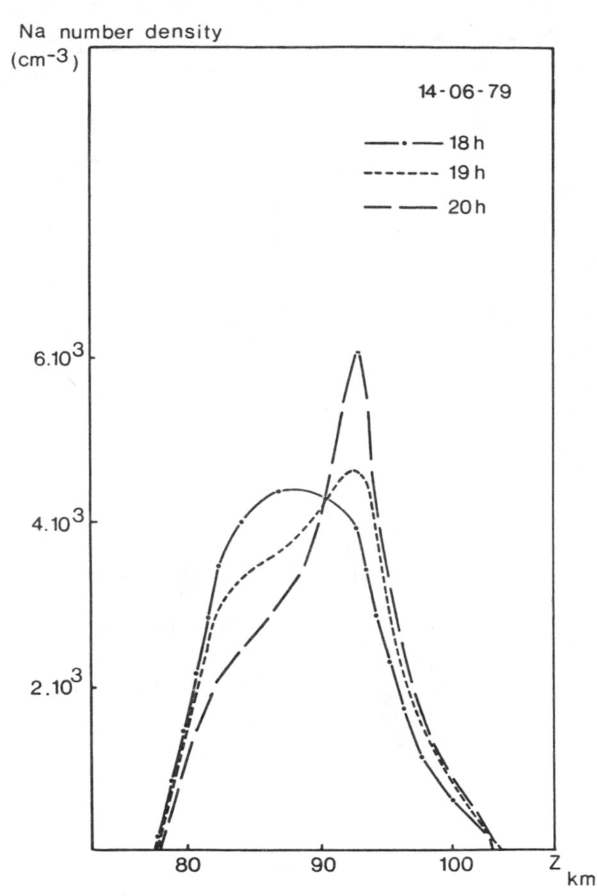

Fig. 9.39. Evolution of the sodium altitude profile during twilight, showing the influence of a wavelike propagating perturbation (Granier and Megie, 1982).

atmosphere, and to provide synoptic coverage with high resolution. In order to develop the rationale for a Shuttle atmospheric lidar system, an international working group of scientists was convened by the U.S. National Aeronautics and Space Administration (NASA) in 1977 for the purpose of (1) identifying the major goals of a spaceborne lidar system, (2) proposing a set of experiments that could be undertaken by a lidar system mounted aboard the Space Shuttle, and (3) providing an assessment of the technology available for embarking upon this research program. The final report of this committee was released as a NASA report (SP-433) in 1979, and Abreu (1980) has discussed some of the considerations that led this committee to its conclusions.

The primary goals of the Shuttle atmospheric lidar program are to contribute to an understanding of the processes governing the Earth's atmosphere and to evaluate the susceptibility of the atmosphere to both man-made and natural

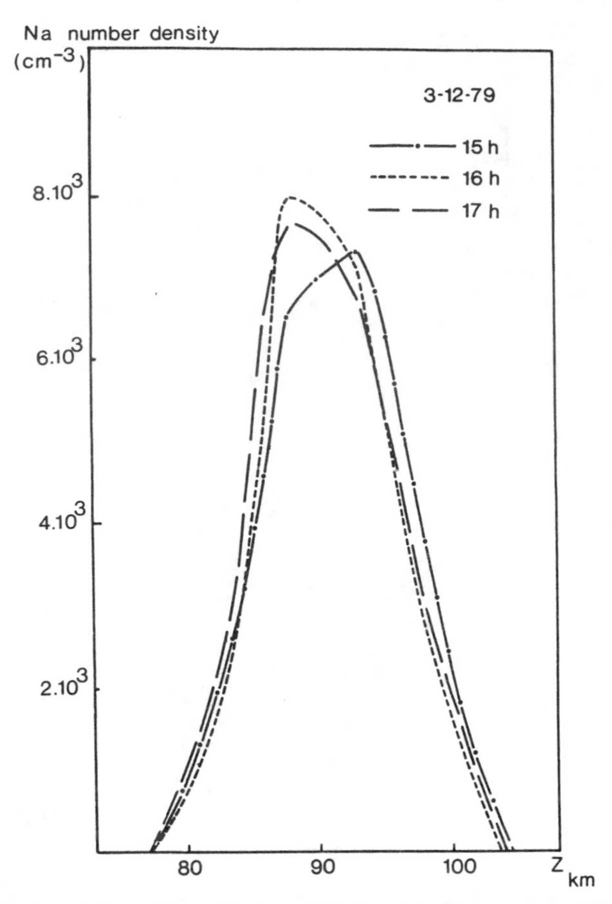

Fig. 9.40. Evolution of the sodium altitude profile during twilight, corresponding to a stable situation as opposed to the evolution described in Fig. 9.39 (Granier and Megie, 1982).

perturbations. It is hoped that the unique attributes of a Shuttle-borne lidar system will enable it to make an important contribution to seven major science and applications objectives, identified by the NASA lidar working group as:

1. Determination of the global flow of water vapor and pollutants in the troposphere and lower stratosphere.
2. Improvement of chemical and transport models of the stratosphere and mesosphere.
3. Evaluation of radiative models of the atmosphere.
4. Augmentation of the meteorological data base.
5. Investigation of excitation, propagation, and dissipation of wave motions in the upper atmosphere.

6. Investigation of the chemistry and transport of thermospheric atomic species.
7. Investigation of magnetospheric aspects of the sun–weather relationships.

In addition a Shuttle lidar could participate in coordinated measurement programs with other instrument packages mounted aboard the Shuttle. Such an arrangement would provide the opportunity for synergistic experiments involving the lidar system with one or more other kinds of remote sensor located on the same (or even another) spacecraft. As an example, simultaneous lidar and passive infrared measurements of cloud-top heights would provide both a means of testing conventional measurement techniques and the necessary data for improving passive-analysis algorithms. There is even the possibility of undertaking active perturbation experiments involving excitation by the laser beam and detection by other sensors.

9.2.1. Shuttle Lidar Facility

Abreu (1980) suggests that the evolutionary modular approach to the lidar hardware represents one of the most important proposals made by the NASA Atmospheric Lidar Working Group. This idea involves (1) making the lidar components readily interchangeable so that they can be replaced as demanded by technological developments or changing measurement needs, and (2) allowing the relatively simple initial hardware complement of the early experiments to be augmented by better or more sophisticated technology as it develops. The modular approach is based on the fact that a lidar system can naturally be divided into components linked by well-defined interfaces. These components are the laser, the receiving optics, the spectral analyzer (or filter), and the detector.

In order to cover the full range of lidar activities it will be necessary to operate with several lasers that will range in wavelength from the ultraviolet through to the infrared. The most promising choices at present appear to be the Nd–YAG and CO_2 lasers, since they have both been shown to be reliable in extensive flight trials. The Nd–YAG laser can be frequency-doubled, -tripled, and even -quadrupled to produce outputs in the UV, and it has become one of the standard pump sources for tunable dye lasers. The CO_2 laser can operate on many lines in the 9- to 11-μm range and will possibly operate in a heterodyne configuration for maximum sensitivity.

The lidar equation (Chapter 7) shows that the received signal is proportional to the output energy of the laser and the area of the receiver optics. Consequently, in addition to employing lasers with highest output that is practical, it is also necessary to use as large a telescope as feasible. Weight, power, and space limitations aboard the Shuttle have led to the choice of a 1.25-m-diameter Cassegrain telescope, and the system arrangement shown in Fig. 9.41. The output energy of the Nd–YAG laser is likely to be around 1 J, while that of the

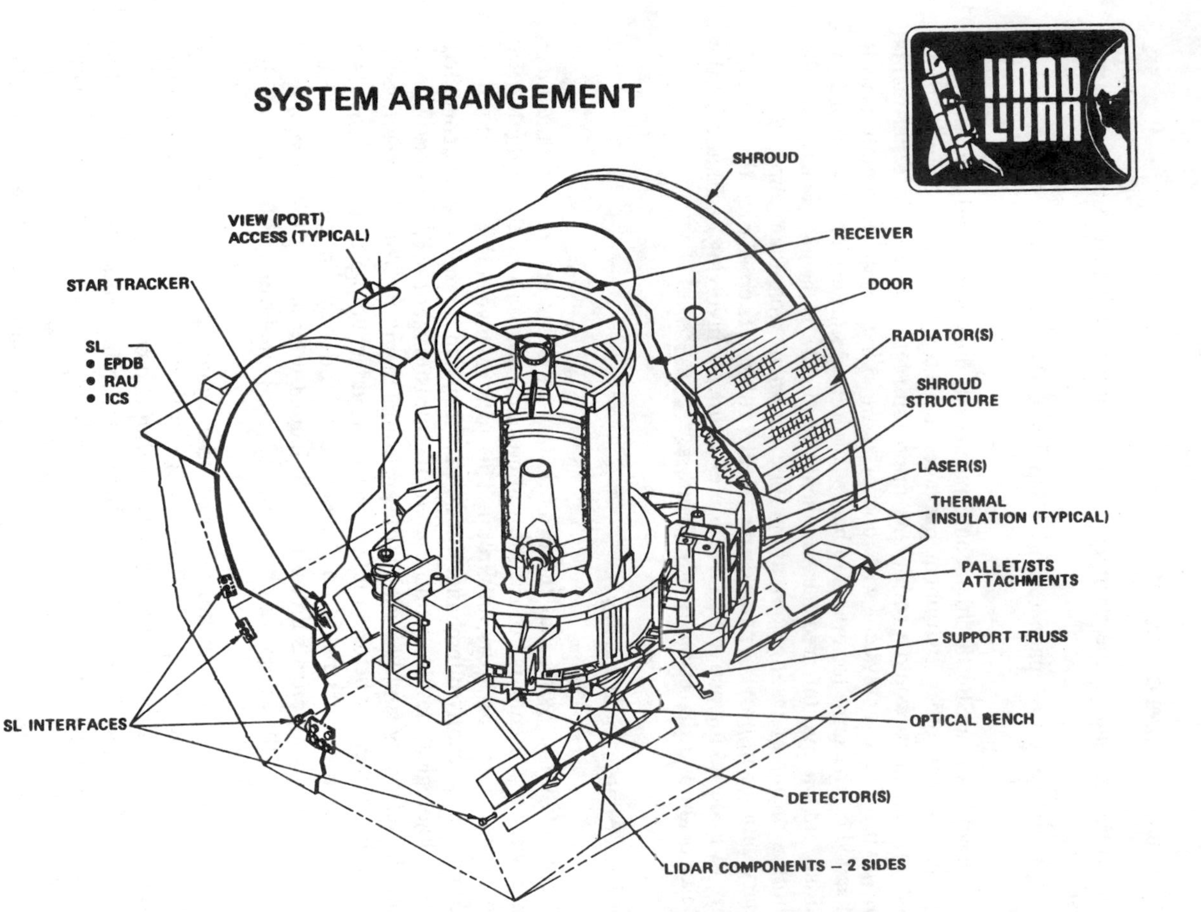

Fig. 9.41. Proposed lidar system arrangement in the Shuttle (Greco, 1979).

CO_2 might be closer to 0.1 J. Interference filters will provide the required spectral discrimination, since they are useful over the wavelength range 0.2 to 30 μm and are rugged, compact, and light-weight. For certain experiments where better spectral resolution or some degree of tunability is required, Fabry–Perot interferometers will probably be utilized.

An important design consideration discussed by Abreu (1980) was the coupling of the telescope to the filter–interferometer detector system. As mentioned in Section 6.5, the parameter used to describe this condition is the *étendue* or light-gathering power of the system. The étendue of the telescope is defined as the telescope area times the solid angle determined by its field of view, that is, $A_T\Omega_T$. For maximum efficiency the étendue of the telescope should be matched to that of the filter or interferometer. The étendue of a Fabry–Perot interferometer with an on-axis aperture is $2\pi A_{FP}\Delta\lambda/\lambda_0$, where A_{FP} is the area of the Fabry-Perot and $\Delta\lambda$ is its bandpass for wavelength λ_0. It follows that maximum light collection is achieved when

$$A_T\theta_T^2 = \frac{2A_{FP}\Delta\lambda}{\lambda_0}$$

where θ_T represents the telescope's angular field of view. Evidently a bandpass-limited device may dictate the field of view of the receiver and therefore the laser's divergence angle.

It is expected that in the UV, visible, and near IR the primary detection devices will be photomultipliers, while in the mid IR semiconductor devices will be used. It is also possible that for certain experiments detector arrays will be employed; these could include photodiode arrays, multichannel plates, vidicons, charge-coupled devices, and coupled photomultipliers (Abreu, 1980).

9.2.2. Planned Lidar Observations from Space

The NASA Atmospheric Lidar Working Group identified 26 candidate experiment classes which would significantly contribute to the seven science objectives discussed in Section 9.2.1. These are listed in Table 9.5 with the altitude of interest, the lidar principle involved, the type of laser required, and the scientific objectives to which each pertains.

The capability of an orbiting lidar to probe between clouds with very good vertical and horizontal resolution could provide global measurements that might revolutionize investigations of the troposphere. Some of the principal measurements would include monitoring the distribution and total abundance of aeroticulates; studying the optical thickness and height of clouds; and evaluating such meteorological parameters as humidity, winds, temperature, and pressure. In addition a DIAL system could be used to measure water-vapor profiles in the troposphere and lower stratosphere; ozone in the stratosphere; and the tropospheric profiles of gases such as O_3, NH_3, CH_4, CO, CO_2, and C_2H_4 among others. It is expected that these differential-absorption-based

TABLE 9.5. SET OF SHUTTLE-LIDAR CANDIDATE EXPERIMENTS IDENTIFIED BY THE NASA ATMOSPHERIC LIDAR WORKING GROUP[a]

No.	Description	Altitude Region (km)	Principle	Laser	Scientific Objectives[b]
1	Cloud top heights	0–15	Elastic backscattering	Any	3, 4
2	Profiling of tropospheric clouds and aerosols	0–15	Elastic backscattering	Any (0.5–2 μm)	1, 3, 4
3	Cirrus ice–water discrimination	5–15	Polarization-sensitive elastic backscattering	Any	1, 3
4	Profiles of noctilucent clouds and circumpolar particulate layers	60–80	Elastic backscattering	Any (0.5–2 μm)	3, 5
5	Surface reflectance	Ground	Surface scattering	Any	3
6	Stratospheric aerosol backscattering profiles	10–50	Elastic backscattering	Any (0.5–2 μm)	1, 2, 3,
7	Alkali-atom density profiles	80–120	Resonant scattering	Tuned dye	5, 6
8	Ionospheric metal-ion distributions	80–600	Resonant scattering	Tuned dye	6, 7
9	Water-vapor profiles	0–20	DIAL	Tuned dye	1, 3, 4, 5
10	Atmospheric species measurements using cw IR laser ground and cloud returns	0–30	Long-path absorption (column content)	Line-tunable cw CO_2	1, 3, 4
11	Chemical-release diagnosis	90–50,000	Resonant scattering	Tuned dye	7
12	Stratospheric ozone concentration profiles	20–60	Differential-range absorption	Nd × 4, and/or dye	1, 2, 3, 7
13	Upper-atmospheric trace-species measurements using two-satellite occultation	10–50	Long-path absorption	Tunable, mainly IR	2, 3, 7
14	Sodium-layer temperature and winds	80–110	Doppler-sensitive resonant scattering	Tuned dye	5, 6, 7
15	Surface pressure and cloud-top pressure and height measurements	0–10	O_2 absorption (column content)	Tuned dye	4
16	Vertical profiles of atmospheric pressure	0–10	O_2 absorption (range-resolved)	Tuned dye	4

17	Temperature profile	0–10	Temperature-sensitive O_2 absorption	Tuned dye	1, 2, 3, 4
18	Altitude distribution of atmospheric constituents using IR DIAL	0–15	DIAL	Line-tunable pulsed CO_2	1, 3, 4
19	Cloud-top winds	0–15	Doppler-sensitive elastic backscattering	Any narrowband	1, 2, 4
20	Aerosol winds	0–25	Doppler-sensitive elastic backscattering	Any narrowband	1, 2, 4, 5
21	OH density profile between 35- and 100-km altitude	35–100	Resonance fluorescence	Tuned dye	2, 5
22	Simultaneous measurement of metallio-atom, ion, and oxide profiles	80–600	Resonant scattering	Tuned dye	6
23	Tropospheric NO_2 concentration profile and total burden of NO_2	0–15	DIAL	Tuned dye	1
24	Stratospheric aerosol composition	15–30	Differential scattering (DISC)	Line-tunable pulsed CO_2	1, 3
25	NO density profiles between 70- and 150-km altitude	70–150	Resonance fluorescence	Tuned dye	2, 7
26	Abundance and vertical profiles of atomic oxygen	80–500	Two-photon fluorescence	Tuned dye	6, 7

[a] From Shuttle Atmospheric Lidar Research Program NASA SP-433, 1979, and Abreu (1980).

[b] Objective numbers denote the following: 1, global flow of water vapor and pollutants; 2, stratospheric and atmospheric chemistry and transport; 3, radiative models; 4, meteorological data; 5, upper-atmospheric waves; 6, thermospheric chemistry and transport; 7, magnetospheric sun and weather relationships.

measurements will involve the use of a tunable dye laser or a pulsed tunable CO_2 laser.

As we have seen (Section 9.1.7), ground-based lidar systems using resonance fluorescence have been able to map the distributions of sodium and potassium in the mesosphere. An orbital lidar would significantly augment this research, for it would not only provide continuous global coverage, but would also permit the laser resonance-fluorescence technique to be extended to evaluating the distributions of metal ions in the ionosphere and the density profiles of nitric oxide molecules and hydroxyl free radicals in the mesosphere. Such measurements are important in studies of the chemistry and transport of the upper atmosphere. For the most part a tunable dye laser is expected to be the source of excitation for this work.

The hydroxyl radical plays a central role in the chemistry of the atmosphere. In the stratosphere OH takes part in catalytic ozone destruction processes, which regulate the concentration of O_3 in both the upper stratosphere and the lower mesosphere. For example,

$$OH + O_3 \rightarrow HO_2 + O_2$$

$$HO_2 + O \rightarrow OH + O_2$$

OH also serves to regulate similar catalytic ozone destruction cycles involving chlorine and nitrogen oxides: the reaction

$$OH + NO_2 + M \rightarrow HNO_3 + M$$

removes NO_2 from the nitrogen oxide catalytic cycle, and

$$OH + HCl \rightarrow H_2O + Cl$$

releases atomic chlorine, which can then destroy ozone in a catalytic destruction cycle (see Chapter 1). Measurements of the diurnal variation in the OH concentration by a Shuttle-borne lidar system would permit a detailed test of the chemical models that have recently been developed to understand these processes.

Heaps (1980) has undertaken an analysis of the expected resonance fluorescence return for a Shuttle-mounted lidar system that is optimized for the detection of the hydroxyl radical. This system involves a laser operating at either 282 or 308 nm with an output energy of about 1 J. The cross section, averaged over the typical Doppler width and a laser linewidth of 0.001 nm, is about 2×10^{-16} cm^2. Atmospheric transmission for this situation is primarily determined by ozone absorption and Rayleigh (molecular) scattering. They have been calculated using the U.S. Standard Atmosphere, giving a transmission factor of close to unity down to 60 km, and about 0.013 at 30 km for 282 nm. A 1-m-diameter Cassegrain telescope (with a 20% obscuration) and a 10-nm-broad bandpass filter having a transmission of 10% were assumed.

Simulations using these values and a spatial resolution of 5 km led to the results presented in Table 9.6. The OH concentration and the expected number of resonance-fluorescence photons returned is shown as a function of altitude for four different times of the day.

The largest signals arise at noon (1200) for mid altitudes where the OH concentration reaches a reasonably high value and the ozone attenuation is modest. These daytime signals cannot be interpreted in terms of a hydroxyl measurement without consideration of the scattered solar flux, which is the principal source of noise during the day. For nighttime operations the values presented in Table 9.6 can be taken as representative of the measurement capability of the system. The poor photon returns from the lower altitudes derive from quenching and from attenuation of the 282-nm laser beam through absorption by ozone. If narrowband detection, sufficient to differentiate between fluorescence and scattered radiation, were available, then fluorescence could be excited by direct (0, 0) pumping at 308 nm. This wavelength is attenuated less by the atmosphere and consequently would lead to larger return signals from the low altitudes. Unfortunately the poor efficiency of such a narrowband filter would considerably reduce the number of detected photons. In summary, Heaps (1980) contends that measurement of mesospheric and upper stratospheric hydroxyl concentration by a Shuttle-borne lidar should be feasible.

Yeh and Browell (1982, I and II) have also undertaken calculations regarding the detection capabilities of the Shuttle lidar. Their analysis was concerned with the resonance-fluorescence measurements of the sodium and potassium number density in the upper atmosphere (80 to 110 km) and the magnesium-ion (Mg^+) number density in the ionosphere (80 to 500 km). In the case of the sodium and potassium measurements, Yeh and Browell (1982, I) show that

TABLE 9.6. HYDROXYL CONCENTRATION AND ANTICIPATED PHOTON RETURN AS A FUNCTION OF ALTITUDE FOR FOUR LOCAL TIMES[a]

Local time:	1200		2000		0000		0400	
Altitude (km)	[OH] (cm^{-3})	N_ν (photons)	[OH] (cm^{-3})	N_ν (photons)	[OH] (cm^{-3})	N_ν (photons)	[OH] (cm^{-3})	N_ν (photons)
90	7×10^3	0.26	4×10^4	1.5	4×10^4	1.5	4×10^4	1.5
80	4×10^5	12.4	2×10^6	62	4×10^6	124	3×10^6	93
70	4×10^6	106	5×10^6	127	4×10^6	106	1.5×10^6	40.2
60	8×10^6	182	6×10^6	136	2×10^6	45	3×10^5	6.8
50	2×10^7	315	2×10^6	32	1.5×10^5	2.7	1×10^5	1.5
40	3×10^7	128	4×10^5	1.6	1×10^5	0.43	6×10^4	0.25
30	1×10^7	0.29	1×10^5	0.003	5×10^3	—	2×10^3	—

[a] From Heaps (1980). For this example fluorescence is excited by a 1-J pulse at 282 nm, and fluorescence is detected from (1, 1) and (0, 0) bands by a 1-m-diameter telescope and 10%-efficient detection optics. The shuttle is assumed to operate at an altitude of 200 km.

saturation effects are negligible for nighttime operation, as the laser-beam divergence can be large. This is no longer true for daytime operation, where the optimal laser-beam divergence is determined by a tradeoff between the reduction of signal return (due to saturation) and the reduction of background light (due to narrowing the receiver field of view). Their calculations suggest that a minimum signal error will necessitate a laser-beam divergence of 0.18 mrad for sodium and 0.108 mrad for potassium. For sodium this leads to a saturation correction factor of 0.65 [Section 4.3, equation (4.55)], while for potassium it leads to 0.57.

The results of these simulations suggest that the performance of the Shuttle lidar system should be adequate to meet most of the identified scientific requirements and thereby provide new insights into the stability and dynamics of the upper atmosphere.

9.3. ATMOSPHERIC POLLUTION SURVEILLANCE

The adverse effect of air pollution on human health is well established (Stern, 1968; Williamson 1973), and many government agencies have established maximum levels of exposure for a wide range of atmospheric contaminants. Although there are many methods of monitoring the degree of atmospheric pollution, those based on lasers are in a special class. In the following section we shall look at the techniques that are capable of providing information on the general level of pollution, while in the subsequent section we shall focus our attention on those techniques that are better suited for effluent-source monitoring.

9.3.1. Long-Path Resonance Absorption

The possibility of using multiple-line gas lasers to detect pollution over an extended path was first considered towards the end of the sixties by Hanst and Morreal (1968). They showed that it was possible to detect gaseous pollutants, such as CO, NO, SO_2, and O_3, down to concentrations of a few ppm over a 1-km path using either a CO_2 or an I_2 laser and a retroreflector. In order to avoid an absolute calibration of the system, a differential-absorption approach was adopted. Under these circumstances the mean level of pollution that can be detected over a range increment ΔR is given by an equation that is similar to (7.31), namely,

$$\langle C \rangle = \frac{1}{2\kappa_A(\lambda_0)\,\Delta R} \ln\left\{ \frac{E_s(\lambda_w)}{E_s(\lambda_0)} \right\} \tag{9.47}$$

where $\langle C \rangle$ represents the mean concentration of the constituent of interest in ppm, and $\kappa_A(\lambda_0)$ represents the corresponding absorption coefficient in

$(\text{ppm cm})^{-1}$ at STP, see [equation (8.33)]. $E_s(\lambda_\alpha)$ represents the received laser energy at wavelength λ_α, where α is identified with ω for the wing of the line and 0 at the line center. Equation (9.47) is based on the assumption that $\sigma^A(\lambda_0) \gg \sigma^A(\lambda_w)$ and $\bar{\kappa}(\lambda_0) \approx \bar{\kappa}(\lambda_w)$. If we now introduce y for the fractional change in the received laser energy between the "on" and "off" (wing) laser wavelengths,

$$y \equiv \frac{E_s(\lambda_w) - E_s(\lambda_0)}{E_s(\lambda_w)} \tag{9.48}$$

then we can write

$$\langle C \rangle^{\min} = \frac{1}{2\kappa_A(\lambda_0)\,\Delta R} \ln\left\{\frac{1}{1 - y_{\min}}\right\} \tag{9.49}$$

Earlier, in Section 8.2.3, we developed an expression for the minimum density that could be detected for a DIAL system, equation (8.24). In terms of concentration [see equation (8.34)], this detection limit takes the form

$$C^{\min} \approx \frac{1}{4\kappa_A(\lambda_0)\,\Delta R} \tag{9.50}$$

Consequently, the relative improvement in the detection limit attained by use of a retroreflector can be determined from the ratio of (9.50) and (9.49), i.e.,

$$\frac{C^{\min}}{\langle C \rangle^{\min}} = \frac{1}{2\ln\{1/(1 - y_{\min})\}} \tag{9.51}$$

assuming that ΔR is the same. For $y_{\min} \approx 0.05$—corresponding to a 5% change in the received laser energy between λ_0 and λ_w—we have

$$\frac{C^{\min}}{\langle C \rangle^{\min}} = 9.75$$

We see that the use of a retroreflector can lead to nearly an order-of-magnitude improvement in the detection limit of the differential-absorption technique. This greater sensitivity is possible because the return from a retroreflector is large enough that it is only the difference in the signal returns at λ_0 and λ_w that determines the detection limit. It should not be forgotten, however, that this improvement is made at the expense of spatial resolution.

Table 3.6 provides both the absorption wavelength and the absorption coefficient at STP for a large number of atmospheric pollutants. From this table and equation (9.49) it is possible to estimate the minimum concentration (in ppm) that can be detected for a given range increment and the minimum fractional change observable in the received laser signals between the wing and line-center wavelengths. For example, suppose that $y_{\min} = 0.05$ (5%) and $\Delta R = 100$ m; then we can write

$$\langle C \rangle^{\min} = \frac{5 \times 10^{-5}}{\kappa_A(\lambda_0)} \ln\left\{ \frac{1}{1 - 0.05} \right\} = \frac{2.565 \times 10^{-6}}{\kappa_A(\lambda_0)} \tag{9.52}$$

Consequently, if we operate on the 4-μm band of SO_2, then $\langle C \rangle^{\min} \approx 4.77$ ppm, while if we choose the 300-nm band of SO_2, then $\langle C \rangle^{\min} \approx 0.096$ ppm. Clearly a lower detection limit is expected for operation around the 300-nm band, all other things being equal. In general a correction has to be made to these figures to allow for the actual atmospheric density, since $\kappa_A(\lambda_0)$ was based on STP conditions. A more realistic density to use in evaluating the absorption coefficient would have been 2.55×10^{19} cm^{-3} (corresponding to sea level at 0°C). This correction is accomplished by multiplying $\kappa_A(\lambda_0)$, obtained from Table 3.6, by the ratio 2.55/2.69 (or 0.948).

9.3.2. Differential Absorption and Topographical Scattering

If we consider the 9.5-μm band of O_3, we estimate a detection limit of 0.18 ppm. This compares well with the 0.2 ppm measured in a photochemical smog by Asai and Igarashi (1975). Their observations involved differential absorption between the $R(14)$ and $R(16)$ lines of a CO_2 laser and a retroreflector that was positioned 80 m from the laser.

Tunable infrared lasers clearly increase the scope of pollution detection, since they provide a wider choice of wavelengths. If this tuning can be accomplished rapidly, additional opportunities present themselves. Ku et al. (1975) demonstrated that fast modulation of the laser frequency about an absorption line can lead to the elimination of signal noise arising from atmospheric turbulence. This *fast-derivative spectroscopy* was discussed in Section 6.7 and involved the use of a cryogenically cooled $PbS_{0.82}Se_{0.18}$ semiconductor diode laser operating around the CO 4.7-μm absorption band with an output power of less than 1 mW. The laser frequency was modulated over a band of about 0.5 cm^{-1} at 10 kHz by a 10% oscillation in the injection current. Detection was accomplished with a liquid-nitrogen-cooled InSb infrared detector. In order to select the best CO line within the atmospheric transmission window, Ku et al. computed the expected atmospheric transmission spectrum around 4.7 μm for a 10-km path. Their results are presented as Fig. 9.42. The heavy vertical lines show the locations and relative strengths of the strong CO lines in this region. In order to discover if any spectroscopic interference was likely to arise from other atmospheric pollutants, they scanned the CO lines in samples of raw automobile exhaust. A number of the CO lines appeared to be free of this kind of interference, and for the field measurements the CO $P(4)$ line at 2127.69 cm^{-1} was chosen. The absorption coefficient for this line was determined to be 3.8 cm^{-1} ppm^{-1}.

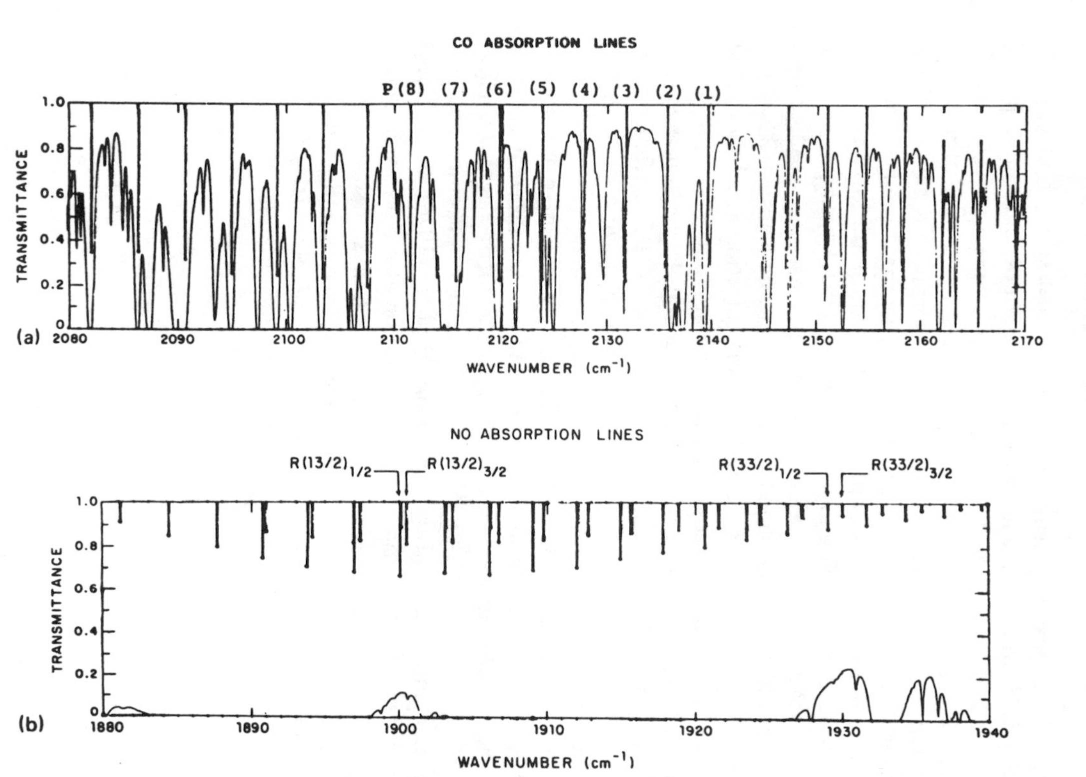

Fig. 9.42. Atmospheric transmission computed for a 10-km path at sea level during midwinter humidity conditions. The upper spectrum (a) shows the 4.7-μm region for CO detection, where the solid vertical lines designate CO transitions. The lower spectrum (b) is centered around the 5.2-μm region, for NO detection, and the NO lines are similarly identified (Hinkley, 1976).

An example of the sensitivity achieved with this system over a 0.61-km path is shown in Fig. 9.43, where the laser was tuned to the $P(4)$ absorption line of CO. The system was calibrated within the first 10 min by means of a cell which was filled with 82 ppb and then 170 ppb of CO and by the use of a second retroreflector that eliminated the background CO absorption. A small van was parked near the laser path, and its engine started at the 10-min point and allowed to run for 4 min. During this period the CO concentration was seen to fluctuate above the ambient 400-ppb level. Ku et al. (1975) speculate that the large CO signal associated with starting the engine may have been caused by a rich fuel mixture, which then became leaner as the engine warmed. They also suggest that the noticeable increase in the CO concentration at 37 min could have been due to a group of cars leaving a nearby parking lot.

Hinkley (1976) reported on a U.S. Environmental Protection Agency's Regional Air Pollution Study that involved the use of this technique for monitoring the CO concentration over a 1-km path in downtown St. Louis. Because of the high concentration and long path, the diode laser had to be tuned further off the line center than normal for the first derivative. Figure 9.44 shows a representative result of this study. Bag samples were taken at 1:35 and 1:45 P.M. at the monitoring van and at the retroreflector respectively. These values were 0.5 ppm at the van and 1.9 ppm at the retroreflector versus 1.1 ppm for the laser integrated path measurement. Hinkley (1976) also presents the results of monitoring the nitric oxide concentration at a busy traffic rotary in Cambridge, Massachusetts using resonance absorption and second-derivative spectroscopy. Figure 9.45 illustrates this work and shows both calibrations and sudden increases in NO concentration due to individual vehicles.

In principle, very much greater sensitivity can be obtained by either a multipath technique or positioning the retroreflector much further away.

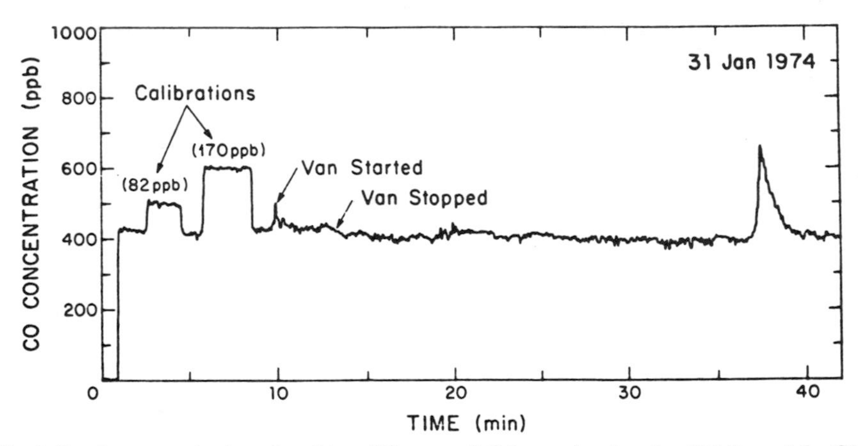

Fig. 9.43. Laser monitoring of ambient CO over a 0.61-km path using the $P(4)$ line of the CO_2 laser, illustrating effects caused by intentionally adding CO to a portion of the path. Integration time was 1 s (Ku et al., 1975).

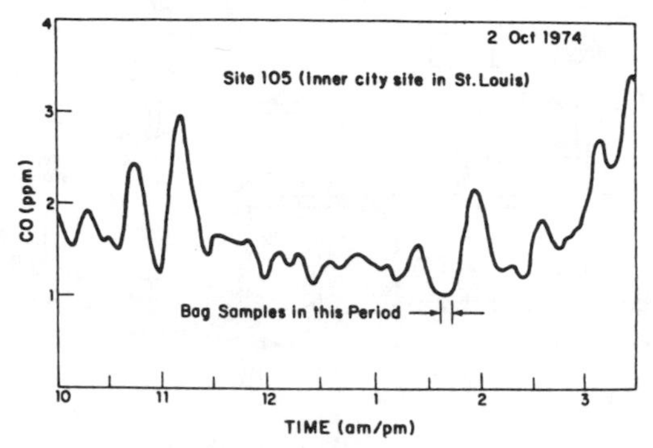

Fig. 9.44. Monitoring of ambient CO in St. Louis, Missouri, over a total (round-trip) path of 2 km. Comparative bag samples were analyzed off line by a chromatographic technique, as indicated. Averaging time for the laser measurements was 10 min (Hinkley, 1976).

Indeed, the limit of sensitivity could easily be extended by a factor of 50, which would mean in some instances a sensitivity of better than a few ppb (Reid et al., 1978). Additional improvement can be achieved if heterodyne detection is employed (Menzies, 1972, 1976). A further point worth making stems from the very modest laser energy required for long-path absorption. This allows the use of a broad spectrum of lasers, including tunable infrared laser diodes and Raman spin-flip lasers (Hinkley, 1972; Nill, 1974). Unfortunately, the range of situations amenable to installation of a retroreflector will invariably be restrictive. Topographical targets may be used to relax this constraint, but at a cost of

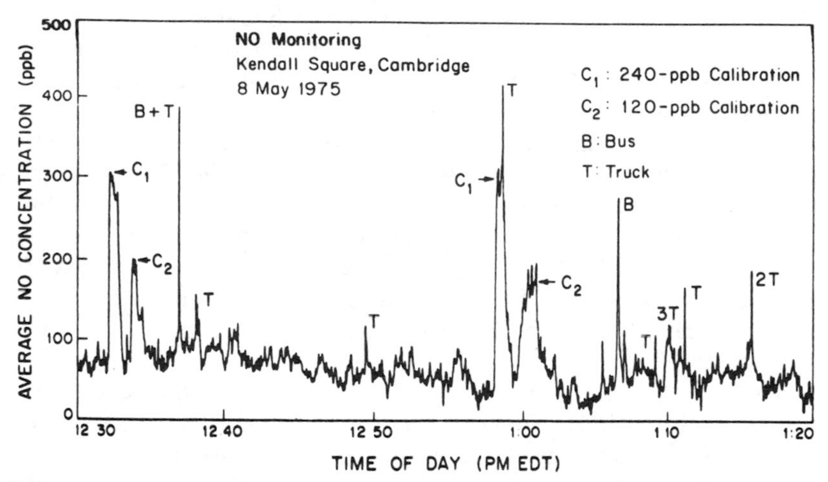

Fig. 9.45. Nitric oxide monitoring by resonance absorption (second derivative) at a busy traffic rotary in Cambridge, Massachusetts. Calibrations are shown, and large increases due to diesel-fueled buses (B) and trucks (T) are evident. The time constant was 1 s (Hinkley, 1976).

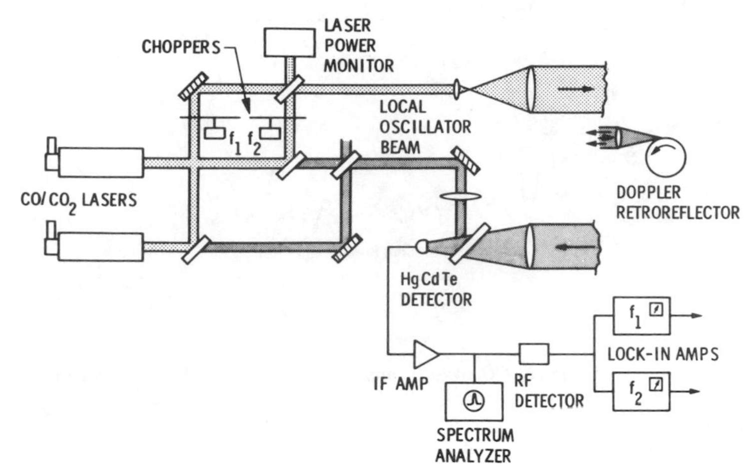

Fig. 9.46. Schematic diagram of the laser differential-absorption apparatus of Menzies and Shumate, (1976).

increases laser energy due to the return signal's range dependence [see equation (7.30)].

Menzies and Shumate (1976) measured the average concentrations of nitric oxide, ozone, and ethylene over a 0.8- and a 3.75-km path using differential absorption with either a retroreflector or a solid scattering surface to provide the return signals. CO and CO_2 lasers supplied the respective 5.2-, 9.5-, and 10.5-μm-wavelength radiation. Direct detection of the laser radiation was employed with cube-corner retroreflectors, while heterodyne detection was employed when the return signal was laser radiation scattered from the rough surface. In principle heterodyne detection of laser radiation can be several orders of magnitude more sensitive than direct photodetection in the 5- to 12-μm infrared wavelength region. This holds out the prospect of detection from topographical surfaces at a range of several kilometers, even when the emitted laser power is less than one watt. However, as pointed out by Menzies and Shumate (1976), proper operation of the heterodyne receiver requires a means of providing a suitable frequency displacement between the return signal and the local oscillator. A Doppler shift, generated by the rotation of the backscattering surface, provided this frequency displacement in their experiments. This simulates an airborne system that is directed at a small forward angle to the local vertical.

A schematic of the system developed by Menzies and Shumate (1976) is presented as Fig. 9.46. The cw lasers were mechanically chopped in order to allow synchronous detection. Ratioing the signal returns from two simultaneously transmitted wavelengths proved to be successful in reducing the effects of turbulence-induced fluctuations and permitted an accuracy of about 0.2% to be attained over a 1-km path. This corresponded to a minimum detectable mixing ratio of about 2 ppb in the case of ozone. In general they observed that when the concentration of NO (and C_2H_4) increased, due to heavy traffic, there was a noticeable decrease in the O_3 concentration. This can be understood in terms of the mutual annihilation of NO and O_3,

$$NO + O_3 \rightarrow NO_2 + O_2$$

More recently, Menzies and Shumate (1978) measured tropospheric ambient ozone concentrations using an airborne differential-absorption lidar system.

Henningsen et al. (1974) and Guagliardo and Bundy (1974, 1975) were amongst the first to demonstrate that single-ended resonance absorption using topographical backscattering was practical. Henningsen et al. (1974) were able to detect CO over a range of 107 m using backscattered radiation at 2.3 μm from foliage. Guagliardo and Bundy (1974, 1975) developed an airborne system to measure the column concentration of O_3 using the earth as a topographical target. They employed two grating-tuned TEA CO_2 lasers that were fired within a 20-μs interval in order to both minimize atmospheric scintillation and ensure that both beams struck, as far as possible, the same spot on the ground.

Ethylene is an important urban air pollutant in that it is directly emitted in the exhaust of motor vehicles, yet many plants exhibit symptoms of ethylene toxicity at concentrations as low as 10 ppb. Laser backscattering from foliage on foothills located at a range of around 5 km has been used to monitor the ambient concentrations of ethylene (Murray and van der Laan, 1978). A representative example of the backscattered signal is presented as Fig. 9.47. The signal backscattered from aerosols decreases with increasing range until it fades into the system noise. The large signal at the end of the oscilloscope trace, corresponding to a range of about $5\frac{1}{2}$ km, represents laser radiation backscattered from coastal hills.

These experiments were conducted with 1-J (100-ns) pulses on the $P(14)$ and $P(16)$ lines of a CO_2 TEA laser with a beam divergence of around 1.8 mrad. The receiver system comprised a 31.75-cm-diameter telescope, having a 3.3-mrad field of view, and a HgCdTe detector with a detectivity D^* of 1.1×10^{10} cm $Hz^{1/2}$ W^{-1}. Two extreme examples of ethylene measurements undertaken by Murray and van der Laan (1978) are presented as Fig. 9.48. The spectral interference from water vapor was estimated to be equivalent to about 7.6 ppb of ethylene. The results shown in Fig. 9.48 were corrected for this interference.

Killinger et al. (1980) have also used a line-tunable CO_2 TEA laser to monitor the concentration of carbon monoxide near a major traffic roadway. They frequency-doubled the CO_2 output in a crystal of $CdGeAs_2$ to provide pulses of about 1 mJ at around 4.65 μm. The return radiation was collected by a 30-cm-diameter Cassegrain telescope and directed onto a InSb pyroelectric detector. The frequency-doubled $R(18)$ line is fairly strongly absorbed by CO and is relatively free from spectral interference of other atmospheric constituents. The diurnal variation of the average atmospheric CO concentration

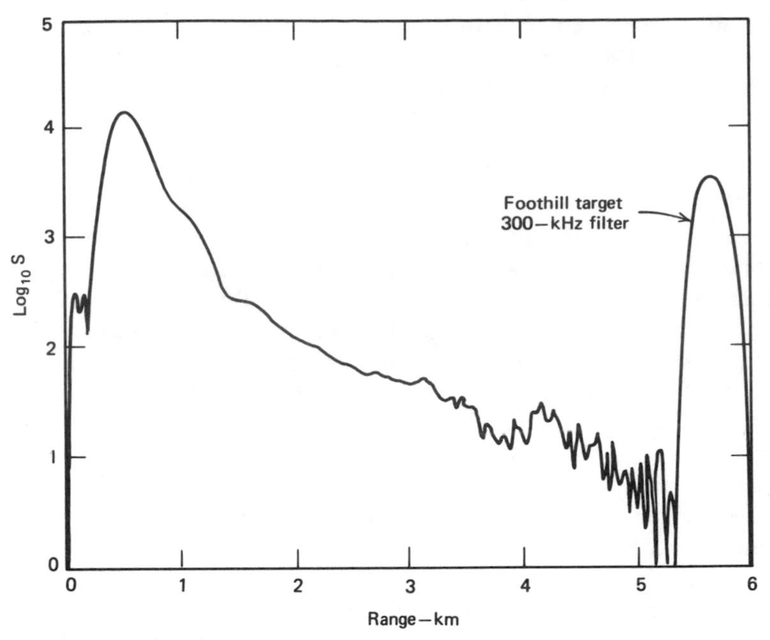

Fig. 9.47. Backscattered signal measured with an infrared lidar system (Murray and van der Laan, 1978).

over a 13-hr period for a 500-m path was observed to range between 0.35 and 1.0 ppm. The peaks tended to correlate with periods of heavy traffic on a nearby major roadway. Furthermore, pseudo-range-resolved measurements obtained through the use of several targets reinforced this connection, revealing concentrations as high as 1.8 ppm at a range corresponding to the location of the roadway.

This same system was also used to monitor the concentration of nitric oxide over a well-traveled road (Menyuk et al., 1980). These measurements involved the spectral coincidence between the frequency-doubled CO_2 laser radiation and the NO absorption lines near 5.3 μm. Return signals were obtained from a topographical target at a range of 1.4 km, and significant NO concentrations (of about 250 ppb) above ambient were recorded even under high-humidity conditions. Recently, Menyuk et al. (1982) have reported that this system has also been used to detect about 100 ppb of highly toxic rocket fuels over ranges between 0.5 and 5 km using a topographical backscattering target (see Table 9.7).

The first remote measurement of the ambient background concentration of N_2O using a DF laser-based lidar was reported by Altmann et al. (1980b). They used the coincidence of the DF $P_3(7)$ laser line, at around 3.8903 μm, with a strong absorption (1.33×10^{-6} cm^{-1} ppm^{-1}) in N_2O. The adjacent $P_3(6)$ laser line at 3.855 μm provided the reference signal, having a much

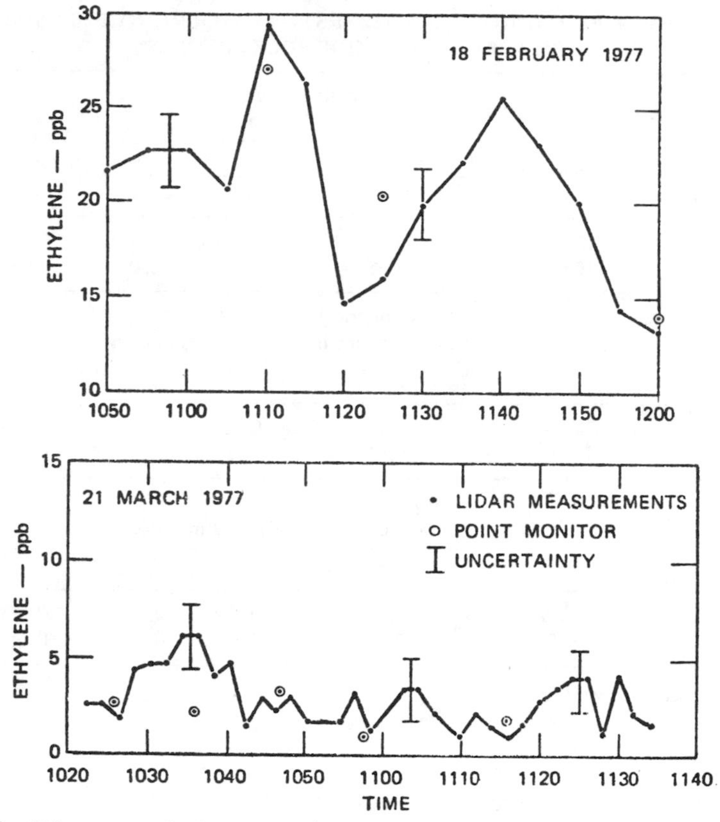

Fig. 9.48. Lidar-measured ethylene concentrations corrected for water-vapor interference and compared with a point monitor (Murray and van der Laan, 1978).

weaker absorption (8.1×10^{-8} cm^{-1} ppm^{-1}). Return signals from topographical targets at distances up to 8 km were detected, and their N_2O concentration measurement of about 290 ppb was found to be in reasonable agreement with *in situ* measurements. Altmann and Pokrowsky (1980) measured the absorption coefficient of SO_2 for twenty of the DF laser lines. Only the $P_4(6)$ line at 3.9843 μm was found to have a strong absorption coefficient (4.4×10^{-7} cm^{-1} ppm^{-1}).

A Nd–YAG-pumped $LiNbO_3$ optical parametric oscillator (OPO) represents an important alternative to line-tunable gas lasers as a source of intense tunable infrared radiation (Yariv, 1976; Baumgartner and Byer, 1979). Baumgartner and Byer (1978a, b) have developed a differential-absorption lidar based on such a transmitter. Their $LiNbO_3$ parametric oscillator provided an output of about 20 mJ over the range 1.4 to 4.2 μm with a 1.0-cm^{-1} linewidth. A 41-cm-diameter telescope collected the return signal and focused it onto a 1-mm^2 InSb photovoltaic detector (cooled to 77 K). The output of the

TABLE 9.7. ABSORPTION PARAMETERS FOR LASER REMOTE SENSING OF SEVERAL HYDRAZINE ROCKET FUELS[a]

CO_2-Laser Transition	Wavelength λ (μm)	Absorption Coefficients κ_A (cm^{-1} ppm^{-1})			Atmospheric Attenuation κ_e (km^{-1})
		Hydrazine	Interfering Species		
		N_2H_4	NH_3	C_2H_4	
P(22)	10.611	4.77	0.045	1.09	0.1142
P(28)	10.675	2.06	0.36	1.30	0.0976
		Unsymmetrical Dimethylhydrazine	Interfering Species		
		$(CH_3)_2N_2H_4$	NH_3	C_2H_4	
P(30)	10.696	2.22	0.86	1.63	0.0907
R(10)	10.318	0.18	0.78	1.51	0.1142
		Monomethyl-hydrazine	Interfering Species		
		$CH_3N_2H_4$	NH_3	C_2H_4	
R(30)	10.182	1.69	0.029	0.56	0.1137
R(18)	9.282	0.31	0.13	0.61	0.1418

[a] Menyuk et al. (1982).

detector was amplified through a computer-controlled gain-switchable amplifier having a 1-MHz bandwidth.

Preliminary observations were undertaken with this system using a topographic reflector for SO_2 at 4.0 μm, CH_4 at 3.3 μm (and 1.66 μm), and H_2O at 1.7 μm (Baumgartner and Byer, 1978a, b). During the CH_4 measurements at 3.3 μm, absorption scans of the atmosphere were made to determine the possible interference effects of water vapor (Murray and Byer, 1980). These spectra are shown in Fig. 9.49. From this observation the $P(10)$ transition of the CH_4 3.4-μm band was selected, and a representative concentration-versus-time measurement is displayed as Fig. 9.50. These results are seen to be in good agreement with hourly point-source measurements of CH_4.

Recently, Aldén et al. (1982) have reported the first lidar measurements of atomic mercury in the atmosphere. Differential absorption and topographical scattering were employed. An anti-Stokes-shifted, frequency-doubled Nd–YAG-pumped dye laser with an output of about 0.7 mJ at the "on" and "off" wavelengths (253.65 and 253.68 nm, respectively) of the resonance transition of mercury was used in conjunction with a 25-cm, $f/4$ Newtonian telescope. A detection limit of about 8 μg m^{-3} m was achieved with a laser linewidth of around 0.15 cm^{-1}. This corresponds to an average concentration of 4 ng m^{-3} over a path length of 2×1 km, which is representative of

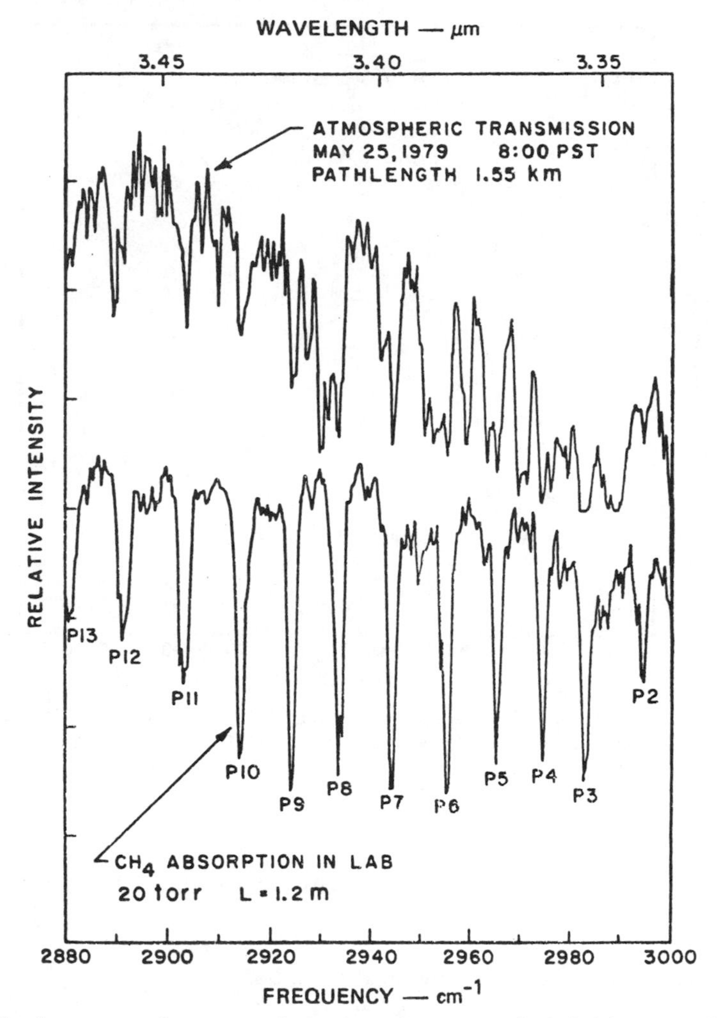

Fig. 9.49. A segment of an atmospheric absorption scan and an in-laboratory absorption spectrum near the CH_4 fundamental absorption band at 3.4 μm (Murray and Byer, 1980).

background concentrations and typically one-tenth of that expected near chloroalkali plants.

Probably one of the most ambitious schemes for mapping the spatial distribution of gaseous pollutants over an extensive area using resonance absorption of laser radiation was proposed by Byer and Shepp (1979) and Wolfe and Byer (1982). This scheme can be thought of as the optical analogue of X-ray tomography, whereby the absorption of a laser beam that is made to cross the specified area from many directions is deconvolved through an

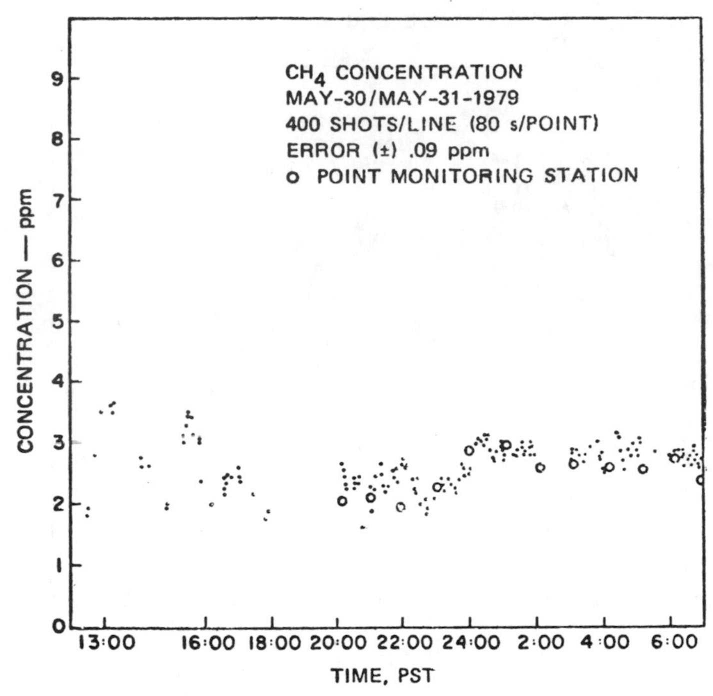

Fig. 9.50. A part of the CH_4 measurement showing CH_4 fluctuations over the path being monitored. The statistical rms noise of the measurement is 0.03 ppm, which is less than the CH_4 fluctuation level (Murray and Byer, 1980).

elaborate computer program into a concentration map of the constituent for which the laser is tuned.

9.3.3. Differential Absorption and Scattering Lidar

Occasionally the ambient concentration of the constituent of interest can be large enough to limit the range of observations based on resonance absorption through excessive attenuation of the laser radiation. This was first pointed out by Measures and Pilon (1972) and is illustrated for the case of SO_2 in Fig. 8.1. Detuning of the "on"-wavelength laser can relax this constraint. To show this we use equation (8.23) and draw upon the same assumptions as used to obtain (8.24). Then we can write

$$E(\lambda_0, R) - E(\lambda_0, R + \Delta R) \approx E_L H \mathrm{e}^{-B\sigma}(1 - \mathrm{e}^{-C\sigma})$$

where

$$B = \int_0^R N_i(R)\, dR, \qquad C = N_i(R)\,\Delta R, \qquad \sigma = \sigma_i^A(\lambda)$$

and H is an unimportant parameter for this analysis. For detection we invoke (8.19) and thereby deduce that the laser energy required can be expressed in the form

$$E_L = \frac{De^{B\sigma}}{1 - e^{-C\sigma}}$$

If we set $dE_L/d\sigma = 0$, then E_L is minimum† for

$$\sigma_i^A(\lambda) = \frac{1}{2N_i(R)\,\Delta R} \ln\left\{1 + \frac{N_i(R)\,\Delta R}{\int_0^R N_i(R)\,dR}\right\} \tag{9.53}$$

which is well approximated by

$$\sigma_i^A(\lambda) \approx \frac{1}{2N_i(R)R} \tag{9.54}$$

for $\Delta R < R/10$, assuming that $N_i(R)$ is reasonably independent of R. Consequently, in order to detect 1 ppm of *any* species at a range of, say, 5 km with minimum laser energy, the optimum cross section is approximately 3.92×10^{-20} cm^{-2} or, put another way, $\kappa_A^{min}(\lambda) \approx 10^{-6}$ $(cm\ ppm)^{-1}$. Byer and Garbuny (1973) have shown by similar arguments that where an excess density of some pollutant, ΔN_i, is to be detected against a background density N_i of the same constituent, then the cross section appropriate to a minimum in the laser energy is given by

$$\sigma_i^A(\lambda) = \frac{1}{2L\,\Delta N_i} \ln\left\{1 + \frac{L\,\Delta N_i}{RN_i}\right\} \tag{9.55}$$

where L is the extent of the pollutant plume and R its range.

†If

$$E_L = \frac{De^{B\sigma}}{1 - e^{-C\sigma}}$$

then

$$\frac{dE_L}{d\sigma} = E_L\left\{B - \frac{Ce^{-C\sigma}}{1 - e^{-C\sigma}}\right\}$$

Clearly E_L is a minimum (exclusive of zero) for $dE_L/d\sigma = 0$, that is,

$$\sigma = \frac{1}{C} \ln\left\{1 + \frac{C}{B}\right\}$$

As we have seen from Table 3.6, most pollutants have absorption bands in the infrared. However, a few—notably O_3, NO_2, SO_2, and a number of metals—have healthy absorption features that lie in the visible or near ultraviolet part of the spectrum. Several groups have undertaken studies of SO_2, O_3, and NO_2 using these shorter wavelengths and differential absorption and scattering. Grant and Hake (1975) and Grant et al. (1974) used a 2.5-m sample cell at a range of 306 m to demonstrate that the kind of sensitivity indicated in Table 3.6 could in fact be achieved in practice for all three pollutants. Hoell et al. (1975) and Thompson et al. (1975) attained a measurement sensitivity of 10 ppb at a range of 0.8 km for SO_2 in actual field experiments. A frequency-doubled, flashlamp-pumped tunable dye laser having an output of 100 μJ, in a spectral bandwidth of less than 0.03 nm, and a duration of 1.3 μs was used for these observations. More recently, Baumgartner et al. (1979) have also used a flashlamp-pumped dye laser to measure the ambient NO_2 concentration over Redwood City, California. Their laser emits a 10-mJ, 700-ns pulse that alternates in wavelength between 448.1 and 446.5 nm (to correspond to the "on" and "off" absorption bands of NO_2) with a divergence of about 1.3 mrad. The laser repetition rate was 5 pps, and the output linewidth was 0.2 nm. A 51-cm-diameter Newtonian telescope was used in conjunction with a set of narrowband interference filters. An example of the results obtained with this system is presented as Fig. 9.51. It can be seen by reference to this figure that good agreement was obtained between the lidar measurements and those attained from a standard instrument provided the wind speed was less than 3 mi/hr.

Similar measurements for SO_2 over the city of Goteborg (Sweden) have been undertaken by Fredriksson et al. (1979). Their system employed a frequency-doubled flashlamp-pumped dye laser with an output of 0.4 mJ in a 1-μs pulse around 300 nm. The laser repetition rate was 25 pps, and a 25-cm-diameter $f/4$ Newtonian telescope served as the optical receiver. An example of their results is displayed in Fig. 9.52. A comparison with the day

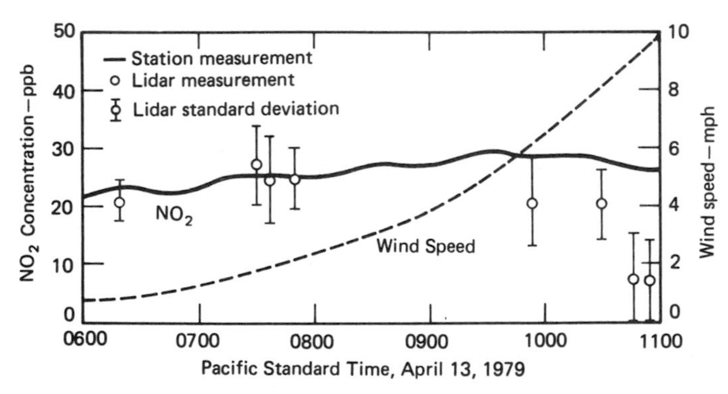

Fig. 9.51. Ambient NO_2 concentration and average wind speed at Redwood City, California, as determined by lidar and station measurements (Baumgartner et al., 1979).

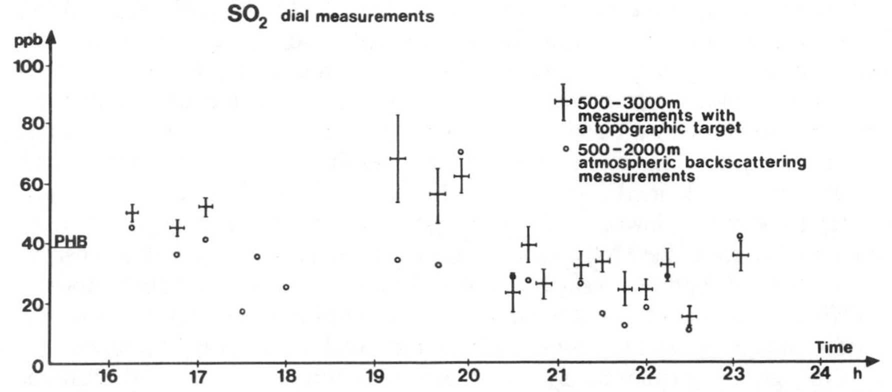

Fig. 9.52. Time variation of the SO_2 concentration over the city of Goteborg, Sweden. The average concentration over a distance of 3 km, as determined by DIAL measurements against a topographic target, is shown with indications of the time periods of measurement and the estimated maximal error. In addition, mean values over a distance of 2 km are given by (circles), as obtained by DIAL measurements using atmospheric backscattering. The day-average value as measured by the local Public Health Board (PHB) with conventional chemical methods is also given. This value was obtained at a point along the measurement path (Fredriksson et al., 1979).

average SO_2 concentration obtained by the local Public Health Board, using conventional chemical methods, is also provided.

9.4. POLLUTION-SOURCE MONITORING

As we have seen, the differential-absorption approach is unlikely to be rivaled for range and sensitivity. Nevertheless, the sophistication of these DIAL systems and the complexity of their signal interpretation provides considerable incentive to develop alternative approaches where the bounds on range and concentration are conducive. Such a situation is likely to be encountered when the spatial and temporal distributions of specific gaseous contaminants are to be monitored as they emerge from a pollution source.

9.4.1. Fluorescence and Raman Measurements

An approach based upon laser-induced fluorescence may appear to be attractive for a number of constituents such as SO_2, NO_2, I_2, O_3, various hydrocarbon vapors, and certain kinds of aerosol pollutants. An early analysis of the fluorescence return signal expected from a localized source of pollution was undertaken by Measures and Pilon (1972). The results of this study revealed that above a certain peak concentration a distortion of the returned signal could lead to a misinterpretation of both the range and concentration of the source. An example of the kind of distortion predicted for a Lorentzian distribution of NO_2 was presented as Fig. 8.2.

Unfortunately, in the case of atmospheric work, collision quenching and the broadband nature of the emission, combined with the concomitant high aeroticulate background associated with such sources of pollution, tend to restrict the remote-sensing potential of this approach. In spite of this, fluorescence has been used by Tucker et al. (1975) to measure the NO_2 concentration within a sampled volume of urban air with a sensitivity of less than 1 ppb, and calculations based on the experimental work of Gelbwachs and Birnbaum (1973) have led Gelbwachs (1973) to predict that on clear nights remote sensing of laser-induced NO_2 fluorescence could be undertaken with a sensitivity of about 50 ppb at a range of 1 km. Schuster and Kyle (1980) proposed studying the transport and diffusion of pollution plumes through the use of a fluorescence tracer and a low-power laser mounted aboard a light aircraft.

Although the extraordinary small cross section associated with Raman scattering represents a considerable impediment to its use in remote sensing, it possesses several desirable characteristics which make it very attractive for pollution-source monitoring:

1. The spectral shift of the Raman-backscattered radiation is specific to each molecule; see Fig. 3.23 and Tables 3.3 and 3.4.
2. The intensity of a given Raman signal is directly proportional to the density of the appropriate scattering molecule and independent of the others. Consequently, a direct measure of the concentration of a pollutant relative to nitrogen can be obtained without calibration problems.
3. The narrow spectral width and shift of the Raman return are conducive to spectral discrimination against both solar background radiation and elastically scattered laser radiation.
4. The inherent short duration of the Raman process can also be used to discriminate against solar background radiation when a small range interval is of interest.
5. Only a single, fixed-frequency laser is required to produce the simultaneous Raman spectra of all the pollutants within the region being probed. The advantage of this for multiplexing is obvious.
6. Good spatial and temporal resolution is possible, since a backscattering process is involved.

There is one additional problem, alluded to earlier, that can arise in pollution monitoring. The Raman signal from a trace constituent of a plume could be masked by the *O*- or *S*-branch Raman signal from a major component. This can be appreciated by reference to Fig. 3.22. Inaba and Kobayasi (1969, 1972), who were the first to undertake a comprehensive study of Raman scattering for pollution monitoring, addressed this problem and found that in most instances spectral interference can be avoided by the use of narrow spectral filters. An excellent example of the difference in the Raman-backscattered spectra observed at a range of 20 m between ordinary air, an oil-smoke plume, and the exhaust gas of an automobile was presented by Inaba and Kobayasi (1972) and is reproduced here as Fig. 9.53. The corresponding molecular concentrations relative to atmospheric N_2 are presented in Table 9.8; the numbers inside the parentheses are deemed to be less reliable. These results were obtained using a N_2 laser emitting 0.2 mJ at 337 nm in a 10-ns pulse with a 50-pps repetition rate. The receiver contained a 30-cm-diameter Newtonian telescope, and spectral discrimination was achieved using an $f/8.5$, 0.5-m single-grating monochromator and a short-wave blocking filter having a transmission of 10^{-3} at 337 nm. The integration time was 5-ns, and the 50-nm scans shown in Fig. 9.53 were accumulated in 40 min.

An early foretaste of what can be achieved was provided by Hirschfeld et al. (1973), who designed and built one of the most powerful lidars to date. This system incorporated a frequency-doubled, *Q*-switched, 2-J (2-pps) ruby laser, a 91.4-cm-diameter $f/6.8$ Dall–Kirkham Cassegrain telescope, a polychromator, and an array of photomultipliers to provide multiplex detection. The high sensitivity achieved with this lidar is attested by the daylight Raman spectra obtained from a controlled plume of SO_2 and kerosene at a range of 200 m, with a 10-m range resolution. An example of the results obtained with this system are presented as Fig. 9.54. DeLong (1974) extended these measurements and demonstrated that sensitivities of the order of 100 ppm for a wide variety of constituents should be achieved, even under unfavourable daytime weather conditions for a range of several hundred meters. Melfi et al. (1973) also reported the detection of Raman scattering by SO_2 in the plume of a 200-MW coal-burning electrical generating plant at a slant range of 210 m. An example of the correlation observed between the output power of the electrical plant and the SO_2 return signal is presented as Fig. 9.55. The lidar system used in this study comprised a *Q*-switched ruby laser operating at 694.3 nm with an output of about 1.5 J in 25 ns. A 61-cm-diameter $f/4$ Newtonian telescope collected the backscattered laser energy and focused it onto a stop which limits the field of view to 12 mrad. Typically two 3-nm-bandpass (45% transmission) interference filters were used in conjunction with a 694.3-nm 10^5 blocking filter and a photon-counting detection system containing extended-red-sensitive RCA-8852 photomultipliers.

Poultney et al. (1977) demonstrated that an improved version of this system was capable of measuring SO_2 at a concentration of about 10^3 ppm with 12% accuracy at a range of 300 m in about 15 min with a range resolution of 6 m. In all of this Raman work the pollutant signals are normalized by the N_2 (or sometimes O_2) Raman signals. This serves to provide a direct measurement of the relative concentration of the pollutant and thereby eliminate uncertainties and variations in both laser energy and atmospheric (including plume) transmission. There is also no need for calibration with this technique. Poultney et al. used three photodetector channels: one for the Raman signal from nitrogen at 828.3 nm (or oxygen at 778.4 nm), another for the pollutant Raman signal

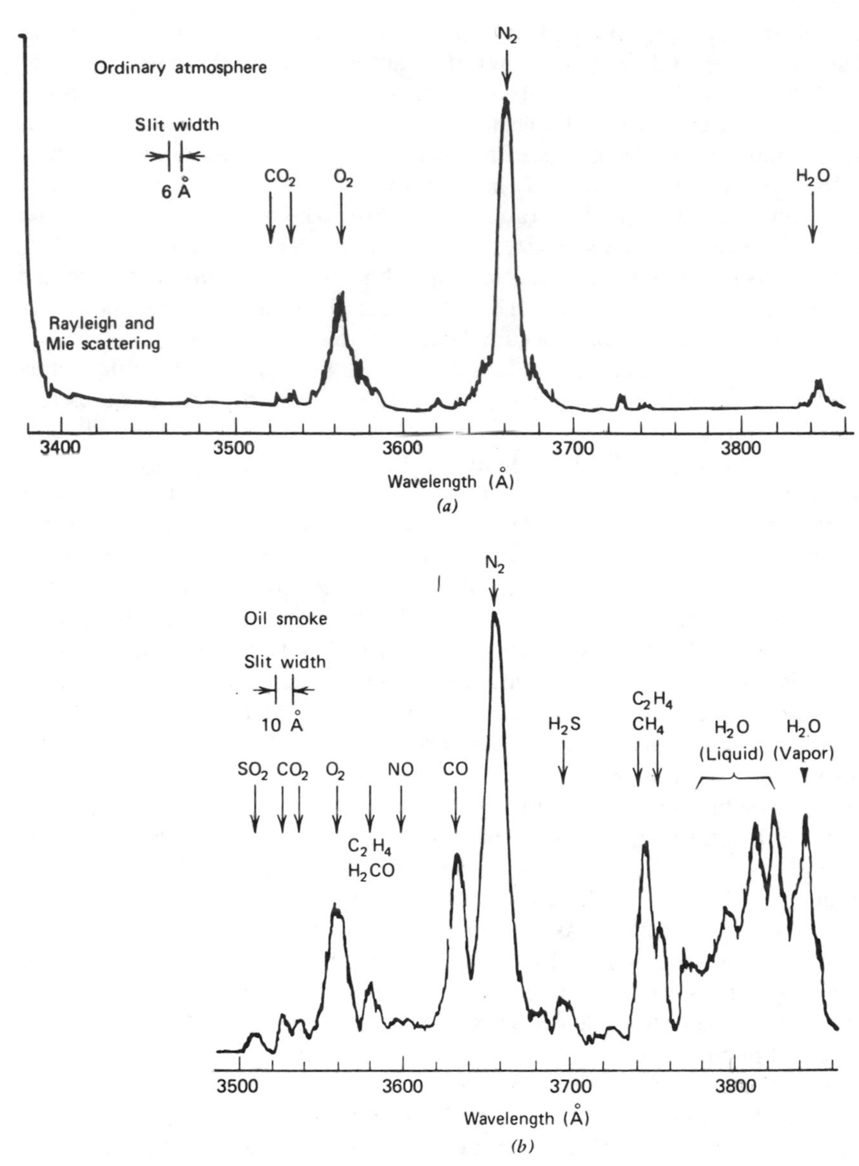

Fig. 9.53. (a) Lidar-measured spectrum of Raman-shifted and unshifted backscattering from the ordinary atmosphere. Lidar-measured spectrum of Raman-shifted backscattering from (b) an oil-smoke plume and (c) the exhaust gas from an automobile (Inaba and Kobayasi, 1972).

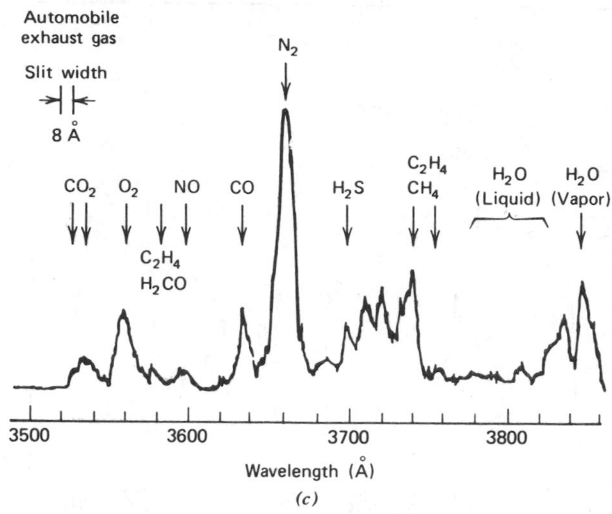

(c)

Fig. 9.53. (*Continued*)

(754.6 nm in the case of SO_2), and an auxiliary channel for viewing the 694.3-nm elastically backscattered signal. The last signal was used to locate the plume quickly in range and time and to estimate its diameter.

9.4.2 Differential-Absorption Measurements

In spite of the complexity of both equipment and data processing required for the differential-absorption and scattering technique, its sensitivity enables it to

TABLE 9.8. EXPERIMENTAL ESTIMATION OF MOLECULAR CONCENTRATIONS[a] RELATIVE TO THAT OF ATMOSPHERIC N_2 MOLECULES FROM RAMAN SPECTRA MEASURED BY RAMAN LIDAR TECHNIQUES

	Molecular Species										
				H_2CO +					CH_4 +	H_2O	
Observed Spectrum	SO_2	CO_2	O_2	C_2H_4	NO	CO	N_2	H_2S	C_2H_4	Liquid	Vapor
Ordinary atmosphere, Fig. 9.53(a)	—	0.02	0.26	—	—	—	1	—	—	—	0.016
Oil smoke plume, Fig. 9.53(b)	0.01	0.06	0.24	(0.12)	(0.03)	0.43	1	0.015	(0.069)	(0.11)	0.10
Automobile exhaust gas, Fig. 9.53(c)	—	0.11	0.21	(0.04)	(0.09)	0.28	1	0.024	(0.057)	(0.05)	0.071

[a] Values in parentheses are less accurate because of large experimental errors or uncertainties in Raman cross sections, Inaba and Kobayasi (1972).

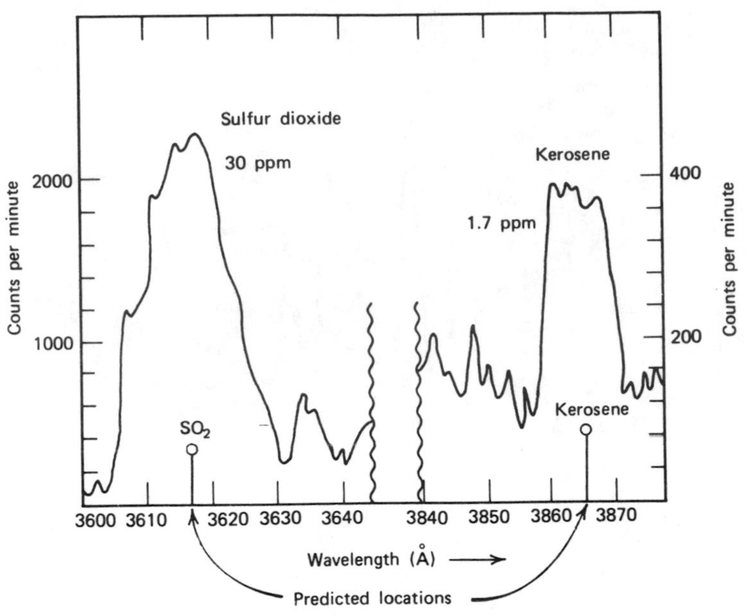

Fig. 9.54. Remote Raman lidar returns for SO_2 and kerosene air pollution at a range of 200 m (Hirschfeld et al., 1973).

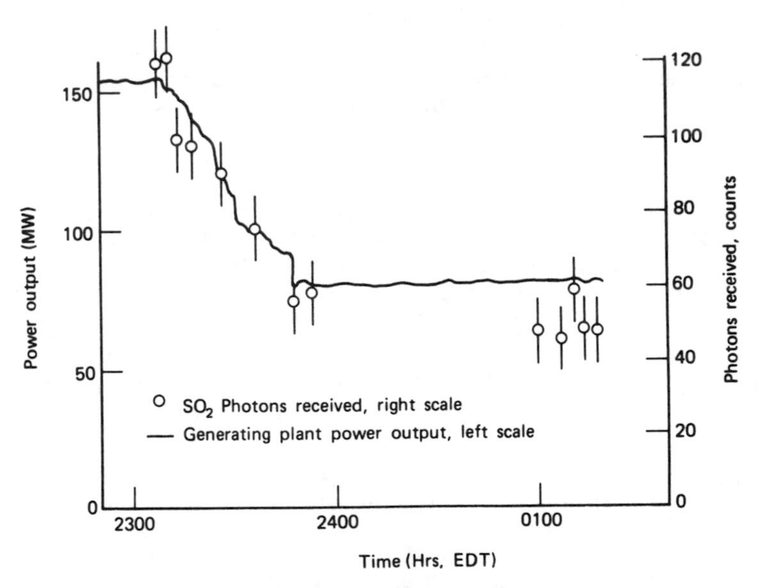

Fig. 9.55. Comparison of generating-plant power output with the observed SO_2 Raman scattering from the plume. Slant range for the lidar system was 210 m (Melfi et al., 1973).

do more than just monitor the pollutant concentration within the plume at the source of emission. In particular, a DIAL system can be used to map the dispersion of these pollutants, day or night. This, combined with the growing concern regarding the international problem of acid rain, has prompted the development of many operational DIAL systems throughout the industrialized world. We shall examine four representative systems.

One of the first such operational DIAL systems was developed by Rothe et al. (1974a, b) in Germany. They used a 1-mJ, 300-ns, flashlamp-pumped, tunable dye laser operating between 455 and 470 nm with a 1-pps repetition rate to demonstrate that NO_2 concentrations down to 200 ppb could be detected at night up to a range of 4 km over the city of Cologne. Their map of the NO_2 spatial distribution over a chemical factory obtained with the lidar situated some 750 m from the chimney stack is reproduced here as Fig. 9.56 and

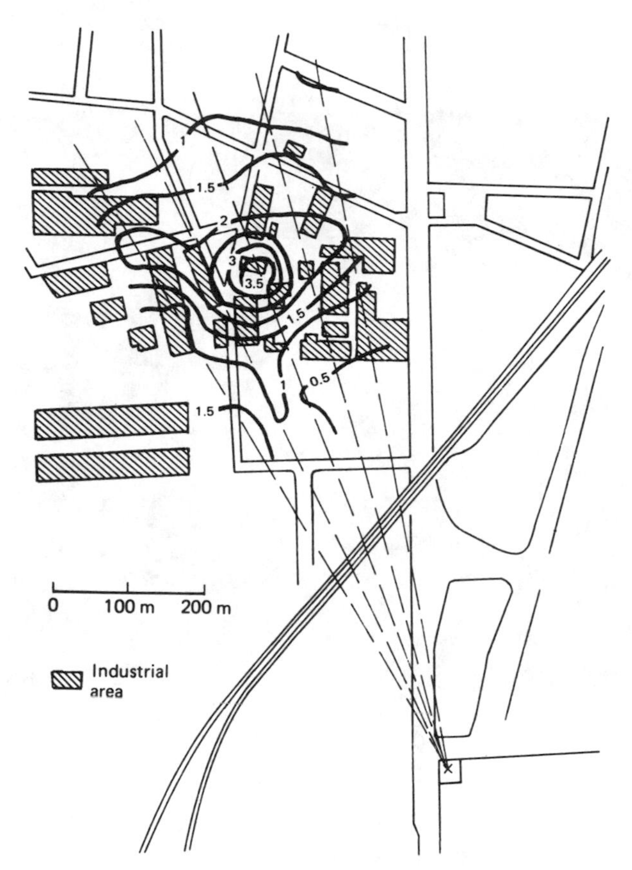

Fig. 9.56. NO_2 distribution over a chemical factory as derived from DAS-lidar measurements performed in the indicated directions at an altitude of about 45 m. The concentrations are given in ppm (Rothe et al., 1974).

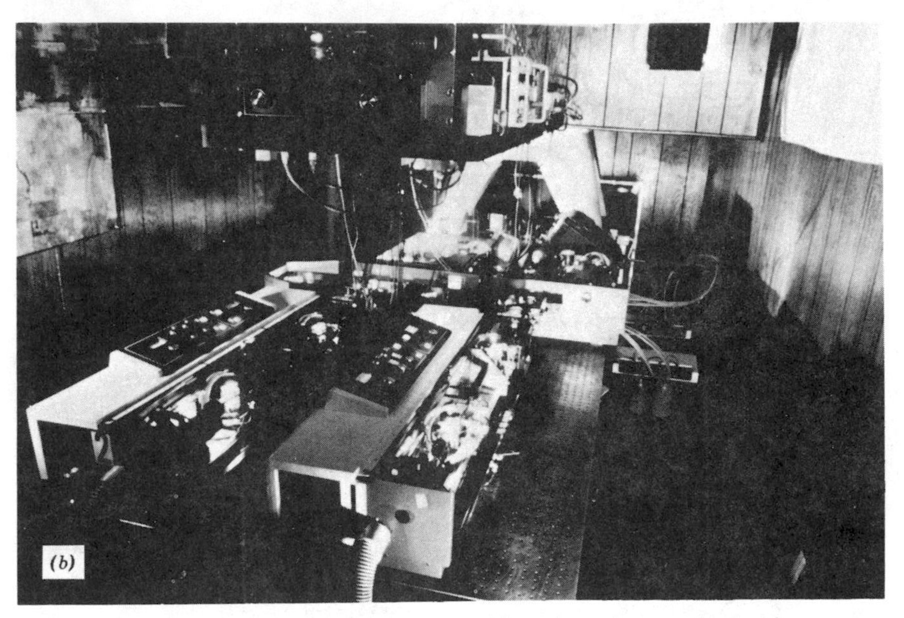

Fig. 9.57. SRI International mobile lidar: (a) van with receiver telescope; (b) dual laser system and receiver electronics within the van (Hawley, 1981).

represents one of the most striking examples of the potential of DAS. To obtain these results signals from about 8000 shots were averaged along each of the five indicated directions.

In the United States the Electric Power Research Institute (EPRI) has promoted the development of a mobile DIAL system for monitoring the emission and dispersion of SO_2 and NO_2 released from the tall stacks commonly used by utilities burning fossil fuels. At the heart of this system, which was designed and built at SRI International, California, lie two independently fired Nd–YAG lasers (Hawley, 1981). Their outputs are frequency-doubled and used to pump two dye lasers, which are in turn frequency-doubled to provide outputs at 300.0 and 299.5 nm for SO_2. Detuning of the second laser (the "off" line for SO_2) to 291.4 nm will also provide information on O_3. The 448.1- and 446.5-nm wavelengths for NO_2 detection can be obtained by pumping a pair of coumarine dye lasers with the third-harmonic outputs of the Nd–YAG lasers.

The sensitivity of this DIAL system for SO_2 was about 2 parts per million meter (ppm m) when the output energy was around 10 mJ and a 51-cm-diameter telescope was employed. The integration period was around 2 min. This means that a concentration of only 220 ppb of SO_2 could be detected for a smokestack plume of 10-m diameter. Moreover, such measurements could be undertaken at a range of 3 km. A photograph of the van, laser, and receiver telescope is shown in Fig. 9.57(a). The twin lasers and other electro-optical equipment that constitute this DIAL system are dramatically illustrated in Fig. 9.57(b).

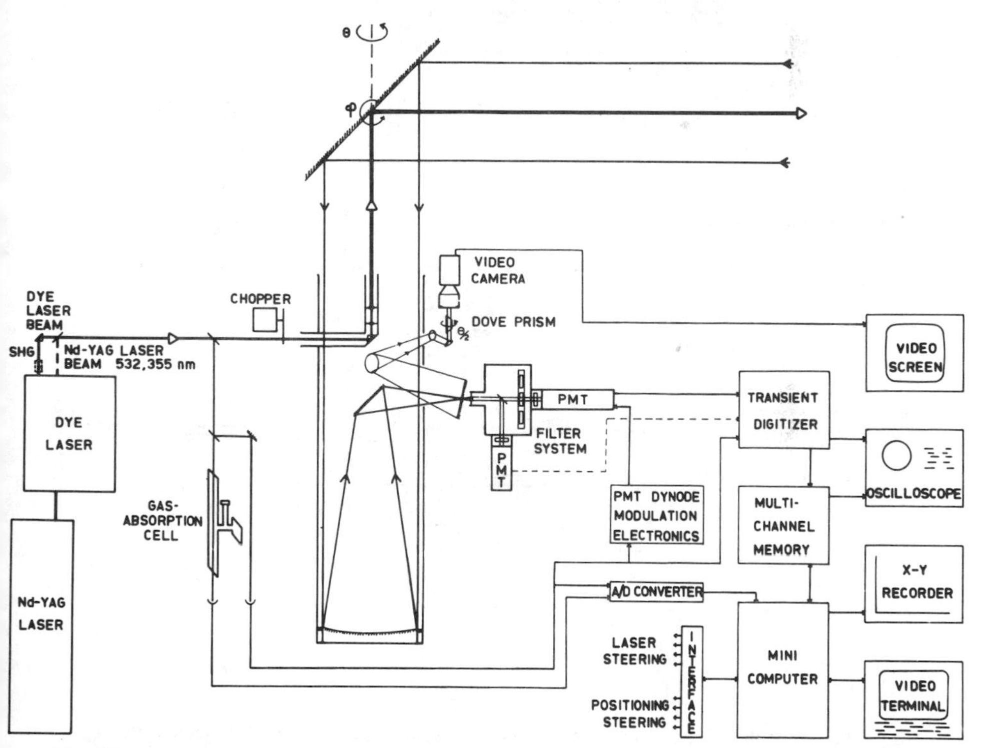

Fig. 9.58. Optical and electronic arrangement of a lidar system operated by the National Swedish Environmental Protection Board (Frederiksson et al., 1981).

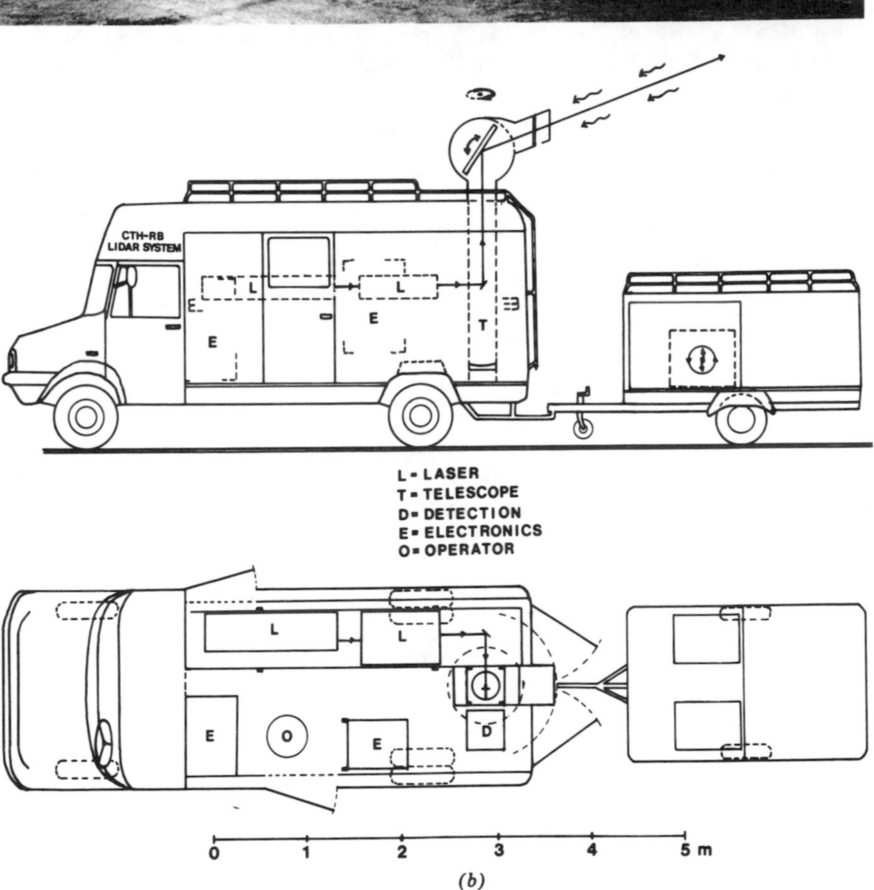

Fig. 9.59. (a) Mobile lidar system of the National Swedish Environmental Protection Board; (b) schematic view of the system (Frederiksson et al., 1981).

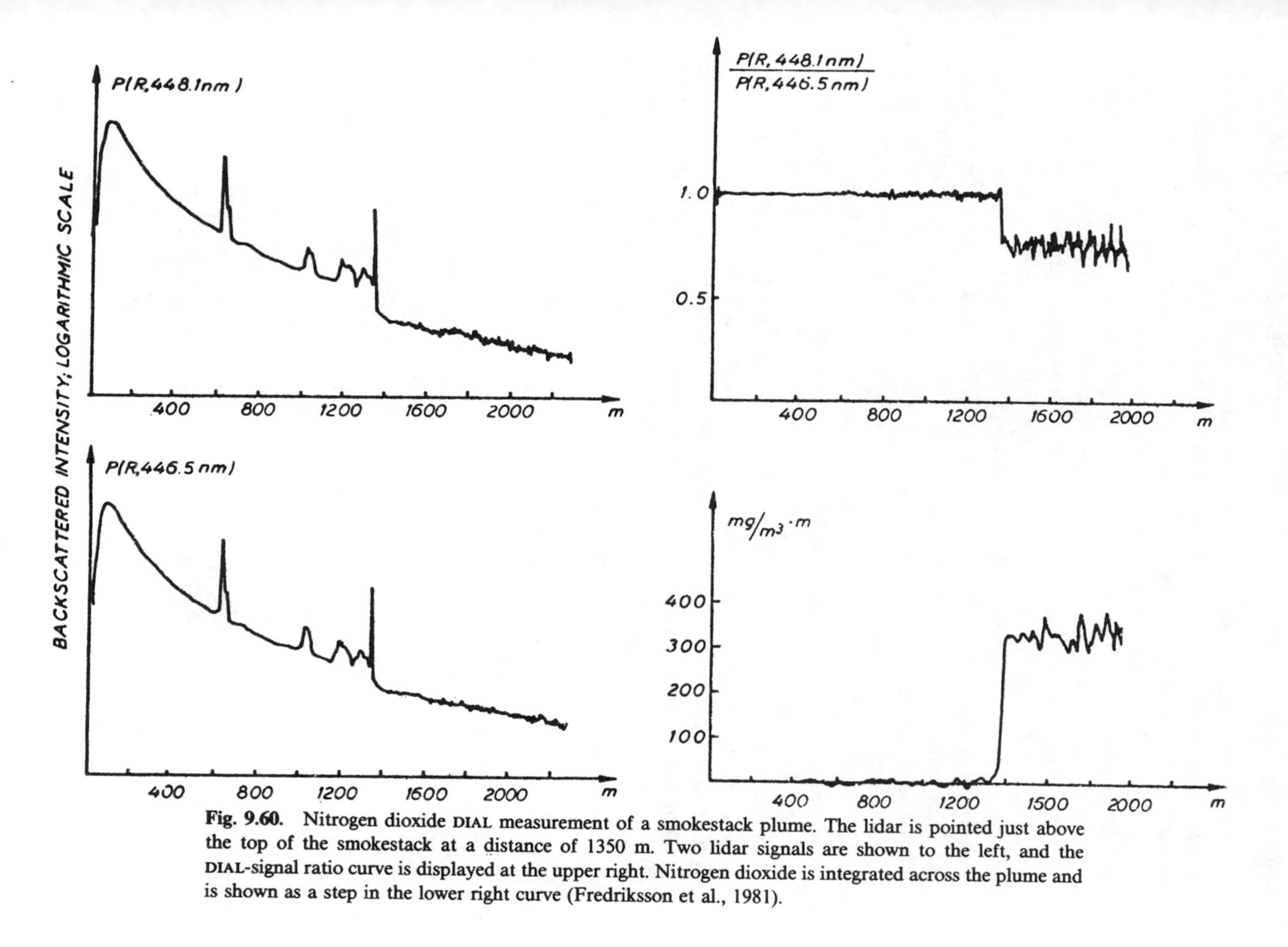

Fig. 9.60. Nitrogen dioxide DIAL measurement of a smokestack plume. The lidar is pointed just above the top of the smokestack at a distance of 1350 m. Two lidar signals are shown to the left, and the DIAL-signal ratio curve is displayed at the upper right. Nitrogen dioxide is integrated across the plume and is shown as a step in the lower right curve (Fredriksson et al., 1981).

In early 1979 the Swedish Space Corporation funded the development of a powerful, fully mobile lidar system that would be used for both research and routine atmospheric monitoring. The lidar was built around a 250-mJ, 7-ns, 10-pps Nd–YAG laser that could be frequency-doubled or -tripled to yield 100 and 50 mJ, respectively. These shorter-wavelength outputs are used to pump a tunable dye laser which consists of a grating-tuned oscillator and one or two amplifier stages (Fredriksson et al., 1981). For NO_2 measurements 448.1 and 446.5 nm were again chosen as the "on" and "off" wavelengths, while for SO_2 the dye laser is frequency-doubled to provide 300.05 and 299.30 nm respectively. The receiver system used a 30-cm $f/3.3$ Newtonian telescope to focus the backscattered laser radiation onto an EMI 9817 photomultiplier after transmission through a set of narrowband optical filters. The first 20 μs of signal (corresponding to backscattered radiation out to 3 km) is captured with a Biomation 8100 transient digitizer. A schematic of this DIAL system is presented as Fig. 9.58, while a photograph and schematic of the van containing it are displayed as Fig. 9.59(a) and (b).

It is worth noting that Fredriksson et al. (1981) avoided having to match the large dynamic range of the return signal (arising primarily from its $1/R^2$ falloff) to the limited dynamic range of the fast-transient digitizer through a combination of applying a synchronized ramp voltage to the dynode chain of

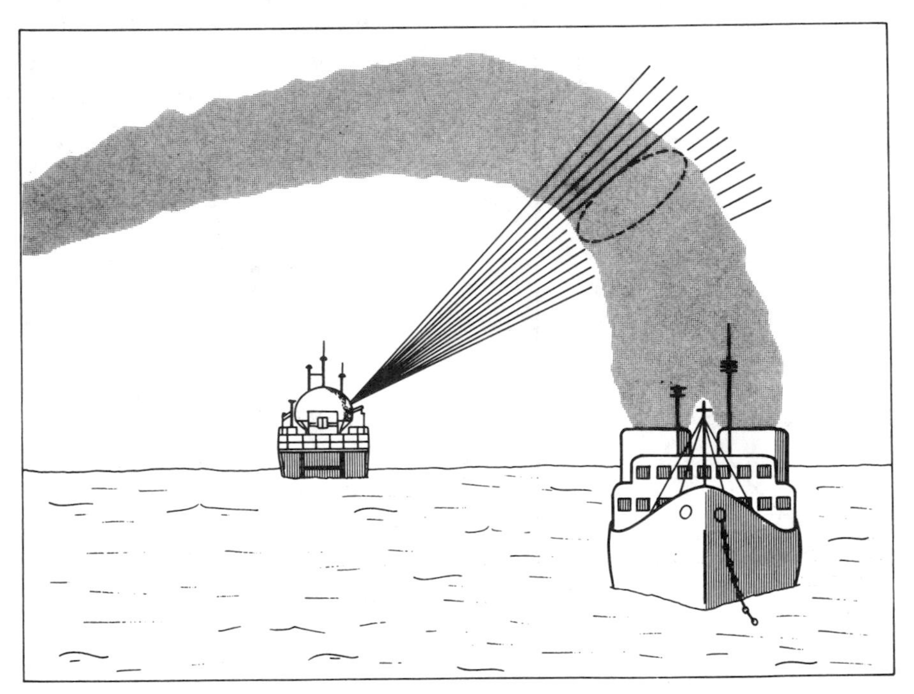

Fig. 9.61. Pictorial illustration of the remote monitoring of an incineration ship plume by a ship-mounted lidar (Weitkamp et al., 1981).

the photomultiplier and using geometrical compression in the optical design (Harms et al., 1978); see Section 7.4. A representative set of backscattered signals, their appropriate signal ratio, and the resultant pollutant mass-density–range profile are presented as Fig. 9.60 for NO_2. As can be seen, the atmosphere was very inhomogeneous between the lidar and the plume of interest. The aeroticulate signals have been eliminated in the DIAL-signal ratio curve. From this DIAL curve the integrated NO_2 concentration has been evaluated. These NO_2 measurements were recorded from the chimney of a saltpeter plant at a range of 1350 m and required averaging over 1000 laser-pulse pairs of rather low-energy (2-mJ) pulses. The NO_2 concentration in the plume was estimated to be 122 mg m^{-3} (which corresponds roughly to 100 ppm) and agreed well with the value determined by conventional means.

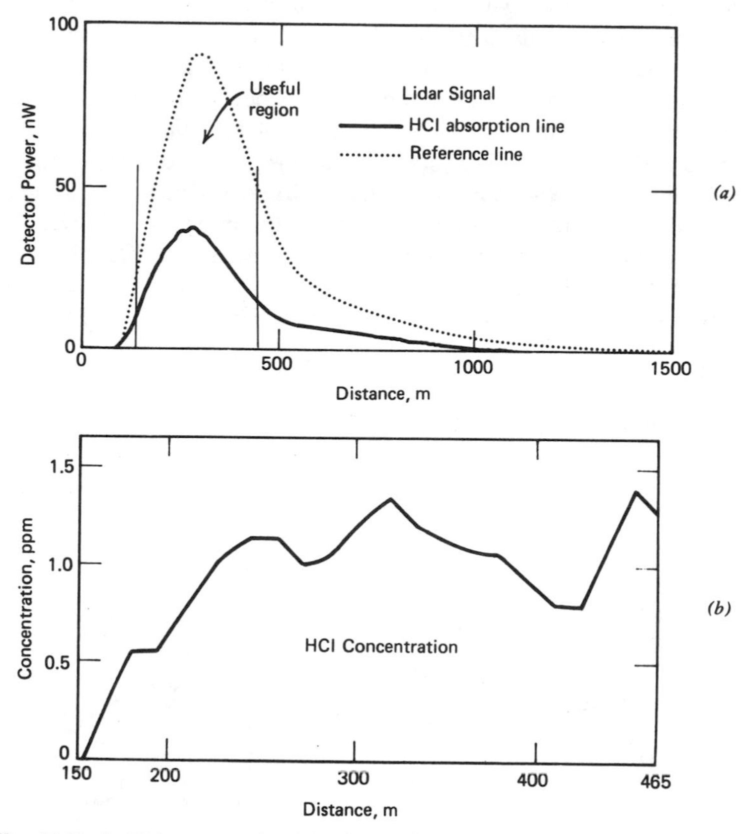

Fig. 9.62. (a) Typical lidar return signal for $P_2(3)$ HCl absorption and $P_2(5)$ reference lines (top) of a DF laser. To reduce noise, signals are averaged over 100 [$P_2(3)$] and 50 [$P_2(5)$] laser shots, respectively. Electromagnetic interference from the laser and partial beam overlap determine the short-distance limit, while photon statistics are responsible for the long-distance limit to the useful region. In the above case this extends from 150 to 465 m. (b) The corresponding lidar-determined HCl concentration profile (Weitkamp et al., 1980).

In closing this chapter I should like to present one last example of pollution-source monitoring based on differential absorption and scattering. This application is not only somewhat different from those previously considered but turns out to be fairly appropriate for interfacing this chapter on atmospheric lidar applications with the subsequent chapter on hydrographic lidar applications.

Chlorinated hydrocarbons are today extensively used in a variety of industrial processes. Indeed, organochloride wastes in excess of 100,000 tons are produced annually by the industrialized nations. Because of the established carcinogenic, mutagenic, teratogenic, and acute toxic effects of certain components of this waste, its disposal presents considerable problems. Destruction by incineration is regarded as the best means of disposing of these substances with minimal impact on the biosphere. If this is done at sea, expensive scrubbers (for the removal of the hydrogen chloride produced by combustion) are not required, provided the incineration is conducted far enough from populated areas. However, the further out to sea, the greater the cost, so the question arises—how far and in what direction? What is needed to answer this question is an investigation of the plume concentration of HCl under different weather conditions and a mathematical model for describing its disappearance.

Weitkamp et al. (1980) and Weitkamp (1981) describe a DIAL system based on a pulsed TEA deuterium fluoride laser for this purpose; see Fig. 9.61. The relevant data pertaining to this system are presented in Table 9.9 (Weitkamp, 1981), and some of the results obtained by Weitkamp et al. (1980) are displayed as Figs. 9.62 and 9.63. Unfortunately, the need for sequential

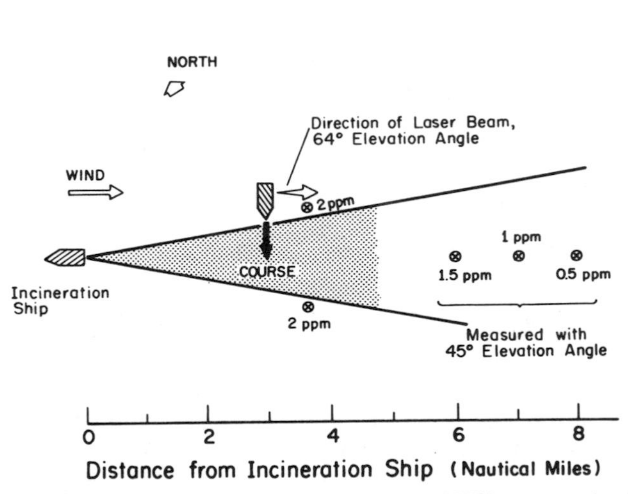

Fig. 9.63. Schematic representation of the lidar evaluation of HCl concentration in the plume of an incineration ship, indicating specific measurements undertaken (Weitkamp, 1981).

TABLE 9.9. DATA FOR HCL DIFFERENTIAL ABSORPTION AND SCATTERING LIDAR[a]

Laser	DF, pulsed	
	Signal	Reference
Lines	$P_2(3)$	$P_2(5)$
Wavelength (μm)	3.6363	3.6983
Pulse energy (mJ)	30	10
HCl absorption coefficient (cm^{-1} atm^{-1})	5.64	0.02
Pulse duration	300 ns	
Repetition frequency	1 Hz	
Wavelength selection	Grating	
Receiver optics	50-cm afocal Cassegrain	
Detector	InSb 1-mm-diameter LN_2-cooled	
Compression of signal dynamics	Geometric	
Transmitter–receiver geometry	Noncoaxial	
Amplifier bandwidth	500 kHz to 2 MHz	

[a] Weitkamp (1981).

measurements at the signal and reference ("on" and "off") wavelengths precludes the evaluation of data from those parts of the plume that appear opaque, because there are rapid and large changes in the elastic backscattering coefficient between the individual laser shots. A system that involves near-simultaneous "on" and "off" laser pulses would do much to alleviate this problem. Accurate and extensive measurements of HCl diminution through diffusion, chemical reaction, absorption within aerosols, and washout by rain will help in testing mathematical models that will aid in determining the right balance between safe incineration at sea and the economics of this disposal method (Weitkamp, 1981).

In summary, we see that laser remote sensing is likely to have a significant impact on atmospheric investigations. One of the most striking features of this impact is its diversity, for it appears that almost every facet of atmospheric study can benefit from lidar measurements. We have seen that lidar systems can provide important local or global meteorological information on such features as cloud structures and humidity and temperature profiles. Lidars can also be used to probe both the stratosphere and the mesosphere, and they can be mounted aboard ships, vans, helicopters, aircraft, and space platforms. Lastly, we have seen that they can also be used to undertake pollution surveillance measurements as well as map the dispersion of trace pollutants from various kinds of emission sources.

10

Hydrographic Lidar Applications

Despite the fact that lasers have been used to probe the atmosphere almost since their inception, a number of years passed before they were mounted in aircraft or ships and directed downwards to investigate the hydrosphere. Remote sensing of the oceans, lakes, and rivers of our planet are possible with visible, infrared, and microwave radiation. Indeed, an enormous wealth of data has been collected from ships, aircraft, and satellites. Most of this information was derived passively (by other than radar observations) prior to the introduction of the laser into hydrographic work. As we shall see in this chapter, not only does the laser complement and extend the kinds of measurements undertaken; it can truly be said to add another dimension to hydrographic research, for it permits a degree of depth resolution and subsurface interrogation that is unattainable through other remote-sensing techniques. The primary reasons for this are that infrared and microwave radiation have negligible penetration in water (see Fig. 10.1), and visible observations were essentially passive in nature prior to the introduction of the hydrographic lidar systems.

10.1. BATHYMETRIC SURVEY AND TURBIDITY MEASUREMENTS

Towards the end of the sixties Hickman and Hogg (1969) demonstrated the feasibility of deploying an airborne blue–green laser for the purpose of mapping subsurface topography using the time interval between the surface and subsurface returns as a measure of the depth (see Fig. 10.2). A 60-μJ pulsed neon laser was found adequate to record depths of close to 8 m from an altitude of 150 m over the shores of Lake Ontario at night. Although the 3-ns duration of their pulsed neon ion laser enabled them to achieve a spatial resolution of about 0.34 m, they pointed out that scattering and beam spreading could ultimately limit the depth resolution attainable.

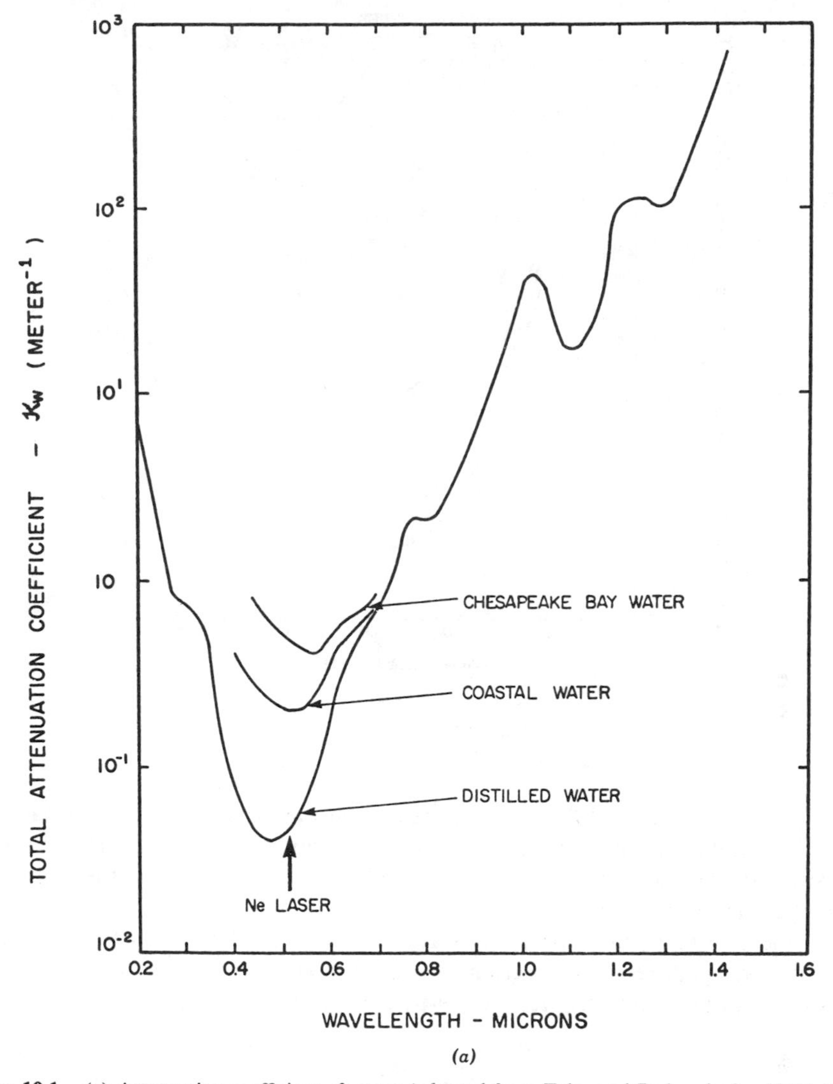

(a)

Fig. 10.1. (a) Attenuation coefficient of water (adapted from Tyler and Preisendorfer, 1962). (b) Downward irradiance attenuation coefficient measured by Jerlov (1976) in the first 10 m of depth as a function of wavelength for a variety of deep ocean and coastal water types (Northam et al., 1981).

Coastal waters, which are likely to be prime targets for laser bathymetry, are often rich in microscopic marine life. Thus not only is there likely to be enhanced scattering in such areas, but also absorption due to phytoplankton. This is reflected in the different spectral profiles of the attenuation coefficient $\kappa_W(\lambda)$ seen in Fig. 10.1(a) and (b). The so-called *window on sea* can be seen to lie in the blue–green region of the spectrum. Extensive studies by Hickman

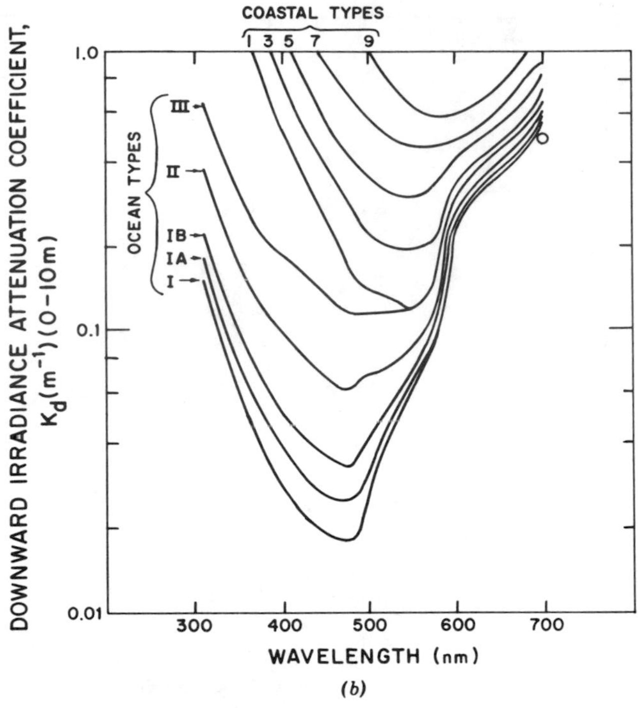

(*b*)

Fig. 10.1. (*Continued*)

(1973) and Levis et al. (1973) have revealed that the depth-measuring capability of a bathymetric lidar depends upon the attenuation ratio for absorption versus scattering for water (κ_W^A/κ_W^s) as well as the reflectivity of the bottom and the overall attenuation coefficient ($\kappa_W = \kappa_W^A + \kappa_W^S$).

In the case of hydrographic applications several modifications to the lidar equation are required. The fraction of radiation emanating from a depth R_W below the surface and accepted by the lidar is given by

$$\frac{\Omega_S}{4\pi} = \frac{A_S}{4\pi R_W^2} \tag{10.1}$$

where A_S represents the surface area within the field of view of the receiver optics. However, as seen from the geometry of the situation (Fig. 10.3),

$$\frac{A_S}{(R_W/n)^2} = \frac{A_0}{(R_A + R_W/n)^2} \tag{10.2}$$

where n is the refractive index of the water, A_0 is the effective area of the

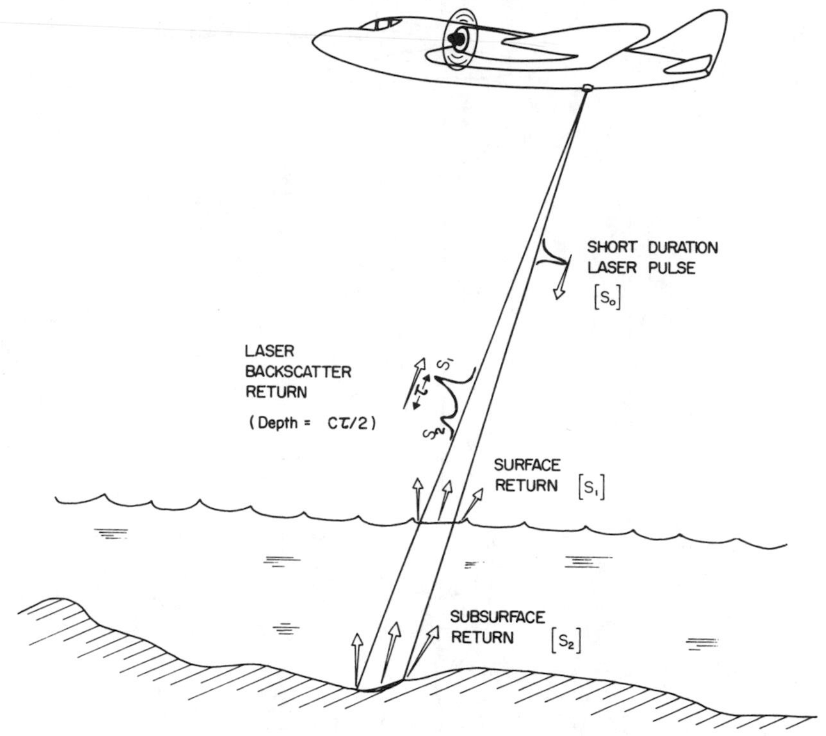

Fig. 10.2. Pictorial representation of airborne laser bathymetry.

collection optics, and R_A is the altitude of the lidar system. Thus

$$\frac{\Omega_S}{4\pi} = \frac{A_0}{n^2(R_A + R_W/n)^2 4\pi} \tag{10.3}$$

and the energy received at wavelength λ from depth R_W and arising from an interaction with a species having an appropriate cross section $\sigma_S(\lambda, \lambda_L)$ can be derived from equation (7.15) by substitution of Ω_S for A_0/R^2:

$$E_S(\lambda) = E_L T(R_A)\xi(R_A)\xi(\lambda)\frac{c\tau_d A_0 \phi_W N_S \sigma_S(\lambda, \lambda_L)}{8\pi n^2(R_A + R_W/n)^2}$$

$$\times \exp\left(-\int_0^{R_W}\{\kappa_W(\lambda_L) + \kappa_W(\lambda)\}\, dR\right) \tag{10.4}$$

where N_S represents the number density of the scattering species at depth R_W, $\xi(\lambda)$ is the receiver's spectral transmission factor and $\xi(R_A)$ can be taken as

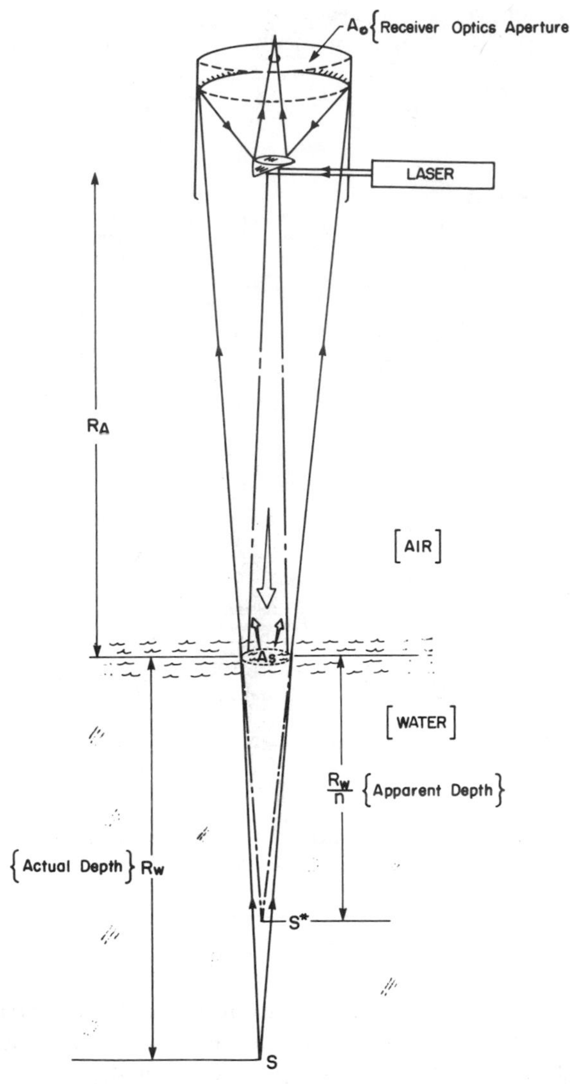

Fig. 10.3. Schematic of optical geometry of an airborne oceanographic lidar, illustrating how the refractive index affects the range factor in the lidar equation, namely, $A_s/(R_W/n)^2 = A_0/(R_A + R_W/n)^2$.

the geometrical overlap factor, $T(R_A)$ accounts for the two-way attenuation within the atmosphere, and ϕ_W is termed the two-way (air–water and water–air) *transmission factor* for the air–water interface. ϕ_W not only includes the transmission factors for a smooth interface (see Section 2.5), but should also allow for the influence of surface roughness, by taking a sea-slope probability factor (Levis et al., 1973) into account. In the case of bathymetry we may write for the surface return (essentially Fresnel reflection)

$$E_1(\lambda_L) = E_L \xi(\lambda_L) \xi(R_A) \frac{A_0}{R_A^2} \frac{\rho_W(\lambda_L)}{\pi} e^{-2\kappa_A(\lambda_L) R_A} \quad (10.5)$$

and for the subsurface return

$$E_2(\lambda_L) = E_1(\lambda_L) \frac{R_A^2 \phi_W}{(R_A + R_W/n)^2 n^2} \times \frac{\rho_B(\lambda_L)}{\rho_W(\lambda_L)} e^{-[\kappa_W^-(\lambda_L) + \kappa_W^+(\lambda_L)] R_W} \quad (10.6)$$

where we have assumed a common overlap factor and have introduced the water and subsurface backscattering efficiencies, $\rho_W(\lambda_L)$ and $\rho_B(\lambda_L)$, respectively.

The attenuation coefficient used in equation (10.6) for the downward $\kappa_W^-(\lambda_L)$ and upward $\kappa_W^+(\lambda_L)$ paths can be different due to beam spreading. Duntley (1963) has shown that beyond a few exponential attenuation lengths, forward scattering effectively reduces the rate of attenuation. In bathymetry the downward beam will initially be well collimated, and so the monopath attenuation coefficient might be expected to apply. However, as the beam spreads and diffuses, forward scattering tends to replenish the beam, and a multipath attenuation coefficient might be more appropriate. Levis et al. (1973), using a frequency-doubled *Q*-switched neodymium–glass laser mounted in a tower 14 m above the water surface, were able to detect targets to a depth of 26 m with a considerable sensitivity margin. In general their measured returns fell between the values predicted on the basis of monopath and multipath attenuation. Based on this, one might expect that the maximum depth-measuring capability would lie around 8 to 10 monopath attenuation lengths, although Hickman (1973) has indicated that values closer to 15 might be achieved under favorable conditions. If Δt is the time interval between the two returns, then a monopath assumption leads to

$$R_W = \frac{c\,\Delta t}{2n} \quad (10.7)$$

Obviously, where the subsurface return is observed by virtue of a multipath attenuation coefficient for part of the path, some error will be incurred in using this relation.

Kim et al. (1975) and Kim (1977) developed and flight-tested a prototype operational airborne laser bathymeter (ALB) over Chesapeake Bay and over Key West, Florida. Although, the primary objective of these tests was to evaluate ALB technology, a subsurface contour map of the test site was created and is illustrated in Fig. 10.4. An example of the comparison between the subsurface topography observed with the airborne lidar and that obtained from a sonar system aboard a surface ship is shown as Fig. 10.5. Although reasonably good agreement was observed for the general features, the relative inaccuracy of the sonar equipment prevented a more detailed comparison of the two approaches. Water depths of 10 ± 0.25 m were recorded in waters having an effective attenuation coefficient of $0.175\ \mathrm{m}^{-1}$.

Although some of the preliminary runs of this system were conducted with a 2-kW, 6-ns, pulsed neon ion laser operating at 540 nm, the major portion of this bathymetric work was undertaken with a 2-MW, 8-ns, frequency-doubled Nd–YAG laser having an output at 532 nm and a repetition rate up to 50 Hz. The receiver system involved a 28-cm-diameter $f/1$ Cassegrain telescope, a 0.4-nm interference filter centered at 532 nm, and an RCA 8575 photomulti-

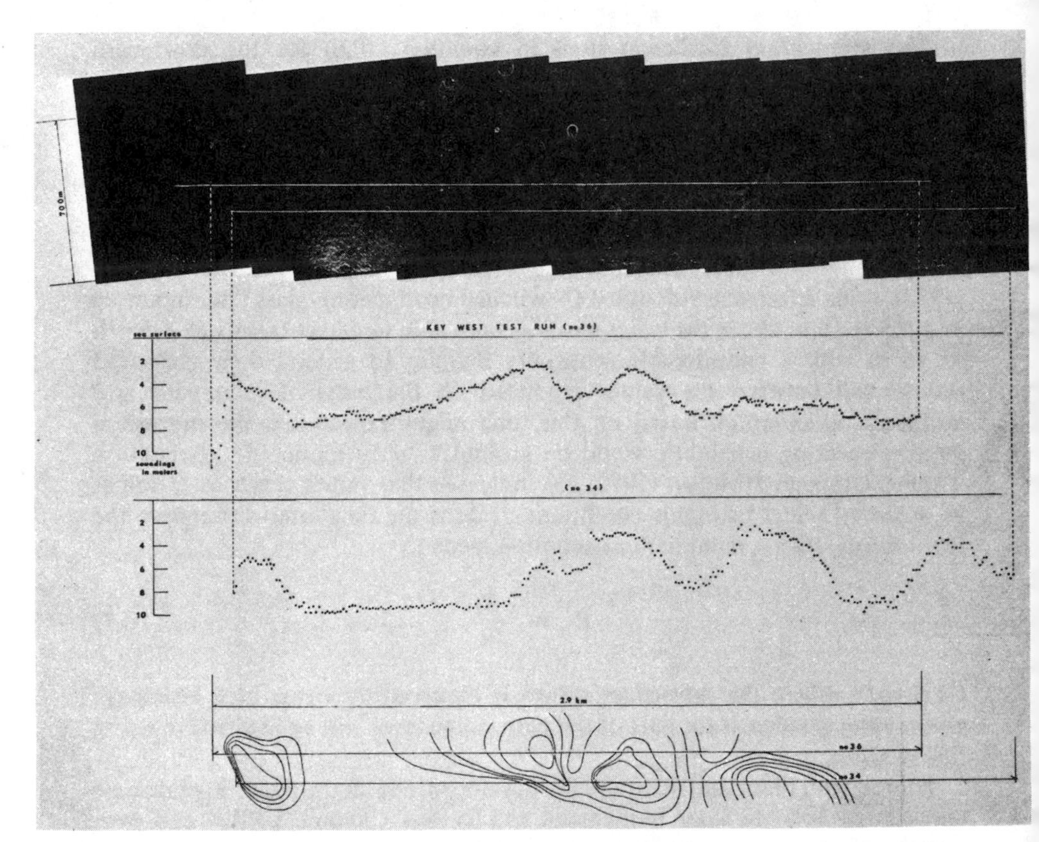

Fig. 10.4. Airborne lidar bathymetric survey results of Key West, Florida, undertaken by NASA (Kim et al., 1975).

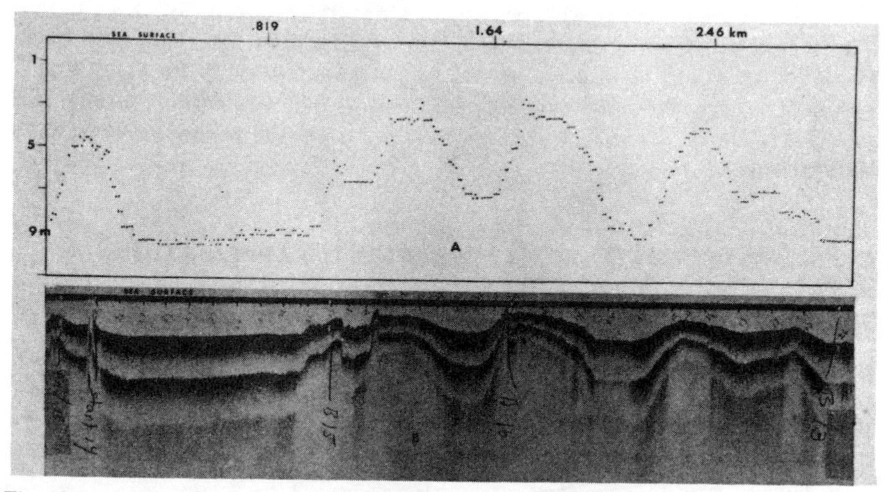

Fig. 10.5. Ocean floor profile: A, determined by an airborne bathymetric lidar; B, a matching sonar chart (Kim et al., 1975).

plier. The data were recorded onto 9-track magnetic tape from a waveform digitizer and the system was flown on a military version of a DC-4 at an altitude between 150 and 600 m while flying at about 150 knots (Kim, 1977).

A true operational airborne laser bathymeter would require two improvements over the system described above. First, the lidar would have to scan rapidly through an appreciable nadir angle in order to extract bathymetric information from a reasonable width of the overflight, and second, to facilitate good lateral resolution a laser with a higher repetition rate would be needed. Hoge et al. (1980) reported the initial baseline performance of a new NASA airborne oceanographic lidar (AOL) system that includes a state-of-the-art conical-type scanning pulsed laser. As shown in Fig. 10.6, the 400-Hz, 7-ns, 2-kW output of this neon ion laser at 540.1 nm is folded into the adjustable-beam-divergence collimating lens, and then directed downward through the main receiver folding flat and onto the 56-cm, round, nutating scanner mirror, which directs the beam towards the water's surface. Flight tests over the Atlantic Ocean yielded water depth measurements to 10 m, while water depths to 4.6 m were measured in the more turbid Chesapeake Bay. To aid in evaluating the lidar system's operational performance, water-truth measurements of depth and beam attenuation coefficients were determined from a boat at the same position as the aircraft overflights. Recently, Northam et al. (1981) have reported on the development of a high-repetition-rate frequency-doubled Nd–YAG laser that has been optimized for airborne bathymetry. This system has an output of 0.4 MW in 7 ns at 400 Hz and represents an improvement of more than two orders of magnitude over the laser used by Hoge et al. (1980).

Turbidity measurements have formed a natural extension of these bathymetric studies. Hickman et al. (1974) have shown that the volumetric backscattered signal can be directly related to the water turbidity for $\kappa_W \leq 6\ \mathrm{m}^{-1}$ and that the greatest sensitivity was achieved at 440 nm. Although only 10% accuracy may be feasible, use of several wavelengths would improve such measurements.

10.2. THE LASER FLUOROSENSOR AND ITS APPLICATIONS

Towards the end of the sixties concern over the environmental impact of oil spills on the oceans, lakes, and rivers, combined with inadequate means of

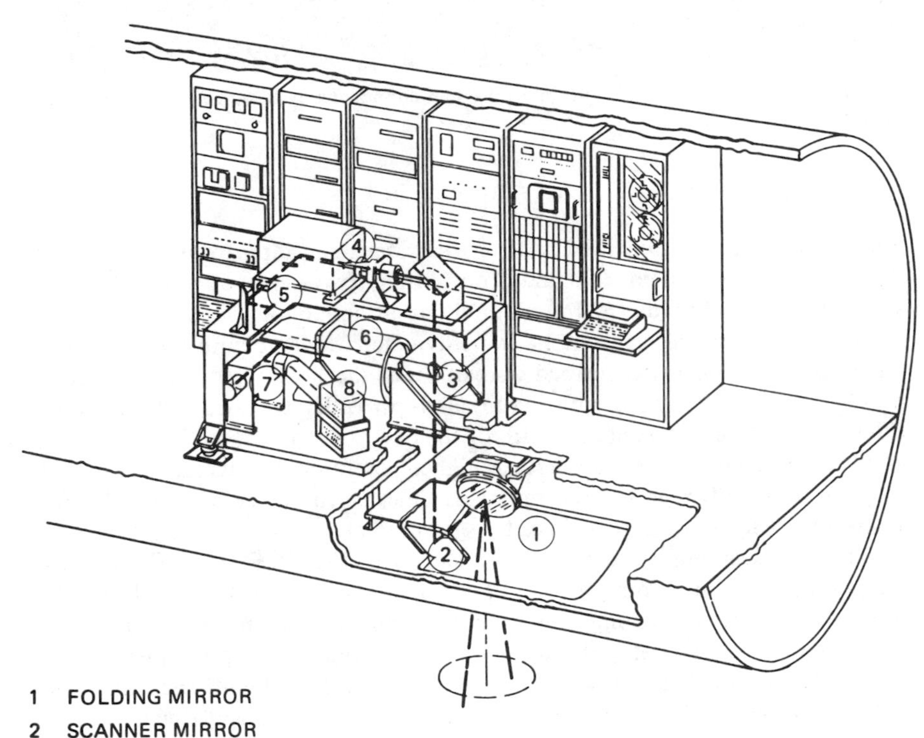

Fig. 10.6. Cutaway illustration of the AOL installed in the NASA Wallops Flight Center C-54 aircraft (Hoge et al., 1980).

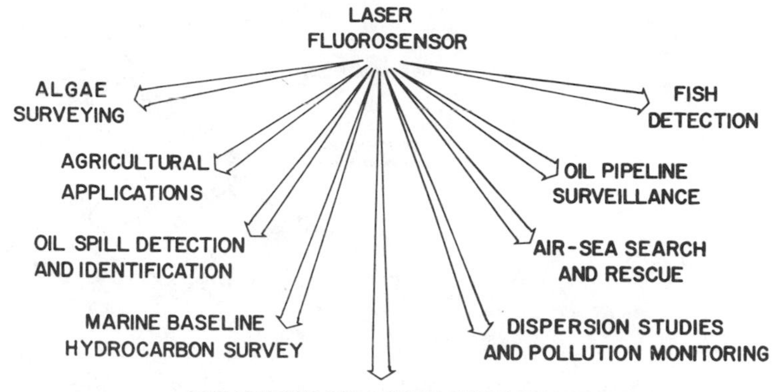

Fig. 10.7. Range of application amenable to laser-induced fluorescence studies of the environment (Measures et al., 1975).

Fig. 10.8. (a) University of Toronto, Institute for Aerospace Studies (UTIAS) laser fluorosensor at field test site (1972); (b) principle of operation of the laser fluorosensor (Measures et al., 1975).

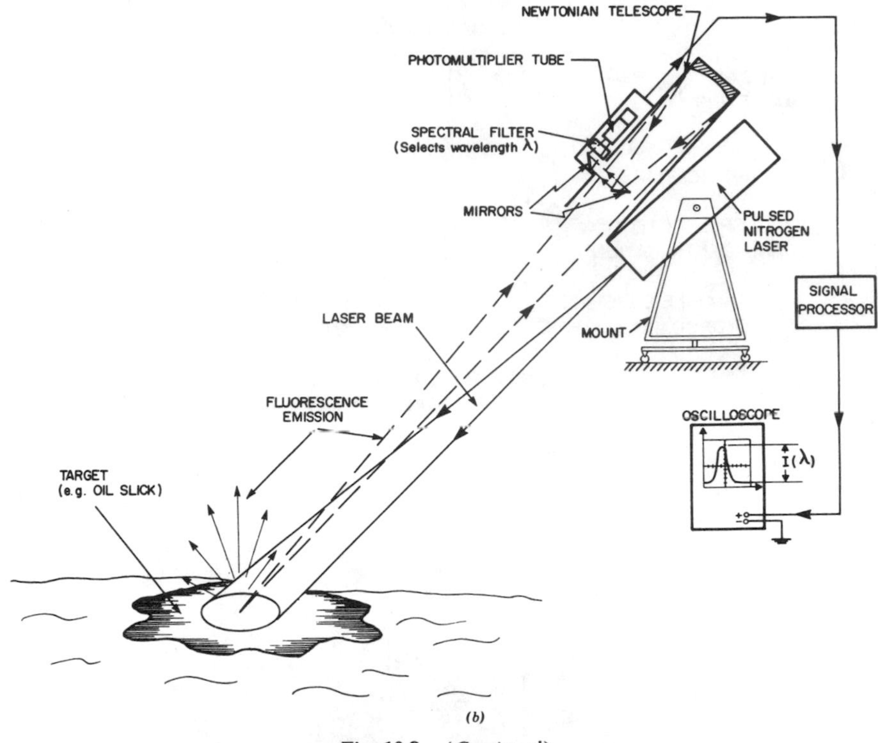

Fig. 10.8. (*Continued*)

detecting intentional spills at night, provided the incentive for the development of a new class of remote airborne sensor termed the *laser fluorosensor* (Measures and Bristow, 1971; Fantasia et al., 1971). Fluorescence had been used in many disciplines as a powerful analytical tool enabling substances to be detected down to the few parts per million in the laboratory. The invention of the high-power laser operating in the near ultraviolet made it possible to extend this capability to the field of remote sensing. Indeed, we saw earlier that laser-induced fluorescence had an important role in studies of the upper atmosphere.

Although the laser fluorosensor was originally conceived for airborne oil-spill detection, Measures et al. (1975) indicated that from the onset this new form of active remote sensor was conceived to be capable of undertaking a broad class of missions, some of which had never previously been considered to be within the realm of airborne surveillance. An appreciation for this spectrum of potential application can be obtained by reference to Fig. 10.7. The prototype laser fluorosensor developed by Measures et al. (1973) involved a 100-kW, 10-ns, 337-nm nitrogen laser and is shown in Fig. 10.8(a). Its principle of operation is illustrated in Fig. 10.8(b). The airborne version is shown, mounted

Fig. 10.9. Canada Centre for Remote Sensing DC-3-mounted airborne version of the UTIAS laser fluorosensor. (Courtesy M. Bristow, private communication.)

within a DC-3 of the Canada Centre for Remote Sensing, in Fig. 10.9. A more sophisticated version was developed by Fantasia and Ingrao (1974) for the U.S. Coast Guard.

10.2.1. Oil-Spill Detection and Identification

In the case of oil pollution at sea, or on lakes and rivers, there are many different kinds of sensor that are capable of indicating the presence of some anomaly. These include multiband cameras, radar mappers, infrared scanners, and microwave radiometers. However, none of these can unequivocally recognize the presence of an oil patch on a 24-hour basis or offer any possibility of classifying the type of oil. Fantasia et al. (1971), in the U.S. Coast Guard, studied the fluorescence characteristics of some 29 samples of crude oil. Their results suggested that each sample could be uniquely characterized by a measurement of its peak emission wavelength, lifetime, and fluorescence efficiency [equation (7.52)]. Moreover, the magnitude of this fluorescence efficiency was sufficient to make airborne measurements possible. This conclusion was also reached independently by Measures and Bristow (1971).

As a result of this work the role of the laser fluorosensor was considered to be complementary to that of other, high altitude oil detection sensors. Once one of these sensors detected the presence of some anomaly, it was thought that the reconnaissance aircraft would reduce altitude in order that the laser fluorosensor could probe the target to determine if the anomaly arose from an oil spill, and if so, characterize the oil. An advanced airborne version of the U.S. DOT–Coast Guard oil-sensing lidar was mounted inside a Sikorsky CH53 U.S. marine helicopter and demonstrated the feasibility of this approach (Fantasia and Ingrao, 1974). This laser fluorosensor employed a 10^5-W, 10-ns nitrogen laser operating at 337 nm with a 500-Hz repetition rate. A 30-cm-diameter, $f/3.5$ Newtonian telescope was used with an image dissector and an optical multichannel analyzer (OMA) to record and process the complete fluorescence spectrum. This system was arranged to operate in two modes. In the detection mode two of the 35 OMA channels were calibrated for the ambient sea-water fluorescence and were used to indicate the presence of an anomaly in the fluorescence return. Once such an anomaly was monitored, the operator was alerted and the system was switched to the classification mode in which all channels were activated.

A representative set of fluorescence spectra acquired with the above helicopter-mounted laser fluorosensor during controlled oil spills are presented as Fig. 10.10. All measurements were made from an altitude that ranged from 31 to 46 m and at a forward speed of about 70 km hr^{-1}. The background sea-water fluorescence profiles were obtained from the vicinity of the test site prior to the oil spills. Some difference is observed between the day and the night fluorescence spectrum for each target. In particular, the fluorescence peak at night appears at a slightly shorter wavelength than the peak during the day. Fantasia and Ingrao (1974) provided no explanation of this observation.

Laboratory studies (Fantasia et al., 1971; Rayner and Szabo, 1978) have shown that although the fluorescence spectrum of a mineral oil would not in itself permit identification of that sample of oil at sea, it would allow the oil to be classified within one of three groups, namely, *light refined products* such as diesel fuel, *crude oils*, and *heavy residual products* such an Bunker C fuel oil. A representative spectrum from each group is displayed for comparison purposes as Fig. 10.11.

In most oil-spill situations, the fluorescence from the oil film will be much stronger than the background water fluorescence, so that there is unlikely to be any problem in detecting the oil. As we have seen above, the class of oil may also be characterized from the general shape of the fluorescence spectra, but in order to be more specific, additional information is required. The fluorescence conversion efficiency represents one such parameter, according to the laboratory studies. Unfortunately, this parameter appears to be of limited value under the real conditions of an oil spill, where the oil film may not be optically thick over part or all of the laser footprint, or the oil has been weathered or become dispersed through a water column. O'Neil et al. (1980) have proposed that one way of improving the characterization of the oil is to use the Pearson linear correlation coefficient ρ, which is sensitive to the spectral shape and not the integrated fluorescence signal. When the correlation coefficient of a given oil with each oil in a number of samples is calculated, the oil groups are found

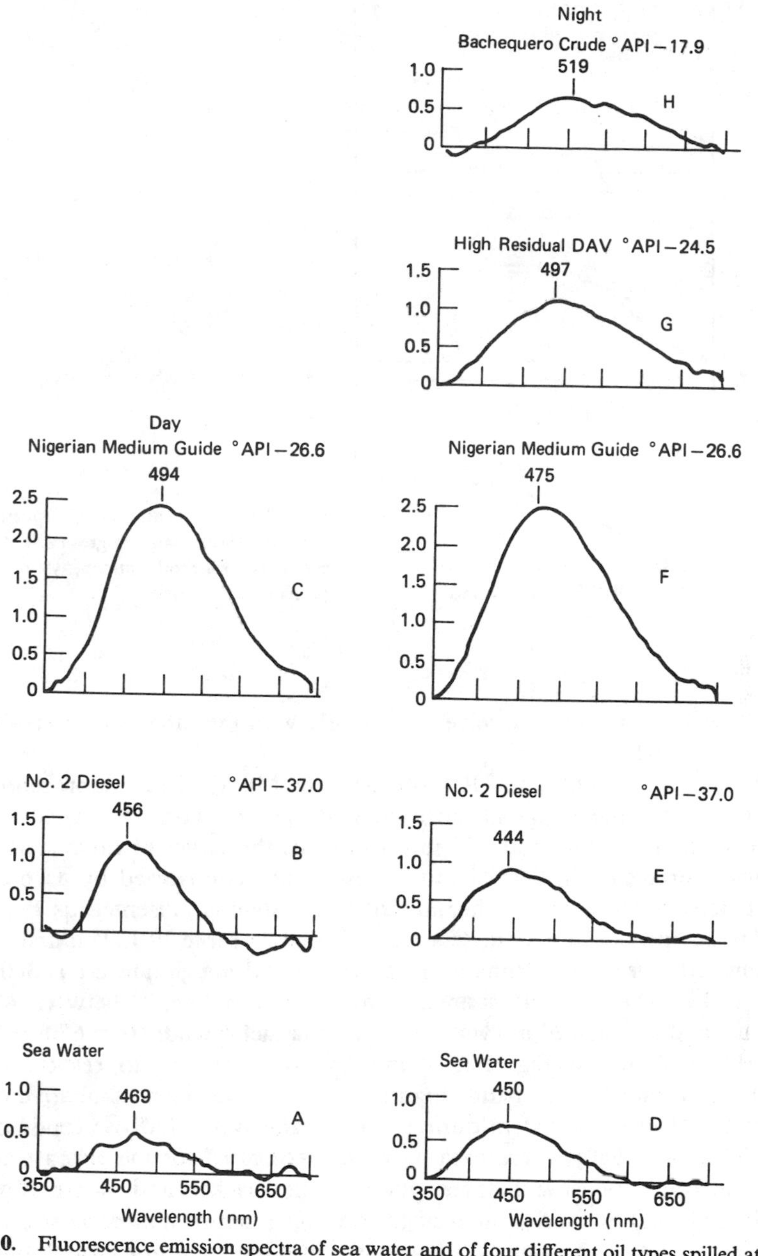

Fig. 10.10. Fluorescence emission spectra of sea water and of four different oil types spilled at sea as measured by the Transport Systems Centre (Cambridge, Massachusetts experimental airborne laser oil-spill remote-sensing system during field test (Fantasia and Ingrao, 1974).

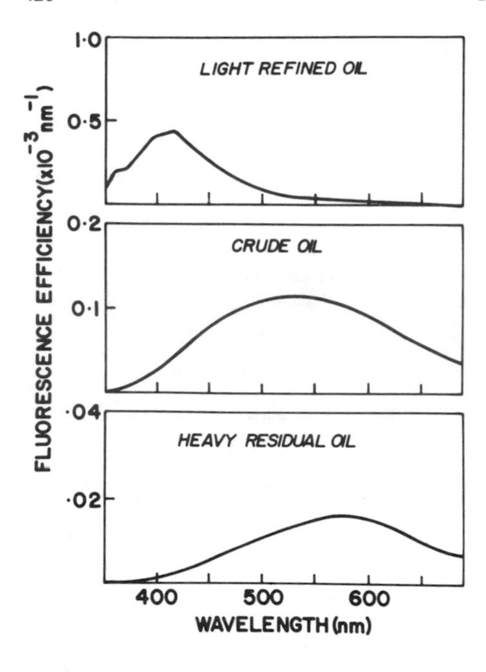

Fig. 10.11. Typical emission spectra of oils from the three main oil groups: light refined products, crude oils, and heavy residual oils (O'Neil et al., 1978).

to correlate well with themselves but poorly with the other two (O'Neil et al., 1981).

The Canada Centre for Remote Sensing Mk III laser fluorosensor is a representative state-of-the-art airborne hydrographic lidar. A rugged and reliable N_2 laser, operating at 337 nm, irradiates the target of interest while the return fluorescence in the 380- to 700-nm range is observed by a 16-channel photodetection system. A schematic of this system is presented as Fig. 10.12. The basic operational parameters are provided in Table 10.1. The first channel is centered at the water Raman line at 381 nm (although labeled as 380 nm in Fig. 10.13). The next 14 channels have center wavelengths between 400 and 660 nm and are each 20 nm wide. The last channel extends from 670 to 720 nm (labeled 680 nm in Fig. 10.13) and has been chosen to respond to the chlorophyll fluorescence band. Figure 10.13 presents a set of relative outputs from the 16-channel system during a test flight over a LaRosa crude oil spill (O'Neil et al., 1980). These data have been corrected for the aircraft altitude, photodetector gain, and laser power. Solar-background subtraction was achieved by gating the first intensifier "on" again between laser pulses.

The oil slick is apparent—at about 16 s into the flight line. The fluorescence of the oil is most apparent in the wavelength interval from 520 to 640 nm. Suppression of the water Raman return is seen to occur in the 380-nm channel at the location of the oil spill. In an operational laser fluorosensor, the

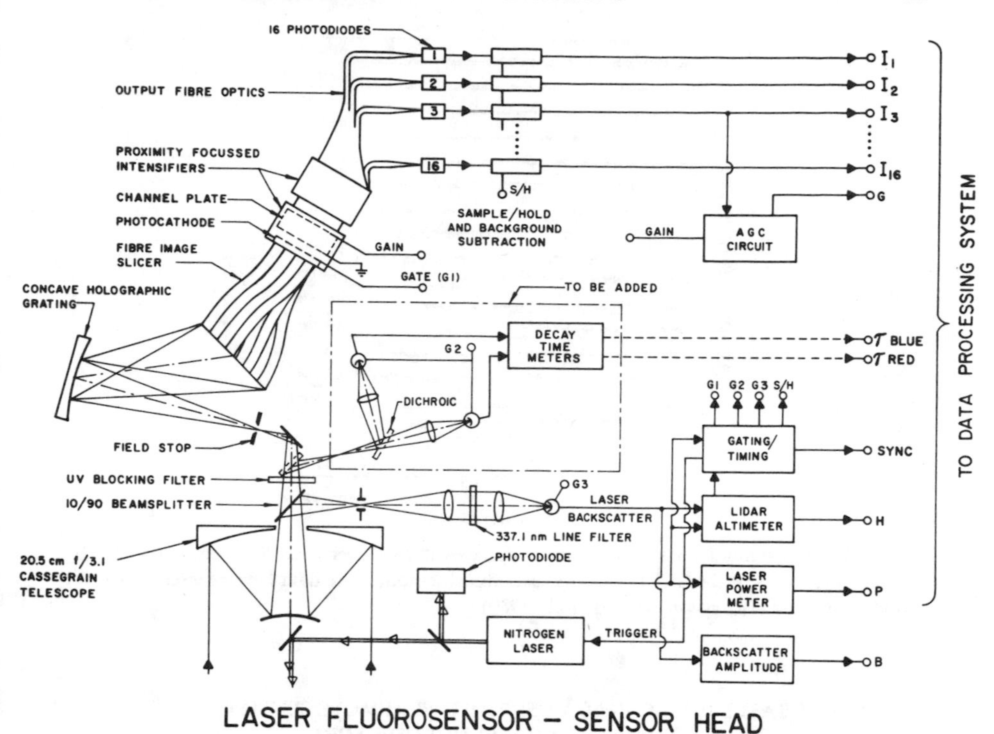

Fig. 10.12. Schematic diagram of the Canada Centre for Remote Sensing (CCRS) airborne laser fluorosensor showing the major electrooptical components. In addition, the entire fluorosensor system requires a laser support pallet (consisting of a power supply, a nitrogen supply, and a vacuum pump) and a microprocessor-based data-acquisition system that includes a real-time display and a computer-compatible tape transport.

correlation technique provides a better signal-to-noise ratio than the spectra in any one channel, as all of the radiation returned from the target contributes to the correlation signal. This can be seen by reference to Fig. 10.14, where the upper trace represents the fluorescence signal in the 540-nm channel while the lower trace displays the computed correlation of the data from all of the spectral channels (with the exception of the Raman channel labeled 380 nm) with the laboratory fluorescence spectra of a La Rosa crude oil.

We see that the airborne laser fluorosensor appears ideally suited for detecting, characterizing, and (as we shall see shortly) mapping the thickness of oil spills. There is also evidence that these sensors can detect oil that is dispersed within a water column, and in addition some tests have suggested that a laser fluorosensor could be unique in having an ability to detect oil in water with a significant degree of ice coverage (O'Neil et al., 1981). This could be quite important with regard to oil spills in the arctic regions.

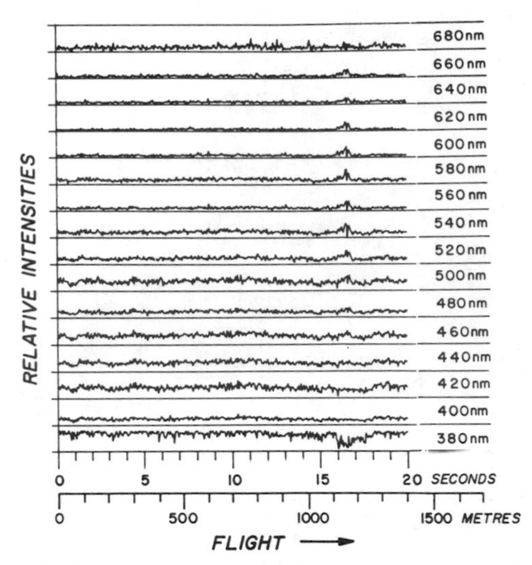

Fig. 10.13. Representative laser-fluorosensor signals gathered over a La Rosa crude oil spill by the sixteen channels of the CCRS airborne laser fluorosensor. These data have been corrected for altitude, gain, and laser power (O'Neil et al., 1980).

TABLE 10.1. CANADA CENTRE FOR REMOTE SENSING AIRBORNE LASER FLUOROSENSOR[a]

Transmitter: N_2 laser	
Wavelength	337.1 nm
Bandwidth	$\leq$ 0.1 nm
Pulse width	3 ns
Repetition rate	100 Hz
Peak output power (max)	300 kW
Beam divergence	3 $\times$ 1 mrad
Receiver: *f*/3.1 Dall–Kirkham telescope with a clear aperture of 0.0232 m^2	
Nominal spectral range	386–690 nm
Nominal spectral bandpass (2 to 15 channels)	20 nm/channel
Field of view	3 $\times$ 1 mrad
Intensifier on-gate period	70 ns
Noise-equivalent energy[b]	$\approx 4.8 \times 10^{-17}$ J
Experiment:	
Aircraft altitude	75–750 m
Aircraft velocity	$\approx$ 100 m/s
Measurement background	Day and night

[a] O'Neil et al. (1980).

[b] This is the apparent fluorescence energy (after background subtraction) collected by the receiver in one wavelength channel for a single laser pulse that equals the noise in that channel.

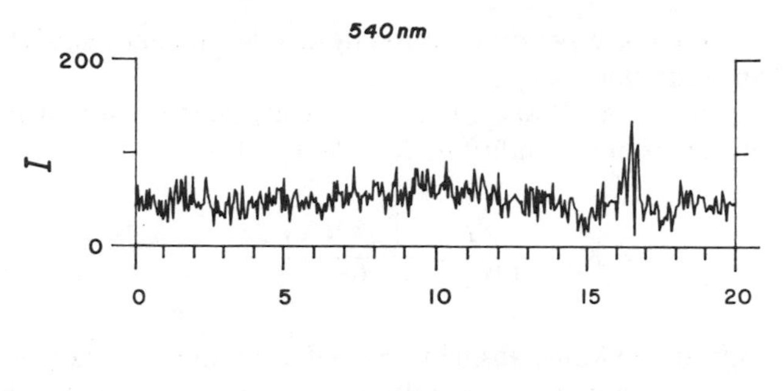

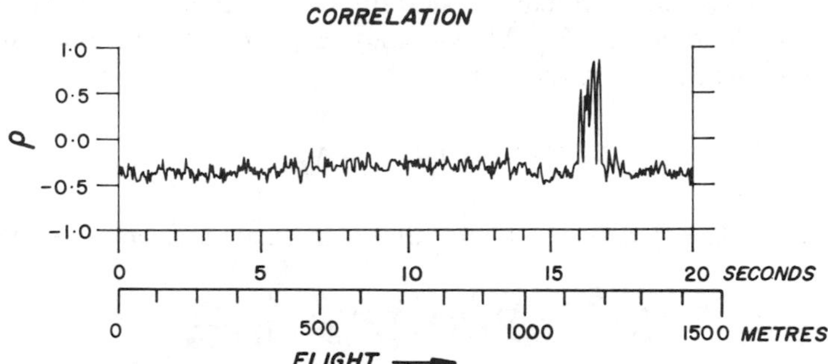

Fig. 10.14. Identification and classification of a crude-oil spill by means of an airborne laser fluorosensor. The upper trace shows a single channel of the raw data, while the lower trace shows the result of a correlation of the data in all spectral channels (with the exception of the Raman channel labeled 380 nm) with the fluorescence emission spectrum of La Rosa crude oil. The correlation procedure enhances the oil slick 16 s into the flight line and increases the signal-to-noise ratio (O'Neil et al., 1981).

10.2.2. Oil-Film Thickness Measurement

The theory of oil-film thickness measurement by a laser fluorosensor has been given by Kung and Itzkan (1976). We present here a somewhat simplified version. In the case of a dense medium comprising many different fluorescent molecules, laser-induced fluorescence can be a very complex interaction. If we assume steady-state conditions, we may introduce the fluorescence power radiated per unit volume per unit solid angle per unit wavelength interval, when excited by laser radiation at wavelength λ_L:

$$P^F(\lambda) = \sum_i \frac{N_i^* \mathscr{L}_i^F(\lambda, \lambda_L) hc}{4\pi\lambda\tau_i^{\mathrm{rad}}} \tag{10.8}$$

where N_i^* represents the excited-state number density of the ith molecular species, $\mathscr{L}_i^F(\lambda, \lambda_L)$ the corresponding emission profile function, and τ_i^{rad} the appropriate radiative lifetime.

According to equations (4.24) and (4.35) the steady-state density of excited molecules under nonsaturating conditions (i.e., $S_I \ll 1$) is

$$N_i^* = N_i B_{12}^i \tau_i \int \frac{I^L(\lambda) \mathscr{L}_i^A(\lambda)\, d\lambda}{4\pi} \tag{10.9}$$

where B_{12}^i is the appropriate Milne absorption coefficient and τ_i the excited-state lifetime. For laser radiation that is spectrally narrow compared to the absorption-line profile function $\mathscr{L}_i^A(\lambda)$, we may approximate the laser spectral irradiance $I^L(\lambda)$ by a delta function:

$$I^L(\lambda) = I_L \delta(\lambda - \lambda_L) \tag{10.10}$$

where I_L is the laser irradiance. Under these circumstances we can write

$$P^F(\lambda) = I_L \sum_i \kappa_i^A(\lambda_L) \mathscr{L}_i^F(\lambda, \lambda_L) \frac{\tau_i}{\tau_i^{\text{rad}}} \tag{10.11}$$

where we have introduced the volume absorption coefficient at the laser wavelength,

$$\kappa_i^A(\lambda_L) = \frac{N_i B_{12}^i \mathscr{L}_i^A(\lambda_L) hc}{4\pi\lambda} \tag{10.12}$$

By analogy with (3.182) we can write

$$P^F(\lambda) = I_L \sum_i \kappa_i^A(\lambda_L) F_i(\lambda, \lambda_L) \tag{10.13}$$

where $F_i(\lambda, \lambda_L)$ represents the fluorescence efficiency of the ith species.

In a complex mixture of fluorescent molecules, such as found in a mineral oil, there is also an exchange of energy possible between the various hydrocarbon molecules. The best we can do under these circumstances is consider the *bulk phenomenological* properties of the material. In that case we may write the fluorescence in terms of the total absorption coefficient $\kappa^A(\lambda_L)$ at the laser wavelength, and the total fluorescence efficiency $F(\lambda, \lambda_L)$:

$$P^F(\lambda) = I_L \kappa^A(\lambda_L) F(\lambda, \lambda_L) \tag{10.14}$$

If this is used in conjunction with equation (10.4), the hydrographic lidar equation appropriate to an oil film on the water surface can be written in the form

$$\Delta E_S^H(\lambda) = E_A \kappa_H^A(\lambda_L) F_H(\lambda, \lambda_L)\, \Delta R_H e^{-(\kappa_H^A(\lambda_L)+\kappa_H^A(\lambda))(R_H - R_A)} \tag{10.15}$$

with

$$E_A \equiv E_L T(R_A) \xi(R_A) \xi(\lambda) \frac{c \tau_d A_0 \phi_H}{8 \pi n_H^2 R_A^2} \tag{10.16}$$

$\Delta E_S^H(\lambda)$ represents the increment of fluorescence energy received by the laser fluorosensor arising from a thickness ΔR_H of oil located at a depth $R_H - R_A$ within the oil film. We have assumed that the range of this hydrocarbon layer R_H is approximately the atmospheric range to the surface of the oil R_A. The subscript H has been used to designate hydrocarbon quantities, and we have set the fluorescence correction factor $\gamma(R)$, introduced in Section 7.3, equal to unity. This is based on the assumption that the detector integration time is greater than or comparable to the laser pulse duration and that both are larger than the fluorescence lifetime (Measures, 1977). This may not always be justified, and substantial deviations from unity could lead to errors in estimating the oil-film thickness.

On integrating equation (10.15) over the oil layer of thickness d, we arrive at the total fluorescence received by the laser fluorosensor from the oil film alone:

$$E_S^H(\lambda) = E_A F_H(\lambda, \lambda_L)\left[1 - e^{-\kappa_H^A d}\right] \frac{\kappa_H^A(\lambda_L)}{\kappa_H^A} \tag{10.17}$$

where we have introduced

$$\kappa_H^A = \kappa_H^A(\lambda_L) + \kappa_H^A(\lambda) \tag{10.18}$$

If we chose to observe the return radiation at the wavelength corresponding to the water Raman signal, then there are likely to be two additional terms to consider in equation (10.17): one arising from water-induced fluorescence (actually fluorescence from dissolved organic contamination of the water) and the other from Raman scattering by the OH-stretch band of the water molecules. The total radiation received can then be expressed in the form

$$E_S^H(\lambda_R) = E_A \Big[F_H(\lambda_R, \lambda_L)\{1 - e^{-\kappa_H^A d}\} + F_W(\lambda_R, \lambda_L) e^{-\kappa_H^A d} + R_W(\lambda_R, \lambda_L) e^{-\kappa_H^A d} \Big] \tag{10.19}$$

where $F_W(\lambda_R, \lambda_L)$ represents the water fluorescence efficiency of an optically thick column at the Raman wavelength λ_R, and $R_W(\lambda_R, \lambda_L)$ represents the Raman conversion efficiency of such a water column. It should be noted that in writing this relation we have assumed that $\kappa_H^A(\lambda_L) \gg \kappa_H^A(\lambda_R)$ and that the laser irradiance incident upon the oil–water interface has been reduced by the factor $\mathbf{e}^{-\kappa_H^A d}$ due to absorption within the oil film. In the absence of an oil layer the return signal is

$$E_S^W(\lambda_R) = E_A\{F_W(\lambda_R, \lambda_L) + R_W(\lambda_R, \lambda_L)\} \tag{10.20}$$

It is evident that if there is negligible oil fluorescence at the water Raman wavelength, then division of equation (10.19) by (10.20) leads to a simple relation for the oil thickness, namely,

$$d = -\frac{1}{\kappa_H^A} \ln\left\{\frac{E_S^H(\lambda_R)}{E_S^W(\lambda_R)}\right\} \tag{10.21}$$

This means that if the total attenuation coefficient κ_H^A for the oil is known, the thickness of the layer can be evaluated from the ratio of the laser fluorosensor signal obtained while over the oil to that obtained from an adjacent water mass where no oil slick is present (O'Neil et al., 1980). In essence it is the *suppression* of the return at the water Raman wavelength resulting from attenuation within the oil film that is used to determine the thickness of the oil. If the oil fluorescence at the water Raman wavelength is not negligible, then its contribution can be taken into account by using the laser-fluorosensor signal at another wavelength (where the water Raman contribution is negligible) (Hoge and Swift, 1980).

An example of this suppression of the Raman signal resulting from a controlled spill of LaRosa crude and observed by an airborne laser fluorosensor can be seen in Fig. 10.15(a). The corresponding oil thickness computed from this information is shown in Fig. 10.15(c). Throughout the near-totally depressed area, the oil film can only be characterized as being greater than 4 μm in thickness. It is interesting to observe that since the flight path was parallel to the wind direction, the eye (or head) of the oil slick is encountered gradually, as shown by the smaller slope of the Raman suppression in region A of Fig. 10.15(a). After the head of the slick is traversed, the Raman signal increases rapidly (region B).

This work was undertaken with a new, state-of-the-art scanning lidar having a multispectral time-gated receiver capability (Hoge et al., 1980). As discussed earlier, this NASA AOL was also designed to perform bathymetric surveys. When used as a laser fluorosensor it utilizes a high-repetition-rate nitrogen laser and a bank of optically coupled RCA C71042 photomultipliers to cover the spectrum from 350 to 800 nm. The basic characteristics of the system are indicated in Table 10.2. To remove solar background radiance the AOL system is triggered at 200 Hz with the laser being fired every alternate pulse. In the

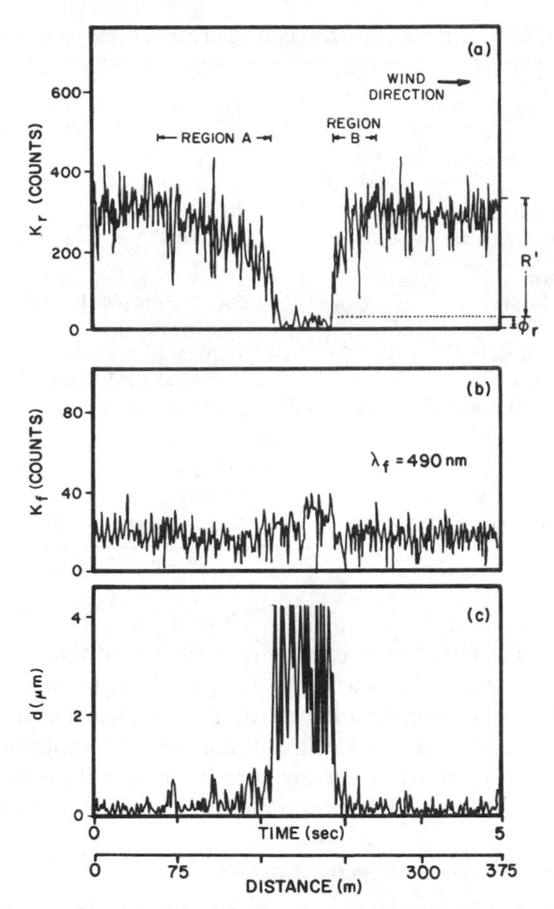

Fig. 10.15. (a) Near-total suppression of the Raman peak signal K_r as the La Rosa crude oil slick was traversed. (b) Peak fluorescence emission at $\lambda_f = 490$ nm obtained from the La Rosa crude during the same pass as (a). (c) Oil thickness computed from the data in (a) by using equation (10.21) together with the laboratory-measured extinction coefficients (Hoge and Swift, 1980).

scanning mode the oil slicks were often only viewed for a portion of the sweep cycle. This imposes a roughly 5-Hz modulation on the received signals. An example of an oil-thickness map produced by this NASA AOL is presented as Fig. 10.16. The width of the observed track was 26.5 m and was determined by the 10° full-cone off-nadir angle scan and the 150-m altitude sustained during this flight test.

One of the largest sources of error associated with this oil-thickness technique is the unknown influence of weathering and aging on the attenuation coefficient κ_H^A. Also an underestimate of the thickness is likely if the oil coagulates into small, optically opaque regions (thickness $\gtrsim 20$ μm) due to spatial averaging within the 3-mrad laser footprint. In spills such as the one

TABLE 10.2. AOL OIL FLUOROSENSING MODE PARAMETERS[a]

Transmitter: N_2 laser	
Wavelength	337.1 nm
Bandwidth	0.1 nm
Pulse width	10 ns
PRF	≤ 100 Hz
Peak output power (max)	100 kW
Beam divergence	2–20mrad
Receiver:	
Bandwidth	350–800 nm (40 channels)
Spectral resolution (min)	11.25 nm
Field of view	1–20 mrad, variable, vertical and horizontal
Temporal resolution	8–150 ns, variable
Experiment:	
Aircraft altitude	150 m
Aircraft velocity	75 m/s
Measurement background	Day and night

[a] Hoge et al. (1980).

reported above it has been found that more than 50% of the oil is contained in films having a thickness of from 0.5 to 1.0 μm. Measurements with a dual-frequency microwave radiometer indicate that this instrument responds only to thick layers of oil, and so the two kinds of instrument complement each other. Shorter-wavelength lidars based on excimer lasers could be used to monitor nanometer-thick oil films, and development along this line is in progress.

10.2.3. Fluorescence-Decay Spectroscopy

As we have seen above, the state-of-the-art laser fluorosensors are at best only able to classify the source of an oil slick as being light, crude, or heavy. In principle better characterization could be achieved through the use of many more spectral channels. Unfortunately, this would not only substantially increase the cost and complexity of the system, it would also significantly degrade the signal-to-noise ratio by reducing the signal energy available to each channel. Furthermore, it may not be possible to compensate for this loss of signal per channel by increasing the laser energy, for reasons of eye safety. One way to improve the specificity of the laser fluorosensor is to excite at two or more wavelengths, as shown by Houston et al. (1973). As we shall see later, this approach turns out to be quite successful for classifying algae by means of remote laser-induced fluorescence.

An alternative method of economically enhancing the identification capability of a laser fluorosensor involves making use of the time domain. Fantasia et al. (1971) were the first to suggest that fluorescence lifetime measurements might serve as an additional technique for characterizing crude oils. Measures

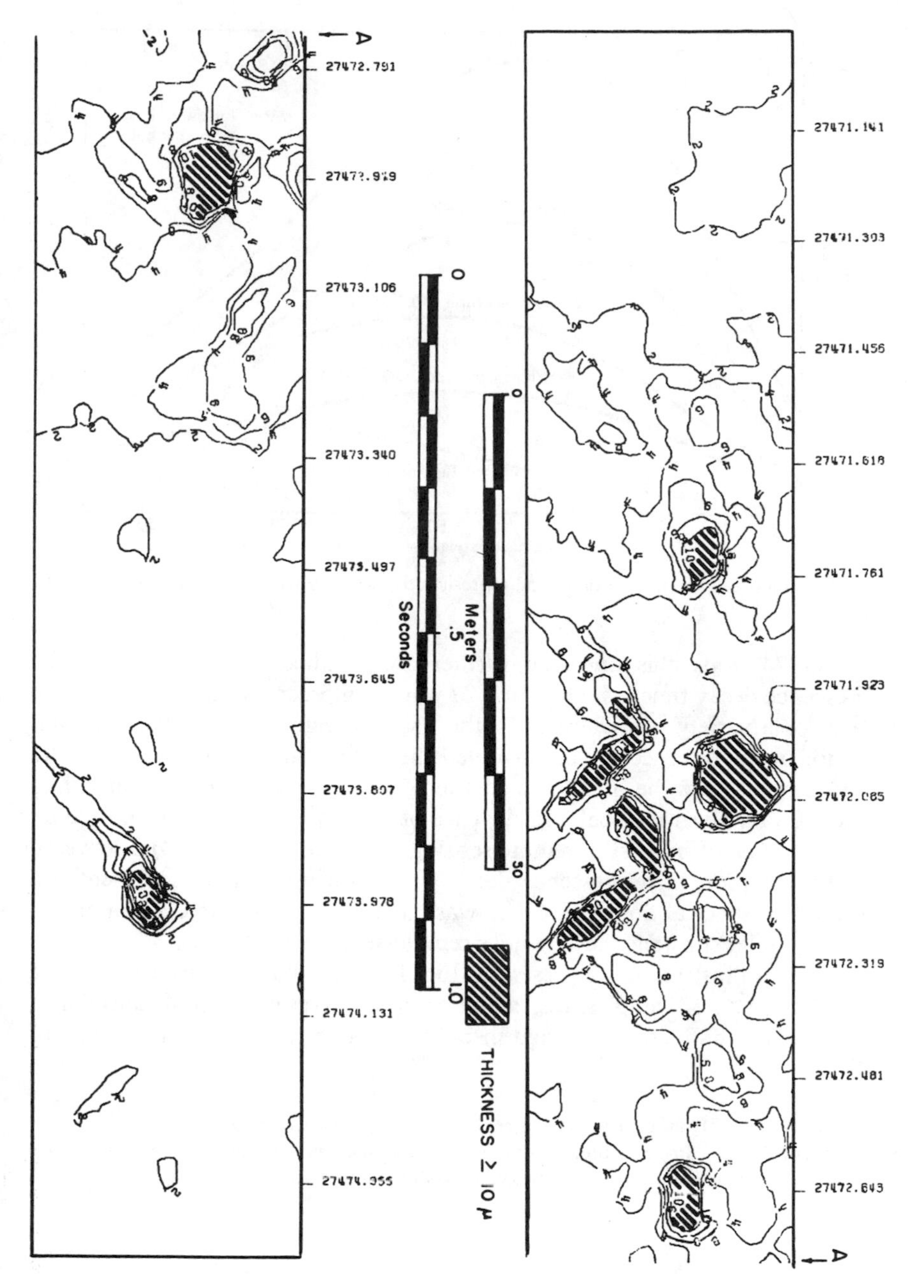

Fig. 10.16. Map of oil-film thickness obtained from airborne oceanographic lidar operated in a scanning mode. The important longitudinal features of the slick may be seen in proper perspective by joining the two segments at the points labeled A (O'Neil et al., 1981).

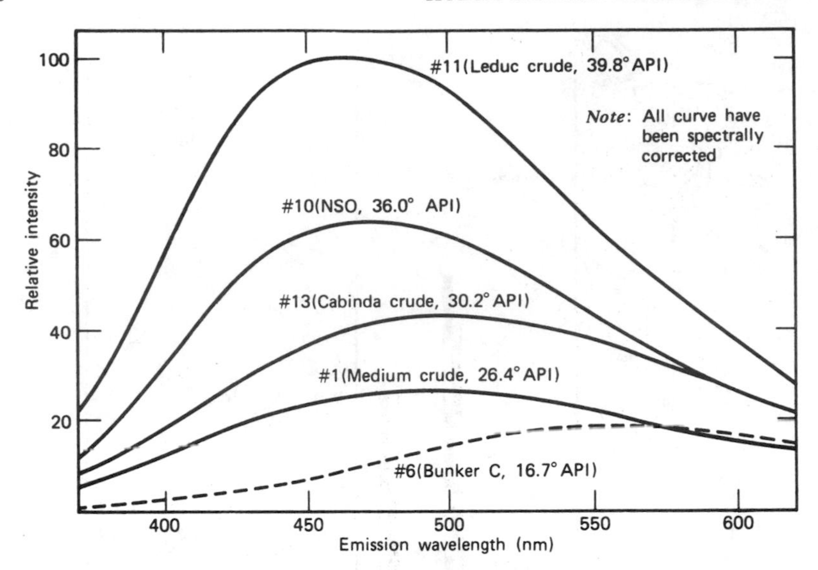

Fig. 10.17. Fluorescence spectra of several representative samples of crude oil (Measures et al., 1974).

et al. (1974) took this one step further and studied the variation of the fluorescence decay time[†] as a function of wavelength across the entire emission profile for a variety of materials. In the past, measurements of the temporal behavior of fluorescence usually involved the entire emission band and could, for pure substances, be attributed to the lifetime of the excited manifold of levels within a given molecule. In the case of a complex mixture of molecules, such as is often of interest in remote sensing, Measures et al. (1974) discovered that the variation in fluorescence decay time with wavelength represented a new kind of spectral fingerprint that was capable of very fine discrimination between similar substances even under remote-sensing conditions.

By way of illustration, we present in Fig. 10.17 the fluorescence spectra of a number of crude-oil samples. Inspection of these curves clearly indicates that it would be very difficult to distinguish between these samples by the shape of

[†] Measures et al. (1974) introduced the concept of *spectral fluorescence decay time* $\tau(\lambda)$, defined as the exponential decay time measured per unit wavelength interval. With this in mind, the *lifetime* of the entire emission band, τ_F, can be defined according to the relation

$$\frac{1}{\tau_F} \equiv \frac{\int \frac{\mathscr{L}^F(\lambda)}{\tau(\lambda)} d\lambda}{\int \mathscr{L}^F(\lambda)\, d\lambda}$$

where $\mathscr{L}^F(\lambda)$ is the fluorescence profile. Clearly, if $\tau(\lambda)$ is independent of wavelength, then the spectral fluorescence decay time and the transitional lifetime are indistinguishable.

their spectra. The magnitude of the signal, which measures the fluorescence efficiency of each sample, would aid in this purpose. However, the absolute signal amplitude of an airborne laser fluorosensor can be seen, from equations (10.16) and (10.19), to depend upon many variables, including the laser energy, the various transmission losses, the overlap between the laser beam and the area of observation, the extent to which the oil slick fills the receiver field of view, the altitude and inclination of the laser beam relative to the water's surface, and the nature of the target (i.e., whether optically thick or thin, patchy, etc.); see Fig. 10.18.

The fluorescence-decay spectra (FDS) for the same set of crude-oil samples is presented as Fig. 10.19. It is immediately apparent that these FDS possess far better discriminatory power than the traditional fluorescence spectra. The primary reason for this is that the ordinate of the FDS represents a time measurement whose magnitude, within the remote-sensing context, is much more closely related to the properties of the target media than is the magnitude of the fluorescence return. Of particular significance is the fact that the decay times tend to increase with increasing density (or API) of the oil, a trend first noted by Fantasia et al. (1971). To reinforce the identification potential of FDS, Measures et al. (1974) also compared the normal fluorescence spectra

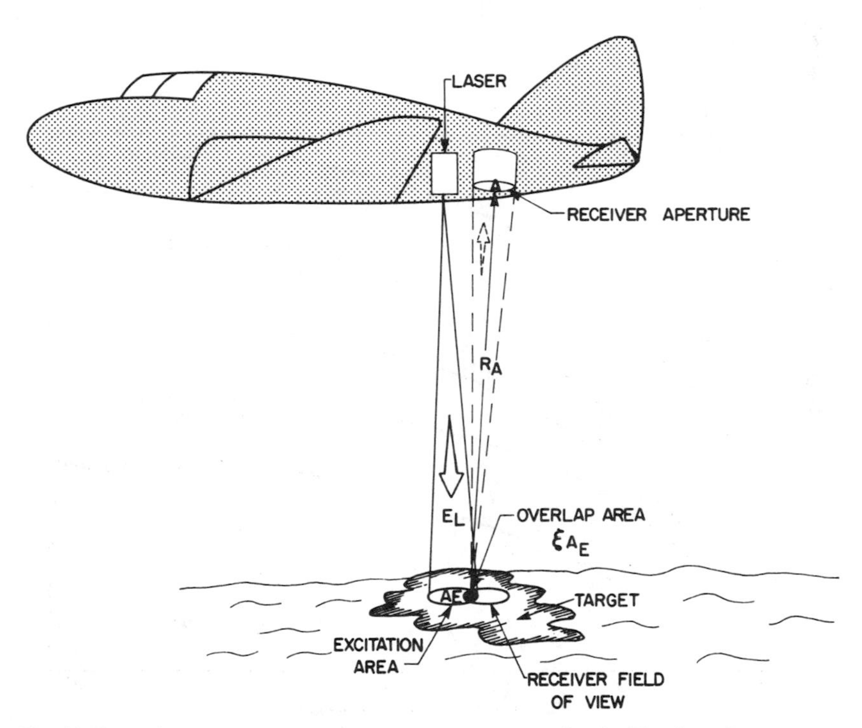

Fig. 10.18. Schematic of the geometrical arrangement associated with a laser fluorosensor.

with the fluorescence decay-time spectra for a selection of refined petroleum products. This comparison is presented as Fig. 10.20(a) and (b). In general it appears that the decay time tends to increase with increasing wavelength, the exception being stove oil. Although not shown, the FDS was even capable of distinguishing gasolines of different octane values and oil films from different types of fish.

To make an absolute measurement of a FDS, an allowance has to be made for the finite duration of the laser pulse and the response time of the photodetector. In addition, the thickness of the target should be small to avoid depth effects interfering with the decay-time measurements. This is virtually always true in the case of oil spills, as the optical thickness of such materials tends to be of order 10^{-3} cm (Fantasia et al. (1971; Measures and Bristow, 1971). However, variations in the height of the oil film due to wave action could lead to some error in the absolute measurements of the lifetime due to the finite speed of light. This kind of systematic error may not detract too much from the characterization capability of the FDS.

The recorded fluorescence temporal profile $f(t)$ can be expressed in the form of a convolution integral

$$f(t) = \int_0^t i(t')\phi(t - t')\, dt' \tag{10.22}$$

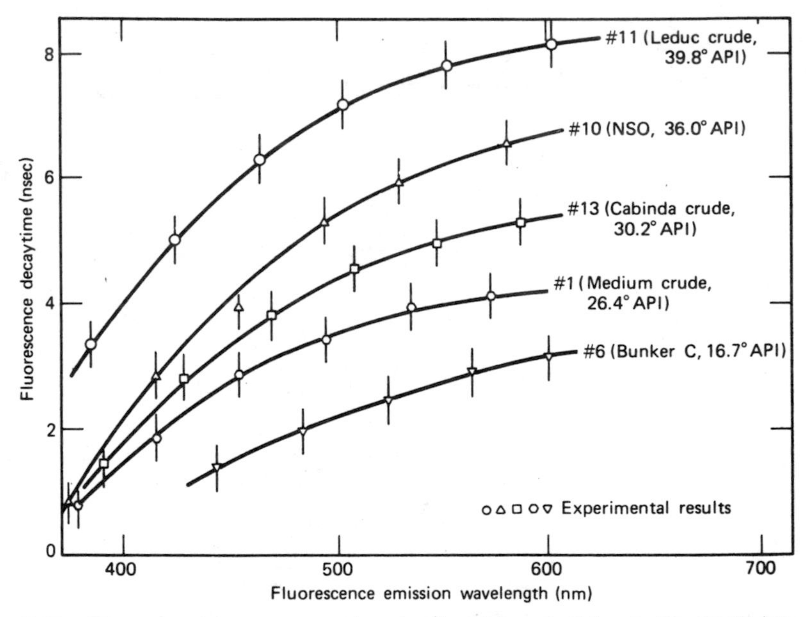

Fig. 10.19. Fluorescence-decay spectra of crude-oil samples referred to in Fig. 10.17 (Measures et al., 1974).

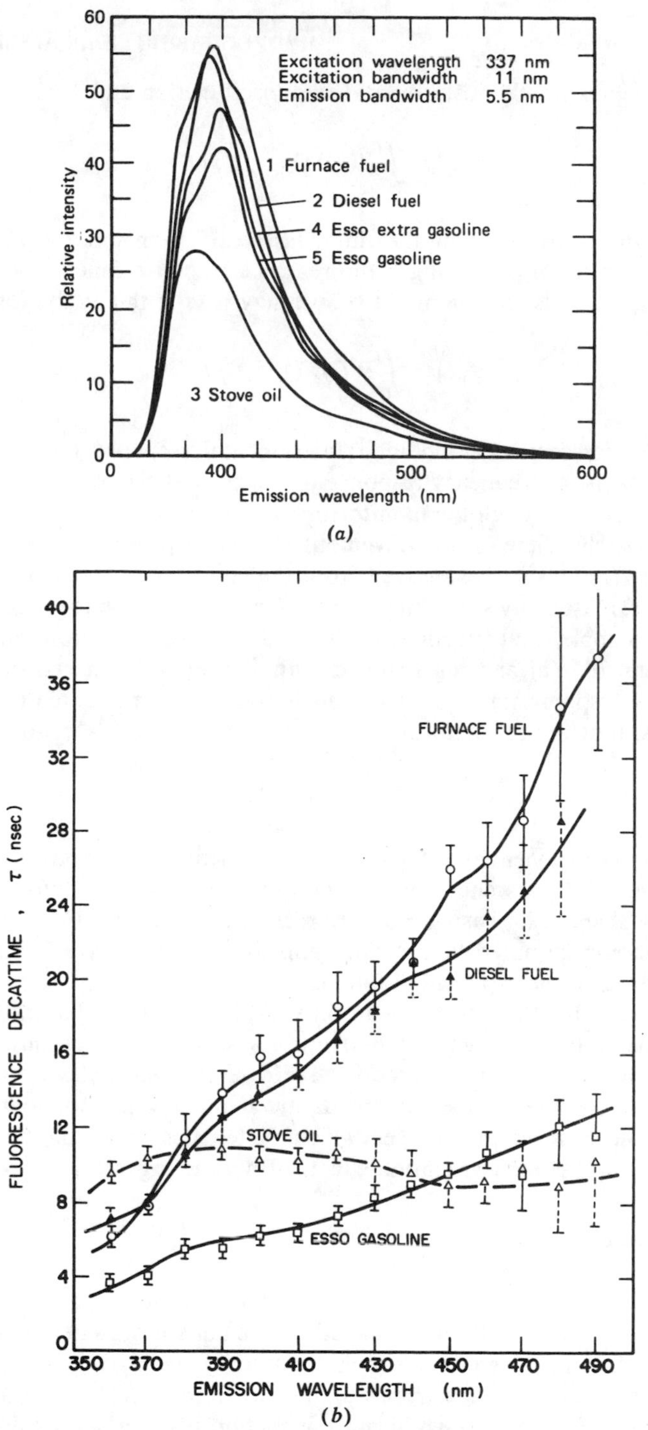

Fig. 10.20. (a) Fluorescence spectra of several refined petroleum products; (b) corresponding fluorescence-decay spectra of four out of the five products (Measures et al., 1974).

where $i(t)$ represents the instrumental response function and

$$\phi(t) = \int_0^t P(t')F(t - t')\, dt' \tag{10.23}$$

represents the convolution of the finite laser excitation pulse $P(t)$ with the fluorescence medium possessing a fluorescence response function $F(t)$. Since convolution integrals are commutative, we may rewrite this in the form

$$f(t) = \int_0^t I(t')F(t - t')\, dt' \tag{10.24}$$

where $I(t)$ represents the modified instrumental response function, taking account of both the frequency response of the photodetector and the shape of the exciting pulse through the monitoring system.

Unfortunately, there is no convenient analytical procedure for evaluating $F(t)$, given $I(t)$ and $f(t)$ as derived from the convolution integral. There are, nevertheless, reasonably simple techniques for obtaining the required information. For example, trial functions of $f(t)$ can be computed from an accurate measurement of $I(t)$ and an assumed form for $F(t)$. In most instances the fluorescence response function of the medium can be represented by a simple exponential function of the form

$$F(t) = \mathrm{e}^{-t/\tau}$$

In this case the exponential decay time τ can be determined from a comparison of the observed fluorescence pulse shape $f_0(t)$ with a set of computer-synthesized pulse shapes $f_c(t)$, using τ as the matching parameter (Measures et al., 1974). It should be noted that Rayner and Szabo (1978) have intimated that sometimes a multiple exponential function may be required.

In the event that the target, say an oil slick, is simply to be characterized in order to match it to a sample taken from a suspected ship, deconvolution of the FDS may not even be necessary. However, before laser-fluorosensor tagging of oil spills becomes accepted, much more will have to be known regarding the changes in the spectral characteristics of an oil slick due to weathering and the biodegradation effects of marine organisms.

10.2.4. Algae Mapping

Photosynthesis is the ultimate source of all food and the oxygen we breathe. It is the process by which all plants, including the aquatic varieties, convert solar energy into chemical energy. Chlorophyll *a* plays a leading role in this process, and so it should come as no surprise to find that Sorenzen (1970) and El-Sayed (1970) discovered a highly significant correlation between chlorophyll *a* concentration and the primary productivity in marine surface waters. Yentsch and

Menzel (1963) were able to show that a fluorescence technique could be used to determine the concentration of chlorophyll *a* with phytoplankton, and Sorenzen (1966) developed an *in vivo* fluorescence approach to evaluating chlorophyll concentrations.

Hickman and Moore (1970) first suggested, on the basis of laboratory studies, that chlorophyll measurements with a sensitivity of better than 10 mg m^{-3} might be undertaken using an airborne pulsed neon laser (output of about 20 kW) from an altitude of 100 m. An improved system, incorporating a 250-mJ, 300-ns dye laser, was shown by Kim (1973) to be capable of detecting chlorophyll *a* concentrations down to a fraction of a mg m^{-3} at an altitude of 30 m. Kim (1973) also drew attention to the difference expected in the fluorescence profiles of algae and plant foliage, and reported that excitation at 590 nm yielded the largest fluorescence signal. In the pier-mounted test of a prototype chlorophyll laser fluorosensor, he obtained fairly close agreement between laser measurements and those based on wet chemistry. These results are presented here as Fig. 10.21. Helicopter flight trials of this system revealed concentrations varying between 4 and 12 mg m^{-3} across Lake Ontario.

Mumola et al. (1973) stated that accurate measurements of the chlorophyll *a* concentration can only be undertaken if allowance is made for the different color groups that may be present in any natural mixture of algae. Indeed, they showed that additional fluorescence due to other pigments may often appear, as indicated in Fig. 10.22, even though the characteristic chlorophyll *a* fluorescence peak at 685 nm is always present. Furthermore, the excitation spectrum was found to vary considerably from one phytoplankton color group to another, as seen in Fig. 10.22. In spite of this, the excitation and emission spectra for different species within one color group are observed (see Fig. 10.23) to be remarkably similar. Mumola et al. (1973) proposed a four-wave-

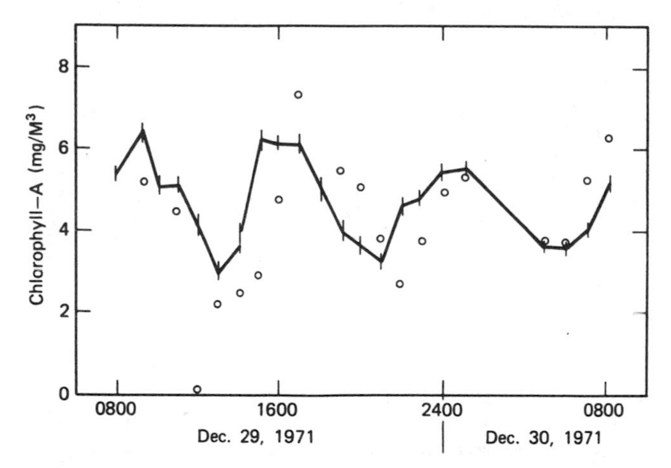

Fig. 10.21. Comparison of hourly averaged chlorophyll *a* readings as determined by laser chlorophyll analysis (solid line) and wet chemistry chlorophyll analysis (circles) (Kim, 1973).

length excitation scheme that would exploit these features in order to evaluate the concentration of chlorophyll *a* in an arbitrary mixture of phytoplankton color groups.

The optically thick form of the lidar equation (7.50) is appropriate to this situation, and so we can write the fluorescence return received in a band centered on the chlorophyll wavelength λ_f (685 nm in all cases) when excited by a laser of wavelength λ^i_L as

$$E_S(\lambda^i_L) = \frac{E_L K_0(\lambda_f) T(R_A) \xi(R_A) A_0 \phi_W}{4\pi R_A^2 \left[\kappa(\lambda^i_L) + \kappa(\lambda_f)\right] n^2} \sum_j N_j \sigma_j^F(\lambda^i_L) \qquad (10.25)$$

where N_j is the number density of chlorophyll *a* molecules in phytoplankton color group j; $\sigma_j^F(\lambda^i_L)$ is the *in vivo* fluorescence cross section of the color-group-j

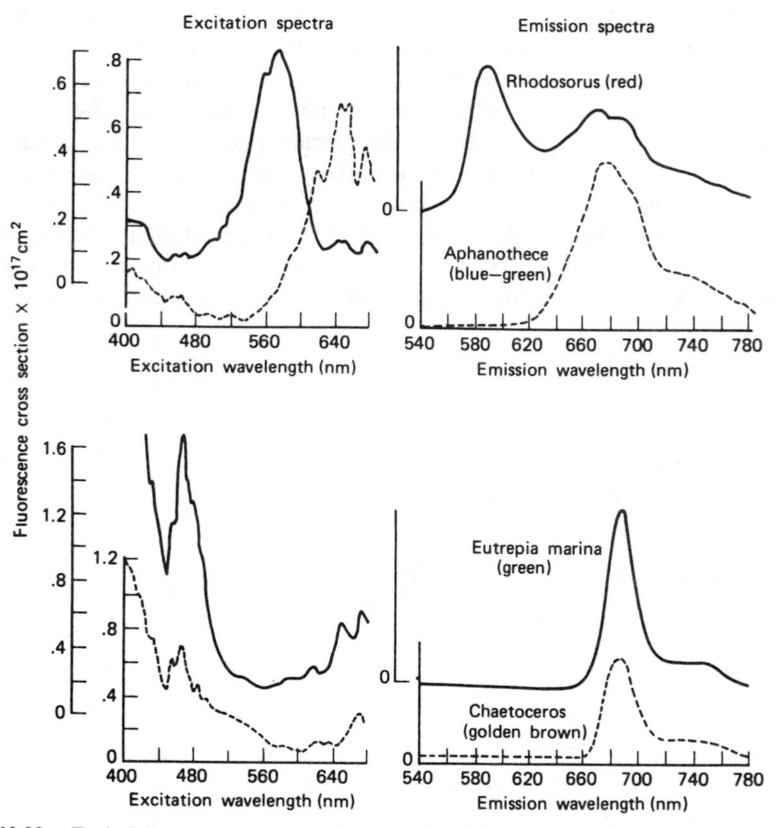

Fig. 10.22. Typical fluorescence cross sections and emission spectra of red, blue–green, green, and golden–brown algae samples (Mumola et al., 1973).

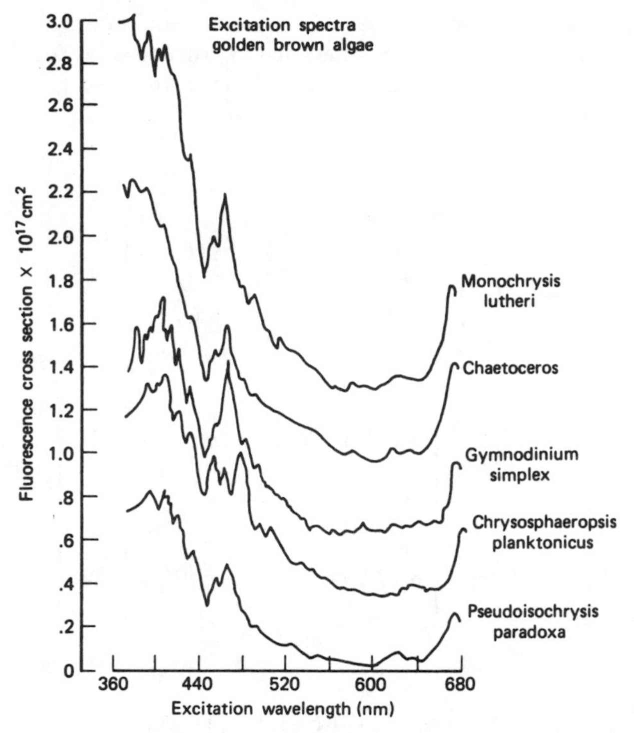

Fig. 10.23. Effective fluorescence cross sections of chlorophyll *a* in golden–brown algae. Data shown are normalized to single-molecule values with a 5-nm resolution (Mumola et al., 1973).

molecules, for emission per unit wavelength interval centered about λ_f (the 685-nm chlorophyll *a* band) with excitation at λ^i_L; ϕ_w is the two-way transmission factor for the air–water interface; $K_0(\lambda_f)$ is the filter function of the laser fluorosensor; $T(R_A)$ is the two-way atmospheric transmission factor; A_0 is the receiver area; R_A is the altitude of the aircraft; E_L is the transmitted laser energy; and $\kappa(\lambda^i_L)$ and $\kappa(\lambda_f)$ are the effective attenuation coefficients at the laser and fluorescence wavelengths, λ^i_L and λ_f, respectively.

The form of equation (10.25) inherently assumes that the altitude of the laser fluorosensor is far greater than the effective penetration depth of the laser beam. This would be reasonable in any operational airborne mission. We have also assumed that τ_d (the detector integration time) $\gtrsim \tau_L$ (the laser pulse duration) in order to capitalize on the return signal (at the expense of spatial resolution), and that the fluorescence lifetime is much less than the laser pulse duration. This is a reasonable assumption, since the lifetime of the chlorophyll *a* molecule is known to be very short [close to one nanosecond (Houston et al., 1973)] compared to the pulse duration of the flashlamp-pumped dye lasers used to ensure a large value for E_L and provide the several wavelengths of excitation.

Under these circumstances the set of four lidar equations, $j = 1$ to 4 in (10.25) (one for each laser wavelength), needed in order to solve for the density of chlorophyll a as distributed amongst the four main color groups can be expressed in the matrix form

$$\begin{bmatrix} E_S(\lambda_L^1) \\ \cdot \\ \cdot \\ E_S(\lambda_L^4) \end{bmatrix} = \begin{bmatrix} \frac{E_A^*}{\kappa_1 + \kappa_f} & 0 & 0 & 0 \\ 0 & \cdot & 0 & 0 \\ 0 & 0 & \cdot & 0 \\ 0 & 0 & 0 & \frac{E_A^*}{\kappa_4 + \kappa_f} \end{bmatrix} \times \begin{bmatrix} \sigma_1(\lambda_L^1) & \cdot & \cdot & \sigma_4(\lambda_L^1) \\ \cdot & \cdot & \cdot & \cdot \\ \cdot & \cdot & \cdot & \cdot \\ \sigma_1(\lambda_L^4) & \cdot & \cdot & \sigma_4(\lambda_L^4) \end{bmatrix} \begin{bmatrix} N_1 \\ \cdot \\ \cdot \\ N_4 \end{bmatrix} \tag{10.26}$$

where

$$E_A^* \equiv \frac{E_L K_0(\lambda_f) T(R_A) \xi(R_A) A_0 \phi_W}{4\pi R_A^2 n^2} \tag{10.27}$$

and we have written κ_i for $\kappa(\lambda_L^i)$ and κ_f for $\kappa(\lambda_f)$. It should be pointed out that in the case of chlorophyll fluorescence the signal window of the photodetector is often sufficient to include almost the entire fluorescence band at 685 nm. In that case, the total fluorescence cross sections would have to be used in the above equations and an appropriate change made in the filter function $K_0(\lambda)$.

In principle, the densities N_j can be evaluated by matrix inversion, given all of the other factors. A prototype four-wavelength (454.4, 539.0, 598.7, and 617.8 nm) laser fluorosensor was developed by Mumola et al. (1973). This system operated with a minimum energy of 0.6 mJ per pulse and was able to detect < 1 mg m^{-3} of chlorophyll a from an altitude of 100 m during daytime operating conditions. It is also worth noting that at this altitude the system met the eye-safety limits for a water-level observer.

Browell (1977) undertook an analysis of the errors likely to be incurred in such measurements. His results indicate that remote quantification of chlorophyll a requires optimum excitation wavelengths and careful measurement of the marine attenuation coefficients. Difficulties in determining these attenuation coefficients constituted a major limitation on the implementation of this form of remote sensing.

Fortunately, in most situations of interest this does not occur, and the solution to the problem of the unknown attenuation coefficients lies in the recognition that the laser-excited water Raman signal could serve as an indicator of changes in optical attenuation. This Raman-scattered signal is a property of the water alone, and consequently its magnitude will be primarily determined by the attenuation coefficients at the laser and Raman wavelengths, $\kappa(\lambda_L)$ and $\kappa(\lambda_R)$, respectively.

If we use (10.27) with (10.25), we can write, for a given laser wavelength λ_L, the fluorescence signal observed in the 685-nm chlorophyll band:

$$P_F = \frac{E_A^* K_0(\lambda_f) N_c \sigma_c^F(\lambda_L)}{K_0(\lambda_f)[\kappa_L + \kappa_f]} \tag{10.28}$$

where we have introduced the total chlorophyll *a* molecular density N_c and the appropriate mean fluorescence cross section $\sigma_c^F(\lambda_L)$. In a similar manner, the laser fluorosensor signal observed in the Raman band at wavelength λ_R (displaced 3418 cm^{-1} to the longer-wavelength side of λ_L) can be written

$$P_R = \frac{E_A^* K_0(\lambda_R) N_W \sigma_W^R(\lambda_L)}{K_0(\lambda_f)[\kappa_L + \kappa_R]} \tag{10.29}$$

where N_W represents the density of water molecules and σ_W^R is the Raman-scattering cross section for the OH stretching vibrational mode of liquid water. Division of the fluorescence signal by the Raman signal eliminates uncertainties and fluctuations associated with E_L, R_A, $\xi(R_A)$, ϕ_W, n, and $T(R_A)$ and yields

$$\frac{P_F}{P_R} = \frac{K_0(\lambda_f)}{K_0(\lambda_R)} \left\{ \frac{\kappa_L + \kappa_R}{\kappa_L + \kappa_f} \right\} \frac{N_c \sigma_c^F(\lambda_L)}{N_W \sigma_W^R(\lambda_L)} \tag{10.30}$$

Clearly, for equation (10.30) to be useful for evaluating the chlorophyll *a* concentration, the attenuation ratio $(\kappa_L + \kappa_R)/(\kappa_L + \kappa_f)$ must remain constant, or at most change slowly for significant changes in κ_L, κ_R, and κ_f. This appears to be true for a water body in which dissolved and particulate materials change only in concentration but not in character, according to the observations of Bristow et al. (1981). Their relatively simple laser fluorosensor employed at 200-kW, 250-ns (FWHM) flashlamp-pumped dye laser, a 30-cm-diameter $f/1.8$ refracting telescope (using an acrylic Fresnel collector lens), and the two-channel photodetection system shown in Fig. 10.24. The repetition rate for the laser was 1 Hz, and its wavelength was chosen to be 470 nm. This wavelength was selected on the basis of a compromise. In order to minimize the background water fluorescence, ensure that the attenuation ratio was reasonably constant, and maximize both the penetration depth of the laser pulse and the output energy available, a long wavelength of excitation was desired, while to maximize both the Raman and fluorescence signals, a short wavelength was desired. An example of the results obtained by Bristow et al. (1981) from a single laser shot and displayed on the sweep of an oscilloscope through the use of gated photomultipliers and an electronic delay are presented as Fig. 10.25.

Results of an airborne flight test conducted over the Las Vegas Bay are shown in Fig. 10.26 in the form of smoothed profiles of the chlorophyll *a* fluorescence pulse energy P_F, the water Raman pulse energy P_R, and P_F/P_R as functions of flight time. In addition, the location of the sampling sites and an approximate horizontal distance scale are indicated. Also shown in Fig. 10.26 are the ground truth measurements of the extracted chlorophyll *a* concentration for grab samples obtained at the 28 sampling sites at the time of the laser-fluorosensor overflights. From this figure it is immediately apparent that P_F/P_R and the chlorophyll *a* data are highly correlated over the extent of the flight path.

A note of caution should be sounded at this point. If the attenuation is due primarily to the presence of phytoplankton and the attenuation coefficients at the laser wavelengths are greater than that at the chlorophyll fluorescence wavelength, then

$$\kappa_i \approx \sum_j N_j \sigma_j^A(\lambda_L^i) \tag{10.31}$$

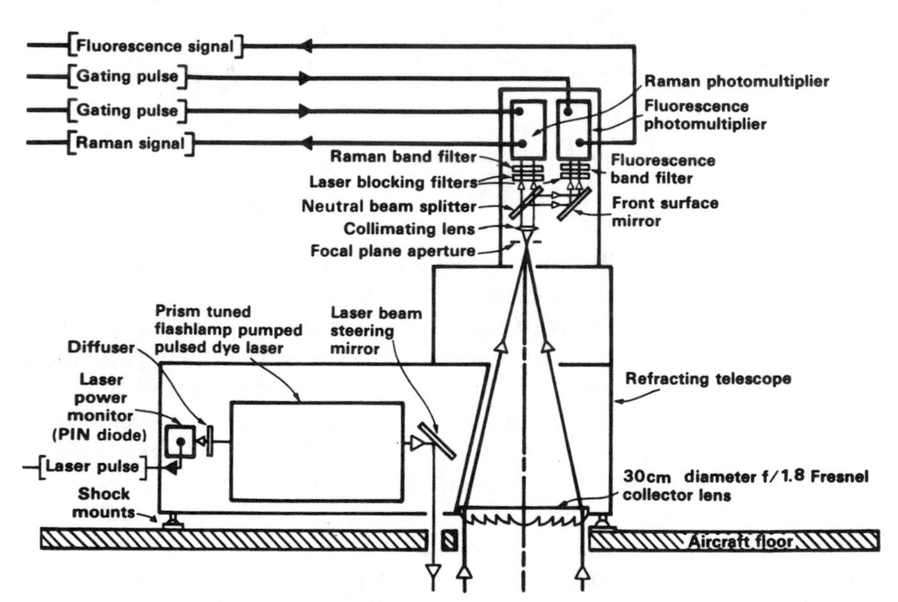

Fig. 10.24. Optical diagram of airborne laser fluorosensor for monitoring chlorophyll *a* fluorescence and water Raman signals (Bristow et al., 1981).

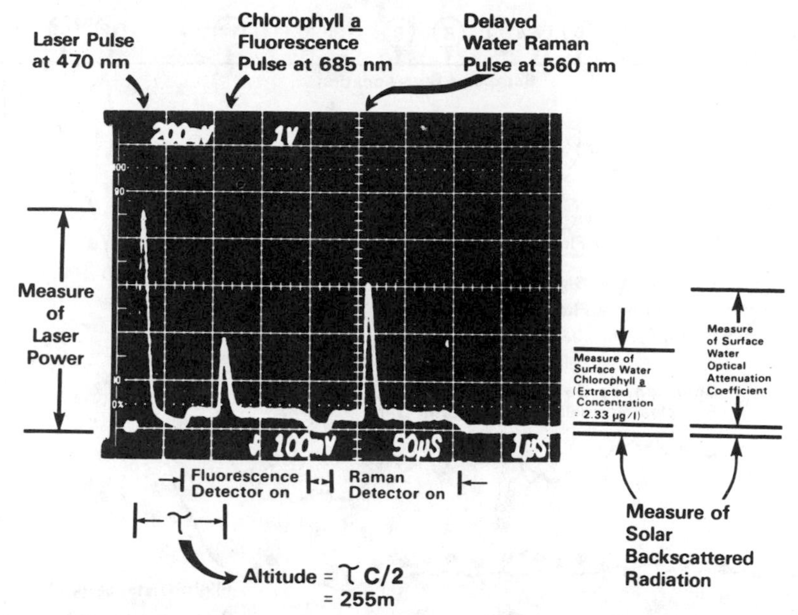

Fig. 10.25. Oscillogram showing time sequence of airborne-laser fluorosensor signals obtained over buoy 11 in the Las Vegas Bay region of Lake Mead, Nevada on 24 January 1979 for a measured chlorophyll *a* concentration of 2.33 μg/l (Bristow et al., 1981).

If, in addition, the quantum yield for each color group has a similar value, it follows from equation (10.25) that the return signal becomes independent of the density of chlorophyll *a* molecules. Under these circumstances it becomes impossible to measure the chlorophyll *a* concentration; see equation (7.51). This has also been confirmed by Bristow et al. (1981).

The NASA airborne oceanographic lidar referred to earlier has also been used for the remote sensing of chlorophyll *a* in the oceans (Hoge and Swift, 1981). Unfortunately, the low spectral resolution of their 40-channel receiver, combined with a laser wavelength of 532 nm, led to problems of deconvolution between the chlorophyll fluorescence at 685 nm and the water Raman band, which in this instance occurred at 645 nm and often appeared to ride on the shoulder of the chlorophyll fluorescence profile. Another problem caused by the poor choice of laser wavelength was competitive absorption and fluorescence by organic material (often referred to as *Gelbstoff*) carried to the sea by rivers (Yentsch, 1973). We shall return to this subject in our next section. In spite of these difficulties, Hoge and Swift (1981) were able to map the relative concentration of chlorophyll *a* over an extensive area of the North Sea and at the intersection of Chesapeake Bay with the Potomac River in the United States.

It is possible that the ultimate limit on the precision of chlorophyll *a* measurements undertaken from an airborne laser fluorosensor may well be

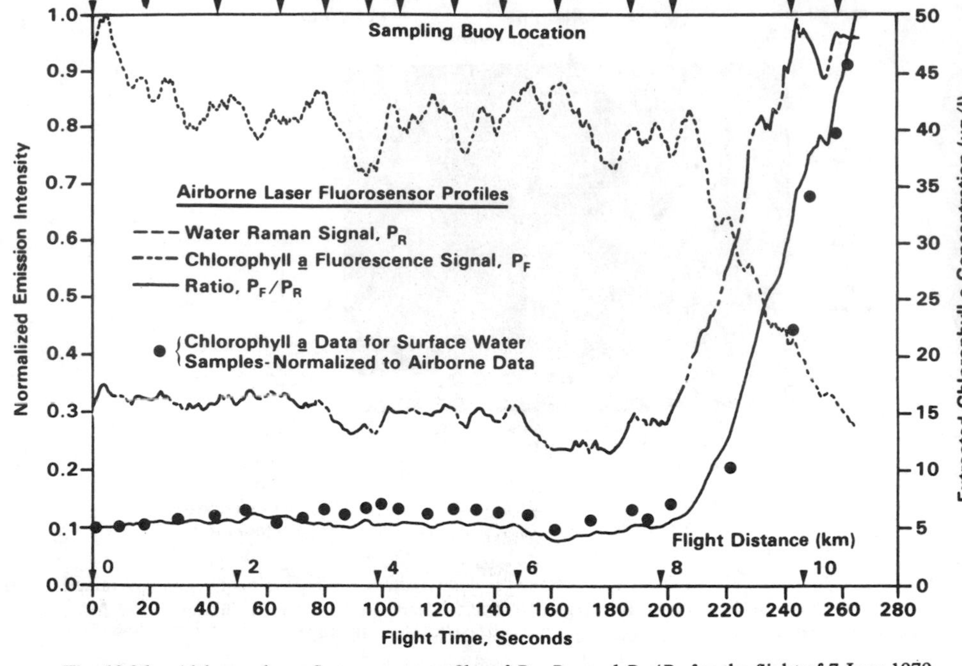

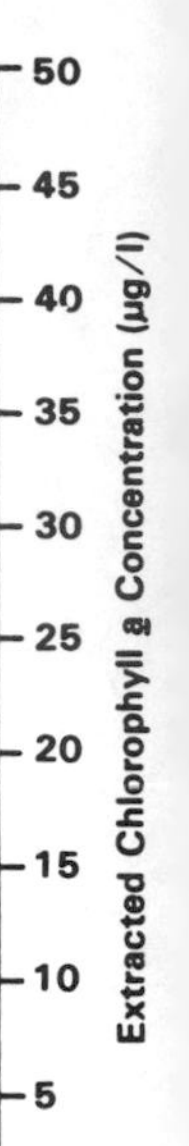

Fig. 10.26. Airborne-laser-fluorosensor profiles of P_F, P_R, and P_F/P_R for the flight of 7 June 1979 over Las Vegas Bay. Profiles smoothed using a moving five-point average scheme. Also shown are the chlorophyll *a* values for the 28 ground truth samples, which have been normalized to the laser-fluorosensor profile for P_F/P_R (Bristow et al., 1981).

determined by the fact that the quantum efficiency of phytoplankton fluorescence depends upon the ambient light level (photo-inhibition). This could mean that accurate measurements can only be made at night and with very short pulse lasers.

10.2.5. Water-Pollution Studies

It has long been recognized that all water samples (except those of the very highest purity) can be made to fluoresce by exposure to ultraviolet radiation. As mentioned earlier, this so-called "blue" water fluorescence has represented a considerable source of annoyance to those involved in fluorescence studies (Parker, 1968). From the remote-sensing point of view this fluorescence might constitute a useful parameter by which to study water quality. Indeed, the effluent from the pulp and paper industry has long been recognized as susceptible to measurement by fluorescence techniques (Thurston, 1970), since it contains high concentrations of lignin sulphonate. This has been corroborated in flight trials conducted over settling ponds and effluent streams of the CIP Pulp Mill at Hawkesbury on the Ottawa River. Figure 10.27 is an aerial photograph of this region showing the approximate location of the three flight lines used in these experiments. The fluorescence profile, made on a northwesterly heading corresponding to flight line B, is shown in Fig. 10.28(a), while the profile for line A, made on a northeasterly heading, is shown in Fig. 10.28(b). An interesting feature observed in Fig. 10.28(b) is the large, well-defined fluorescence anomaly occurring in the effluent channel region. Its magnitude is seen to be comparable to that from the settling pond and therefore suggests that little plume dispersion has occurred at this location. These observations

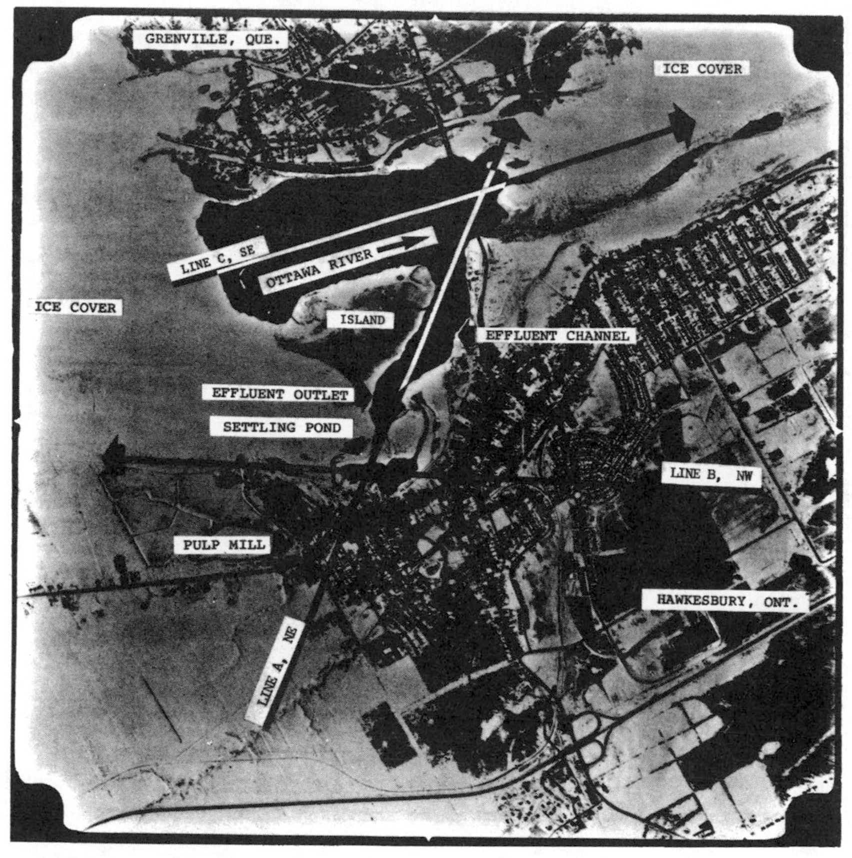

Fig. 10.27. Aerial photograph of Hawkesbury, Ontario, indicating flight lines over a pulp mill, March–April 1974 (Bristow, 1978).

were undertaken with the airborne version of the prototype laser fluorosensor shown earlier in Fig. 10.8 (Bristow, 1978).

An examination of the fluorescence of several different kinds of effluent was undertaken by Measures et al. (1975) and revealed that the spectral form of this fluorescence may not be as informative as its amplitude. In order to facilitate such amplitude measurements, Bristow et al. (1973) suggested that the OH-stretch water Raman band could be employed as a built-in reference signal

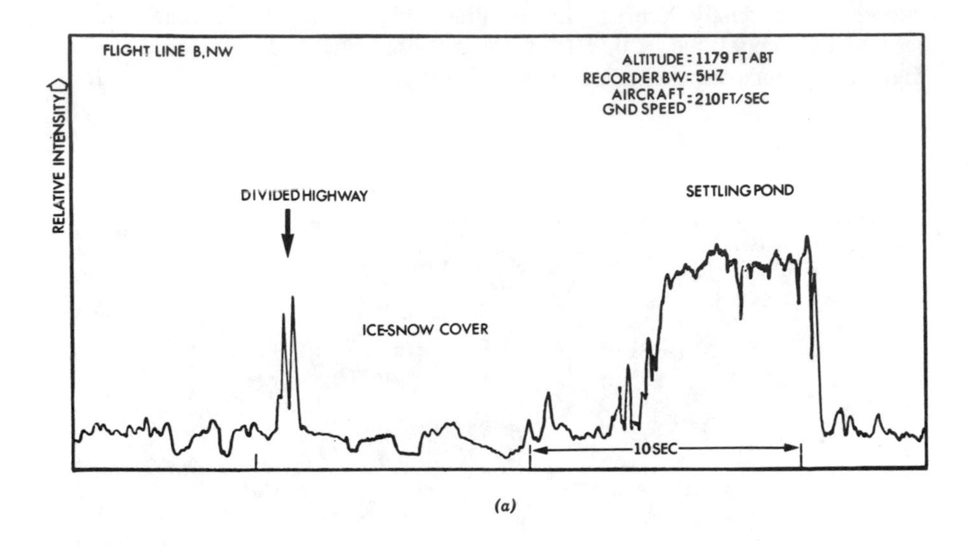

(a)

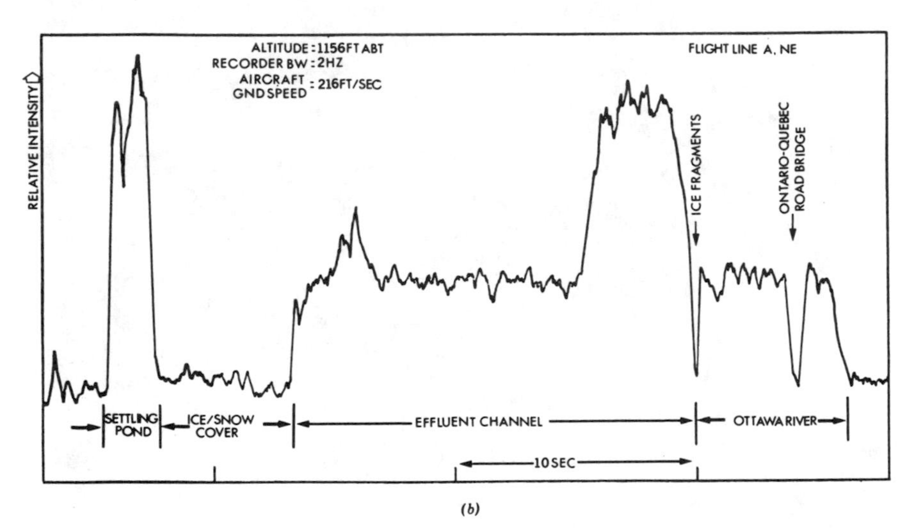

(b)

Fig. 10.28. Laser fluorosensor profiles obtained during overflights (a) of a pulp-mill settling pond, Hawkesbury, Ontario, line B, NW, on 2 March 1974, and (b) of the same settling pond and effluent channel, Line A, NE, on 28 March 1974; see Fig. 10.27 (Bristow, 1978).

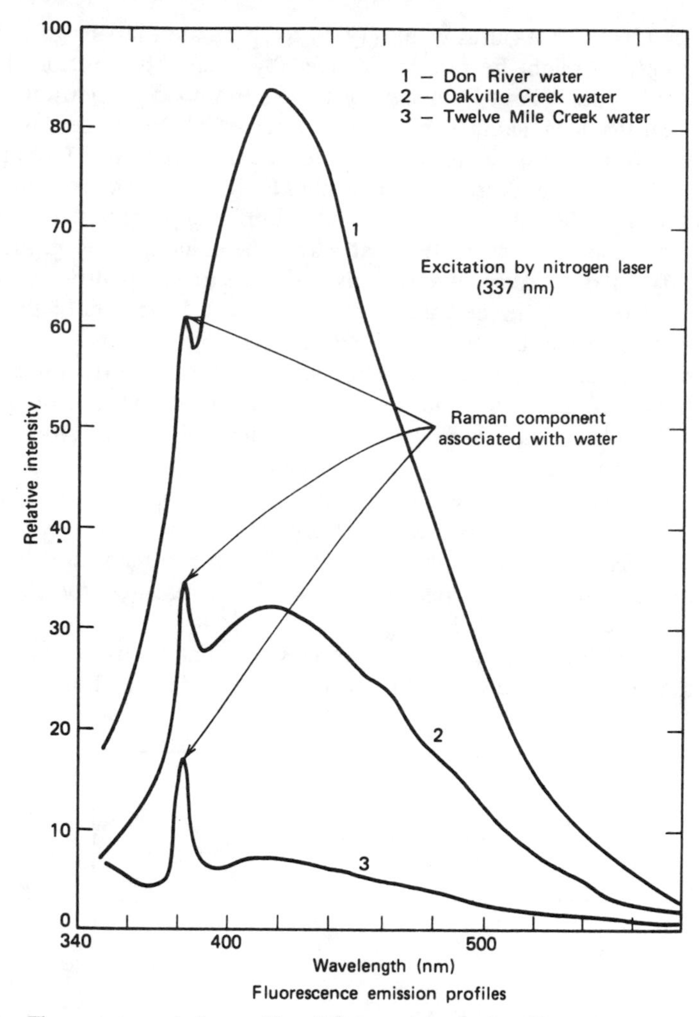

Fig. 10.29. Fluorescence emission profiles of three water samples, illustrating wide variation of fluorescence and Raman signals (Measures et al., 1975).

by which to normalize the fluorescent return. In this way many of the difficulties of amplitude measurement, discussed above in detail with respect to chlorophyll measurements, could be avoided. The wide variation observed in this fluorescence-to-Raman (F/R) ratio† for different water samples is illustrated in Fig. 10.29.

† The fluorescence-to-Raman (F/R) ratio used in these measurements was defined as the ratio of the peak amplitude of the fluorescence signal to the peak magnitude of the Raman component. The latter was obtained by subtracting the fluorescence component from the total signal at the Raman wavelength.

These results led Measures et al. (1975) to speculate that the dispersion of an effluent plume might be rapidly mapped by a suitably positioned (tower-mounted or airborne) laser fluorosensor that is equipped to monitor the F/R ratio. To test this idea, samples of various kinds of effluent were diluted (with water taken from an upstream location for each source) and measurements made of the corresponding change in the F/R ratio. An example of the variation in this F/R ratio as a function of dilution is presented as Fig. 10.30. The results of this work indicated that where the effluent of an industry or a municipality enters a river of relatively good upstream quality, it might be possible to make fairly precise measurements of the dispersion of the effluent plume based upon the change in the F/R ratio; see Fig. 6.2(d).

In addition, Measures et al., 1975 found evidence of a correlation between the magnitude of this F/R ratio and the total organic burden of the water samples. This observation is in general agreement with other studies (Kalle, 1966; Christman and Ghasseni, 1966; Sylvia et al., 1974) and suggests the possibility of rapidly surveying the organic loading of natural bodies of water by means of an airborne laser fluorosensor. High concentrations of organic materials in the water can render the water harmful to man and the aquatic ecosystems, as well as reducing the value of the resource for recreational purposes, (Donaldson, 1977; Keith, 1979).

A balance in the energy and carbon cycles of aquatic ecosystems requires a certain quantity of natural organic material. Bacteria degrade both natural and

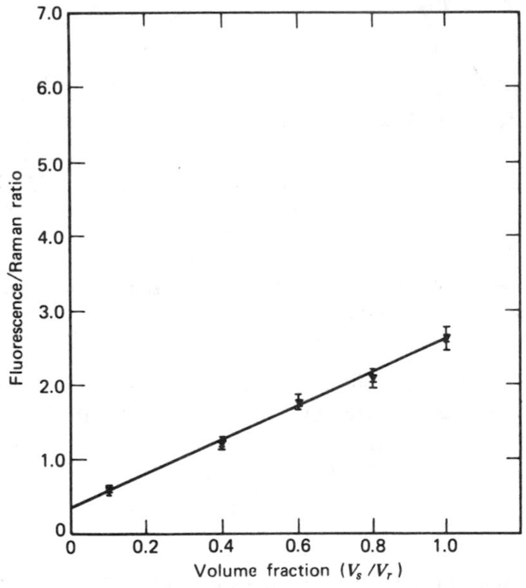

Fig. 10.30. Variation of the fluorescence/Raman ratio with dilution of effluent sample concentration in upstream Twelve Mile Creek water (Measures et al., 1975).

man-made organics through the use of dissolved oxygen. Consequently, an excess of organic material could lead to a depletion of this oxygen, which would lead to undesirable effects on other aquatic organisms. Both the concentration of chlorophyll *a* and the concentration of dissolved organic carbon (DOC) (Bristow and Nielsen, 1981) are two accepted indicators of water quality which can be measured remotely by means of a laser fluorosensor.

Fluorescent dyes have long been used to study water movement and dispersal rates for natural bodies of water. The airborne laser fluorosensor represents a natural method of extending the range and capability of such measurements. Hickman (1973), in an extensive study of suitable dyes, revealed that Acridine red and Rhodamine B have the highest product of optical density and quantum efficiency and are therefore likely to be the best candidates for detecting dye cloud movement and dispersion close to the surface. For studies that involve a greater penetration his work indicates that 3,6-dichlorofluorescein (absorption maximum at 432 nm) might have some advantage.

An interesting outgrowth of this investigation has been the suggestion that differences in the fluorescence sensitivity of dyes to the environment might be exploited to measure various parameters of importance. These include temperature, salinity, and pH. The approach to be adopted, in each instance, is to disperse a calibrated mixture of two appropriate dyes in the area of interest, then probe the resultant cloud with an airborne laser fluorosensor and evaluate the local condition by comparison of the fluoroescence returns from the two dyes. The laboratory work of Hickman (1973) has suggested that temperature evaluation to an accuracy of $\pm 0.5°C$ might be possible if the ratio of the fluorescence return from a mixture of Rhodamine B with Eosin Y is measured to within 2%. A very attractive feature of this technique is that almost all of the factors which would lead to uncertainty in an amplitude measurement cancel in such a ratio provided the wavelength separation needed for spectral discrimination is not too large. Nevertheless, questions of differential dispersion and sensitivity to more than one parameter remain to be carefully investigated.

As in the previous laser-fluorosensor applications, the radiation inelastically backscattered from the water can again be used to provide a reference signal that eliminates most of the uncertainties and fluctuations in both the system and environmental parameters. In this instance the ratio of the fluorescence to Raman signals from flight runs is compared with the same ratio found in the prepared laboratory dilutions to determine the field concentrations (Hoge and Swift, 1981). A representative laser-induced fluorescence profile at 590 nm obtained with the NASA AOL (using the conical scanner) over a Rhodamine WT dye spill in the ocean is displayed as Fig. 10.31(a). In this experiment about 19 liters of 20% solution of the dye was deployed and a 100-kW nitrogen laser (at 337 nm) was used to excite both the dye fluorescence and the water Raman signal. The dye-concentration map produced from such data is reproduced here as Fig. 10.31(b), and the results were found to be in fairly close agreement with the maximum plume concentration of 13 ppb measured in the laboratory from grab samples (O'Neil et al., 1981).

Other dye-tracing experiments have been conducted with the AOL reconfigured to use a neon laser having only 3% of the output energy of the nitrogen laser and operating at 540 nm. With this system a concentration of only 4 ppb could be detected with Rhodamine 6G dye. In clear ocean waters it is expected that concentrations down to 0.1 ppb should be detectable with a single laser pulse from an altitude of 150 m (Hoge and Swift, 1981). Furthermore, this might eventually be undertaken with time resolution in order to provide full 3-D dye-plume mapping.

Laser fluorosensors have clearly evolved to the point where they could integrate into a global environmental monitoring scheme. In particular, airborne laser fluorosensors could serve as the intermediary between satellite reconnaissance and grab sampling. They can cover areas of moderate size rapidly and

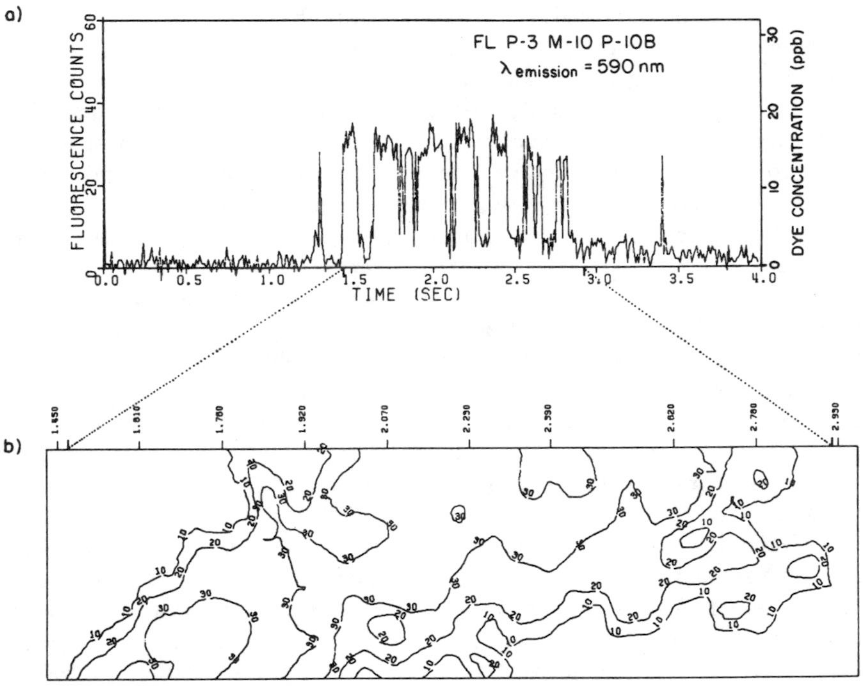

Fig. 10.31. Map of Rhodamine WT dye. (a) The raw fluorescence emission signal at a wavelength of 590 nm, near the peak of the Rhodamine WT dye emission, is plotted as a function of time along the flight line over an ocean dye spill. This trace shows the effect of the conical scan as the footprint passes in and out of the regions of high concentration. (b) The data in (a) have been unwound to produce a contour map of dye concentration. The contours are labeled according to signal levels in the fluorosensor: a level of 10 corresponds to a dye concentration of 3.3 ppb, a level of 20 to 6.7 ppb, and so on. These data were acquired with the AOL (O'Neil et al., 1981).

provide data that would enable more accurate interpretation of the images received from space. It is also likely that laser fluorosensors will shortly be used in a routine matter for water-quality surveillance.

10.3. SUBSURFACE WATER-TEMPERATURE MEASUREMENTS

The oceans of our planet are recognized as one of our greatest resources, not only for the food they yield, the oxygen they provide, and their possible future use as a source of energy, but also for the influence they have on our climate. Basic among the ocean's properties, certainly from these points of view, is its thermal structure down to about 100 m. Currently there is no operational means of remotely monitoring the subsurface temperature profile. At present satellites can map the surface temperature globally. A remote subsurface-temperature measurement capability would greatly benefit weather forecasting and aid in our understanding of the climate and the air–sea interface. It would also facilitate the development of more reliable models for thermal waste-disposal studies and for the design of ocean thermal-energy conversion (OTEC) power plants.

There are three approaches to remote sensing of subsurface temperatures under consideration. As we have seen earlier (Fig. 10.1), appreciable penetration of natural bodies of water is possible only for a narrow portion of the spectrum, and consequently all three methods involve some form of laser backscattering—Rayleigh, Raman, and Brillouin. Of these the Raman approach appears to be the most important at present, having undergone an initial sea trial.

Chang and Young (1973) were amongst the first to consider using Raman backscattering for ocean temperature measurements, and the experimental work of Slusher and Derr (1975) led to a Raman differential scattering cross section of 4.5×10^{-29} cm^2 sr^{-1} for the O—H stretching band of liquid water. Their work also indicated that ice temperatures might be determined by measuring the ratio of Stokes to anti-Stokes Raman lines.

More recently, Leonard et al. (1979) have undertaken the first remote ocean-temperature measurements based on Raman backscattering from a ship-mounted lidar system. The physical basis of the Raman temperature measurement stems from the coexistence of both monomer and polymer forms of liquid water. The relative concentration of these two species depends primarily on the temperature, and since the O—H Raman stretching frequency is different for each, it is possible to infer the water temperature from the resulting Raman backscattering spectra (Walrafen, 1967). This model is shown in a heuristic manner by means of the Raman spectrum displayed in Fig. 10.32 (Leonard et al., 1979). As is seen, each species has its individual Raman spectrum, proportional to its concentration, so that a simultaneous measurement of the Raman spectra at the two wavelengths λ_1 and λ_2 provides a measure of the concentration ratio and, through an equilibrium constant, the temperature.

The preliminary ocean tests of this technique were undertaken by Leonard et al. (1979) using the Raman lidar system originally designed and used by Leonard and Caputo (1974) for airport transmissometer work. A 100-kW, 10-ns, 337-nm pulsed nitrogen laser with a repetition rate of 500 Hz and a 2-mrad divergence was used as the transmitter. The output of the laser is passed through an interference filter, which passes the 337.1-nm laser line but blocks associated broadband spontaneous emission. A water solution of 2,7-dimethyl-3,6-diazacyclohepta-1,6-dieneperchlorate in a quartz cell acts as a very effective filter for blocking backscattered UV laser radiation, while permitting the Raman signal at about 375 nm to be transmitted quite efficiently into a double 0.25-m-focal-length scanning spectrometer having a 0.5-nm spectral resolution. This system was mounted aboard the afterdeck of the unique twin-hulled research vessel *Hayes* shown in Fig. 10.33.

In light of the well-known fluorescence induced in natural bodies of water by short-wavelength lasers, initial work involved broad spectral scans to determine the extent of this interference. An example of such a scan (on a logarithmic scale) obtained from a clear ocean location in the Mediterranean is presented as Fig. 10.34. It is quite evident from this observation that the 3400-cm^{-1} liquid-water Raman band is by far the most prominent feature of the spectrum and that laser-induced fluorescence is less than 1% of the Raman signal.

The laser Raman spectra of the ocean waters clearly confirmed the temperature-dependent effects (Leonard et al., 1979). As an example, the Raman spectra from Labrador Current water is compared with that obtained from the

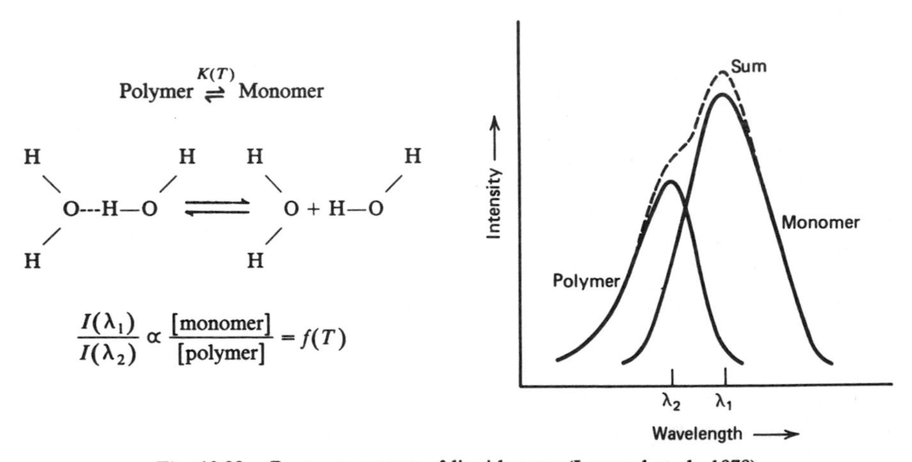

Fig. 10.32. Raman spectrum of liquid water (Leonard et al., 1979).

Fig. 10.33. USNS Hayes, used for laser oceanographic probing (Leonard et al., 1979).

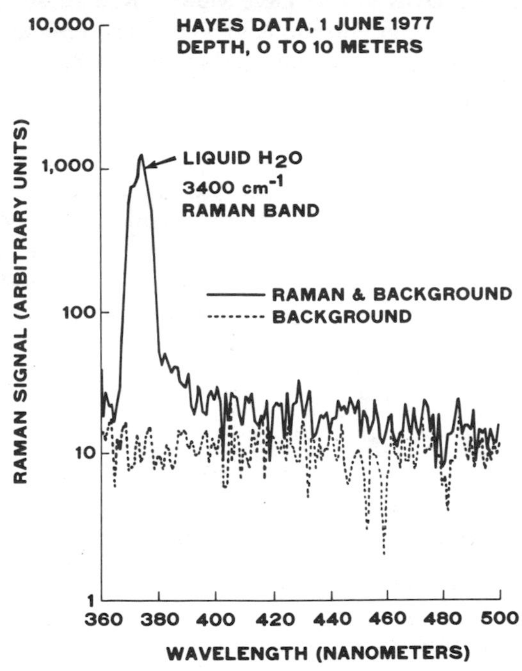

Fig. 10.34. Raman water band plus laser-induced background obtained from Raman lidar system mounted aboard the USNS Hayes (Leonard et al., 1979).

Gulf Stream in Fig. 10.35. The ground truth measurements were 4.7 and 15.7°C, respectively. Both spectra were obtained for water in the first 10 m of depth. Analysis of the two band shapes with the standard two-color technique reveals that the polymer component decreased by 13% between the Labrador Current and the Gulf Stream. This suggests a temperature difference of 13°C, compared to the measured difference of 11°C. In attempting to undertake depth-resolved Raman temperature measurements, differential attenuation was observed to cause an apparent increase in the spectral temperature as the depth increases. Analysis using a Gaussian deconvolution technique appeared to account for this effect and could therefore be used to correct for it.

The conclusions reached by Leonard et al. (1979) with regard to these preliminary experiments are: that there is negligible interference from laser-induced fluorescence—at least in the open ocean waters; that Raman spectral temperature measurements were possible to an accuracy of ±1°C while under way; that measurement of the subsurface temperature gradient to a depth of 30 m was possible, and that this could possibly be extended to 100 m if a more penetrating laser wavelength were chosen.

This section would be incomplete without some reference to the interest in Raman studies of water pollution. Bradley (1970) and Davis et al. (1973) have considered the possibility of using laser Raman spectroscopy for the detection and identification of molecular water pollutants, while Walrafen (1970), Houghton (1973), and Dylis (1974) have viewed Raman spectroscopy as a means of measuring ionic concentrations in water. The remote-sensing potential of these approaches seems at present to be a little remote!

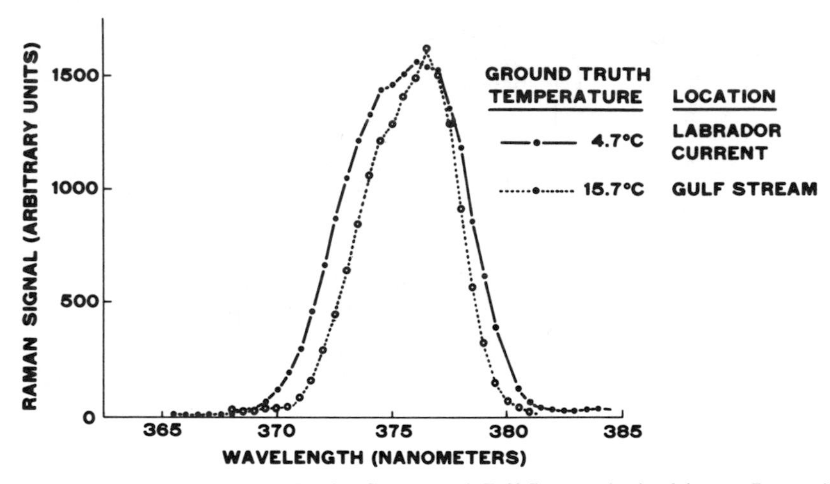

Fig. 10.35. Raman spectra of Labrador Current and Gulf Stream obtained from a Raman lidar system mounted aboard the USNS Hayes (Leonard et al., 1979).

11

Concluding Remarks

We have seen that *lidar* constitutes an important new weapon in the arsenal of those who would study the environment. The range of this active probe extends from the tenuous outer regions of the atmosphere to below the surface of the oceans. Atomic densities in the 10^3-cm^{-3} range have been observed at altitudes of about 90 km by ground-based systems, while oil films only a few micrometers thick have been measured by airborne lidars. Lidar systems can be mounted on the ground or in towers; or they can be operated from mobile platforms such as vans, aircraft, helicopters, and ships; or, under very special circumstances, they can be directed towards the earth from space. They may be used in a variety of roles, from simply evaluating the spatial distribution of a specific component within the environment, to identifying the constituents of a complex target by essentially performing a spectral analysis at a distance.

By and large we have concentrated on lidars that involve pulsed lasers, as they possess the capability of making both spatial and temporal measurements and in certain situations are capable of greater sensitivity and specificity than their cw counterparts. Nonetheless, there are certain classes of applications where cw systems might have some advantage in sensitivity, cost, or size.

To speculate on the future of laser environmental sensing would be rather naive in the light of past dramatic advances. Nevertheless, it is clear that changes in four areas will play a major role in shaping the destiny of this subject: advances in laser technology; improvements in detector sensitivity, particularly in the infrared; developments in data-processing technology; and changes in eye-safety regulations. One thing, however, appears sure: laser environmental sensing is likely to blossom into an important branch of remote sensing. Indeed, the real-time remote analysis capability of lidars raise the fascinating prospect of immediate detection and location of any excessive release of pollution over an extensive or sensitive area. Not only will this prevent intentional violation of emission-control standards, but in the event of an accidental discharge, this radarlike pollution early-warning system could ensure a rapid shutdown of the offending source and possibly provide a real time map of the dispersing plume.

11

Concluding Remarks

References

Abreau, V. J. (1980), "Lidar from Orbit," *Optical Engineering*, **19**, 489–493.

Ahmed, S. A. (1973), "Molecular Air Pollution Monitoring by Dye Laser Measurement for Differential Absorption of Atmospheric Elastic Backscatter," *Appl. Optics*, **12**, 901–903.

Aldén, M., H. Edner, and S. Svanberg (1982), "Remote Measurements of Atmospheric Mercury Using Differential Absorption Lidar," *Optics Letters*, **7**, 221–223.

Allen, R. J., and C. M. Platt (1977), "Lidar for Multiple Backscattering and Depolarization Observations," *Appl. Optics*, **16**, 3193–3199.

Aller, L. H. (1963), *Astrophysics—The Atmospheres of the Sun and Stars*, 2nd ed., The Ronald Press Co., New York.

Altmann, J., and P. Pokrowsky (1980), "Sulphur Dioxide Absorption at DF Laser Wavelengths," *Appl. Optics*, **19**, 3449–3452.

Altmann, J., S. Kohler, and W. Lahmann (1980a), "Fast Current Amplifier for Background-Limited Operation of Photovoltaic InSb Detectors," *J. Physics. 6: Sci. Instrum.* **13**, 1275–1279.

Altmann, J., W. Lahmann, and C. Weitkamp (1980b), "Remote Measurements of Atmospheric N_2O with a DF Laser Lidar," *Appl. Optics*, **19**, 3453–3457.

American National Standards Institute (1973), *Laser Standard Designated Z-136.1-173*, ANSI, New York.

Armstrong, R. L. (1975), "Rotational Raman Interferometric Technique to Measure Gas Temperatures," *Appl. Optics*, **14**, 383–387.

Aruga, T., H. Kamiyama, M. Jyumonji, T. Kobayashi, and H. Inaba (1974), "Laser Radar Observation of the Sodium Layer in the Upper Atmosphere," *Report of Ionosphere and Space Research in Japan*, **28**, 65–68.

Asai, K., and T. Igarashi (1975), "Detection of Ozone by Differential Absorption Using CO_2 Laser," *Optical and Quantum Electronics*, **7**, 211–214.

Asai, K., T. Itabe, and T. Igarashi (1979), "Range-Resolved Measurements of Atmospheric Ozone Using a Differential-Absorption CO_2 Laser Radar," *Appl. Phys. Lett.*, **35**, 60–62.

Baardsen, E. L., and R. W. Terhune (1972), "Detection of OH in the Atmosphere Using a Dye Laser," *Appl. Phys. Lett.*, **21**, 209–211.

Barber, P., and C. Yeh (1975), "Scattering of Electromagnetic Waves by Arbitrarily Shaped Dielectric Bodies," *Appl. Optics*, **14**, 2864–2872.

Barrett, J. J. (1974), "The Use of a Fabry–Perot Interferometer for Studying Rotational Raman Spectra of Gases," *Laser Raman Gas Diagnostics* (M. Lapp and C. M. Penny, Eds.), Plenum Press, New York, 63–85.

Barrett, J. J. (1976), "Gas Temperature and Centrifugal Distortion Constant Determination from Profiles of Rotational Raman Interferometric Spectra," *J. Opt. Soc. Am.*, **66**, 801–812.

Barrett, J. J. (1977), "Interferometric Investigations of Rotational Raman Spectra of Gases," *Optical Engineering*, **16**, 85–106.

Barrett, J. J., and S. A. Myers (1971), "New Interferometric Method for Studying Periodic Spectra Using Fabry–Perot Interferometer," *J. Opt. Soc. Am.*, **61**, 1246–1251.

Basting, O., F. P. Schafer, and B. Steyer (1972), "A Simple High Power Nitrogen Laser," *Opto-Electronics*, **4**, 43–44.

Baumgartner, R. A., and R. L. Byer (1978a), "Remote SO_2 Measurements at 4 m with a Continuously-Tunable Source," *Optics Letters*, **2**, 163.

Baumgartner, R. A., and R. L. Byer (1978b), "Continuously-Tunable IR Lidar with Applications to Remote Measurements of SO_2 and CH_4," *Appl. Optics*, **17**, 3555.

Baumgartner, R. A., and R. L. Byer (1979), "Optical Parametric Amplification," *IEEE J. Quant. Electr.*, **QE-15**, 432.

Baumgartner, R. A., J. G. Depp, W. E. Evans, W. B. Grant, J. G. Hawley, R. G. March, E. R. Murray, and E. K. Proctor (1979a), *Characterization of the EPRI Differential Absorption Lidar* (DIAL) *System*, EA-1267, Project Report 862-14.

Baumgartner, R. A., L. D. Fletcher, and J. G. Hawley (1979b), "A Comparison of Lidar and Air Quality Station NO_2 Measurements," *ARCO Journal*, **29**, 1162–1165.

Bevington, P. R. (1969), *Data Reduction and Error Analysis for the Physical Sciences*, McGraw-Hill, New York.

Bhaumik, M. L., R. S. Bradford and A. R. Roit (1976), "High Efficiency KrF Excimer Laser," *Appl. Phys. Lett.*, **28**, 23–24.

Bilbro, J. W. (1980), "Atmospheric Laser Doppler Velocimetry: An Overview," *Optical Engineering*, **19**, 533–542.

Blamont, J. E., M. L. Chanin, and G. Megie (1972), "Vertical Distribution and Temperature Profile of the Nighttime Atmospheric Sodium Layer Obtained by Laser Backscatter," *Appl. Optics Ann. Geophys.*, **28**, 833–838.

Bloembergen, N. (1965), *Nonlinear Optics*, Benjamin, New York; 3rd printing, Addison-Wesley, Reading, Mass., 1978.

Bloembergen, N. (1980), "Conservation Laws on Nonlinear Optics," *J. Opt. Soc. Am.*, **70**, 1429–1436.

Born, M., and E. Wolf (1964), *Principles of Optics*, 2nd ed., Pergamon Press, New York.

Boudreau, R. D. (1970), "On the Use of Ultraviolet Lidar for Observing Atmospheric Constituents by Raman Scattering," *J. Appl. Meterol.*, **9**, 316–317.

Bowman, M. R., A. J. Gibson, and M. C. W. Sandford (1969), "Atmospheric Sodium Measured by a Tuned Laser Radar," *Nature*, **221**, 456.

Boyd, G. O., and J. P. Gordon, (1961), "Confocal Multimode Resonator for Millimeter through Optical Wavelength Masers," *Bell Syst. Tech. J.*, **40**, 489.

Bradley, E. (1970), "On the Exploitation of Laser Raman Spectroscopy for Detection and Identification on Molecular Water Contaminants," *Water Research*, **4**, 125.

Bradley, D. J. (1974), "Generation and Measurement of Frequency-Tunable Picosecond Pulses from Dye Lasers," *Opto-electronics*, **6**, 25–42.

Bradley, D. J., and G. J. C. New (1974), "Ultrashort Pulse Measurements," *Proc. IEEE*, **62**, 313–345.

Brassington, D. J. (1978), "Alternate-Pulse Wavelength Switching of a Dye Laser," *J. Phys. E: Sci. Instrum.*, **2**, 119–120.

Brassington, D. J. (1981), "Sulfur Dioxide Absorption Cross Section Measurements from 290 nm to 713 nm," *Appl. Optics*, **20**, 3774–3779.

Bridges, W. B., A. N. Chester, A. S. Halsted and J. V. Parker (1971), "Ion Laser Plasmas," *Proc. IEEE*, **59**, 724–737.

Bristow, M. P. F. (1978), "Airborne Monitoring of Surface Water Pollutants by Fluorescence Spectroscopy," *Remote Sensing of Environment*, **7**, 105–127.

Bristow, M. P. (1979), "Fluorescence of Short Wavelength Cutoff Filters," *Appl. Optics*, **18**, 952–955.

Bristow, M. P. F., W. R. Houston, and R. M. Measures (1973), "Development of a Laser Fluorosensor for Airborne Surveying of the Aquatic Environment," *NASA Conference on the Use of Lasers for Hydrographic Studies, Wallops Island, Sept. 1973*, SP-375, 197–202.

Bristow, M., D. Nielsen, D. Bundy, and R. Furtek (1981), "Use of Water Raman Emission to Correct Airborne Laser Fluorosensor Data for Effects of Water Optical Attenuation," *Appl. Optics*, **20**, 2889–2906.

Bristow, M., and D. Nielsen (1981), *Remote Monitoring of Organic Carbon in Surface Waters*, Report No. EPA-80/4-81-001, U.S. Environmental Protection Agency, Las Vegas, Nevada.

Browell, E. V. (1977), *Analysis of Laser Fluorosensor Systems for Remote Algae Detection and Quantification*, NASA TN D-8447.

Browell, E. V. (1982a), "Lidar Measurements of Tropospheric Gases," *Optical Eng.*, **21**, 128–132.

Browell, E. V. (1982b), "Remote Sensing of Tropospheric Gases and Aerosols with an Airborne Dial System," *Proc. of Workshop on Optical and Laser Remote Sensing, Monterey, Feb. 9–11, 1982.*

Browell, E. V., T. D. Wilkerson, and T. J. McIlrath (1978), "Water Vapor Differential-Absorption Lidar Development and Evaluation," *Appl. Optics*, **18**, 3474–3483.

Bücher, H., and W. Chow (1977), "A 1 MW *p*-Terphenyl Dye Laser," *Appl. Phys.*, **13**, 267–269.

Burgess, S., and I. W. Shepherd (1977), "Fluorescence Suppression in Time-Resolved Raman Spectra," *J. Physics E: Sci. Instrum.*, **10**, 617–620.

Burnham, R., and E. J. Schimitschek (1981), "High-Power Blue–Green Lasers," *Laser Focus*, 54–61.

Byer, R. L. (1975), "Remote Air Pollution Measurement," *Optical and Quantum Electronics*, **7**, 147–177.

Byer, R. L., and M. Endermann (1982), "Remote Measurements of Trace Species in the Troposphere," *AIAA J.*, **20**, 395–403.

Byer, R. L., and M. Garbuny (1973), "Pollutant Detection by Absorption Using Mie Scattering and Topographic Targets as Retroreflectors," *Appl. Optics*, **12**, 1496–1505.

Byer, R. L., and L. A. Shepp (1979), "Two-Dimensional Remote Air-Pollution Monitoring via Tomography," *Optics Letters*, **4**, 75–77.

Cahen, C., and G. Megie (1981), "A Spectral Limitation of the Range Resolved Differential Absorption Lidar Technique," *J. Quant. Spectrosc. Radiat. Transfer*, **25**, 151–157.

Chandrasekhar, S. (1960), *Radiative Transfer*, Dover, New York.

Chang, C. H., and L. A. Young (1973), "Remote Measurement of Ocean Temperature from Depolarization in Raman Scattering," *NASA Conference on the Use of Lasers for Hydrographic Studies*, NASA SP-375, 105–112.

Christman, R. F., and M. Ghasseni (1966), "Chemical Nature of Organic Color in Water," *J. AWWA*, 723–731.

Chylek, P., G. W. Grams, and R. G. Pinnick (1976), "Light Scattering by Irregular Randomly Oriented Particles," *Science*, **193**, 480–482.

Clay, M. R., and A. P. Lenham (1981), "Transmission of Electromagnetic Radiation in Fogs in the 0.53 to 10.1 Micron Wavelength Range," *Appl. Optics*, 3831–3832.

Cohen, A., and M. Graber (1975), "Double-Scattering Calculations and Laboratory Dye-Laser Multiple Scattering Measurements," *Optical and Quantum Electronics*, **7**, 221–228.

Cohen, A., J. Cooney, and K. N. Geller (1976), "Atmospheric Temperature Profiles from Lidar Measurements of Rotational Raman and Elastic Scattering," *Appl. Optics*, **15**, 2896–2901.

Cohen, A., M. Kleiman, and J. Cooney (1978), "Lidar Measurements of Rotational Raman and Double Scattering," *Appl. Optics*, **17**, 1905–1910.

Collis, R. T. H., and P. B. Russell (1976), "Lidar Measurement of Particles and Gases by Elastic Backscattering and Differential Absorption," *Laser Monitoring of the Atmosphere* (E. D. Hinkley, Ed.), Springer-Verlag.

Collis, R. T. H., and E. E. Uthe (1972), "Mie Scattering Techniques for Air Pollution Measurement with Lasers," *Opto-Electronics*, **4**, 87–99.

Condon, E. U., and H. Odishaw, Eds. (1967), *Handbook of Physics*, McGraw-Hill, New York.

Cooney, J. A. (1968), "Measurements on the Raman Component of Laser Atmospheric Backscatter," *Appl. Phys. Lett.*, **12**, 40–42.

Cooney, J. A. (1970), "Comparisons of Water Vapor Profiles Obtained by Radiosonde and Laser Backscatter," *J. Appl. Meteor.*, **9**, 182–184.

Cooney, J. A. (1971), "Remote Measurement of Atmospheric Water Vapor Profiles Using Raman Component of Laser Backscatter," *J. Appl. Meteor.*, **10**, 301–308.

Cooney, J. A. (1972), "Measurement of Atmospheric Temperature Profiles by Raman Backscatter," *J. Appl. Meteor.*, **11**, 108–112.

Cooney, J. A. (1975), "Normalization of Elastic Lidar Returns by Use of Raman Rotational Backscatter," *Appl. Optics*, **14**, 270–271.

Cooney, J. A., J. Orr, and C. Tomasetti (1969), "Measurements Separating the Gaseous and Aerosol Components of Laser Atmospheric Backscatter," *Nature*, **224**, 1098–1099.

Corliss, C. H., and W. R. Bozman (1962), *Experimental Transition Probabilities for Spectral Lines of Seventy Elements Derived from the NBS Tables of Spectral-Line Intensities*, NBS Monograph 53, U.S. Government Printing Office, Washington, D.C.

Corney, A. (1977), *Atomic & Laser Spectroscopy*, Clarendon Press, Oxford.

Davis, P. A. (1969), "The Analysis of Lidar Signatures of Cirrus Clouds," *Appl. Optics*, **8**, 2099–2102.

Davis, A., M. Bristow, and J. Koningstein (1973), "Raman Spectroscopy as a Water Quality Indicator," *Remote Sensing and Water Resources Management Proc.* **17**, 239–246.

Davis, D. D., W. S. Heaps, D. Philen, M. Rodgers, T. McGee, A. Nelson, and A. J. Moriarty (1979), "Air-borne Laser Induced Fluorescence System for Measuring OH and Other Trace Gases in the Parts-per-Quadrillion to Parts-per-Trillion Range," *Rev. Sci. Instrum.*, **50**, 1505–1516.

Davis, D. D., W. S. Heaps, D. Philen, M. Rodgers, T. McGee, A. Nelson, and A. J. Moriarty (1980), "Erratum: Air-borne Laser Induced Fluorescence System for Measuring OH and Other Trace Gases in the Parts-per-Quadrillion to Parts-per-Trillion Range," *Rev. Sci. Instrum.*, **51**, 1584–1585.

Deirmendjian, D. (1964), "Scattering and Polarization Properties of Water Clouds and Hazes in the Visible and Infrared," *Appl. Optics*, **2**, 187–196.

Deirmendjian, D. (1969), "Electromagnetic Scattering on Spherical Polydispersions," American Elsevier, New York.

DeLong, H. P. (1974), "Air Pollution Field Studies with a Raman Lidar," *Optical Engineering*, **13**, 5–9.

Dobrowolski, J. A., G. E. Marsh, D. G. Charbonneau, J. Eng, and P. D. Josephy (1977), "Colored Filter Glasses: An Intercomparison of Glasses Made by Different Manufacturers," *Appl. Optics*, **16**, 1491–1521.

Donaldson, W. T. (1977), "Trace Organics in Water," *Environmental Science & Technology*, **11**, 348–351.

Douglas-Hamilton, D. H. (1978), "Transmission at 10.6 Micrometers Wavelength through the Upper Atmosphere," *Appl. Optics*, **17**, 2316–2320.

Duntley, S. Q. (1963), "Light in the Sea," *J. Opt. Soc. Am.*, **58**, 214–233.

Durst, F., A. Melling, and J. H. Whitelaw (1976), *Principles and Practice of Laser-Doppler Anemomentry*, Academic Press.

Dylis, D. D. (1974), "A Raman Technique for the Remote Measurement of Aqueous Acid Solutions," *Opt. Eng.*, **13**, 502–505.

Eisberg, R., and R. Resnick (1974), *Quantum Physics of Atoms, Molecules, Solids, Nuclei, and Particles*, Wiley, New York.

El-Sayed, S. Z. (1970), "Phytoplankton Production of the South Pacific and the Pacific Sector of the Antarctic," *Scientific Exploration of the South Pacific*, National Academy of Sciences, Washington, D.C.

Elterman, G. (1953), *J. Geophys.*, **58**, 519.

Elterman, G. (1954), *J. Geophys.*, **59**, 351.

Elterman, L. (1968), *UV, Visible and IR Attenuation for Altitudes to* 50 *km*, Air Force Cambridge Research Laboratories, AFCRL-68-0153, Environmental Research Paper No. 285 (Apr.).

Elterman, L. (1970), "Relationships between Vertical Attenuation and Surface Meteorological Range," *Appl. Optics*, **9**, 1804–1810.

Emmons, R. B., S. R. Hawkins, and C. F. Cuff (1975), "Infrared Detectors: An Overview," *Opt. Eng.*, **14**, 21–30.

Endermann, M., and R. L. Byer (1980), "Remote Single-Ended Measurements of Atmospheric Temperature and Humidity at 1.77 Micrometers Using a Continuously Tunable Source," *Optics Letters*, **5**, 452–454.

Endermann, M., and R. L. Byer (1981), "Simultaneous Remote Measurements of Atmospheric Temperature and Humidity Using a Continuously Tunable IR-Lidar," *Appl. Optics*, **20**, 3211–3217.

Ewing, J. J. (1978), "Rare-Gas Halide Lasers," *Physics Today*, May, 32–39.

Fantasia, J. F., and H. C. Ingrao (1974), "Development of Experimental Airborne Laser Remote Sensing System for the Detection and Classification of Oil Spills," *Proc. of the 9th Intern. Symp. on Remote Sensing of the Environment, 15–19 April 1974,* Paper 10700-1-X, 1711–1745.

Fantasia, J. F., T. M. Hard, and H. C. Ingrao (1971), *An Investigation of Oil Fluorescence as a Technique for Remote Sensing of Oil Spills*, Report No. DOT-TSC-USCG-71-7, Transportation Systems Center, Dept. of Transportation, Cambridge, Mass.

Fegley, R. W., and H. T. Ellis (1975), "Lidar Observations of a Statospheric Dust Cloud Layer in the Tropics," *Geophys. Res. Lett.*, **2**, 139–141.

Felix, F., W. Keenliside, G. Kent, and M. C. W. Standford (1973), "Laser Radar Observations of Atmospheric Potassium," *Nature*, **246**, 345–346.

Fenn, R. W. (1966), "Correlation between Atmospheric Backscattering and Meteorological Visual Range," *Appl. Optics*, **5**, 293–295.

Fenner, W. R., H. A. Hyatt, J. M. Kellan, and S. P. S. Porto (1973), "Raman Cross-sections of Some Simple Gases," *J. Opt. Soc. Am.*, **63**, 73–77.

Fiocco, G., and Colombo, (1964), "Optical Radar Results and Meteoric Fragmentation," *J. Geophys. Res.*, **69**, 1795–1803.

Fiocco, G., and G. Grams (1964), "Observations of the Aerosol Layer at 20 km by Optical Radar," *J. Atmos. Sci.*, **21**, 323–324.

Fiocco, G., and L. D. Smullin (1963), "Detection of Scattering Layers in the Upper Atmosphere (60–140 km) by Optical Radar," *Nature*, **199**, 1275–1276.

Foord, R., R. Jones, C. Oliver, and E. Pike (1969), *Appl. Optics*, **8**, 1975.

Fouche, D. G., and R. K. Chang (1971), "Relative Raman Cross-section for N_2, O_2, CO, CO_2, SO_2, and H_2S," *Appl. Phys. Lett.*, **18**, 579–580.

Fouche, D. G., and R. K. Chang (1972a), "Relative Raman Cross-section for O_3, CH_4, C_3H_8, NO, NO_2 and H_2," *Appl. Phys. Lett.*, **20**, 256–257.

Fouche, D. G., and R. K. Chang (1972b), "Observation of Resonance Raman Scattering below the Dissociation Limit in I_2 Vapor," *Phys. Rev. Lett.*, **29**, 536–539.

Fouche, D. G., A. Herzenberg, and R. K. Chang (1972), "Inelastic Photon Scattering by a Polyatomic Molecule: NO_2," *J. Appl. Phys.*, **43**, 3846–3851.

Fox, A. G., and T. Li (1961), "Resonant Modes in a Maser Interferometer," *Bell Syst. Tech. J.*, **40**, 453–488.

Fredriksson, K., I. Lindgren, S. Svanberg, and G. Weibull (1976), *Measurements of the Emission from Industrial Smoke-stacks Using Laser Radar Techniques*, Goteborg Institute of Physics Reports—121 (Jan.).

Fredriksson, K., I. Lindgren, and S. Svanberg (1978), "Measurements of Source Emissions and Ambient Air Quality Using Pulsed Nitrogen and Dye Lasers," *Proceedings, 4th Joint Conference on Sensing Environmental Pollutants*, **113**, 416–419.

Fredriksson, K., B. Galle, K. Nystrom, and S. Svanberg (1979), "Lidar System Applied in Atmospheric Pollution Monitoring," *Appl. Optics*, **18**, 2998–3003.

Fredriksson, K., B. Galle, K. Nystrom, and S. Svanberg (1981), "Mobile Lidar System for Environmental Probing," *Appl. Optics*, **20**, 4181–4189.

Frush, C. L. (1975), "A New Lidar Signal Processing and Display System," *Optical and Quantum Electronics*, **7**, 179–185.

Fymat, A. (1976), *Inversion Methods in Atmospheric Remote Sounding*, NASA CP-004.

Garvey, M. J., and G. S. Kent (1974), "Raman Backscatter of Laser Radiation from the Stratosphere," *Nature*, **248**, 124–125.

Gebbie, H. S., W. R. Harding, C. Hilsum, A. W. Pryce, and V. Roberts (1951), "Atmospheric Transmission in the 1 to 14 Micrometer Region," *Proc. R. Soc. A*, **206**, 87–107.

Gelbwachs, J. (1973), "NO_2 Lidar Comparison: Fluorescence vs. Backscattered Differential Absorption," *Appl. Optics*, **12**, 2812–2813.

Gelbwachs, J., and M. Birnbaum (1973), "Fluorescence of Atmospheric Aerosols and Lidar Implications," *Appl. Optics*, **12**, 2442–2447.

Gibson, F. W. (1976), "*In Situ* Photometric Observations of Angular Scattering from Atmospheric Aerosols," *Appl. Optics*, **15**, 2520–2533.

Gibson, A. J., and M. C. W. Sandford (1971), "The Seasonal Variation of the Night-time Sodium Layer," *J. Atmospheric and Terrestrial Physics*, **33**, 1675–1684.

Gibson, A. J., and M. C. W. Sandford (1972), "Daytime Measurements of the Atmospheric Sodium Layer," *Nature*, **239**, 509–511.

Girard, A., and P. Jacquinot (1967), "Principles of Instrumental Methods in Spectroscopy," *Advanced Optimal Techniques* (A. van Heel, Ed.), North-Holland, Amsterdam, Chapter 3.

Gires, F., and F. Combaud (1965), "Saturation de l'Absorption Optique de Certaines Solutions de Phthalocyanines," *J. Physique*, **26**, 325–330.

Gordon, E. I., E. F. Labuda, and W. B. Bridges (1964), "Continuous Visible Laser Action in Singly Ionized Argon, Krypton and Xenon," *Appl. Phys. Lett.*, **4**, 178.

Grams, G. W., and C. M. Wyman (1972), "Compact Laser Radar for Remote Atmospheric Probing," *J. Appl. Meteor.*, **11**, 1108–1113.

Granier, C., and G. Megie (1982), "Daytime Lidar Measurements of Mesospheric Sodium Layer," *Planet. Space Sci.*, **30**, 169–177.

Grant, W. B. (1982), "Effect of Differential Spectral Refectance on DIAL Measurements Using Topographical Targets," *Appl. Optics*, **21**, 2390–2394.

Grant, W. B., and R. D. Hake, Jr. (1975), "Calibrated Remote Measurements of SO_2 and O_3 Using Atmospheric Backscattering," *Appl. Phys. Lett.*, **17**, 139–141.

Grant, W. B., R. D. Hake, Jr., E. M. Liston, R. C. Robbins, and E. K. Proctor, Jr. (1974), "Calibrated Remote Measurements of NO_2 Using Differential Absorption Backscatter Technique," *Appl. Phys. Lett.*, **24**, 550–552.

Greco, R. V. (1979), *Atmospheric Lidar Multi-User Instrument System Definition Study*, General Electric Final Report to NASA, NAS 1-15476.

Green, A. E. S., and P. J. Wyatt (1965), *Atomic and Space Physics*, Addison-Wesley, Reading.

Green, A. E. S. (1966), *The Middle Ultraviolet*, Wiley, New York.

Griem, H. R. (1974), *Spectral Line Broadening by Plasmas*, Academic Press, New York.

Guagliardo, J. L., and D. H. Bundy (1974), "Differential Monitoring of Ozone in the Troposphere Using Earth Reflected Differential Absorption," International Telemetering Conference (Oct.).

Guagliardo, J. L., and D. H. Bundy (1975), "Earth Reflected Differential Absorption Using TEA Lasers: A Remote Sensing Method for Ozone," 7th International Laser Radar Conference, Palo Alto, Calif.

Gunning, W. J. (1981), "Electro-Optically Tuned Spectral Filters: Review," *Opt. Eng.*, **20**, 837–845.

Hake, R. D., Jr., D. E. Arnold, D. W. Jackson, W. E. Evans, B. P. Ficklin, and R. A. Long (1972), "Dye-Laser Observations of the Nighttime Atomic Sodium Layer," *Geophysical Research*, **77**, 6389–6848.

Hall, F. F., Jr. (1974), "Laser Systems for Monitoring of the Environment," Academic Press, Vol. II.

Halldórsson, T., and J. Langerholc (1978), "Geometrical Form Factors for the Lidar Function," *Appl. Optics*, **17**, 240–244.

Hänsch, T. W. (1972), "Repetitively Pulsed Tunable Dye Laser for High Resolution Spectroscopy," *Appl. Optics*, **11**, 895.

Hanst, P. L. (1976), "Optical Measurement of Atmosphere Pollutants: Accomplishments and Problems," *Optical and Quantum Electronics*, **8**, 87–93.

Hanst, P. L., and J. A. Morreal (1968), "Detection and Measurement of Air Pollutants by Absorption of Infrared Radiation," *J. Air Poll. Control Assoc.*, **18**, 754–759.

Harmony, M. D. (1972), *Introduction to Molecular Energies and Spectra*, Holt, Rinehart & Winston.

Harms, J. (1979), "Lidar Return Signals for Coaxial and Noncoaxial Systems with Central Obstruction," *Appl. Optics*, **18**, 1559–1566.

Harms, J., W. Lahmann, and C. Weirkamp (1978), "Geometrical Compression of Lidar Return Signals," *Appl. Optics*, **17**, 1131–1135.

Hawley, J. G. (1981), "Dual-Wavelength Laser Radar Probes for Air Pollutants," Laser Focus, Mar., 60–62.

Heaps, W. S. (1980), "Measurement of Hydroxyl Radical in the Upper Atmosphere Using Lidar from the Space Shuttle," *Appl. Optics*, **19**, 243–249.

Heaps, W. S. (1981), "Selection of Fluorescence Lidar Operating Parameters for SNR Maximization," *Appl. Optics*, **20**, 583–587.

Heaps, W. S., T. J. McGee, R. D. Hudson, and L. O. Caudill (1982), "Stratospheric Ozone and Hydroxyl Radical Measurements by Balloon-Borne Lidar," *Appl. Optics*, **21**, 2265–2274.

Heard, H. (1963), *Nature*, Nov. 16, 667.

Hecht, E., and A. Zajac (1974), *Optics*, Addison-Wesley.

Heintzenberg, J., H. Muller, H. Quenzel, and E. Thomalla (1981), "Information Content of Optical Data with Respect to Aerosol Properties: Numerical Studies with a Randomized Minimization-Search-Technique Inversion Algorithm," *Appl. Optics*, **20**, 1308–1322.

Heitler, W. (1954), *The Quantum Theory of Radiation*, Clarendon Press, Oxford, England.

Heller, W., and M. Nakagaki (1974), "Light Scattering of Spheroids. III. Depolarization of the Scattered Light," *J. Chemical Physics*, **61**, 3619–3621.

Henningsen, T., M. Garbuny, and R. L. Byer (1974), "Remote Detection of CO by Parametric Tunable Lasers," *Appl. Phys. Lett.*, **24**, 242–244.

Herzberg, G. (1950), *Spectra of Diatomic Molecules*, vol. 1, Van Nostrand, Princeton.

Herzberg, G. (1967), *Electronic Spectra and Electronic Structure of Polyatomic Molecules*, vol. III, Van Nostrand, Princeton.

Hickman, G. D. (1973), "Recent Advances in the Application of Pulsed Lasers to the Hydrosphere," *NASA Conference on the Use of Lasers for Hydrographic Studies*, NASA SP-375, 81–88.

Hickman, G. D., and J. E. Hogg (1969), "Application of an Airborne Pulsed Laser for Near Shore Bathymetric Measurements," *Remote Sensing of Environment*, **1**, 47–58.

Hickman, G. D., and R. B. Moore (1970), "Laser Induced Fluorescence in Rhodamine 3 and Algae," *Proc. 13th Conf. Great Lakes Res.*, Int. Assoc. Great Lakes Res., 1–14.

Hickman, G. D., A. H. Ghovanlou, E. J. Friedman, C. S. Gault, and J. E. Hogg (1974), *A Feasibility Study for a Remote Laser Water Turbidity Meter*, NASA CR-132376.

Hinkley, E. D. (1972), "Tunable Infrared Lasers and Their Applications to Air Pollution Measurements," *Opto-electronics*, **4**, 69–86.

Hinkley, E. D. (1976), "Laser Spectroscopic Instrumentation and Techniques: Long Path Monitoring by Resonance Absorption," *Optical and Quantum Electronics*, **8**, 155–167.

Hinkley, E. D., and P. L. Kelly (1971), "Detection of Air Pollutants with Tunable Diode Lasers," *Science*, **17**, 635–639.

Hinkley, E. D., R. T. Ku, and P. L. Kelley (1976), "Techniques for Detection of Molecular Pollutants by Absorption of Laser Radiation," *Laser Monitoring of the Atmosphere* (E. D. Hinkley, Ed.), Springer-Verlag.

Hirschfeld, T. (1974), "Range Independence of Signal in Variable Focus Remote Raman Spectrometry," *Appl. Optics*, **13**, 1435–1437.

Hirschfeld, T. (1977a), "The Choice between Absorption and Fluorescent Techniques," *Applied Spectroscopy*, **31**, 245.

Hirschfeld, T. (1977b), "On the Nonexistence of Nonfluorescent Compounds and Raman Spectroscopy," *Applied Spectroscopy*, **31**, 328–329.

Hirschfeld, T., and S. Klainer (1970), "Remote Raman Spectroscopy as a Pollution Radar," *Optical Spectra*, 63–66.

Hirschfeld, T., E. R. Schildkraut, H. Tannenbaum, and D. Tannenbaum (1973), "Remote Spectroscopic Analysis of ppm-level Air Pollutants by Raman Spectroscopy," *Appl. Phys. Lett.*, **22**, 38–40.

Hochenbleicher, J. G., W. Kiefer, and J. Brandmuller (1976), "A Laboratory Study for Resonance Raman Lidar System," *Applied Spectroscopy*, **30**, 528–531.

Hodgeson, J. A., W. A. McClenny, and P. L. Hanst (1973), "Air Pollution Monitoring by Advanced Spectroscopic Techniques," *Science*, **182**, 248–258.

Hodgkinson, I. J., and J. I. Vukusic (1978), "Birefringent Filters for Tuning Flashlamp-Pumped Dye Lasers: Simplified Theory and Design," *Appl. Optics*, **17**, 1944–1948.

Hoell, J. M., Jr., W. R. Wade, and R. T. Thompson, Jr. (1975), "Remote Sensing of Atmospheric SO_2 Using the Differential Absorption Lidar Technique," Int. Conf. on Environ. Sens. & Assessment, Las Vegas, 14 Sept.

Hoge, F. E., and R. N. Swift (1981), "Airborne Simultaneous Spectroscopic Detection of Laser Induced Water Raman Backscatter and Fluorescence from Chlorophyll-*a*, and Other Naturally Occurring Pigments," *Appl. Optics*, **20**, 3197–3205.

Hoge, F. E. (1982), "Laser Measurements of the Spectral Extinction Coefficients of Fluorescent, Highly Absorbing Liquids," *Appl. Optics*, **21**, 1725–1729.

Hoge, F. E., and R. N. Swift (1980), "Oil Film Thickness Measurement Using Airborne Laser-Induced Water Raman Backscatter," *Appl. Optics*, **19**, 3269–3281.

Hoge, F. E., and R. N. Swift (1981), "Absolute Tracer Dye Concentration Using Airborne Laser-Induced Water Raman Backscatter," *Appl. Optics*, **20**, 1191–1202.

Hoge, F. E., R. N. Swift, and E. B. Frederick (1980), "Water Depth Measurement Using an Airborne Pulsed Neon Laser System," *Appl. Optics*, **19**, 871–883.

Houghton, W. M. (1973), "Measurement on Raman Spectra of H_2O and SO_4 in Sea Water," *NASA Conference on the Use of Lasers for Hydrographic Studies*, NASA SP-375, 113–118.

Houston, W. R., D. G. Stephenson, and R. M. Measures (1973), "LIFES: Laser Induced Fluorescence and Environmental Sensing," *NASA Conference on the Use of Lasers for Hydrographic Studies*, NASA SP-375, 153–169.

Hurn, R. W. (1968), "Mobile Combustion Sources," *Air Pollution* (A. C. Stern, Ed.), Vol. 3, Academic Press, New York.

Hutcheson, L. D., and R. S. Hughes (1974), "Rapid Acousto-Optic Tuning of a Dye Laser," *Appl. Optics*, **13**, 1395–1398.

Hyatt, H. A., J. M. Cherlow, W. R. Fenner, and S. P. S. Porto (1973), "Cross Section for Raman Effect in Molecular Nitrogen Gas," *J. Opt. Soc. Am.*, **63**, 1604–1606.

Imes, E. S. (1919), "Measurements on the Near Infrared Absorption of Some Diatomic Gases," *Astrophys. J.*, **50**, 251–276.

Inaba, H. (1976), "Detection of Atoms and Molecules by Raman Scattering and Resonance Fluorescence," *Laser Monitoring of the Atmosphere* (E. D. Hinkley, Ed.), Springer-Verlag.

Inaba, H., and T. Kobayasi (1969), "Laser-Raman Radar for Chemical Analysis of Polluted Air," *Nature*, **224**, 170–172.

Inaba, H., and T. Kobayasi (1972), "Laser-Raman Radar," *Opto-electronics*, **4**, 101–123.

Inomata, H., and A. I. Carswell (1977), "Simultaneous Tunable Two-Wavelength Ultra-violet Dye Lasers," *Optics Comm.*, **19**, 5–6.

Jacquinot, P. J. (1954), "The Luminosity of Spectrometers with Prisms, Gratings or Fabry–Perot Etalons," *J. Opt. Soc. Am.*, **44**, 761.

Jerlov, N. G. (1976), *Marine Optics*, Elsevier, New York.

Johnson, W. B., and E. E. Uthe (1971), "Lidar Study of the Keystone Stack Plume," *Arm. Environ.*, **5**, 703–724.

Junge, C. E. (1963), *Air Chemistry and Radioactivity*, Academic Press, New York.

Junge, C. E., and J. E. Manson (1961), "Stratospheric Aerosol Studies," *J. Geophys. Res.*, **66**, 2163–2182.

Kalle, K. (1966), "The Problem of the Gelbstoff in the Sea," *Oceanogr. Mar. Biol. Ann. Rev.*, **4**, 91–104.

Kalshoven, J. E., Jr., C. L. Korb, G. K. Schwemmer, and M. Dombrowski (1981), "Laser Remote Sensing of Atmospheric Temperature by Observing Resonant Absorption of Oxygen," *Appl. Optics*, **20**, 1967–1971.

Kanstad, S. O., A. Bjerkestrand, and T. Lund (1977), "Tunable Dual-Line CO_2 Laser for Atmospheric Spectroscopy and Pollution Monitoring," *J. Physics & Scientific Instrum.*, **10**, 998–1000.

Kattawar, G. W., and G. N. Plass (1976), "Asymptotic Radiance and Polarization in Optically Thick Media: Ocean and Clouds," *Appl. Optics*, **15**, 3166–3178.

Keith, L. H. (1979), "Analysis of Organic Water Pollutants," *Environmental Science & Technology*, **13**, 1469–1471.

Kelley, P. L., R. A. McClatchey, R. K. Long, and A. Snelson (1976), "Molecular Absorption of Infrared Laser Radiation in the Natural Atmosphere," *Optical and Quantum Electronics*, **8**, 117–144.

Kent, G. S., and R. W. H. Wright (1970), "A Review of Laser Radar Measurements of Atmospheric Properties," *J. Atmospheric and Terrestial Physics*, **32**, 917–943.

Kent, G. S., M. C. W. Sandford, and W. Keenliside (1971), "Laser Radar Observations of Dust from Comet Bennett," *J. Atmospheric and Terrestrial Physics*, **33**, 1257–1262.

Kerker, M. (1969), *The Scattering of Light*, Academic Press, New York.

Kerker, M., and D. D. Cooke (1976), "Remote Sensing of Particle Size and Refractive Index by Varying the Wavelength," *Appl. Optics*, **15**, 2105–2111.

Keyes, R. J., and R. H. Kingston (1972), "A Look at Photon Detectors," *Phys. Today*, Mar., 48–54.

Kildal, H., and R. L. Byer (1971), "Comparison of Laser Methods for the Remote Detection of Atmospheric Pollutants," *Proc. IEEE*, **59**, 1644–1663.

Killinger, D. K., and N. Menyuk (1981), "Remote Probing of the Atmosphere Using a CO_2 DIAL System," *IEEE J. Quant. Electr.*, **QE-17**, 1917–1929; "Effects of Turbulence-Induced Correlation on Laser Remote Sensing Errors," *Appl. Phys. Lett.*, **38**, 968–970.

Killinger, D. K., N. Menyuk, and W. E. DeFeo (1980), "Remote Sensing of CO Using Frequency-Doubled CO_2 Laser Radiation," *Appl. Phys. Lett.*, **36**, 402–405.

Kim, H. H. (1973), "New Algae Mapping Technique by the Use of an Airborne Laser Fluorosensor," *Appl. Optics*, **12**, 1454–1459.

Kim, H. H. (1977), "Airborne Bathymetric Charting Using Pulsed Blue–Green Lasers," *Appl. Optics*, **16**, 46–56.

Kim, H. H., and G. D. Hickman (1973), "An Airborne Laser Fluorosensor for the Detection of Oil on Water," *NASA Conference on the Use of Lasers for Hydrographic Studies, Wallops Island, Sept. 1973*, SP-375, 197–202.

Kim, H. H., P. O. Cervenka, and C. B. Lankford (1975), *Development of an Airborne Laser Bathymeter*, NASA TN D-8079, Oct.

Klainer, S. M., T. Hirschfeld, and E. R. Schildkraut (1970), "The Detection of Toxic Contaminants in the Atmosphere Using Single Ended Remote Raman Spectrometric Techniques," The Central States Section on the Combustion Institute, Houston, Texas, 7–8 April.

Klein, M. V. (1970), *Optics*, Wiley, New York.

Klett, J. D. (1981), "Stable Analytical Inversion for Processing Lidar Returns," *Appl. Optics*, **20**, 211–220.

Kobayasi, T., and H. Inaba (1969), "Laser Raman Radar for Chemical Analysis of Polluted Air," *Nature*, **224**, 170–172.

Kobayasi, T., H. Shimizu, and H. Inaba (1974), "Laser Radar Techniques for Remote Measurement of Atmospheric Temperature," *1974 International Laser Radar Conf., Sept. 3–6, Sendai, Japan,* 49.

Koller, L. R. (1969), *Ultraviolet Radiation*, 2nd ed., Wiley, New York.

Kressel, H. (1972), "Semiconductor Lasers," *Laser Handbook* (Ed. F. T. Arecchi and E. O. Schultz-Dubois), North Holland, Amsterdam.

Kruse, P. W., L. D. McGlauchlin, and R. B. McQuistan (1963), *Elements of Infrared Technology*, Wiley, New York.

Ku, R. T., E. D. Hinkley, and J. O. Sample (1975), "Long-Path Monitoring of Atmospheric Carbon Monoxide with a Tunable Diode Laser System," *Appl. Optics*, **14**, 854–861.

Kung, R. T. V., and I. Itzkan (1976), "Absolute Oil Fluorescence Conversion Efficiency," *Appl. Optics*, **15**, 409–415.

LaRocca, A. J. (1975), Laser Focus, 41, Jan. 1976. "Methods of Calculating Atmospheric Transmittance and Radiance in the Infrared," *Proc. IEEE*, **63**, 75–94.

Lefrere, J., J. Pelon, C. Cahen, A. Hauchecorne, and P. Flamant (1981), "Lidar Survey of the Post-Mt. St. Helens Statospheric Aerosol at Haute Provence Observatory," *Appl. Optics*, **20**, A70, 1117.

Leonard, D. A. (1965), "Saturation of the Molecular Nitrogen Second Positive Laser Transition," *Appl. Phys. Lett.*, **7**, 4–6.

Leonard, D. A. (1967), "Observation of Raman Scattering from the Atmosphere Using a Pulsed Nitrogen Ultraviolet Laser," *Nature*, **216**, 142–143.

Leonard, D. A. (1974), "Measurement of Aircraft Turbine Engine Exhaust Emissions," *Laser Raman Gas Diagnostics*, Plenum Press, New York, 45–61.

Leonard, D. A., and B. Caputo (1974), "Single-Ended Atmospheric Transmissometer," *Optical Engineering*, **13**, 10–14.

Leonard, D. A., B. Caputo, and F. E. Hoge (1979), "Remote Sensing of Subsurface Water Temperature by Raman Scattering," *Appl. Optics*, **18**, 1732–1745.

Leskovar, B. (1977), "Microchannel Plates," *Physics Today*, **30**, 42–49.

Levis, C. A., W. G. Swarner, C. Prettyman, and G. W. Reinhart (1973), "An Optical Radar for Airborne Use over Natural Waters," *NASA Conference on the Use of Lasers for Hydrographic Studies*, NASA SP-374, 67–80.

Liboff, R. L. (1980), *Introductory Quantum Mechanics*, Holden-Day, San Francisco.

Ligda, M. G. H. (1963), *Proc. Conf. Laser Technol.*, 1st, San Diego, Calif., 63–72.

Likens, G. E. (1975), "Acid Precipitation," *Conference on Emerging Environmental Problems, Rensselaerville, New York*, EPA Report 902/9-75-001.

Lintz, J., Jr., and D. S. Simonett (1976), *Remote Sensing of the Environment*, Addison-Wesley.

Liou, K. N., and R. M. Schotland (1971), "Multiple Backscattering and Depolarization from Water Clouds for a Pulsed Lidar System," *J. Atmos. Sci.*

Lochte-Holtgreven, W. (1968), *Plasma Diagnostics*, North-Holland, Amsterdam.

Lopez, R. J., and M. A. Rebolledo (1981), "Fatigue in Photomultipliers due to Excitation by Pulsed Light Sources," *Rev. Sci. Instrum.* **52**, 1852–1854.

Loree, T. R., R. C. Sze, D. L. Barker, and P. B. Scott (1979), "New Lines in the UV: SRS of Excimer Laser Wavelengths," *IEEE J. Quant. Elec.* **QE-15**, 337–342.

Loudon, R. (1973), *The Quantum Theory of Light*, Clarendon Press, Oxford, England.

Lussier, F. M. (1976a), "Choosing an Infrared Detector", Oct. Issue of Laser Focus, 66-71.

Lussier, F. M. (1976b), "Guide to IR-Transmissive Materials", Dec. Issue of Laser Focus, 47-50.

Maiman, T. H. (1960), "Stimulated Optical Radiation in Ruby," *Nature*, **187**, 493–494.

Marling, J. B., J. G. Hawley, E. M. Liston, and W. B. Grant (1974), "Lasing Characteristics of Seventeen Visible Wavelength Dyes Using a Coaxial Flashlamp Pumped Laser," *Appl. Optics*, **13**, 2317–2320.

Mason, J. B. (1975), "Lidar Measurements of Temperature: A New Approach," *Appl. Optics*, **14**, 76–78.

Maugh, T. M., II (1980), "Ozone Depletion Would Have Dire Effects," *Science*," **207**, 394–395.

Mavrodineanu, R., and H. Boiteux (1965), *Flame Spectroscopy*, Wiley, New York.

McClatchey, R. A., Fenn, J. E. A. Selby, F. E. Volz, and J. S. Garing (1971), *Optical Properties of the Atmosphere*, Air Force Cambridge Research Laboratories 71-0279; also 72-0497 (1972).

McClatchey, R. A., W. S. Benedict, S. A. Clough, D. E. Burch, R. F. Calfee, K. Fox, L. S. Rothman, and J. S. Garing (1973), *AFCRL Atmospheric Absorption Line Parameters Compilation*, AFCRL-TR-73-0096.

McClung, F. J. and R. W. Hellwarth (1962), "Giant Optical Pulsations from Ruby," *J. Appl. Phys.*, **33**, 828–829.

McCormick, M. P. (1971), "Simultaneous Multiple Wavelength Laser Radar Measurements of the Lower Atmosphere," Electro-Optics International Conference, Brighton, England, 24–26 Mar.

McCormick, M. P. (1982), "Lidar Measurements of Mount St. Helens Effluents," *Opt. Eng.* **21**, 340–342.

McCormick, M. P., and W. H. Fuller, Jr. (1971), "Lidar Applications to Pollution Studies," Joint Conference on Sensing of Environmental Pollutants, Palo Alto, California, 8–10 Nov.

McCormick, M. P., and W. H. Fuller, Jr. (1975), "Lidar Measurements of Two Intense Stratospheric Dust Layers," *Appl. Optics* **14**, 4–5.

McCormick, M. P., S. K. Poultney, V. van Wijk, C. O. Alley, R. T. Bettinger, and J. A. Perschy (1966), "Backscattering from the Upper Atmosphere (75–160 km) Detected by Optical Radar," *Nature*, **209**, 798–799.

McCormick, M. P., J. O. Lawrence, Jr., and F. R. Crownfield, Jr. (1968), "Mie Total and Differential Backscattering Cross Sections at Laser Wavelengths for Jungle Aerosol Modes," *Appl. Optics*, 7, 2424–2425.

McCormick, M. P., T. J. Swissler, W. P. Chu, and W. H. Fuller, Jr. (1978), "Post-volcanic Stratospheric Aerosol Decay as Measured by Lidar," *J. Atmospheric Sciences*, **35**, 1296–1303.

McGee, T. J., and T. J. McIlrath (1979), "Stratospheric Temperature and Pressure Determinations from OH Fluorescence Lidar Instrument," *Appl. Optics*, **18**, 1710–1714.

McIlrath, T. J. (1980), "Fluorescence Lidar," *Optical Engineering*, **19**, 494–502.

McNeil, W. R., and A. I. Carswell (1975), "Lidar Polarization Studies of the Troposphere," *Appl. Optics*, **14**, 2158–2168.

Measures, R. M. (1968), "Selective Excitation Spectroscopy and Some Possible Applications," *J. Appl. Phys.*, **39**, 5232–5245.

Measures, R. M. (1970), "Electron Density and Temperature Elevation of a Potassium Seeded Plasma by Laser Resonance Pumping," *J. Quant. Spectrosc. Radiant Transfer*, **10**, 107–125.

Measures, R. M. (1971), "A Comparative Study of Laser Methods of Air Pollution Mapping," *Can. Aeronaut. Space J.*, **17**, 417–418.

Measures, R. M. (1977), "Lidar Equation Analysis—Allowing for Target Lifetime, Laser Pulse Duration, and Detector Integration Period," *Appl. Optics*, 16, 1092–1103.

Measures, R. M., and M. Bristow (1971), "The Development of a Laser Fluorosensor for Remote Environmental Probing," Joint Conference on Sensing of Environmental Pollutants, Palo Alto, Nov. 1971, AIAAA Paper 71-112; *Can. Aeron. Space J.*, **17**, 421–422.

Measures, R. M., and H. Herchen (1983), "Laser Absorption under Saturating Conditions with Allowance for Spectral Hole Burning," *J. Quant. Spectrosc. Radiat. Transfer*, **29**, 9–18.

Measures, R. M., and G. Pilon (1972), "A Study of Tunable Laser Techniques for Remote Mapping of Specific Gaseous Constituents of the Atmosphere," *Opto-electronics*, **4**, 141–153.

Measures, R. M., W. R. Houston, and M. Bristow (1973), "Development and Field Tests of a Laser Fluorosensor for Environmental Monitoring," *Can. Aeron. Space J.*, **19**, 501–506.

Measures, R. M., H. R. Houston, and D. G. Stephenson (1974), "Laser Induced Fluorescence Decay Spectra-A New Formof Environmental Signature," *Optical Engineering*, **13**, 494–450.

Measures, R. M., J. Garlick, W. R. Houston, and D. G. Stephenson (1975), "Laser Induced Spectral Signatures of Relevance to Environmental Sensing," *Can. J. Remote Sensing*, **1**, 95–102.

Mecherikunnel, A., and C. Duncan (1982), "Total and Spectral Solar Irradiance Measured at Ground Surface," *Appl. Optics*, **21**, 554–556.

Megie, G., and J. E. Blamont (1977), "Laser Sounding of Atmospheric Sodium—Interpretation in Terms of Global Atmospheric Parameters," *Planet. Space Sci.*, **25**, 1093–1109.

Megie, G., and R. T. Menzies (1980), "Complementarity of UV and IR Differential Absorption Lidar for Global Measurements of Atmospheric Species," *Appl. Optics*, **19**, 1173–1183.

Megie, G., F. Bos, J. E. Blamont, and M. L. Chanin (1978a), "Simultaneous Nighttime Lidar Measurements of Atmospheric Sodium and Potassium," *Planet. Space Sci.*, **26**, 27–35.

Megie, G., F. Bos, J. E. Blamont, and M. L. Chanin (1978b), "Simultaneous Nighttime Lidar Measurements of Atmospheric Sodium and Potassium," *Planet. Space Sci.*, **26**, 27–35.

Melchior, H. (1972), "Demodulation and Photodetection Techniques," *Laser Handbook*, (F. T. Arecchi and E. O. Schultz-Dubois, Eds.), North-Holland, Amsterdam, **1**, C7.

Melfi, S. H. (1972), "Remote Measurements of the Atmosphere Using Raman Scattering," *Appl. Optics*, **11**, 1605–1610.

Melfi, S. H., J. D. Lawrence Jr., and M. P. McCormick (1969), "Observation of Raman Scattering by Water Vapor in the Atmosphere," *Appl. Phys. Lett.*, **15**, 295–297.

Melfi, S. H., M. L. Brumfield, and R. M. Storey, Jr. (1973), "Observation of Raman Scattering by SO_2, in a Generating Plant Stack Plume," *Appl. Phys. Lett.*, **22**, 402–403.

Menyuk, N., and D. K. Killinger (1981), "Temporal Correlation Measurements of Pulsed Dual CO_2 Lidar Returns," *Optics Letters*, **6**, 301–303.

Menyuk, N., D. K. Killinger, and W. E. DeFeo (1980), "Remote Sensing of NO Using a Differential Absorption Lidar," *Appl. Optics*, **19**, 3282–3286.

Menyuk, N., D. K. Killinger, and W. E. DeFeo (1982), "Laser Remote Sensing of Hydrazine, MMH and UDMH Using a Differential-Absorption CO_2 Lidar," *Appl. Optics*, **21**, 2275–2286.

Menzies, R. T. (1972), "Remote Sensing With Infrared Heterodyne Radiometers," *Opto-electronics*, **4**, 178–186.

Menzies, R. T. (1976), "Laser Heterodyne Detection Techniques," *Laser Monitoring of the Atmosphere* (Ed. E. O. Hinkley), Springer-Verlag.

Menzies, R. T., and M. S. Shumate (1976), "Remote Measurements of Ambient Air Pollutants with a Bistatic Laser System," *Appl. Optics*, **15**, 2080–2084.

Menzies, R. T., and M. S. Shumate (1978), "Tropospheric Ozone Distributions Measured with an Airborne Laser Absorption Spectrometer," *J. Geophy. Res.*, **83**, 4039.

Middleton, A. (1976), "Sulphur Dioxide Pollution Over Europe," *New Scientist*, 4 Nov., 279.

Mie, G. (1908), *Ann. Physik*, **25**, 377.

Molina, M. J., and F. S. Rowland (1974), "Stratospheric Sink for Chlorofluoromethanes: Chlorine Atom Catalyzed Destruction of Ozone," *Nature*, **249**, 810–812.

Morhange, J. F., and C. Hirlimann (1976), Luminescence Rejection in Raman Spectroscopy," *Appl. Optics*, **15**, 2969–2970.

Morrow, T., and T. W. Price (1974), "A Simple Reliable Coaxial Dye Laser System," *Optics Comm.* **10**, 133–136.

Mumola, P. B., O. Jarrett, Jr., and C. A. Brown, Jr. (1973), "Multiwavelength Lidar for Remote Sensing of Chlorophyll *a* in Algae and Phytoplankton," *NASA Conference on the Use of Lasers for Hydrographic Studies*, NASA SP-375, 137–145.

Murphy, W. F., W. Holzer, and H. J. Bernstein (1969), "Gas Phase Raman Intensities: A Review of 'Pre-laser' data," *Appl. Spectrosc.*, **23**, 211–218.

Murray, E. R. (1977), "Remote Measurement of Gases Using Discretely Tunable Infrared Lasers," *Optical Engineering*, **16**, 284–290.

Murray, E. R., and R. L. Byer (1980), *Remote Measurements of Air Pollutants*, SRI International report, Jan.

Murray, E. R., and J. E. van der Laan (1978), "Remote Measurement of Ethylene Using a CO_2 Differential-Absorption Lidar," *Appl. Optics*, **17**, 814–817.

Murray, E. R., R. D. Hake, Jr., J. E. Van der Laan, and J. G. Hawley (1976), "Atmospheric Water Vapor Measurements with a 10 Micrometer DIAL System," *Appl. Phys. Lett.*, **28**, 542–543.

Murray, E. R., M. F. Williams, and J. E. van der Laan (1978), "Single-Ended Measurement of Infrared Extinction Using Lidar," *Appl. Optics*, **17**, 296–299.

Nakahara, S., K. Ito, S. Ito, A. Fuke, S. Komatsu, H. Inaba, and T. Kobayasi (1972), "Detection of Sulphur Dioxide in Stack Plume by Laser Raman Radar," *Opto-electronics*, **4**, 169–177.

Nicholls, R. W. (1971), "Recent Researches in Shock-Tube Spectroscopy," *Modern Optical Methods in Gas Dynamic Research*, (D. S. Dosanjh, Ed.), Plenum Press, New York.

Nill, K. W. (1974), "Tunable Infrared Lasers," *Opt. Eng.*, **13**, 516–522.

Nilsson, B. (1979), Meterological Influence on Aerosol Extinction in the 0.2–40 Micrometer Wavelength Range," *Appl. Optics*, **18**, 3457–3473.

Northam, G. B., J. M. Rosen, S. H. Melfi, T. J. Pepin, M. P. McCormick, D. J. Hofman, and W. H. Fuller, Jr. (1974), "Dustsonde and Lidar Measurements of Stratospheric Aerosols: A Comparison," *Appl. Optics*, **13**, 2416–2421.

Northam, D. B., M. A. Guerra, M. E. Mack, I. Itzkan, and C. Deradourian (1981), "High Repetition Rate Frequency-Doubled Nd:YAG Laser for Airborne Bathymetry," *Appl. Optics*, **20**, 968–971.

Northend, C. A., R. C. Honey, and W. E. Evans (1966), "Laser Radar (Lidar) for Meteorological Observations," *Rev. Sci. Inst.*, **37**, 393–400.

Oettinger, P. E., and C. F. Dewey, Jr. (1976), "Lasing Efficiency and Photochemical Stability of IR Laser Dyes in the 710–1080 nm Spectral Region," *IEEE J. Quant. Electr.*, **QE-12**, 95–101.

Oliver, B. M. (1965), "Thermal and Quantum Noise," *Proc. IEEE*, **53**, 436–454.

O'Neil, R. A., A. R. Davis, H. G. Gross, and J. Kruus (1973), "A Remote Sensing Laser Fluorometer," *NASA Conference on the Use of Lasers for Hydrographic Studies, Wallops Island, Sept. 1973*, SP-375, 173–195.

O'Neil, R. A., L. Buje-Bijunas, and D. M. Rayner (1980), "Field Performance of a Laser Fluorosensor for the Detection of Oil Spills," *Appl. Optics*, **19**, 863–870.

O'Neil, R. A., F. E. Hoge, and M. P. F. Bristow (1981), "The Current Status of Airborne Laser Fluorosensing," *15th International Symposium on Remote Sensing of Environment, Ann Arbor, Michigan, May*, 379–398.

Oppenheim, U. P., and R. T. Menzies (1982), "Aligning the Transmitter and Receiver Telescopes of an Infrared Lidar: A Novel Method," *Appl. Optics*, **21**, 174–175.

O'Shea, C., and L. G. Dodge (1974), "NO_2 Concentration Measurements in an Urban Atmosphere Using Differential Absorption Technique," *Appl. Optics*, **13**, 1481–1486.

Pal, S. R., and A. I. Carswell (1973), "Polarization Properties of Lidar Backscattering from Clouds," *Appl. Optics*, **12**, 1530–1535.

Pal, S. R., and A. I. Carswell (1976), "Multiple Scattering in Atmospheric Clouds: Lidar Observations," *Appl. Optics*, **15**, 1990–1995.

Pal, S. R., and A. I. Carswell (1978), "Polarization Properties of Lidar Scattering from Clouds at 347 nm and 694 nm," *Appl. Optics*, **17**, 2321–2328.

Parker, C. A. (1968), *Photoluminescence of Solutions*, Elsevier, New York.

Penndorf, R. (1957), "Tables of the Refractive Index for Standard Air and the Rayleigh Scattering Coefficients for the Spectral Region between 0.2 and 20.0 Micrometers and Their Application to Atmospheric Optics," *J. Opt. Soc. Amer.*, **47**, 176–182.

Penney, C. M. "Light Scattering in Terms of Oscillator Strengths and Refractive Indices," *J. Opt. Soc. Amer.*, **59**, 34–42.

Penney, C. M. (1974), "Light Scattering and Fluorescence in the Approach to Resonance—Stronger Probing Techniques," *Laser Raman Gas Diagnostics* (M. Lapp and C. M. Penney, Eds.), Plenum Press, New York, 191–217.

Penney, C. M., R. L. St. Peters, and M. Lapp (1974), "Absolute Relational Raman Cross Sections for N_2, O_2, and CO_2," *J. Opt. Soc. Am.*, **64**, 712–716.

Petri, K., A. Salik, and J. Cooney (1982), "Variable-Wavelength Solar-Blind Raman Lidar for Remote Measurement of Atmospheric Water-Vapour Concentration and Temperature," *Appl. Optics*, **21**, 1212–1218.

Pipes, L. A. (1958), *Applied Mathematics for Engineers and Physicists*, 2nd ed., McGraw-Hill.

Pitts, J. N., Jr. (1977), "Keys to Photochemical Smog Control," *Environmental Sciences & Technology*, **11**, 456–461.

Placzek, G. (1934), "Rayleigh-streuung und Raman-effekt," *Handbuch der Radiolgie*, Akademische Verlag, Leipzig, Vol. VI, Pt. 2, 205–374 (UCRL-Trans-526L).

Plass, G. N., G. W. Kattawar, and J. A. Guinn, Jr. (1976), "Radiance Distribution over a Ruffled Sea: Contributions from Glitter, Sky, and Ocean," *Appl. Optics*, **15**, 3161–3165.

Pomraning, G. C. (1973), *The Equations of Radiation Hydrodynamics*, Pergamon Press, New York.

Post, M. J., F. F. Hall, R. A. Richter, and T. R. Lawrence (1982), "Aerosol Backscattering Profiles at $\lambda = 10.6$ Microns,"*Appl. Optics*, **21**, 2442–2446.

Poultney, S. K. (1972a), "Laser Radar Studies of Upper Atmosphere Dust Layers and the Relation to Temporary Increases in Dust to Cometary Micrometeoroid Streams," *Space Research*, **12**, 403–421.

Poultney, S. K. (1972b), "Single Photon Detection and Timing: Experiments and Techniques," *Advances in Electronics and Electron Physics*, **31**, 39–117.

Poultney, S. K., M. L. Brumfield, and J. H. Siviter, Jr. (1977), "Quantitative Remote Measurements of Pollutants from Stationary Sources Using Raman Lidar," *Appl. Optics*, **16**, 3180–3182.

Pourny, J. C., D. Renaut, and A. Orszag (1979), "Raman-Lidar Humidity Sounding of the Atmospheric Boundary-Layer," *Appl. Optics*, **18**, 1141–1148.

Pratt, W. K. (1969), *Laser Communication Systems*, Wiley, New York.

Pressley, R. J. (1971), *Handbook of Lasers*, Chemical Rubber Co., Cleveland.

Prettyman, C. E., and M. D. Cermak (1969), "Time Variation of the Rough Ocean Surface and its Effect on an Incident Laser Beam," *IEEE Trans. Geoscience Electronics*, **GE-7**, 235–243.

Ramsay, J. V. (1962), "A Rapid-Scanning Fabry-Perot Interferometer with Automatic Parallelism Control," *Appl. Optics*, **1**, 411–413.

Rayner, D. M., and A. G. Szabo (1978), "Time-Resolved Laser Fluorosensors: A Laboratory Study of Their Potential in the Remote Characterization of Oil," *Appl. Optics*, **17**, 1624–1630.

Rayner, D. M., M. Lee, and A. G. Szabo (1978), "Effect of Sea-State on the Performance of Laser Fluorosensors," *Appl. Optics*, **17**, 2730–2733.

Reid, J., J. Schewchun, B. K. Garside, and E. A. Ballik (1978), "High Sensitivity Pollution Detection Employing Tunable Diode Lasers," *Appl. Optics*, **17**, 300–307.

Remsberg, E. E., and L. L. Gordley (1978), "Analysis of Differential Absorption Lidar from the Space Shuttle," *Appl. Optics*, **17**, 624–630.

Renaut, D., J. C. Pourny, and R. Capitini (1980), "Daytime Raman-Lidar Measurements of Water Vapor," *Optics Letters*, **5**, 233–235.

Res, M. A., C. J. Koke, J. Bednarik, and K. Kroger (1977), "Bandpass Filters for Use in the Visible Region," *Appl. Optics*, **16**, 1908–1913.

Revelle, R. (1982), "Carbon Dioxide and World Climate," *Scientific American*, **247**, 35–43.

Roberts, R. E., J. E. A. Selby, and L. M. Biberman (1976), Infrared Continuum Absorption by Atmospheric Water Vapor in the 8–12 Micrometer Window," *Appl. Optics*, **15**, 2085–2090.

Rosen, J. M. (1966), "Correlation of Dust and Ozone in the Stratosphere," *Nature* **209**, 1343.

Rosen, H., P. Robish, and O. Chamberlain (1975), "Remote Detection of Pollutants Using Resonance Raman Scattering," *Appl. Optics*, **14**, 2703–2706.

Röss, D. (1969), *Lasers, Light Amplifiers and Oscillators*, Academic Press, New York.

Ross, M. (1966), *Laser Receivers (Devices, Techniques, Systems)*, Wiley, New York.

Rothe, K. W., U. Brinkman, and H. Walter (1974a), "Applications of Tunable Dye Lasers to Air Pollution Detection; Measurements of Atmospheric NO_2 Concentration by Differential Absorption," *Appl. Phys.*, **3**, 115–119.

Rothe, K. W., U. Brinkman, and H. Walter (1974b), "Remote Sensing of NO_2 Emission from a Chemical Factory by the Differential Absorption Technique," *Appl. Phys.*, **4**, 181–182.

Rothman, L. S. (1981), "AFGL Atmospheric Absorption Line Parameters Compilation: 1980 Version," *Appl. Optics*, **20**, 791–795.

Rudder, R. D., and D. R. Bach (1968), "Rayleigh Scattering of Ruby-Laser Light by Neutral Gases," *J. Opt. Soc. Am.*, **58**, 1260–1266.

Russell, P. B., and R. D. Hake, Jr. (1977), "The Post-Fuego Stratospheric Aerosol: Lidar Measurements, with Radiative and Thermal Implications," *J. Atmos. Sci.*, **345**, 163–177.

Russell, P. B., and B. M. Morley (1982), "Orbiting Lidar Simulations. 2: Density, Temperature, Aerosol and Cloud Measurements by a Wavelength-Combining Technique," *Appl. Optics*, **21**, 1554–1563.

Russell, P. B., T. J. Swissler, and M. P. McCormick (1979), "Methodology for Error Analysis and Simulation of Lidar Aerosol Measurements," *Appl. Optics*, **18**, 3783–3797.

St. Peters, R. L., and S. D. Silverstein (1973), "Manifestations of Pressure Broadening on Tuned Resonance Raman Fluorescence," *Opt. Comm.*, **7**, 193–196.

St. Peters, R. L., S. D. Silverstein, M. Lapp, and C. M. Penney (1973), "Resonant Raman Scattering of Resonance Fluorescence in I_2 Vapor," *Phys. Rev. Lett.*, **30**, 191–192.

Salzman, J. A. (1974), "Low Temperature Measurements by Rotational Raman Scattering," *Laser Raman Gas Diagnostics* (Ed. M. Lapp and C. M. Penney), Plenum Press, New York, 179–188.

Sandercock, J. R. (1970), "Brillouin Scattering Study of SbSI using a Double-Pass Stabilized Scanning Interferometer," *Opt. Comm.*, **2**, 73–76.

Sandford, M. C. W. (1967), "Laser Scatter Measurements in the Mesosphere and Above," *J. Atm. and Terres. Phys.*, **29**, 1657–1662.

Sato, T., Y. Suzuki, H. Kashiwagi, M. Nanjo, and Y. Kakui (1978), "Laser Radar for Remote Detection of Oil Spills," *Appl. Optics*, **17**, 3798–3803.

Schäfer, F. P. (1973), *Dye Lasers*, Topics in Applied Physics, Springer-Verlag, New York.

Schawlow, A. L., and C. H. Townes (1958), "Infrared and Optical Masers," *Phys. Rev.*, **112**, 1940–1949.

Schnell, W., and G. Fischer (1975), "Carbon Dioxide Laser Absorption Coefficients of Various Air Pollutants," *Appl. Optics*, **14**, 2058–2059.

Schofield, K. (1977), "Atomic and Molecular Fluorescence as a Stratospheric Species Monitor," *J. Quant. Spectosc. Radiat. Transfer*, **17**, 13–51.

Schotland, R. M. (1966), "Some Observation of the Vertical Profile of Water Vapor by a Laser Optical Radar," *Proc. 4th Symposium on Remote Sensing of the Environment 12–14 April 1966*, Univ. of Michigan, Ann Arbor, 273–283.

Schotland, R. M. (1974), "Errors in Lidar Measurements of Atmospheric Gases by Differential Absorption," *J. Appl. Meteorology*, **13**, 71–77.

Schottky, W. (1918), *Ann. Phys. Leipzig*, **57**, 541.

Schriever, B. A. (1960), *Handbook of Geophysics*, Macmillan, New York.

Schuster, B. G., and T. G. Kyle (1980), "Pollution Plume Transport and Diffusion Studies Using Fluorescence Lidar," *Appl. Optics*, **19**, 2524–2528.

Schwemmer, B. K., and T. O. Wilkerson (1973), "Lidar Temperature Profiling: Performance Simulations of Mason's Method," *Appl. Optics*, **18**, 3539–3541.

Schwiesow, R. L., and N. L. Abshire (1973), "Relative Raman Cross Section of O_3 for Ar Ion Laser Frequencies," *J. Appl. Phys.*, **44**, 3808–3809.

Schwiesow, R. L., and V. E. Derr (1970), "A Raman Scattering Method for Precise Measurements of Atmospheric Oxygen Balance," *J. Geophys. Res.*, **75**, 1629–1632.

Seals, R. K., Jr., and C. H. Bair (1973), *Analysis of Laser Differential Absorption Remote Sensing Using Diffuse Reflection from the Earth*, ISA, JSP 6675.

Selby, J. E. A., and R. A. McClatchey (1975), *Atmospheric Transmittance from 0.25 Microns to 28.5 Microns: Computer Code III*, AFCRL-TR-0255.

Sepucha, R. C. (1977), "N_2 Rayleigh Scattering at 10.6 Micrometers," *J. Opt. Soc. Am.*, **67**, 108–114.

Shapiro, S. L., Ed. (1977), *Ultrashort Light Impulses*, Springer-Verlag, New York.

Shardanand and A. D. Prasad Rao (1977), *Absolute Rayleigh Scattering Cross Sections of Gases and Freons of Stratospheric Interest in the Visible and Ultraviolet Regions*, NASA TN O-8442.

Shen, Y. R. (1976), "Recent Advances in Nonlinear Optics," *Rev. Mod. Phys.*, **48**, 1–32.

Shewchun, J., B. K. Garside, E. A. Ballik, C. C. Y. Kwan, M. M. Elsherbiny, G. Hogenkamp, and A. Kazandjian (1976), "Pollution Monitoring Systems Based on Resonance Absorption Measurements of Ozone with a 'Tunable' CO_2 Laser; Some Criteria," *Appl. Optics*, **15**, 340–346.

Shimizu, H., Y. Sasano, N. Takeuchi, O. Matsudo, and M. Okuda (1980), "A Mobile Computerized Radar System for Observing Rapidly Varying Meteorological Phenomena," *Optical and Quantum Electronics*, **12**, 159–187.

Shipley, S. T., E. W. Eloranta, and J. A. Weinman (1974), "Measurement of Rainfall Rates by Lidar," *J. Appl. Meteorology*, **13**, 800–807.

Shumate, M. S., S. Lundqvist, V. Persson, and S. T. Eng (1982), "Differential Reflectance of Natural and Man Made Materials at CO_2 Laser Wavelengths," *Appl. Optics*, **21**, 2386–2389.

Siegman, A. E. (1971), *An Introduction to Lasers and Masers*, McGraw-Hill.

Singer, F. (1968), "Measurements of Atmospheric Surface Pressure with a Satellite-Borne Laser," *Appl. Optics* **7**, 1125–1127.

Slusher, R. B., and V. E. Derr (1975), "Temperature Dependence and Cross Section of Some Stokes and Anti-Stokes Raman Lines in Ice," *Appl. Optics*, **14**, 2116–2120.

Smiley, V. N., and B. M. Morley (1981), "Lidar Depolarization Studies of the Atmosphere at the South Pole," *Appl. Optics*, **20**, 2189–2195.

Smith, R. A., F. E. Jones, and R. P. Chasmar (1968), *The Detection and Measurements of Infrared Radiation*, Oxford London. Univ. Press.

Smith, W. M. H. (1972), "A New Method for the Detection of Raman Scattering from Atmospheric Pollutants," *Opto-electronics*, **4**, 161–167.

Sorenzen, C. J. (1966), "A Method for the Continuous Measurement of '*In Vivo*' Chlorophyll Concentration," *Deep Sea Research*, **13**, 223–227.

Sorenzen, C. J. (1970), "The Biological Significance of Surface Chlorophyll Measurements," *Limnology and Oceanography*, **15**, 479–480.

Stedman, D. H. (1976), "Measurement Techniques for the Ozone Layer," *Research Development*, Jan., 22–26.

Steinfeld, J. I. (1974), *Molecules and Radiation, An Introduction to Modern Molecular Spectroscopy*, Harper and Row, New York.

Stephenson, D. A. (1974), "Raman Cross Sections of Selected Hydrocarbons and Freons," *J. Quant. Spectrosc. Radiat. Transfer*, **14**, 1291–1301.

Stern, A. C. (1968), *Air Pollution*, Academic Press, New York.

Stewart, R. W., and J. L. Bufton (1980), "Development of a Pulsed 9.5 Micrometer Lidar for Regional Scale O_3 Measurement," *Optical Engineering*, **19**, 503–507.

Strauch, R. G., V. E. Derr, and R. E. Cupp (1971), "Atmospheric Temperature Measurements Using Raman Backscatter," *Appl. Optics*, **10**, 2665–2669.

Strauch, R. G., V. E. Derr, and R. E. Cupp (1972), "Atmospheric Water Vapor Measurements by Raman Lidar," *Remote Sensing of Envir.*, **2**, 101–108.

Sylvia, A. E., D. A. Bancroft, and J. D. Miller (1974), "Detection and Measurements of Microorganics in Drinking Water by Fluorescence," *Am. Water Works Assoc. Tech. Conf. Proc. Dec. 2-3, 1974, Dallas, Texas.*

Sze, R. C. (1979), "Rare Gas Halide Avelanche Discharge Lasers," *IEEE J. Quant. Electr.* **QE-15**, 1338–1347.

Takeuchi, N., H. Shimizu, and M. Okuda (1978), "Detectivity Estimation of the DAS Lidar for NO_2," *Appl. Optics*, **17**, 2734–2738.

Tam, W. G., and A. Zardecki (1980), "Off-axis Propagation of a Laser Beam in Low Visibility Weather Conditions," *Appl. Optics*, **19**, 2822–2827.

Tam, W. G., and A. Zardecki (1982), "Multiple Scattering Corrections to the Beer-Lambert Law I. Open Detector," *Appl. Optics*, **21**, 2405–2412.

Thompson, R. T., Jr., J. M. Hoell, Jr., and W. R. Wade (1975), "Measurements of SO_2 Absorption Coefficients Using a Tunable Dye Laser," *J. Appl. Phys.*, **46**, 3040–3043.

Thrush, B. A. (1977), "The Chemistry of the Stratosphere and its Pollution," *Endeavor, New Series*, **1**, 3–6.

Thurston, A. D., Jr. (1970), "A Fluormetric Method for the Determination of Lignin Sulfonates in Natural Waters," *J. WPCF*, **42**, 1551–1555.

Timothy, J. G. (1981), "Curved-Channel Microchannel Array Plates," *Rev. Sci. Instrum.*, **52**, 1131–1142.

Timothy, J. G., and R. L. Bybee (1975), "Two Dimensional Photon-Counting Detector Arrays Based on Microchannel Array Plates," *Rev. Sci. Instrum.*, **46**, 1615–1623.

Title, A. M., and W. J. Rosenberg (1981), "Tunable Birefringent Filters," *Optical Eng.*, **20**, 815–823.

Tittel, F. K., M. Smayling, W. L. Wilson, and G. Markowsky (1980), "Blue Laser Action by The Rare Gas Halide Trimer Kr_2F," *Appl. Physics. Lett.*, **37**, 862–864.

Topp, J. A., H. W. Schrotter, H. Hacker, and J. Brandmuller (1969), "Improvement of the Signal-to-Noise Ratio of Photomultipliers for Very Weak Signals," *Rev. Sci. Instrum.*, **40**, 1164–1169.

Tucker, A. W., M. Bimbaum, and C. L. Fincher (1975), "Atmospheric NO Determination by 442-nm Laser Induced Fluorescence," *Appl. Optics*, **14**, 1418–1422.

Twomey, S., and H. B. Howell (1965), "The Relative Merit of White and Monochromatic Light for the Determination of Visibility by Backscattering Measurements," *Appl. Optics*, **4**, 501–505.

Tyler, J. E., and R. W. Preisendorfer (1962), *The Sea* (M. N. Hill, Ed.), Wiley-Interscience, New York.

Uchino, O., M. Maeda, J. Kohno, T. Shibata, C. Nagasawa, and M. Hirono (1978), "Observation of Stratospheric Ozone Layer by a XeCl Laser Radar," *Appl. Phys. Lett.*, **33**, 807–809.

Uchino, O., M. Maeda, and M. Hirono (1979), "Applications of Excimer Lasers to Laser-Radar Observations of the Upper Atmosphere," *IEEE J. Quantum Electronics*, **QE-15**, 1094–1107.

Uchino, O., M. Maeda, T. Shibata, M. Hirono, and M. Fujimara (1980), "Measurements of Stratospheric Vertical Ozone Distribution with Xe–Cl Lidar; Estimated Influence of Aerosols," *Appl. Optics*, **19**, 4175–4181.

Uthe, E. E., and R. J. Allen (1975), "A Digital Real-Time Lidar Data Recording Processing and Display System," *Optical and Quantum Electronics*, **7**, 121–129.

Valley, S. L., Ed. (1965), *Handbook of Geophysics and Space Environments*, McGraw-Hill, New York.

Van de Hulst, H. C. (1957), *Light Scattering by Small Particles*, Wiley, New York.

Varanasi, P., and F-K. Ko (1977), "Intensity Measurements in Freon Bands of Atmospheric Interest," *J. Quant. Spectosc. Radiat. Transfer*, **17**, 385–388.

Vassiliadis, A. (1974), *Laser Applications in Medicine and Biology*, Vol. 2 (Ed. M. L. Wolbarsht), Plenum Press, New York.

Veldkamp, W. B., and C. J. Kastner (1982), "Beam Profile Shaping for Laser Radars That Use Detector Arrays," *Appl. Optics*, **21**, 345–356.

Verdeyen, J. T. (1981), *Laser Electronics*, Prentice-Hall, Englewood Cliffs, N.J.

Vergez-Deloncle, M. (1964), "Absorption des Radiations Infrarouges par les Gaz Atmospheriques," *J. Physique* **25**, 773–788.

Viezee, W., J. Obianas, and R. T. H. Collis (1973), *Evaluation of the Lidar Technique of Determining Slant Range Visibility for Aircraft Landing Operations*, SRI Report AFCRL-TR-0708.

Wallenstein, R., and T. W. Hänsch (1975), "Powerful Dye Laser Oscillation–Amplifier System for High Resolution Spectroscopy," *Optics Comm.*, **14**, 353-357.

Walling, J. C., D. G. Peterson, H. P. Jenssen, R. C. Morris, and E. W. O'Dell (1980), "Tunable Alexandrite Lasers," *IEEE J. Quant. Elec.* **QE-16**, 1302–1315.

Walrafen, G. E. (1967), "Raman Spectral Studies of the Effects of Temperature on Water Structure," *J. Chem. Phys.*, **47**, 114–126.

Walrafen, G. E. (1970), "Raman Spectral Studies of the Effects of Perchlorate Ion on Water Structure," *J. Chem. Phys.*, **52**, 4176–4198.

Walther, H., and J. L. Hall (1970), "Tunable Dye Laser with Narrow Spectral Output," *Appl. Phys. Lett.*, **17**, 239.

Wang, C. P. (1974), "Application of Lasers in Atmospheric Probing," *Acta Astronaut.*, **1**, 105–123.

Weitkamp, C. (1981), "The Distribution of Hydrogen Chloride in the Plume of Incineration Ships: Development of New Measurements Systems," *Wastes in the Ocean*, Vol. 3, Wiley; also GKSS 81/E/57.

Weitkamp, C., H. J. Heinrich, W. Herrmann, W. Michaelis, V. Lenhard, and R. N. Schindler (1980), "Measurement of Hydrogen Chloride in the Plume of Incineration Ships," 5th International Clean Air Congress, 20–26 Oct., Buenos Aires, Argentina; also GKSS 80/E/55.

Wells, W. C., G. Gal, and M. W. Munn (1977), "Aerosol Distributions in Maritime Air and Predicted Scattering Coefficients in the Infrared," *Appl. Optics*, **16**, 654–659.

White, M. B. (1977), "*Blue-Green Lasers for Ocean Optics*,"*Opt. Eng.*, **16**, 145–151.

Williams, P. F., D. L. Rousseau, and S. H. Dworetsky (1973), "Resonance Fluorescence and Resonance Raman Scattering; Lifetimes in Molecular Iodine," *Phys. Rev. Lett.*, **32**, 196–199.

Williamson, S. J. (1973), *Fundamentals of Air Pollution*, Addison-Wesley, Toronto.

Wiscombe, W. J. (1980), "Improved Mie Scattering Algorithms," *Appl. Optics*, **19**, 1505.

Wolbarsht, M. L., and D. H. Sliney (1974), *Laser Applications in Medicine and Biology*, Vol. 2 (Ed. M. L. Wolbarsht), Plenum Press, New York.

Wolber, W. (1968), *Res./Develop.*, **19**, 18.

Wolfe, D. C., Jr., and R. L. Byer (1982), "Model Studies of Laser Absorption Computed Tomography for Remote Air Pollution Measurement," *Appl. Optics* **21**, 1165–1178.

Wolfe, W. L., Ed. (1966), *Handbook of Military Infrared Technology*, ONR Cat. No. 65-62266, U.S. Government Printing Office.

Wood, O. R., II (1974), "High Pressure Pulsed Molecular Lasers," *Proc. IEEE*, **62**, 355–397.

Woodman, D. P. (1974), "Limitations in Using Atmospheric Models for Laser Transmission Estimates," *Appl. Optics*, **13**, 2193–2195.

Woodward, L. A. (1967), *Raman Spectroscopy* (H. A. Szymanski, ed.), Plenum Press, New York.

Wright, M. L., E. K. Proctor, L. S. Gasiorek, and E. M. Liston (1975), *A Preliminary Study of Air-Pollution Measurement by Active Remote Sensing Techniques*, NASA CR-132724.

Yariv, A. (1976), *Introduction to Optical Electronics*, 2nd ed., Holt, Rinehart, and Winston.

Yeh, S. O., and E. V. Browell (1982), "Shuttle Lidar Resonance Fluorescence Investigations, (I) Analysis of Na and K Measurements, (II) Analysis of Thermospheric Mg Measurements," *Appl. Optics*, **21**, 2365–2372, 2373–2380.

Yentsch, C. S. (1973), "The Fluorescence of Chlorophyll and Yellow Substance in Natural Waters: A Note on the Problems on Measurements and Their Importance to Remote Sensing," *NASA Conference on the Use of Lasers for Hydrographic Studies*, *Wallops Island*, NASA SP-375, 147–151.

Yentsch, C. S., and D. W. Menzel (1963), "A Method for the Determination of Phytoplankton, Chlorophyll, and Phaeophytin by Fluorescence," *Deep Sea Research*, **10**, 221–231.

Zuchlich, J. R., and J. S. Connolly (1976), "Ocular Hazards of Near UV Laser Radiation," *J. Opt. Soc. Am.*, **66**, 79.

Zuev, V. E. (1976), "Laser-Light Transmission through the Atmosphere," *Laser Monitoring of the Atmosphere* (E. D. Hinckley, ed.), Springer-Verlag.

Zuev, V. E. (1982), *Laser Beams in the Atmosphere*, Transl. from Russian by J. S. Wood, Consultants Bureau, New York.

Index